Biomimetic Microengineering

Biomimetic Microengineering

Edited by

Hyun Jung Kim

CRC Press
Taylor & Francis Group
Boca Raton London New York

CRC Press is an imprint of the
Taylor & Francis Group, an **informa** business

CRC Press
Taylor & Francis Group
6000 Broken Sound Parkway NW, Suite 300
Boca Raton, FL 33487-2742

First issued in paperback 2023

ISBN-13: 978-1-138-03913-1 (hbk)
ISBN-13: 978-1-03-265252-8 (pbk)
ISBN-13: 978-0-367-81480-9 (ebk)

DOI: 10.1201/9780367814809

Visit the Taylor & Francis Web site at
http://www.taylorandfrancis.com

and the CRC Press Web site at
http://www.crcpress.com

Acknowledgment

Taekyung Yoo is an assistant professor in the College of Art & Technology at The Chung-Ang University in South Korea. He has spearheaded the visual effects for motion-picture venues, immersive media, and digital human. He contributed to edit the cover image.

Editor

Hyun Jung Kim is an assistant professor in the Department of Biomedical Engineering at The University of Texas at Austin. After receiving his Ph.D. degree at Yonsei University in the Republic of Korea, he did extensive postdoctoral research at both the University of Chicago and the Wyss Institute at Harvard University. These efforts resulted in cutting-edge breakthroughs in synthetic microbial community research and organomimetic human Gut-on-a-Chip microsystem. His research on Gut-on-a-Chip technology leads to the creation of a microfluidic device that mimics the physiology and pathology of the living human intestine. Since 2015, he has explored novel human host-microbiome ecosystems to discover the disease mechanism and new therapeutics in inflammatory bowel disease and colorectal cancer at UT Austin. In collaboration with clinicians, his lab is currently developing disease-oriented, patient-specific models for the advancement in pharmaceutical and clinical fields.

Contributors

Song Ih Ahn
George W. Woodruff School of
   Mechanical Engineering
Georgia Institute of Technology
Atlanta, Georgia

Philippa Pribul Allen
GlaxoSmithKline
Collegeville, Pennsylvania

Aaron B. Baker
Department of Biomedical Engineering
The University of Texas at Austin
Austin, Texas

Ali Khodayari Bavil
Department of Mechanical
   Engineering
Texas Tech University
Lubbock, Texas

Kambez H. Benam
Department of Medicine &
   Bioengineering
University of Colorado Denver
Denver, Colorado

James Q. Boedicker
Department of Physics
University of Southern California
Los Angeles, California

Daniel Chavarria
Department of Biomedical
   Engineering
The University of Texas at Austin
Austin, Texas

Dongwei Chen
Institute of Microbiology
Chinese Academy of Sciences
Beijing, China

Seung-Woo Cho
Department of Biotechnology
Yonsei University
Seoul, South Korea

Yi Sun Choi
Department of Biotechnology
Yonsei University
Seoul, South Korea

Julianna Deakyne
GlaxoSmithKline
Collegeville, Pennsylvania

Wenbin Du
Institute of Microbiology
Chinese Academy of Sciences
Beijing, China

Jason Ekert
GlaxoSmithKline
Collegeville, Pennsylvania

Zachary Estlack
Department of Mechanical Engineering
University of Utah
Salt Lake City, Utah

Spiro Getsios
Aspect Biosystems
Vancouver, British Columbia, Canada

Landon A. Hackley
Department of Biomedical Engineering
The University of Texas at Austin
Austin, Texas

Stephen Hong
Department of Chemical and
   Biomedical Engineering
Cleveland State University
Cleveland, Ohio

Beiyu Hu
Institute of Microbiology
Chinese Academy of Sciences
Beijing, China

Ran Hu
Institute of Microbiology
Chinese Academy of Sciences
Beijing, China

Jinah Jang
Department of Creative IT Engineering
Pohang University of Science and
    Technology
Pohang, South Korea

Bosu Jeong
Department of Creative IT Engineering
Pohang University of Science and
    Technology
Pohang, South Korea

Claire Jeong
GlaxoSmithKline
Collegeville, Pennsylvania

Alexander J. Kaiser
Department of Medicine &
    Bioengineering
University of Colorado Denver
Denver, Colorado

Hyun Jung Kim
Department of Biomedical Engineering
The University of Texas at Austin
Austin, Texas

Jihoon Kim
Institute of Molecular Biotechnology of
    the Austrian Academy of Sciences
Vienna, Austria

Jin Kim
Department of Biotechnology
Yonsei University
Seoul, South Korea

Jungkyu Kim
Department of Mechanical Engineering
University of Utah
Salt Lake City, Utah

YongTae Kim
George W. Woodruff School of
    Mechanical Engineering
Georgia Institute of Technology
Atlanta, Georgia

Bon-Kyoung Koo
Institute of Molecular Biotechnology of
    the Austrian Academy of Sciences
Vienna, Austria

Esak Lee
Nancy E. and Peter C. Meinig School of
    Biomedical Engineering
Cornell University
Ithaca, New York

Hyunjung Lee
George W. Woodruff School of
    Mechanical Engineering
Georgia Institute of Technology
Atlanta, Georgia

Jason Lee
Department of Biomedical Engineering
The University of Texas at Austin
Austin, Texas

Moo-Yeal Lee
Department of Chemical and
    Biomedical Engineering
Cleveland State University
Cleveland, Ohio

Se-Hwan Lee
Department of Creative IT Engineering
Pohang University of Science and
    Technology
Pohang, South Korea

Nikki Marshall
GlaxoSmithKline
Collegeville, Pennsylvania

Lei Mei
Department of Biomedical
   Engineering
The University of Texas at Austin
Austin, Texas

Sunish Mohanan
GlaxoSmithKline
Collegeville, Pennsylvania

Hyoryung Nam
Department of Creative IT
   Engineering
Pohang University of Science and
   Technology
Pohang, South Korea

Duc-Huy T. Nguyen
Department of Medicine
Weill Cornell Medical College
New York, New York

Brian F. Niemeyer
Department of Medicine
University of Colorado
Boulder, Colorado

Hyun-Ji Park
George W. Woodruff School of
   Mechanical Engineering
Georgia Institute of Technology
Atlanta, Georgia

Yuxin Qiao
Institute of Microbiology
Chinese Academy of Sciences
Beijing, China

Alexander D. Roth
Department of Chemical Engineering
Rochester Institute of Technology
Rochester, New York

Woojung Shin
Department of Biomedical
   Engineering
The University of Texas at Austin
Austin, Texas

Tiger H. Tao
Shanghai Institute of Microsystem and
   Information Technology
Chinese Academy of Sciences
Beijing, China

Jian Wang
Institute of Microbiology
Chinese Academy of Sciences
Beijing, China

Szu-Hsien (Sam) Wu
Institute of Molecular Biotechnology of
   the Austrian Academy of Sciences
Vienna, Austria

Kisuk Yang
Department of Biotechnology
Yonsei University
Seoul, South Korea

Jeong-Kee Yoon
George W. Woodruff School of
   Mechanical Engineering
Georgia Institute of Technology
Atlanta, Georgia

Juanli Yun
Institute of Microbiology
Chinese Academy of Sciences
Beijing, China

Zhitao Zhou
Shanghai Institute of Microsystem and
   Information Technology
Chinese Academy of Sciences
Beijing, China

Part I

Emulating the Microenvironment of a Living System

1 Emulating Biomechanical Environments in Microengineered Systems

Jason Lee, Lei Mei, Daniel Chavarria,
and Aaron B. Baker

University of Texas at Austin

CONTENTS

1.1 INTRODUCTION

Cells in the body are exposed to mechanical forces through a variety of processes. These range from the motions of the body through the contraction of muscles and rotation around the joints to fluidic shear stresses from the flow of blood, lymph, or other body fluids. Mechanobiological processes are being increasingly recognized to be powerful regulators of many organ systems and in pathophysiological processes including those involved in cardiovascular disease (Koskinas et al. 2009), cancer (Jain, Martin, and Stylianopoulos 2014), and stem cell biology (Henderson et al. 2017). On a fundamental scale, forces can alter the rate and equilibrium of biochemical reactions, leading to a broad shift in biological processes that is not easily reproducible by other means. Thus, many processes that are not obviously affected by macroscopic level forces also have a mechanobiological component. Some examples of these include the activation of T cell receptors (Kim, Shin et al. 2012), histone binding to DNA (Miroshnikova, Nava, and Wickstrom 2017), or receptor binding kinetics (Walton, Lee, and Van Vliet 2008).

A major goal of biomimetic systems is to accurately reproduce important aspects of the target systems biology on an *in vitro* system. Some body systems inherently interact with and generate mechanical forces as part of their normal function. The cardiovascular system is a prime example of this in which the physical pumping of the heart moves blood through the elastic arteries of the body. In both the heart and the blood vessels, mechanical forces provide signals for homeostasis or adaptation and are a key part of reproducing the biology ex vivo to model disease or mimic normal physiology.

In this chapter, we will review the methods and techniques for incorporating mechanical stimuli into biomimetic models. We will first examine using patterning and microfabrication techniques to mimic the passive mechanical aspects of an organ system. These include stimuli such as extracellular matrix (ECM) stiffness, patterning of attachment sites or micro- and nanotopographical queues that arise from the fibrillar structure of the ECM. We will next examine methods for simulating mechanical stretch in biomimetic systems to simulate biological processes like arterial stretch due to the cardiac cycle, lung expansion during breathing, or peristalsis of the intestine. Finally, we will examine systems to simulate flow in biomimetic systems.

1.2 METHODS FOR STUDYING SUBSTRATE-MEDIATED MECHANICAL CUES

1.2.1 ECM-Mediated Mechanical Cues Are Important in Physiological and Pathological Processes

In the body, cells exist in a three-dimensional (3D) environment that applies a variety of dynamic and passive mechanical stimuli. Mechanical cues are very important for the hierarchical cells and tissues to maintain particular phenotypes and perform physiological functions. Local sensing of mechanical cues is converted to biochemical signals that results in regulation of cell motility through adhesion, spreading, and migration (Pelham and Wang 1997) modifying cell phenotype (Guilak et al. 2009)

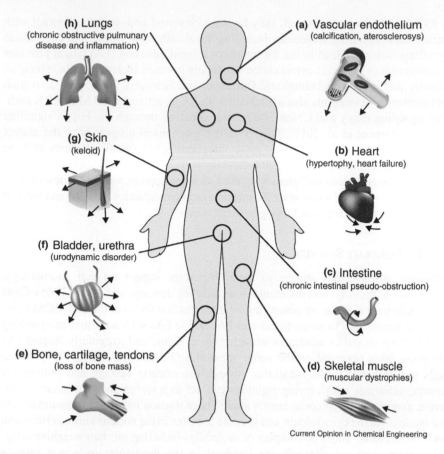

(h) Lungs
(chronic obstructive pulmonary disease and inflammation)

(a) Vascular endothelium
(calcification, aterosclerosys)

(g) Skin
(keloid)

(b) Heart
(hypertophy, heart failure)

(f) Bladder, urethra
(urodynamic disorder)

(c) Intestine
(chronic intestinal pseudo-obstruction)

(e) Bone, cartilage, tendons
(loss of bone mass)

(d) Skeletal muscle
(muscular dystrophies)

Current Opinion in Chemical Engineering

FIGURE 1.1 Human tissue and organs under mechanical stimuli (arrows show force directions) and diseases associated with mechanical forces (Giulitti et al. 2016).

and function (Vogel and Sheetz 2006). Since mechanotransduction is present on all cells, mechanical changes that cells perceive from ECM may impair cellular processes, causing cellular dysfunction and diseases.

Recent investigations have shown the mechanism of common forces, stresses, loads, and deformations on disease that has a component regulated by mechanical forces, including vascular calcification, heart failure, chronic intestinal pseudo-obstruction, and muscular dystrophies among others diseases (Figure 1.1) (Giulitti et al. 2016). Mimicking physiological mechanics exerted from ECM is needed for the creation of accurate biomimetic assays.

The ECM is mainly composed of collagen fibers, proteoglycans, glycoproteins, and other matrix proteins, that bind to integrin receptors on cell surface to form cell–matrix attachments. Most human cells are adherent and must be anchored to an appropriate ECM to survive. The integrin family of receptors are capable of mediating signals from the ECM to regulate cellular behaviors, including proliferation, differentiation, and apoptosis (Giancotti and Ruoslahti 1999). As integrins bind to ECM and become activated, they become clustered and, in turn, interact with intracellular signaling molecules including focal adhesion kinase and Src. These signaling molecules lead to the formation of a focal adhesion (FA) which provides a connection to the actin cytoskeleton via linker proteins including talin, vinculin, filamin, and α-actinin (Humphrey, Dufresne, and Schwartz 2014). The activation and clustering of integrins also associated with RhoA activation, MAPK/ERK pathway signaling (Shyy and Chien 2002), and signaling through the Hippo signaling pathway (Dupont et al. 2011). Therefore, the mechanical properties of the matrix that cells adhere to can influence these integrin-elicited signaling events and are important in regulating various cell behaviors as well as processes such as regeneration, inflammation, and malignancy. In the following section, we will review current research methods and techniques for mimicking the mechanics of ECM and we will mainly focus on stiffness and topographical cues.

1.2.2 SUBSTRATE STIFFNESS

Substrate stiffness has shown to have important impact on cell morphology, proliferation, motility, and differentiation (Discher, Janmey, and Wang 2005). Cells detect substrate stiffness by generating internal traction forces at the cell–ECM interface via actomyosin in stress fibers and FA. Stable FAs with a normal morphology tend to form on stiffer substrates while highly dynamic and irregularly shaped FAs are more often observed on the softer substrates (Pelham and Wang 1997). When cells are anchored to the substrates through the integrin linkage and pulling the matrix, substrates with varying rigidities can act as a spring and thus generate different magnitudes of forces at anchor sites. These traction forces are transmitted into the nucleus via the cytoskeleton and specific proteins in the nuclear lamina (Isermann and Lammerding 2013). A complex of molecules including mechanosensitive adaptor proteins and ion channels are involved in the mechanotransduction process (Kobayashi and Sokabe 2010). It has been observed that cells migrate preferentially towards stiffer substrates in a process known as durotaxis. During the process cytoskeletal tension induces actin fibers to orient along the direction of traction force in response to substrate stiffness (Trichet et al. 2012). Myosin II molecular motors are vital for the cells to sense substrate stiffness as they are key to the contraction phase of the mechanosensing. ECM stiffness and topographical cues contribute to the overall control of stem cell activity. For example mesenchymal stem cells (MSCs) were found to differentiate based on the stiffness of the underlying substrate (Engler et al. 2006). Moreover, cell responses and function are synergistically regulated by substrate rigidity and adhesive pattern (Weng and Fu 2011), and cell cortical stiffness increases as a function of both substrate stiffness and spread area (Tee et al. 2011). These recent studies show the enormous potential for engineering both rigidity and topographical cues of ECM for tissue engineering and regeneration purpose.

1.2.2.1 Traditional Hydrogel System

In recent years, hydrogels have attracted great research interest due to its excellent biocompatibility, synthesis ease, and physical tunability to mimic 3D *in vivo* environment. A major focus of tissue engineering and regeneration is to mimic the *in vivo* cellular environment by 3D gel models. Hydrogels are polymeric materials that can be engineered to achieve this goal. Cell-compatible hydrogels are capable of accommodating cells *in vitro* and for transplantation to injured organs to induce *in vivo* genesis. Hydrogels are often preferred over other biomaterials due to their high water content, diverse physical properties, and their high porosity to provide sufficient oxygen and other nutrients for the encapsulated cells. Traditional hydrogel systems consist of natural hydrogels including collagen, fibrin, hyaluronic acid, Matrigel, and derivatives of natural materials such as chitosan and alginate, as well as the synthetic hydrogels such as polyethylene glycol (PEG), polyacrylamide, and poly(vinyl alcohol) (Sun, Weng, and Fu 2012).

Natural hydrogels are bioactive and usually have high affinity to seeded cells. However, it is difficult to control the composition as they have to be isolated from the body, and their mechanical properties and polymerization conditions are poorly understood. In contrast, the synthetic hydrogel's chemical compositions and mechanical properties are more easily tunable despite the fact that synthetic hydrogels usually need to be modified with proteins and ligands to adhere cells. The mechanical properties of a hydrogel can convey important physical cues to cells via mechanotransduction pathways that mediate cell proliferation, morphogenesis, contractility, differentiation, and other important cell behaviors (Seliktar 2012). The hydrogel can be produced in a wide range of stiffnesses by varying the concentration or the molecular weight of the polymer, or by altering the ratio of crosslinkers, UV irradiation dose, and time. The general techniques used for hydrogel microfabrication are emulsification, photolithography, microfluidic synthesis, and micromolding (Khademhosseini and Langer 2007). Recent research successfully fabricated hydrogel arrays with stiffness gradient and found that cell spreading on these hydrogels linearly correlates with hydrogel stiffness (Sunyer et al. 2012). Hydrogels with different stiffness has also shown to influence stem cell morphology and differentiation (Justin and Engler 2011).

Apart from stiffness, 3D topographical cues can also be used to engineer cell synthetic hydrogels. They have been used to engineer structures with dimensions comparable with those of the cellular-scale functional units of organs via a variety of microfabrication techniques (Table 1.1; Verhulsel et al. 2014). Notably, cell responses to physical environment and stem cell lineage-commitment decisions depend on the mechanical memory from the past physical environment. For example, cultured vascular smooth muscle cells (vSMCs) were sensitive to and more differentiated under physiological waveform strain that is similar to *in vivo* environment (Lee et al. 2013). Another example is that YAP/TAZ and pre-osteogenic transcription factor RUNX2 stayed activated when human mesenchymal stem cells (hMSCs) were cultured initially in stiff environment and then in soft environment (Yang et al. 2014). Hence, a more dynamic hydrogel system is needed to study the consequences of mechanical changes in the cell microenvironment.

1.2.2.2 Photoactivated Hydrogel Systems

Cells are sensitive to their surroundings and making dynamic changes to the microenvironment features constantly. For example, during the wound healing process, fibroblasts are known to differentiate into myofibroblasts and secrete more

TABLE 1.1

Microfabrication Techniques used for 3D Hydrogel Synthesis

Technique	Description	Resolution	Restriction
Soft lithography	Hydrogel molding with a master prepared by microfabrication techniques (e.g., photolithography)	~100 nm	Complex steps and requiring specific equipments
Stereolithography	Structuration of photopolymerizable hydrogel layer by layer with a laser	~1 μm	Requires photopolymerizable hydrogels
Laser microstructuration	Structuration of hydrogel blocks by laser ablation	~1 μm	Depth of structure limited by the penetration of light into the hydrogel
Bioprinting	Precise deposition of various hydrogels using a printer	~10 μm	Limited to specific hydrogels and expensive
Generating fibers using fluid co-flow	Co axial flow of crosslinkable polymer(s) in solution with cells	~10 μm	Geometry limited to linear fibers

Source: Modified from Verhulsel et al. (2014). 3D, three-dimensional.

ECM components to repair damaged ECM around injured tissue and promote fibrogenesis while increasing the matrix modulus (Hinz 2007). Therefore, to emulate the dynamic microenvironment, achieving the real-time control of ECM elasticity is important for unraveling the effect of environment stiffness during various cellular processes. One general approach is to soften hydrogel by degrading it hydrolytically or enzymatically (Metters and Hubbell 2005). This method lacks spatial precision and may have an undesired influence on the encapsulated cells. Recently, great efforts have been focused on temporally modulating hydrogel stiffness by light-triggered reactions. For example, a platform of photodegradable PEG-based hydrogels was designed by attaching photolabile moiety to PEG base bone during synthesis, resulting in the softening of the gel under ultraviolet, visible, or two-photon irradiation (Kloxin et al. 2009). Other studies also demonstrated using UV light to trigger in situ secondary crosslinking, causing temporal stiffening of hyaluronic acid hydrogels in the presence of cells (Guvendiren and Burdick 2012). However, these UV light–triggered softening or stiffening reactions are not powerful enough to penetrate thick hydrogels or deep tissues during *in vivo* experiments.

More recently, an innovative hydrogel system was reported in which alginate gel stiffness can be temporally modulated by near-infrared light–triggered release of calcium or a chelator from temperature-sensitive liposomes. Liposomes loaded with either $CaCl_2$ (for stiffening) or diethylenetriaminepentaacetic acid (for softening) release cargo under irradiation and therefore change the substrate stiffness (Stowers, Allen, and Suggs 2015). This system showed a method to create a hydrogel with user-controlled stiffness that are highly promising for *in vivo* applications. Apart from the photoactivated hydrogel systems, there are also pH- and temperature-responsive hydrogel systems in which stiffness changes with pH or temperature changes (Cha, He, and Ni 2012).

1.2.3 MICRO- AND NANOPATTERNED SYSTEMS

1.2.3.1 Micropost Arrays

Cells apply mechanical forces to their environment during adherence and migration. Arrays of microfabricated silicone elastomeric post arrays have been designed to measure the forces applied directly by living cells at adhesion sites in real time (Balaban et al. 2001). Forces are calculated based on the measured displacements of micropatterned elastomers and the corresponding locations of the FAs visualized via green fluorescent protein-tagged vinculin. Unlike using continuous surfaces for measuring the forces exerted by the cell, the micropost arrays allow researchers to easily compare the forces applied by the specific FA site of a cell. Improving upon this concept, a new method was reported to vary the compliance of the substrate by controlling the geometry of the posts while holding other surface properties constant (Tan et al. 2003). In this approach, when the traction force from the attached cell bends the posts, they behave like simple springs and the deflection is directly proportional to the force, which is measurable on the nanonewton scale. The microfabricated post arrays were fabricated by replica-molding polydimethylsiloxane (PDMS) with different dimensions corresponding to varied stiffnesses that cause traction forces from nanonewton to micronewton scales (Figure 1.2). The system demonstrated a positive correlation between the size of the FAs and the traction force generated on site.

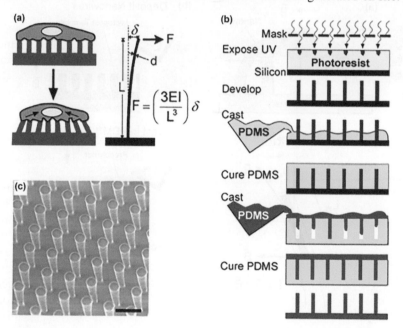

FIGURE 1.2 Fabrication of micropost arrays. (a) Showed the relationship between traction forces F and the physical properties of posts including height L, Young's modulus of the post E, moment of inertia I, and resulting deflection of the post δ. As the equation shows, the deflection is proportional to the force applied by the attached cell. (b) Schematic drawing of the fabrication method. (c) SEM image of the micropost arrays (Tan et al. 2003). SEM, scanning electron microscope.

Traction forces activate integrin-associated pathways, further mediating cytoskeletal tension and transmitting the mechanical cues unto cells (Moore, Roca-Cusachs, and Sheetz 2010). Cells are able to sense the stiffness of their substrates and regulate proliferation, morphogenesis, malignancy, and differentiation to specific lineages (Engler et al. 2006). Therefore, controlling a cell's spreading area is critical in investigating the correlation between traction forces and substrate stiffness. When the effects of stiffness, area, and FA growth were decoupled, results suggesting that the post density, rather than cell spreading, determines the total force and average force exerted by a cell (Han et al. 2012). Using replica-molded arrays of PDMS microposts with identical surface geometry but different post height to control substrate rigidity, it was possible to regulate hMSCs cytoskeletal contractility and lineage differentiation (Fu et al. 2010).

Apart from the elastomeric posts, magnetic posts have also been employed to sense cellular traction forces. This method utilizes a magnetic field to induce torque and deflection in the cobalt nanowires entrapped in the post which in turn impart forces unto the individual adhesion sites of the cells attached to the array (Figure 1.3; Sniadecki et al. 2007). By interspersing magnetic posts among nonmagnetic silicone elastomeric posts, this strategy can apply an external step force to cells by

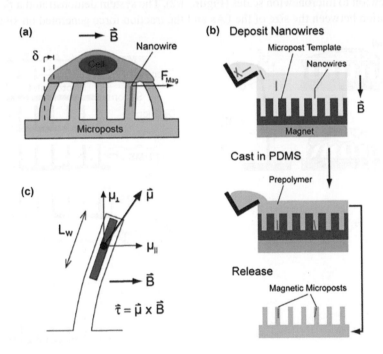

FIGURE 1.3 Fabrication of magnetic post arrays. (a) Magnetic microposts under the influence of a magnetic field $\vec{B}$. A force F_{Mag} is applied to the adherent cell through the magnetic post, that is placed among nonmagnetic silicone elastomeric posts. Nonmagnetic posts reflect traction forces through post deflection δ. (b) Schematic drawing of the fabrication method. (c) Relationship between magnetic torque τ, dipole moment μ, and applied field $\vec{B}$. L_w is the nanowire length (Sniadecki et al. 2007).

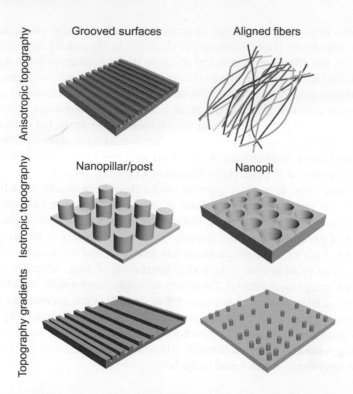

Anisotropic topography

Grooved surfaces   Aligned fibers

Isotropic topography

Nanopillar/post   Nanopit

Topography gradients

FIGURE 1.4 Schematics of representative nanotopography geometries used as substrates: nanogrooved arrays, engineered fibrillar matrices, nanopillar arrays, and nanopit arrays (Kim, Provenzano et al. 2012).

polycarbonate and PEG. Many studies have demonstrated the reduction of cell spreading on nanopits, increased or constant filopodia formation, and reduction in FA formation (Bettinger, Langer, and Borenstein 2009). Other studies also showed changes in cell morphology and cytoskeletal organization when cultured on nanopits (Biggs et al. 2009). These results show great promise for nanofibrillar surfaces to become "smart" biomaterial candidate in tissue engineering and regenerative medicine.

1.3 MICROMECHANICAL DEVICES FOR STUDYING MECHANICAL STRETCH

1.3.1 MECHANICAL STRETCH AS A BIOLOGICAL REGULATOR

Many organs and tissues experience mechanical strain during their normal function or as a result of developing disease. For example, large mechanical strains are introduced to tissues and cells as the lungs expand during breathing or the contraction of the heart. On a smaller scale, mechanical distension is present in capillaries and the small-scale deformation of the ECM. These dynamic mechanical strains drive various processes, including tissue remodeling, cellular signaling pathways, proliferation, migration, phenotype changes (especially in stem cells), and even chromatin remodeling (Lee, Henderson et al. 2018). In mimicking *in vivo* organ or cell behavior, the effects of mechanical strain on the molecular, cellular, and tissue levels is crucial in achieving biologically accurate engineered microsystems.

Many mechanical devices have been developed to induce mechanical strain on cell and tissue cultures. Not only did these technologies reveal ways to investigate the fundamental cell and tissue mechanisms in response to strain, but it also opened up ways to investigate treatment efficacies that would have otherwise been limited to those from a static culture. As mechanical forces can dramatically alter the biological response, it is particularly important to include these forces in assays of drug screening. Most devices for applying mechanical strain to cells work by applying mechanical load to flexible substrate on which the cells are cultured. Forces are applied to the flexible substrate through different mechanisms including the application of pneumatic pressure, stepper, or voice coil motors or electromagnetic mechanisms. Each of these mechanisms has strengths and weaknesses, including aspects such as relative cost, complexity, generation of heat, accuracy in applying complex mechanical profiles, the ability to apply large and/or high-frequency strains. For each application, it is important to consider their accuracy in mimicking the physiological strain, their ease in assembly, and their compatibility to other biological platforms and assays. In this chapter, we will discuss several device types for applying mechanical stretch through different mechanisms and can be used to mimic the mechanical forces found in the body.

1.3.2 Clamp-Based Stretch

Clamp-based methods are one of the most common and early examples of devices that introduced strain to tissue cell cultures. In two-dimensional (2D) cell culture studies, samples are cultured on thin flexible membranes. These can be made of commercially available silicone sheets or custom fabricated thin PDMS sheets (Ahmed et al. 2010). By displacing the motor, strain is introduced to the culture surface (Brown 2000). The motor can be programmed to displace periodically, creating a cyclic and dynamic strain on the culture surface (Houtchens et al. 2008). A single actuator stretching the membrane in one-direction introduces planar uniaxial strain to the cells, while an additional actuator in the perpendicular direction can introduce biaxial strain to samples (Figure 1.5; Win et al. 2017). One of the benefits of using the devices of this type is that the flexible membrane can be fabricated to have grooves or pits, allowing studies that combine stretch and surface effects on cells (Figure 1.6; Ahmed et al. 2010).

In studies with gels, 3D samples, or tissues, specimens are affixed to the actuator via threading, glue, or clamps. Similar to 2D cell culture devices, displacement of the actuator deforms the tissue sample or the matrix/gels. Devices of this type are more appropriate in tensile strength studies of the ECM (e.g., Instron devices) or mechanical strain studies in soft tissues where both cells and the surrounding matrix are deformed similar to an *in vivo* environment (Garcia-Herrera et al. 2012). Devices that have two perpendicular motors can also be used to study complex biaxial deformation on samples (Figure 1.7; Potter et al. 2018).

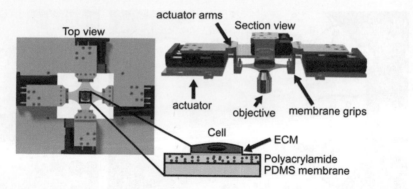

FIGURE 1.5 Example of a clamp-based system for applying mechanical strain. Cells are cultured on a flexible PDMS sheet (blue) fixed into clamps. The sheet is stretched by actuators that control the position of the clamps. Actuators can be placed perpendicular to each other to introduce varying levels of biaxial strain to the cells (Win et al. 2017). PDMS, polydimethylsiloxane.

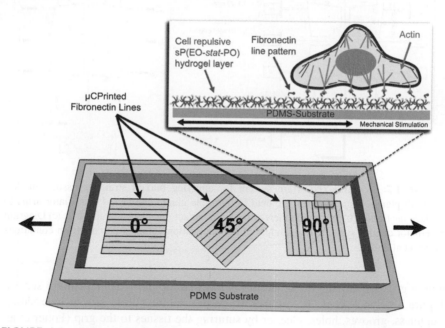

FIGURE 1.6 A thin surface that cells are grown on can be fabricated to have surface grooves and features as shown in this schematic. This allows the study of combined effects of patterned surfaces and mechanical strain on cells (Ahmed et al. 2010).

A limitation of these types of devices is in the grip attachment, where there can be introduction of local strain near the area of the grip or the connection. For applying load to tissues the grip can slip on the tissue or cause local tissue damage. Several variations of the clamp design however exist to circumvent the problem. For example, integrating a "tab" or a "flange" on the flexible membrane provides an extra

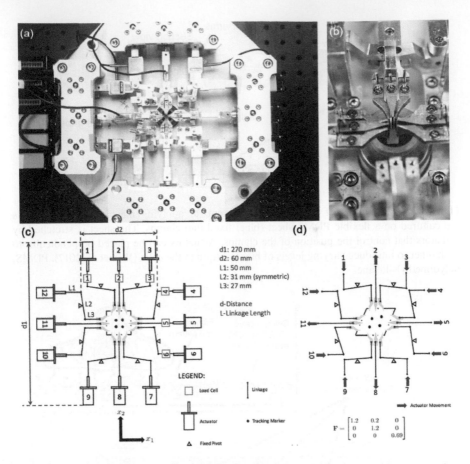

FIGURE 1.7 A multiaxial strain system for applying biaxial strain to tissue samples. (a) Small planar tissue samples are held by multiple clamp connected to actuator arms to displace the tissue. (b) Magnified view of the tissue attachment site and actuator. (c) Diagram of actuator control system. (d) Example of the application complex planar-strain conditions (Potter et al. 2018).

area to grip, trivializing the effect of the local strain on the actual culture surface (Figure 1.8; Rapalo et al. 2015). Other strategies to address these issue include adding roughness, grooves, holes, glue, or by suturing the tissues to the grip (Potter et al. 2018; Houtchens et al. 2008).

The application of uniaxial loads can also lead to transverse strains if the membrane or tissue displacement in the direction perpendicular to the applied strain is not controlled. For studies on cultured cells, clamp-based stretching devices also require some assembly to affix the flexible culture surface to the actuators. Coupled with the fact that each additional axis of stretch requires an additional space for loading devices, clamp-based stretching devices are not suitable for high-throughput studies and are generally incompatible with assays that require standard plate or cell culture dimensions (plate reading, etc). On the other hand, because the actuators are

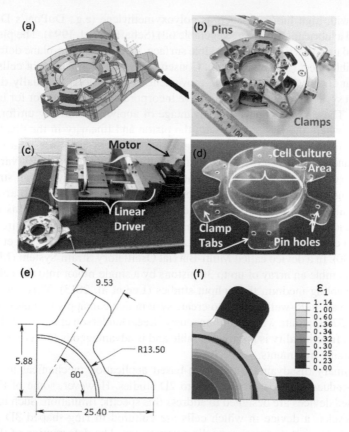

FIGURE 1.8 A device for imaging cells while applying mechanical stretch. (a) Overall CAD model for the device. (b) Photograph of the device after fabrication. (c) The motor and linear driver for the system. These convert rotation of the motor into biaxial stretch. (d) Silicone membrane for cell culture and interfacing with the clamps. (e) Finite element model diagram for the system. (f) Computational model of the strain on the silicone membrane (Rapalo et al. 2015).

not placed above or below the culture surface in 2D studies, live-imaging studies can be conducted relatively easily with proper camera placements (Rapalo et al. 2015). In addition, fabricating patterns on the flexible surface allows studies that combine the effects of surface contour and planar stretch (Ahmed et al. 2010).

1.3.3 Piston-Based Stretch

The application of mechanical stretch to cultured cells can also be achieved by indenting a 2D culture surface with a post or piston. Typically, these devices are used on a monolayer of cells cultured on a thin flexible material such as a silicone sheet. Underneath the flexible surface are posts or pistons. Low friction with the culture surface is needed to apply consistent loads so the indenter is often made of materials with high lubricity, such as polyoxymethylene (e.g., DuPont's Delrin) or PTFE, and a lubricant used (e.g., vegetable oil) (Schaffer et al. 1994). The pistons are then moved upwards towards the flexible surface, which causes in-plane deformation of the flexible sheet (Lee et al. 2013). Consequently, the monolayers of cells experience planar strain (Figure 1.9). The movement of the pistons is usually driven by motor-cam system but newer systems have incorporated a linear motor for improved dynamics. These systems have the advantage of applying relatively uniform strain in the area where there is contact with the piston and linearity in the displacement to strain relationship. Piston- or post-based stretching systems are relatively easy assemble and integrate well with standard multiwell plate formats. By varying the aspect ratio of the piston contact area, the systems can also apply biaxial strain with varying magnitudes in different directions (Chien 2007). One can also arrange the pistons to align with conventional cell culture plate dimensions (such as that of a 6-well), which enables medium or even high-throughput studies that require compatibility with conventional plate readers (Lee and Baker 2015; Lee, Ochoa et al. 2018; Brown 2000). In a device called Multi-Biaxial Oscillatory Strain System (M-BOSS), one can assemble an array of up to 36 pistons by a single motor into the cell culture surface, enabling medium throughput studies (Lee et al. 2013). This was recently extended to a 6 × 96-well format, to create system that can apply complex mechanical load to 576 separate wells simultaneously (Lee, Henderson et al. 2018). The high-throughput functionality is highly desirable and is advantageous for screening drugs and optimizing treatments.

The primary disadvantage of piston-based application mechanical load is that the strain-induced is mostly restricted to 2D studies. However, some of the newer piston-based devices are designed to address this specific limitation. Such is the case for EQUIcycler, a device in which cells are cultured in ring-shaped 3D gels surround an elastic post undergoing cyclic compression. The deformation of the elastic post in turn deforms the 3D gel and induces mechanical strain on the 3D construct (Figure 1.10; Elsaadany, Harris, and Yildirim-Ayan 2017). Another drawback of piston-based devices lies in the heterogeneous strain profile on the culture surface outside of where the piston contacts the flexible membrane. While uniform strain can be applied in the culture area directly in contact with the piston, the edges of culture surface that are not in contact with the piston normally experience higher degree of deformation, introducing shear stress as well as higher strain. There are devices that account for this problem by separating out the area of the membrane that does not come in contact with the piston through an additional wall within the plate (Ursekar et al. 2014).

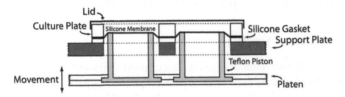

FIGURE 1.9 Generally, piston-based stretching devices have low friction pistons that are driven into a thin flexible culture surface, causing deformation and strain (Lee et al. 2013).

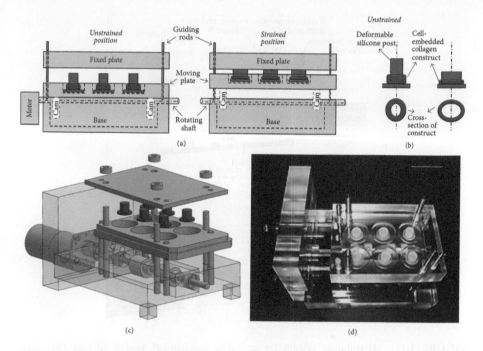

FIGURE 1.10 Diagrams and images of the EQUIcycler, a device that applies mechanical stretch to tissue constructs. (a) Schematic of the EQUIcycler system. (b) Diagram of the mechanism of strain application for the system. (c) Three-dimensional model of the EQUIcycler system. (d) Photograph of the completed system. Bar = 1 in. (Elsaadany, Harris, and Yildirim-Ayan 2017).

1.3.4 PRESSURE-BASED STRETCH

A popular method of applying strain to cell culture surface is using pneumatic pressure to deform a flexible culture surface. These systems work through lowering the pressure underneath a flexible culture surface to cause mechanical strain (Figure 1.11). In the simplest form of these systems, there is a complex strain profile induced in the membrane through combined stretching and bending of the culture membrane (Masuda et al. 2008). Newer systems apply pneumatic pressure around a post/annulus to create an effect similar to piston displacement, which leads to a more uniform application of biaxial strain across the region where the post contacts the membrane.

Devices of this type do require a fabrication step using elastic materials like PDMS, but it also allows flexibility in the shape of the surface where the cells are grown. Many pressure-based stretching devices have different surface contours. For example, one can introduce a "pit" shaped surface with varying depth which enables 3D niche studies, or even introduce a grooved surface to combine the effects of both surface contour and strain on cells or tissues (Figure 1.12; Masuda et al. 2008). The flexibility in design also allows high-throughput functionality much like the piston-based devices. One example is FlexCell, a commercially available device that employs pneumatic pressures to deform the culture surface at a medium/high throughput. It is

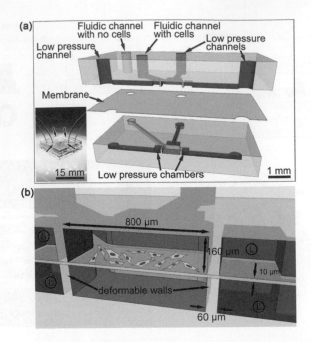

FIGURE 1.11 Microfluidic system for applying mechanical stretch to cultured cells. (a) Diagram of the device and photograph of the completed system. Low- and high pressure can be used to deform elastic structures of the device where the cells are cultured on. (b) An example shown here uses both negative and positive pressure from fluid channels to stretch the walls, which in turn stretches the cell culture surface (Tremblay et al. 2014).

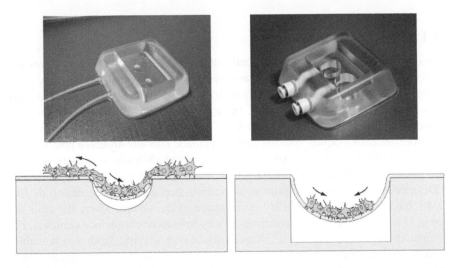

FIGURE 1.12 Pressure-based system are flexible in their design and often require fabrication using elastic material such as PDMS. An example is shown where negative pneumatic pressure via vacuum pump creates a stretchable "pit" of varying sizes which can be controlled during fabrication process (Masuda et al. 2008). PDMS, polydimethylsiloxane.

noteworthy to mention that while the FlexCell uses pneumatic vacuum to induce strain, it also has a post underneath the surface, and thus, introduces both pressure-based and piston-based methods mentioned in this chapter (Corporation 2011).

Pneumatic systems have complex dynamics in terms of the application of complex physiological waveforms. Thus, it should be confirmed that the mechanical loads match those expected in the case of waveforms with high-frequency content strains or sudden changes in strain. The process of causing deformation and returning the flexible surface to the relaxed state requires the creation and removal of pressure in quick succession. The sudden change in the direction of the strain can cause the surface to suddenly vibrate/oscillate due to the momentum much like a stretched rubber band that was suddenly relaxed. Thus, there are challenges in accurately produce dynamic strain waveforms with high frequency and magnitude in pressure-driven devices with pneumatic mechanisms (Lee and Baker 2015).

1.3.5 Devices Using Other Mechanisms to Apply Mechanical Stretch

While conventional stretching devices cause deformation through physical stimulus, there are others that employ other physical mechanisms to apply strain. Examples include thermally- or electrically responsive materials that can create deformation. These methods generally provide an accurate means to control frequency and the area of the strain compared to other methods described in this chapter, as well as having the potential to very small strains for studying molecular level mechanical forces. For example, a device called Hydrogel-Activated Integrated Responsive System (HAIRS) use light as a stimulus to change the temperature of gold rods that are embedded in thermal-responsive hydrogels with cultured cells (Figure 1.13; Sutton et al. 2017). The precision of the light stimulus allows the application of precisely localized small strains in the single cell/sub-area of the cell (<10 μm). This is highly advantageous for studies focusing on cell migration and membrane extension through strain. Another set of devices are known as dielectric elastomer actuators (DEAs), which use electrical current drive the deformation of a cell culture surface (Poulin et al. 2018). In DEAs, the electrodes embedded in an elastomeric material change thickness based on an applied voltage. These changes cause a surrounding gel to deform and apply mechanical deformation to cultured cells (Figure 1.14). While these devices offer precise control and accuracy, they often require relatively complex fabrication processes. In addition, materials that are sensitive to corrosion or

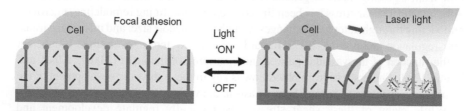

FIGURE 1.13 Example of HAIRS is shown, where thermal energy is transformed into mechanical stretch. The device allows small-scale stretch studies in single cell or sub-cellular regions (Sutton et al. 2017). HAIRS, Hydrogel-Activated Integrated Responsive System.

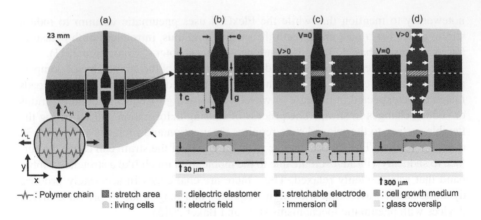

FIGURE 1.14 Example of DEA is shown. Here, electrical stimulation causes deformation of the elastic electrodes, which in turn deforms the cell culture surface (Poulin et al. 2018). Components of the DEAs are labeled in the above diagram wherein the stretchable electrode is shown in black, and the stretched cells from the changes in electric field is shown in green. DEA, dielectric elastomer actuator.

electrolytes from the media need to be shielded, restricting the material and design choices (Poulin et al. 2018). These devices are capable of accurately reproducing small strains but cannot produce large strains such as those produced in piston- or pneumatic-based systems (Figure 1.15; Poulin et al. 2018).

1.4 MICROFLUIDIC AND SIMILAR DEVICES FOR STUDYING SHEAR STRESS

Fluid flow and shear stresses are a normal part of the physiological function of many organ systems, including the cardiovascular system, lungs, kidneys and musculoskeletal system. For example, a large magnitude of shear stress is generated in skeletal system (during daily movement as two large surfaces are moved against each other). Mechanosensing through bones have been linked to osteocyte proliferation and cartilage proteoglycan remodeling (Wang et al. 2013). In the respiratory system, air flow within the lungs generates shear stress on the epithelium of the tracheobronchial tree and alveoli of the lungs (Nucci, Suki, and Lutchen 2003). In gastrointestinal system, fluid flow from digesting material generates shear stress on the intestinal wall and microvilli. Intestinal shear stress has been linked to epithelium lining remodeling, microvilli remodeling through autophagy, changes of epithelial vacuole size, and membrane transport regulation (Kim et al. 2017). In cardiovascular systems, shear stress is inherent to the function of the vascular network and pumping of the heart (Malek, Alper, and Izumo 1999). Shear stress from blood flow is key in the embryonic development of the heart and contributes to normal homeostasis in the vascular system (Baker et al. 2010).

While many of these shear stresses mechanically regulate development and support many functions of organs, altered levels of shear can induce pathophysiological processes and mediate injury and disease progression. For example, shear stresses in joints are cushioned by cartilages, but may lead to osteoarthritis if the

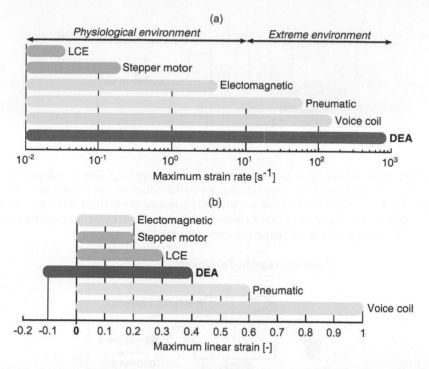

FIGURE 1.15 Stretching devices can induce varying degrees of strain magnitude and frequencies to study both physiological and extreme strains. (a) Plot of different devices and the maximal strain rate they can apply. (b) Plot of different devices and the maximum strain they can apply. DEA, dielectric elastomer actuators; LCE, liquid crystal elastomer actuators (Poulin et al. 2018).

bones are not properly shielded from the shear stress (Wang et al. 2013). In lungs, exposure of high-level shear stress (for example, through the use of ventilator), has been linked to epithelial injury and inflammation (Nucci, Suki, and Lutchen 2003). In vascular systems, studies have shown that abnormal shear stress pattern and low shear stress is implied in tissue calcification, atherosclerosis, and ischemia (Jeong, Lee, and Rosenson 2014). Shear stress has also been shown to regulate proliferation, migration, invasion, and apoptosis of cancer cells, with high implications for directing metastatic behavior (Spencer et al. 2016). As such, understanding how cells and tissues detect shear stress and how they respond to such physical forces is crucial in studying injury development, drug transports, cancer, and cardiovascular diseases. Thus, shear stress is a powerful regulator of biological processes and a key aspect in developing biomimetic systems that can accurate recapitulate the *in vivo* biology of many organ systems.

In general, there are two major approaches to generating shear stress in biomimetic systems. These include is flow chamber or microchips to which a pressure difference is applied to create flow and cone-and-plate devices that drive flow through the motion of a low-angle cone (Figure 1.16). In microfluidic devices, shear stress

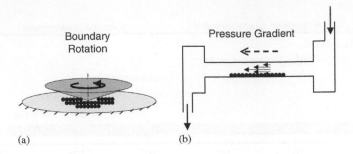

FIGURE 1.16 Diagrams of two common methods for applying flow to cultured cells. (a) The cone-in-plate model generates shear stress on the culture surface by rotating a conical shaft, which in turn creates fluid flow and shear stress in a culture well. (b) In the microchannel model, fluid flow generated through pressure difference between the inlet and the outlet creates shear stress over the cell culture chamber (Brown 2000).

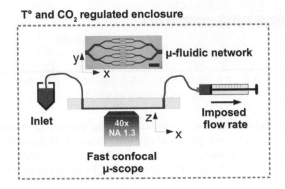

FIGURE 1.17 Microchannel devices use shear stress generated from flow through a narrow passageway or channel walls. Fluid flow is driven by a pump or a vacuum, creating a pressure gradient through the channels (Tsvirkun et al. 2017).

is induced as fluid flows in one direction against the surfaces where the cells are grown (Brown 2000). Fluid flow is generated using a peristaltic or syringe pump, application of pneumatic pressure, or through gravity driven flow using a height difference (Figure 1.17; Dietze et al. 2016). There are also devices that employ capillary actions through absorbent paper to drive flow (Shangguan et al. 2018). Many microfluidics devices fabricate cell culture chambers with grooved surfaces or branched pathways that more closely represent the *in vivo* microenvironment (Figure 1.18). While shear stress is typically generated via fluid flow between two flat parallel surfaces to enable imaging with microscopy, other designs with tubular structures have since emerged to accurately represent the vascular network (Figure 1.19). These microfluidic devices are powerful tools with many flexible designs to produce complex shapes and can be used to study cell response to shear stress as well as protein, particle, or drug adsorption under fluid flow. The strengths of this strategy are the ability to easily create complex flow systems and multichamber tissue models that have flow. While applying steady flow to the systems works well, it is difficult to

FIGURE 1.18 A higher throughput system for performing studies of flow using flow chambers. A multichannel peristaltic pump provides flow in 24 separate flow loops. An optimize pair of pulse dampeners in each flow loop removes the pulsatility and creates steady flow in each flow chamber (Voyvodic, Min, and Baker 2012).

FIGURE 1.19 While most microchannel devices have fluid flow through two parallel flat walls, improvements in microfabrication and 3D printing techniques have enabled studies with more complex tubular channel structures. Here, a microvascular model composed of woven microchannels made of photopolymerizable resin was fabricated using 3D printing. Using this device, channels with different fluorescent dyes were mixed during flow. Scale bay = 0.5 mm (Therriault, White, and Lewis 2003). 3D, three-dimensional.

create more complex flow profiles, such as those found in the artery, accurately in flow chamber/chip type systems.

Another common method to apply shear stress to cultured cells is using a cone-and-plate device. These systems are similar to a cone-and-plate viscometer in which a low-angle cone is rotated close to a flat culture surface. The rotational movement generates fluid flow beneath the cone, creating shear stress on the cultured cells (Figure 1.20; Spencer et al. 2016). A major advantage of these systems is that they can apply temporally changing flow waveforms to mimic the complex flows found in the body. However, these systems inherently require a flat culture surface, limiting their use with 3D tissue constructs or multichamber tissues models. In addition, real-time imaging using these systems is difficult, in contrast to fluidic chips or thin flow chambers.

1.4.1 MICROFLUIDICS

Parallel plates systems have been utilized over 30 years in the application of shear stress in biological systems (Levesque and Nerem 1985). Since then, many improvements in microfabrication have occurred which led to more efficient and tunable

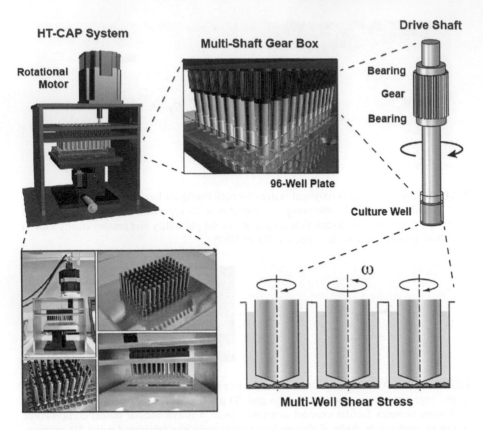

FIGURE 1.20 A high-throughput version of a cone-and-plate device. The system creates flow in the wells of 96-well culture plate through the rotation of cone-tipped shafts. The rotation of the shafts is driven by a single motor linked to the shafts through a 96-shaft gearbox. By modifying the motors rotation, the system can apply continuous or pulsatile flow in each well (Spencer et al. 2016).

devices in the microscopic scale called microfluidics. These microfluidics devices work by creating a difference in pressure between a manifold opening at either end of a chamber while utilizing minute volumes of cell media of approximately 10^{-9}–10^{-18}L (Wu, Huang, and Lee 2010). The difference in pressure creates a uniform laminar flow that applies shear stress to the inner lumen of the chamber where cells adhere. A variety of microfluidic devices have been developed that produce both steady and transient shear stresses in biological applications (Phan et al. 2017). Microfluidics have become a popular technology to integrate shear stress to cell cultures to mimic *in vivo* physiological conditions but like any technology, there are limitations to its implementation.

A wide variety of microfluidic devices have been developed for biological purposes in the past three decades that stem from the principal concepts previously mentioned. In the field of biomedical research, microfluidics has been widely used to study the effects of shear stress on endothelial cells (Butcher and Nerem 2006), specialized cells that line the lumen of the blood vessels. Microfluidics devices are advantageous when recreating specialized 3D cellular architecture such as that of the neurovascular unit in the blood–brain barrier (BBB) (Achyuta et al. 2013), which requires precise layering and culturing of multiple cell types on a microscale. Microfluidics have been utilized to mimic the architecture of the BBB (Figure 1.21) and study neuropathology utilizing a BBB microchip (Figure 1.22). The minimal amount of raw materials needed to create a microfluidic device reduces waste of supplies and the use of PDMS can create optically advantageous properties that enable the microscopic visualization of cultured cells in the microfluidic devices. Microfluidics can often be modified and customized to produce a variety of either steady or pulsatile shear stress applied to cultured cells (Polacheck et al. 2013). These advantages create a favorable microscopic *in vitro* model that can be utilized to study a variety of biological mechanisms such as cellular pathology (Cho et al. 2015), the effects of mechanotransduction on cell behavior (Gray and Stroka 2017) and create organ-on-chips (Huh et al. 2010).

Microfluidic devices have certain limitations when used to mechanically stimulate cultured cells. The scale of microfluidic devices does not allow the culturing of expansive biomass. Careful considerations must be taken with the

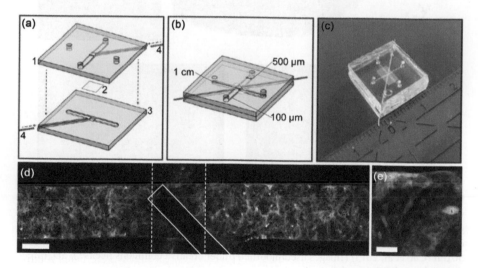

FIGURE 1.21 A chip fabricated to study the mechanical and biochemical modulation of the BBB. The device is crafted from PDMS and incorporates a transwell membrane to separate the vascular and extravascular compartments as well as electrodes to measure the transepithelial electrical resistance. (a and b) Images of the components of the system including: (1) the top plate, (2) Transwell membrane and (3) the bottom plate of the BBB chip, with side-channels to incorporate four platinum electrodes and the assembled device. (c) Photo of the BBB chip after fabrication. (d) Images of the system with human cerebral microvascular endothelial cells (hCMEC/D3 cells) that were stained for DNA (red) and andactin (green). Bar = 250 μm. (e) Higher magnification image of the cells in the BBB device stained for DNA (red) and ZO-1 (green). Bar = 50 μm (Modified and used with permission Griep et al. 2013.). BBB, blood–brain barrier; PDMS, polydimethylsiloxane.

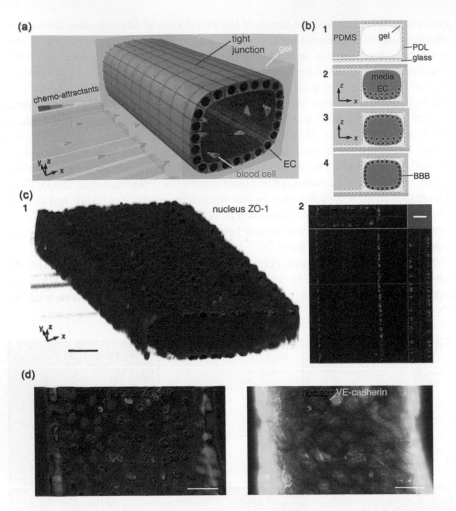

FIGURE 1.22 Diagrams and images of a 3D *in vitro* BBB model. (a) Diagram of the BBB with endothelial cells separating a "blood" compartment from capillaries with chemotactic gradients. (b) Construction of the 3D BBB model. (c) Images of the BBB model with cultured endothelial cells. (d) Immunostaining for ZO-1 and VE-cadherin in the BBB system. Bar = 50 μm (Cho et al. 2015) (Images used/modified under creative commons license.). 3D, three-dimensional; BBB, blood–brain barrier.

material used to fabricate the microfluidic devices as it affects gas exchange rates, osmolarity, and cellular adhesion properties (Halldorsson et al. 2015). More work is needed to characterize the cellular response, such as proliferation and glucose consumption, to the novel materials used to construct microfluidic devices which varies from cell line to cell line (Wu, Huang, and Lee 2010). When these basic criteria are met, intricate constructs can be made to study entire organ systems in microchips such as lung-on-a-chip. This lung-on-a-chip system can reconstitute alveolar–capillary level function and has been utilized to imitate alveolar response to inflammatory cytokines and bacterial infections while integrating biomechanical forces such as shear and strain (Huh et al. 2010). These organ level responses expand the capability of *in vitro* models while reducing the necessity of animal models.

1.4.2 MESOFLUIDIC SYSTEMS

Other research groups overcome the limitations of microfluidic devices by creating mesofluidic systems. Mesofluidic systems bridge the gap between typical macroscopic cell culture, like cell culture plates, and microfluidics devices. The high-throughput cone-and-plate (HT-CAP) system (Spruell and Baker 2013), allows the application of flow in a 96-well plate format that is compatible with commercially available cell culture plates. This allows for the application of shear stress to current commercially available culture plates and the use of multiwell assays read by plate readers or high content imaging systems. The HT-CAP is controlled by an electric motor that rotates a custom gearbox that in turn rotates 96 shafts with a low-angle cone. The position of the 96-well plate is controlled by a microadjustable stage lift that is raised to interface with the low-angle cones. The shear stress applied to cellular monolayers can be adjusted by changing the rotations per minutes of the motor. This allows for high-throughput pulsatile or steady shearing of cells in multiple, separate compartments without interfering with traditional cell culture protocols or vessels.

The HT-CAP system has been successfully utilized to measure MDA-MB-231 and MCF-7 cell adhesion under flow to reveal several candidate pathways to prevent metastatic cancer cell adhesion (Spencer et al. 2016). This system has also been implemented to measure cancer cell adhesion kinetics in varying ECM via a specialized cell adhesion assay in which unlabeled cells are introduced into wells with different ECM coatings followed by the HT-CAP applying shear stress to the wells. In combination with an electric cell–substrate impedance sensing system, the number of unlabeled cancer cells are measured through an impedance measurement by applying an alternating current across the electrode array in the well. The remaining attached cells are then detached by utilizing higher shear stress steps in 1 min intervals while measuring for attached cancer cells in between each interval (Spencer and Baker 2016). Through the use of the HT-CAP and other commercially available assays, significant strides have been made to provide a high-throughput mechanofluidic platform that can be utilized to apply shear stress to traditional cell culture plates. This in turn allows for a more practical device that is compatible with conventional cell culture practices and expedites experimentation and discovery in fields such as cancer research. However, the HT-CAP is also subject to the same limitations that traditional cell culture faces. Traditional cell culture vessels are limited to being 2D and lack the complexity and hierarchy of 3D cell culture. Utilizing the HT-CAP with traditional cell culture techniques adds an extra dimension by integrating physiologically relevant flow to cell culture.

1.5 CONCLUSION

Overall, the inclusion of mechanical forces is key aspect of creating biomimetic systems that recapitulate normal physiologic function and disease processes. Chip-based systems have the advantage of creating multiple compartments to mimic structure. However, these systems can often be limited in creating higher throughput systems. Mesofluidic and well plate-based system for mechanobiology can be readily used for drug screening and high-throughput assays. They however have currently only been used to incorporate 2D culture systems. Thus, there remain opportunities for improvement on both types of assays to create biomimetic systems with 3D architecture than can be used to facilitate drug development and high-throughput science.

REFERENCES

Abagnale, Giulio, Michael Steger, Vu Hoa Nguyen, Nils Hersch, Antonio Sechi, Sylvia Joussen, Bernd Denecke, Rudolf Merkel, Bernd Hoffmann, and Alice Dreser. 2015. Surface topography enhances differentiation of mesenchymal stem cells towards osteogenic and adipogenic lineages. *Biomaterials* 61: 316–326.

Achyuta, Anil Kumar H., Amy J. Conway, Richard B. Crouse, Emilee C. Bannister, Robin N. Lee, Christopher P. Katnik, Adam A. Behensky, Javier Cuevas, and Shivshankar S. Sundaram. 2013. A modular approach to create a neurovascular unit-on-a-chip. *Lab on a Chip* 13(4): 542–553.

Ahmed, Wylie W., Tobias Wolfram, Alexandra M. Goldyn, Kristina Bruellhoff, Borja A. Rioja, Martin Moller, Joachim P. Spatz, Taher A. Saif, Jürgen Groll, and Ralf Kemkemer. 2010. Myoblast morphology and organization on biochemically micro-patterned hydrogel coatings under cyclic mechanical strain. *Biomaterials* 31(2): 250–258.

Baker, Aaron B., Yiannis S. Chatzizisis, Roy Beigel, Michael Jonas, Benjamin V. Stone, Ahmet U. Coskun, Charles Maynard, Campbell Rogers, Konstantinos C. Koskinas, Charles L. Feldman, Peter H. Stone, and Elazer R. Edelman. 2010. Regulation of heparanase expression in coronary artery disease in diabetic, hyperlipidemic swine. *Atherosclerosis* 213(2): 436–442.

Balaban, Nathalie Q., Ulrich S. Schwarz, Daniel Riveline, Polina Goichberg, Gila Tzur, Ilana Sabanay, Diana Mahalu, Sam Safran, Alexander Bershadsky, and Lia Addadi. 2001. Force and focal adhesion assembly: A close relationship studied using elastic micropatterned substrates. *Nature Cell Biology* 3(5): 466.

Bettinger, Christopher J., Robert Langer, and Jeffrey T. Borenstein. 2009. Engineering substrate topography at the micro- and nanoscale to control cell function. *Angewandte Chemie International Edition* 48(30): 5406–5415.

Biela, Sarah A., Yi Su, Joachim P. Spatz, and Ralf Kemkemer. 2009. Different sensitivity of human endothelial cells, smooth muscle cells and fibroblasts to topography in the nano–micro range. *Acta Biomaterialia* 5(7): 2460–2466.

Biggs, Manus J.P., R. Geoff Richards, Nikolaj Gadegaard, Rebecca J. McMurray, Stanley Affrossman, Chris D.W. Wilkinson, Richard O.C. Oreffo, and Mathew J. Dalby. 2009. Interactions with nanoscale topography: Adhesion quantification and signal transduction in cells of osteogenic and multipotent lineage. *Journal of Biomedical Materials Research Part A* 91(1): 195–208.

Brown, Thomas D. 2000. Techniques for mechanical stimulation of cells in vitro: A review. *Journal of Biomechanics* 33(1): 3–14.

Butcher, Jonathan T., and Robert M. Nerem. 2006. Valvular endothelial cells regulate the phenotype of interstitial cells in co-culture: Effects of steady shear stress. *Tissue Engineering* 12(4): 905–915.

Cai, Lei, Li Zhang, Jingyan Dong, and Shanfeng Wang. 2012. Photocured biodegradable polymer substrates of varying stiffness and microgroove dimensions for promoting nerve cell guidance and differentiation. *Langmuir* 28(34): 12557–12568.

Carson, Daniel, Marketa Hnilova, Xiulan Yang, Cameron L. Nemeth, Jonathan H. Tsui, Alec S.T. Smith, Alex Jiao, Michael Regnier, Charles E. Murry, and Candan Tamerler. 2016. Nanotopography-induced structural anisotropy and sarcomere development in human cardiomyocytes derived from induced pluripotent stem cells. *ACS Applied Materials & Interfaces* 8(34): 21923–21932.

Cha, Ruitao, Zhibin He, and Yonghao Ni. 2012. Preparation and characterization of thermal/pH-sensitive hydrogel from carboxylated nanocrystalline cellulose. *Carbohydrate Polymers* 88(2): 713–718.

Chen, Christopher S., Jose L. Alonso, Emanuele Ostuni, George M. Whitesides, and Donald E. Ingber. 2003. Cell shape provides global control of focal adhesion assembly. *Biochemical and Biophysical Research Communications* 307(2): 355–361.

Chen, Christopher S., Milan Mrksich, Sui Huang, George M. Whitesides, and Donald E. Ingber. 1997. Geometric control of cell life and death. *Science* 276(5317): 1425–1428.

Chien, Shu 2007. Mechanotransduction and endothelial cell homeostasis: The wisdom of the cell. *American Journal of Physiology. Heart and Circulatory Physiology* 292(3): H1209–H1224.

Cho, Hansang, Ji Hae Seo, Keith H.K. Wong, Yasukazu Terasaki, Joseph Park, Kiwan Bong, Ken Arai, Eng H. Lo, and Daniel Irimia. 2015. Three-dimensional blood-brain barrier model for in vitro studies of neurovascular pathology. *Scientific Reports* 5: 15222.

Corporation, Flexcell(R) International. 2011. Flexcell® FX-5000™ Tension System.

Dalton, B. Ann, X. Frank Walboomers, Mark Dziegielewski, Margaret D.M. Evans, Sarah Taylor, John A. Jansen, and John G. Steele. 2001. Modulation of epithelial tissue and cell migration by microgrooves. *Journal of Biomedical Materials Research* 56(2): 195–207.

Dietze, Caludia, Sandra Schulze, Stefan Ohla, Kerry Gilmore, Peter H. Seeberger, and Detlev Belder. 2016. Integrated on-chip mass spectrometry reaction monitoring in microfluidic devices containing porous polymer monolithic columns. *Analyst* 141(18): 5412–5416.

Discher, Dennis E., Paul Janmey, and Yu-Li Wang. 2005. Tissue cells feel and respond to the stiffness of their substrate. *Science* 310(5751): 1139–1143.

Downing, Timothy L., Jennifer Soto, Constant Morez, Timothee Houssin, Ashley Fritz, Falei Yuan, Julia Chu, Shyam Patel, David V. Schaffer, and Song Li. 2013. Biophysical regulation of epigenetic state and cell reprogramming. *Nature Materials* 12(12): 1154.

Dupont, Sirio, Leonardo Morsut, Mariaceleste Aragona, Elena Enzo, Stefano Giulitti, Michelangelo Cordenonsi, Francesca Zanconato, Jimmy Le Digabel, Mattia Forcato, and Silvio Bicciato. 2011. Role of YAP/TAZ in mechanotransduction. *Nature* 474(7350): 179.

Elsaadany, Mostafa, Matthew Harris, and Eda Yildirim-Ayan. 2017. Design and validation of equiaxial mechanical strain platform, EQUicycler, for 3D tissue engineered constructs. *BioMed Research International* 2017: 3609703.

Engler, Adam J., Shamik Sen, H. Lee Sweeney, and Dennis E. Discher. 2006. Matrix elasticity directs stem cell lineage specification. *Cell* 126(4): 677–689.

Flemming, Roderick G., Christopher J. Murphy, George A. Abrams, Steven L. Goodman, and Paul F. Nealey. 1999. Effects of synthetic micro- and nano-structured surfaces on cell behavior. *Biomaterials* 20(6): 573–588.

Fu, Jianping, Yang-Kao Wang, Michael T. Yang, Ravi A. Desai, Xiang Yu, Zhijun Liu, and Christopher S. Chen. 2010. Mechanical regulation of cell function with geometrically modulated elastomeric substrates. *Nature Methods* 7(9): 733.

Garcia-Herrera, Claudio, José M. Atienza, Francisco Rojo, Els Claes, Gustavo Guinea, Diego Celentano, Carlos Garcia-Montero, and Raúl L. Burgos. 2012. Mechanical behaviour and rupture of normal and pathological human ascending aortic wall. *Medical & Biological Engineering & Computing* 50(6): 559–566.

Giancotti, Filippo G., and Erkki Ruoslahti. 1999. Integrin signaling. *Science* 285(5430): 1028–1033.

Giulitti, Stefano, Alessandro Zambon, Federica Michielin, and Nicola Elvassore. 2016. Mechanotransduction through substrates engineering and microfluidic devices. *Current Opinion in Chemical Engineering* 11: 67–76.

Gray, Kelsey M., and Kimberly M. Stroka. 2017. Vascular endothelial cell mechanosensing: New insights gained from biomimetic microfluidic models. *Seminars in Cell & Developmental Biology* 71: 106–117.

Griep, Lonneke M., Floor Wolbers, Bjorn de Wagenaar, Paul M. ter Braak, Babette B. Weksler, Ignacio A. Romero, Pierre-Olivier Couraud, Istvan Vermes, Andries Dirk van der Meer, and Albert van den Berg. 2013. BBB ON CHIP: Microfluidic platform to mechanically and biochemically modulate blood-brain barrier function. *Biomedical Microdevices* 15(1): 145–150.

Guilak, Farshid, Daniel M. Cohen, Bradley T. Estes, Jeffrey M. Gimble, Wolfgang Liedtke, and Christopher S. Chen. 2009. Control of stem cell fate by physical interactions with the extracellular matrix. *Cell Stem Cell* 5(1): 17–26.

Guvendiren, Murat, and Jason A. Burdick. 2012. Stiffening hydrogels to probe short-and long-term cellular responses to dynamic mechanics. *Nature Communications* 3: 792.

Halldorsson, Skarphedinn, Edinson Lucumi, Rafael Gómez-Sjöberg, and Ronan M.T. Fleming. 2015. Advantages and challenges of microfluidic cell culture in polydimethylsiloxane devices. *Biosensors and Bioelectronics* 63: 218–231.

Han, Sangyoon J., Kevin S. Bielawski, Lucas H. Ting, Marita L. Rodriguez, and Nathan J. Sniadecki. 2012. Decoupling substrate stiffness, spread area, and micropost density: A close spatial relationship between traction forces and focal adhesions. *Biophysical Journal* 103(4): 640–648.

Henderson, Kayla, Andrew D. Sligar, Victoria P. Le, Jason Lee, and Aaron B. Baker. 2017. Biomechanical regulation of mesenchymal stem cells for cardiovascular tissue engineering. *Advanced Healthcare Materials* 6(22). doi:10.1002/adhm.201700556.

Hinz, Boris. 2007. Formation and function of the myofibroblast during tissue repair. *Journal of Investigative Dermatology* 127(3): 526–537.

Houtchens, Graham R., Michael D. Foster, Tejal A. Desai, Elise F. Morgan, and Joyce Y. Wong. 2008. Combined effects of microtopography and cyclic strain on vascular smooth muscle cell orientation. *Journal of Biomechanics* 41(4): 762–769.

Huh, Dongeun, Benjamin D. Matthews, Akiko Mammoto, Martín Montoya-Zavala, Hong Yuan Hsin, and Donald E. Ingber. 2010. Reconstituting organ-level lung functions on a chip. *Science* 328(5986): 1662–1668.

Humphrey, Jay D., Eric R. Dufresne, and Martin A. Schwartz. 2014. Mechanotransduction and extracellular matrix homeostasis. *Nature Reviews Molecular Cell Biology* 15(12): 802.

Isermann, Philipp, and Jan Lammerding. 2013. Nuclear mechanics and mechanotransduction in health and disease. *Current Biology* 23(24): R1113–R1121.

Jain, Rakesh K., John D. Martin, and Triantafyllos Stylianopoulos. 2014. The role of mechanical forces in tumor growth and therapy. *Annual Review of Biomedical Engineering* 16(16): 321–346.

Jeong, Seul Ki, Jun Young Lee, and Robert S. Rosenson. 2014. Association between ischemic stroke and vascular shear stress in the carotid artery. *Journal of Clinical Neurology* 10(2): 133–139.

Jiang, Xingyu, Derek A. Bruzewicz, Amy P. Wong, Matthieu Piel, and George M. Whitesides. 2005. Directing cell migration with asymmetric micropatterns. *Proceedings of the National Academy of Sciences* 102(4): 975–978.

Justin, R. Tse, and Adam J. Engler. 2011. Stiffness gradients mimicking in vivo tissue variation regulate mesenchymal stem cell fate. *PLoS One* 6(1): e15978.

Khademhosseini, Ali, and Robert Langer. 2007. Microengineered hydrogels for tissue engineering. *Biomaterials* 28(34): 5087–5092.

Kilian, Kristopher A., Branimir Bugarija, Bruce T. Lahn, and Milan Mrksich. 2010. Geometric cues for directing the differentiation of mesenchymal stem cells. *Proceedings of the National Academy of Sciences* 107(11): 4872–4877.

Kim, Deok-Ho, Elizabeth A. Lipke, Pilnam Kim, Raymond Cheong, Susan Thompson, Michael Delannoy, Kahp-Yang Suh, Leslie Tung, and Andre Levchenko. 2010. Nanoscale cues regulate the structure and function of macroscopic cardiac tissue constructs. *Proceedings of the National Academy of Sciences* 107(2): 565–570.

Kim, Deok-Ho, Hyojin Lee, Young Kwang Lee, Jwa-Min Nam, and Andre Levchenko. 2010. Biomimetic nanopatterns as enabling tools for analysis and control of live cells. *Advanced Materials* 22(41): 4551–4566.

Kim, Deok-Ho, Paolo P. Provenzano, Chris L. Smith, and Andre Levchenko. 2012. Matrix nanotopography as a regulator of cell function. *Journal of Cell Biology* 197(3): 351–360.

Kim, Sun Taek, Yongdae Shin, Kristine Brazin, Robert J. Mallis, Zhen Yu Sun, Gerhard Wagner, Matthew J. Lang, and Ellis L. Reinherz. 2012. TCR mechanobiology: Torques and tunable structures linked to early T cell signaling. *Frontiers in Immunology* 3: 76.

Kim, Sun Wook, Jonathan Ehrman, Mok-Ryeon Ahn, Jumpei Kondo, Andres A. Mancheno Lopez, Yun Sik Oh, Xander H. Kim, Scott W. Crawley, James R. Goldenring, Matthew J. Tyska, Erin C. Rericha, and Ken S. Lau. 2017. Shear stress induces noncanonical autophagy in intestinal epithelial monolayers. *Molecular Biology of the Cell* 28(22): 3043–3056.

Kloxin, April M., Andrea M. Kasko, Chelsea N. Salinas, and Kristi S. Anseth. 2009. Photodegradable hydrogels for dynamic tuning of physical and chemical properties. *Science* 324(5923): 59–63.

Kobayashi, Takeshi, and Masahiro Sokabe. 2010. Sensing substrate rigidity by mechano-sensitive ion channels with stress fibers and focal adhesions. *Current Opinion in Cell Biology* 22(5): 669–676.

Koskinas, Konstantinos C., Yiannis S. Chatzizisis, Aaron B. Baker, Elazer R. Edelman, Peter H. Stone, and Charles L. Feldman. 2009. The role of low endothelial shear stress in the conversion of atherosclerotic lesions from stable to unstable plaque. *Current Opinion in Cardiology* 24(6): 580–590.

Lee, Jason, and Aaron B. Baker. 2015. Computational analysis of fluid flow within a device for applying biaxial strain to cultured cells. *Journal of Biomechanical Engineering-Transactions of the ASME* 137(5): 7.

Lee, Jason, Kayla Henderson, Miguel Armenta-Ochoa, Austin Veith, Pablo Maceda, Eun Yoon, Lara Samarneh, Mitchell Wong, Andrew Dunn, and Aaron Baker. 2018. Mechanobiological conditioning of mesenchymal stem cells enhances therapeutic angiogenesis by inducing a hybrid pericyte-endothelial phenotype. *bioRxiv*: 487710.

Lee, Jason, Miguel Ochoa, Pablo Maceda, Eun Yoon, Lara Samarneh, Mitchell Wong, and Aaron B. Baker. 2018. High throughput mechanobiological screens enable mechanical priming of pluripotency in mouse fibroblasts. *bioRxiv*: 480517.

Lee, Jason, Mitchell Wong, Quentin Smith, and Aaron B. Baker. 2013. A novel system for studying mechanical strain waveform-dependent responses in vascular smooth muscle cells. *Lab on a Chip* 13(23): 4573–4582.

Lee, Suk-Won, Su-Yeon Kim, In-Chul Rhyu, Won-Yoon Chung, Richard Leesungbok, and Keun-Woo Lee. 2009. Influence of microgroove dimension on cell behavior of human gingival fibroblasts cultured on titanium substrata. *Clinical Oral Implants Research* 20(1): 56–66.

Levesque, Murina, and Robert Nerem. 1985. The elongation and orientation of cultured endothelial cells in response to shear stress. *Journal of Biomechanical Engineering* 107(4): 341.

Lim, Jung Yul, and Henry J. Donahue. 2007. Cell sensing and response to micro-and nanostructured surfaces produced by chemical and topographic patterning. *Tissue Engineering* 13(8): 1879–1891.

Malek, Adel M., Seth L. Alper, and Seigo Izumo. 1999. Hemodynamic shear stress and its role in atherosclerosis. *JAMA* 282(21): 2035–2042.

Masuda, Taisuke, Ichiro Takahashi, Takahisa Anada, Fumihoto Arai, Toshio Fukuda, Teruko Takano-Yamamoto, and Osamu Suzuki. 2008. Development of a cell culture system loading cyclic mechanical strain to chondrogenic cells. *Journal of Biotechnology* 133(2): 231–238.

McBeath, Rowena, Dana M. Pirone, Celeste M. Nelson, Kiran Bhadriraju, and Christopher S. Chen. 2004. Cell shape, cytoskeletal tension, and RhoA regulate stem cell lineage commitment. *Developmental Cell* 6(4): 483–495.

Metters, Andrew, and Jeffrey Hubbell. 2005. Network formation and degradation behavior of hydrogels formed by Michael-type addition reactions. *Biomacromolecules* 6(1): 290–301.

Miroshnikova, Yekaterina A., Michele M. Nava, and Sara A. Wickstrom. 2017. Emerging roles of mechanical forces in chromatin regulation. *Journal of Cell Science* 130(14): 2243–2250.

Moore, Simon W., Pere Roca-Cusachs, and Michael P. Sheetz. 2010. Stretchy proteins on stretchy substrates: The important elements of integrin-mediated rigidity sensing. *Developmental Cell* 19(2): 194–206.

Nucci, Gianluca, Béla Suki, and Kenneth Lutchen. 2003. Modeling airflow-related shear stress during heterogeneous constriction and mechanical ventilation. *Journal of Applied Physiology* 95(1): 348–356.

Paluch, Ewa, and Carl-Philipp Heisenberg. 2009. Biology and physics of cell shape changes in development. *Current Biology* 19(17): R790–R799.

Pelham, Robert J., and Yu-li Wang. 1997. Cell locomotion and focal adhesions are regulated by substrate flexibility. *Proceedings of the National Academy of Sciences* 94(25): 13661–13665.

Phan, Duc T.T., R. Hugh F. Bender, Jillian W. Andrejecsk, Agua Sobrino, Stephanie J. Hachey, Steven C. George, and Christopher C.W. Hughes. 2017. Blood–brain barrier-on-a-chip: Microphysiological systems that capture the complexity of the blood–central nervous system interface. *Experimental Biology and Medicine* 242(17): 1669–1678.

Polacheck, William J., Ran Li, Sebastien G.M. Uzel, and Roger D. Kamm. 2013. Microfluidic platforms for mechanobiology. *Lab on a Chip* 13(12): 2252–2267.

Potter, Samuel, Jordan Graves, Borys Drach, Thomas Leahy, Christopher Hammel, Yuan Feng, Aaron Baker, and Michael S. Sacks. 2018. A novel small-specimen planar biaxial testing system with full in-plane deformation control. *Journal of Biomechanical Engineering* 140(5). doi:10.1115/1.4038779.

Poulin, Alexandre, Matthias Imboden, Francesca Sorba, Serge Grazioli, Cristina Martin-Olmos, Samuel Rosset, and Herbert Shea. 2018. An ultra-fast mechanically active cell culture substrate. *Scientific Reports* 8(1): 9895.

Rapalo, Gabriel, Josh D. Herwig, Robert Hewitt, Kristina R. Wilhelm, Christopher M. Waters, and Esra Roan. 2015. Live cell imaging during mechanical stretch. *Journal of Visualized Experiments* (102): e52737. doi:10.3791/52737.

Schaffer, Jonathan L., Michael Rizen, Gilbert J. Litalien, Aziz Benbrahim, Joseph Megerman, Louis C. Gerstenfeld, and Martha L. Gray. 1994. Device for the application of a dynamic biaxially uniform and isotropic strain to a flexible cell-culture membrane. *Journal of Orthopaedic Research* 12(5): 709–719.

Schindler, Melvin, Ijaz Ahmed, Jabeen Kamal, Alam Nur-E-Kamal, Timothy H. Grafe, Hoo Young Chung, and Sally Meiners. 2005. A synthetic nanofibrillar matrix promotes in vivo-like organization and morphogenesis for cells in culture. *Biomaterials* 26(28): 5624–5631.

Seliktar, Dror. 2012. Designing cell-compatible hydrogels for biomedical applications. *Science* 336(6085): 1124–1128.

Shangguan, Jin Wen, Yu Liu, Sha Wang, Yun Xuan Hou, Bi Yi Xu, Jing Juan Xu, and Hong Yuan Chen. 2018. Paper capillary enables effective sampling for microfluidic paper analytical devices. *ACS Sensors* 3(7): 1416–1423.

Shyy, John Y.-J., and Shu Chien. 2002. Role of integrins in endothelial mechanosensing of shear stress. *Circulation Research* 91(9): 769–775.

Sniadecki, Nathan J., Alexandre Anguelouch, Michael T. Yang, Corinne M. Lamb, Zhijun Liu, Stuart B. Kirschner, Yaohua Liu, Daniel H. Reich, and Christopher S. Chen. 2007. Magnetic microposts as an approach to apply forces to living cells. *Proceedings of the National Academy of Sciences* 104(37): 14553–14558.

Spencer, Adrianne, Christopher Spruell, Seema Nandi, Mitchell Wong, Mar Creixell, and Aaron Baker. 2016. A high-throughput mechanofluidic screening platform for investigating tumor cell adhesion during metastasis. *Lab on a Chip* 16(1): 142–152.

Spencer, Adrianne, and Aaron B. Baker. 2016. High throughput label free measurement of cancer cell adhesion kinetics under hemodynamic flow. *Scientific Reports* 6(1): 19854.

Spruell, Christopher, and Aaron B. Baker. 2013. Analysis of a high-throughput cone-and-plate apparatus for the application of defined spatiotemporal flow to cultured cells. *Biotechnology and Bioengineering* 110(6): 1782–1793.

Stowers, Ryan S., Shane C. Allen, and Laura J. Suggs. 2015. Dynamic phototuning of 3D hydrogel stiffness. *Proceedings of the National Academy of Sciences* 112(7): 1953–1958.

Sun, Yubing, Shinuo Weng, and Jianping Fu. 2012. Microengineered synthetic cellular microenvironment for stem cells. *Wiley Interdisciplinary Reviews: Nanomedicine and Nanobiotechnology* 4(4): 414–427.

Sunyer, Raimon, Albert J. Jin, Ralph Nossal, and Dan L. Sackett. 2012. Fabrication of hydrogels with steep stiffness gradients for studying cell mechanical response. *PLoS One* 7(10): e46107.

Sutton, Amy, Tanya Shirman, Jaakko V.I. Timonen, Grant T. England, Philseok Kim, Mathias Kolle, Thomas Ferrante, Lauren D. Zarzar, Elizabeth Strong, and Joanna Aizenberg. 2017. Corrigendum: Photothermally triggered actuation of hybrid materials as a new platform for in vitro cell manipulation. *Nature Communications* 8: 15446.

Tan, John L., Joe Tien, Dana M. Pirone, Darren S. Gray, Kiran Bhadriraju, and Christopher S. Chen. 2003. Cells lying on a bed of microneedles: An approach to isolate mechanical force. *Proceedings of the National Academy of Sciences* 100(4): 1484–1489.

Tee, Shang-You, Jianping Fu, Christopher S. Chen, and Paul A. Janmey. 2011. Cell shape and substrate rigidity both regulate cell stiffness. *Biophysical Journal* 100(5): L25–L27.

Therriault, Daniel, Scott R. White, and Jennifer A. Lewis. 2003. Chaotic mixing in three-dimensional microvascular networks fabricated by direct-write assembly. *Nature Materials* 2(4): 265–271.

Tremblay, Dominique, Sophie Chagnon-Lessard, Maryam Mirzaei, Andrew E. Pelling, and Michel Godin. 2014. A microscale anisotropic biaxial cell stretching device for applications in mechanobiology. *Biotechnology Letters* 36(3): 657–665.

Trichet, Léa, Jimmy Le Digabel, Rhoda J. Hawkins, Sri Ram Krishna Vedula, Mukund Gupta, Claire Ribrault, Pascal Hersen, Raphaël Voituriez, and Benoît Ladoux. 2012. Evidence of a large-scale mechanosensing mechanism for cellular adaptation to substrate stiffness. *Proceedings of the National Academy of Sciences* 109(18): 6933–6938.

Tsvirkun, Daria, Alexei Grichine, Alain Duperray, Chaouqi Misbah, and Lionel Bureau. 2017. Microvasculature on a chip: Study of the endothelial surface layer and the flow structure of red blood cells. *Scientific Reports* 7: 45036.

Ursekar, Chaitanya P., Soo Kng Teo, Hiroaki Hirata, Ichiro Harada, Keng Hwee Chiam, and Yasuhiro Sawada. 2014. Design and construction of an equibiaxial cell stretching system that is improved for biochemical analysis. *PLoS One* 9(3): e90665.

Verhulsel, Marine, Maéva Vignes, Stéphanie Descroix, Laurent Malaquin, Danijela M. Vignjevic, and Jean-Louis Viovy. 2014. A review of microfabrication and hydrogel engineering for micro-organs on chips. *Biomaterials* 35(6): 1816–1832.

Vogel, Viola, and Michael Sheetz. 2006. Local force and geometry sensing regulate cell functions. *Nature Reviews Molecular Cell Biology* 7(4): 265.

Voyvodic, Peter L., Daniel Min, and Aaron B. Baker. 2012. A multichannel dampened flow system for studies on shear stress-mediated mechanotransduction. *Lab on a Chip* 12(18): 3322–3330.

Walboomers, X. Frank, Leo Ginsel, and John A. Jansen. 2000. Early spreading events of fibroblasts on microgrooved substrates. *Journal of Biomedical Materials Research* 51(3): 529–534.

Walton, Emily B., Sunyoung Lee, and Krystyn J. Van Vliet. 2008. Extending Bell's model: How force transducer stiffness alters measured unbinding forces and kinetics of molecular complexes. *Biophysical Journal* 94(7): 2621–2630.

Wang, Pu, Pei Pei Guan, Chuang Guo, Fei Zhu, Konstantinos Konstantopoulos, and Zhan You Wang. 2013. Fluid shear stress-induced osteoarthritis: Roles of cyclooxygenase-2 and its metabolic products in inducing the expression of proinflammatory cytokines and matrix metalloproteinases. *FASEB Journal* 27(12): 4664–4677.

Weng, Shinuo, and Jianping Fu. 2011. Synergistic regulation of cell function by matrix rigidity and adhesive pattern. *Biomaterials* 32(36): 9584–9593.

Win, Zaw, Justin M. Buksa, Kerianne E. Steucke, Gu Gant Luxton, Victor H. Barocas, and Patrick W. Alford. 2017. Cellular microbiaxial stretching to measure a single-cell strain energy density function. *Journal of Biomechanical Engineering* 139(7). doi:10.1115/1.4036440.

Wu, Min-Hsien, Song-Bin Huang, and Gwo-Bin Lee. 2010. Microfluidic cell culture systems for drug research. *Lab on a Chip* 10(8): 939.

Yang, Chun, Mark W. Tibbitt, Lena Basta, and Kristi S. Anseth. 2014. Mechanical memory and dosing influence stem cell fate. *Nature Materials* 13(6): 645.

2 Biomimetic Microsystems for Blood and Lymphatic Vascular Research

Duc-Huy T. Nguyen
Weill Cornell Medical College

Esak Lee
Cornell University

CONTENTS

2.1 INTRODUCTION

Blood and lymphatic vasculatures are two major circulatory systems in mammals. Blood vasculatures transport blood throughout the body for delivery of oxygen and nutrients, removal of metabolic wastes, and proper gas exchange. Lymphatic vasculatures do not carry blood; instead, they form "lymph" by draining interstitial fluid into lymphatic vasculatures and transport the lymph that contains immune cells, proteins, and lipids. Blood and lymphatic vessels are separated; however, in most tissues blood capillaries are positioned in close proximity to lymphatic capillaries, where lymph is formed by lymph capillary draining interstitial fluid leaked from blood capillaries. Further, lymphatic trunks are connected to subclavian veins where the lymph goes back to the blood circulation. Thus, these two vascular systems not only play their unique roles, but also are interdependent structurally and functionally, which can involve numerous diseases when they are dysregulated. In this chapter, we discuss fundamental biology of the blood and lymphatic vascular formation, morphogenesis, and homeostasis; pathophysiology of the blood and lymphatic vasculatures in major human diseases; and emerging biomimetic in vitro microsystems to recapitulate key aspects of the vasculatures in normal and disease conditions.

2.2 BLOOD AND LYMPHATIC VASCULAR PHYSIOLOGY

2.2.1 BLOOD VASCULAR PHYSIOLOGY

2.2.1.1 Blood Vascular Development during Embryogenesis

Vasculature is one of the earliest functional organs that form during embryogenesis. The vasculature is comprised of arteries, veins, capillary beds, and lymphatic vessels. These vessels share some common features. The inner most layer of blood vessel is a thin layer of endothelial cells, called endothelium, surrounded by the basement membrane. The arteries, veins, and capillary vessels are also decorated with a layer of perivascular cells. For arteries and veins, the perivascular cells are smooth muscle cells (SMCs) that can constrict or relax to either increase or decrease the pressure, whereas capillary vessels are decorated with pericytes. Due to the proximity to the heart, arteries have thicker wall than veins, to withstand larger blood pressure from the hearts, and they are also surrounded by a denser population of SMCs. The capillary vessels are categorized into three subtypes: continuous, discontinuous, and fenestrated vessels. Continuous capillary vessels have a continuous basement membrane and are decorated with pericytes. Discontinuous capillary vessels are found in sinusoidal blood vessels of the liver and bone marrow where the basement membrane is discontinuous. Fenestrated capillary vessels are characteristic of vessels in tissues that function to filtrate or secrete, such as endocrine and exocrine glands, kidney, and intestine where the endothelium is fenestrated to facilitate exchange and secretion of biomolecules (Potente and Makinen 2017).

Formation of specific vessel subtypes in the vasculature requires the correct spatial and temporal expression of specific genes and transcriptional factors. A close relationship between hematopoiesis and blood vessel formation has led to the notion that hematopoietic cells and endothelial cells originate from a common precursor called hemangioblasts (Park, Kim, and Malik 2013). Brachyury-expressing mesoderm cells appear, and then give rise to FLK1+ cells. FLK1 is preferentially expressed in the endothelial cells in the yolk sac and the embryonic tissues. As a result, deficiency of Flk1 is embryonically lethal due to the failure of vasculature formation in the yolk sac and embryo (Shalaby et al. 1995) (Figure 2.1a).

There are several important transcriptional factors that regulate the emergence of Flk1 cells. One of the many transcriptional factors that regulate the emergence of FLK1+ endothelial cells is the ETS (E-twenty-six specific) transcription factors. ETS factors contain a homologous ETS DNA binding region located at the carboxy terminus. Upon binding to the DNA consensus sequence, ETS regulates the expression of target genes that are critical in multiple biological and pathological processes such as angiogenesis, hematopoiesis, tumorigenesis, and apoptosis. In addition to the ETS factors that regulate endothelial specification, other transcription factors have also been shown to partake in the endothelial lineage specification such as GATA2 and members of the Forkhead transcription factors such as FOXO1 (Lee et al. 1991, Furuyama et al. 2004). Overexpression of GATA2 in embryonic stem cells enhances the population of FLK1+ cells as well as endothelial cells in vitro (Lugus et al. 2007).

Additional transcription factors are also required to ensure proper specification of vessel subtypes, and vascular remodeling. For example, Hey1, Hey2-, Sox7, and Sox18 are important during the specification of arteries and veins (Zhong et al. 2001, Pendeville et al. 2008). COUP-TFII, a member of the nuclear receptor 2F subfamily, is expressed from E8.5 in venous and lymphatic endothelial cells (LECs). Deletion of Coup-TFII in Tie2Cre mice leads to the loss of venous identity as venous endothelial cells gained arterial markers such as Jag1, Notch1, ephrinB2, and Np1 (You et al. 2005) (Figure 2.1a).

2.2.1.2 Signaling Pathways in Blood Vessel Formation

Two of the major processes that shape the vasculature are called vasculogenesis and angiogenesis. The first phase of blood vessel formation is vasculogenesis where endothelial cells or their precursors self-assemble into a network of the primitive vascular plexus (Figure 2.1b). The second phase of blood vessel formation is called angiogenesis where endothelial cells sprout from the existing primitive vascular plexus to form new blood vessels (Figure 2.1c). These new blood vessels undergo extensive remodeling, such as fusion or regression, forming lumens, and recruiting mural cells to form an extensive functional vasculature (Bautch and Caron 2015).

There are several families of proteins, ligands, and receptors that work in junction to tightly regulate the formation of blood vessels (Figure 2.1d). One of the signaling molecules that regulates vessel formation is vascular endothelial growth factor

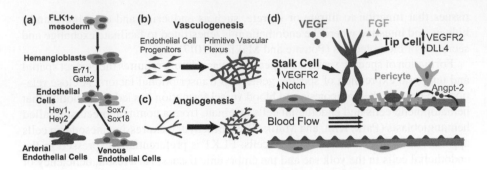

FIGURE 2.1 Formation of the blood vasculature and major signaling pathways. (a) Endothelial cell lineage specification through FLK1+ mesodermal cells to common endothelial progenitor cells to more committed endothelial cells of the arteries and veins. (b) Endothelial progenitor cells self-assemble to form a primitive network of vasculature during early embryo development. (c) Angiogenesis or formation of new blood vessel from the existing vessels occurs during embryogenesis to expand the vasculature and during adult life to support tissue repair and regeneration. (d) A schematic describing the major signaling molecules to trigger angiogenesis (VEGF, FGF), signaling pathways to dictate tip-stalk cell phenotypes during angiogenesis (VEGFR, DLL4, Notch), and key signaling molecule to disrupt perivascular cell attachment during angiogenesis sprouting. FGF, fibroblast growth factor; VEGF, vascular endothelial growth factor.

(VEGF). In mammals, the VEGF family members include VEGF-A, VEGF-B, VEGF-C, VEGF-D, and placental growth factor (PlGF) (Shibuya 2011). Among these, VEGF-A has been shown to be the most important player in blood vessel formation, as loss of a single allele of VEGF-A can cause vascular defects during embryogenesis (Carmeliet et al. 1996). VEGF-A signaling is mediated through the receptor tyrosine kinase VEGFR2 (KDR or FLK1), while the secreted soluble VEGFR1 (FLT1) acts as a ligand trap to antagonize VEGF-A signaling. Another family of proteins that regulates maturation of blood vessel is Angiopoietin (Ang 1 and Ang 2). During angiogenesis and vascular remodeling, Ang 2 is expressed in endothelial cells and disrupts the attachment of perivascular cells to the endothelial cells during angiogenesis (Fagiani and Christofori 2013). The superfamily of fibroblast growth factors (FGFs) exerts several biological functions, including angiogenesis and vessel formation. FGFs can act on endothelial cells directly or activate surrounding cells to promote angiogenesis (Beenken and Mohammadi 2009). FGF-2 in the protein family was first discovered as an angiogenic factor. However, FGF-2 and FGF-1 deficiency does not lead to vascular defects, suggesting that there is a redundancy in the FGF superfamily proteins (Beenken and Mohammadi 2009). Notch and Wnt signaling pathways are also playing essential roles in the formation of blood vessels. As mentioned earlier, tip-stalk cell formation and shuffling are mediated through the VEGF and DLL4-Notch signaling axis (Jakobsson et al. 2010). Notch can also activate Wnt signaling in proliferating stalk cells, and at the same time, the Wnt can reciprocally activate Notch (Corada et al. 2010, Phng et al. 2009).

During the formation of blood vessels, and especially during angiogenic sprouting, the endothelial cells migrate and engage with the surrounding environment. A significant part of this migration process involves matrix degradation and remodeling. For example, the tip cells of the sprouts protrude their filopodial extensions and digest the surrounding matrices to pull the multicellular structures forward, while the stalk cells migrate but at the same time deposit basement membrane proteins along the migrating trail. Among the enzymatic proteases, matrix metalloproteinases (MMPs) have been shown to play important roles in angiogenesis. These MMPs not only degrade several matrix components, but they also liberate other tethered growth factors or proteins that are important for angiogenesis. For instances, MMPs degrade proteoglycan perlecan in the basement membrane to release FGFs or activate TGFβ ligands by cleaving the latent TGFβ binding proteins (Stamenkovic 2003).

During angiogenesis, formation of tip and stalk cells is modulated through VEGF signaling. In response to exogenous VEGF, tip cells express higher level of VEGF receptor 2 (VEGFR2) and delta-like ligand 4 (DLL4) than stalk cells. However, this process of tip-stalk cell selection is a dynamic process that involves the shuffling between tip and stalk cells. At any given time, a stalk cell with higher expression of VEGFR2 and DLL4 can overtake the tip cell position (Jakobsson et al. 2010). Similarly, a stalk cell can initiate a new branch of angiogenic sprout by switching to a tip cell phenotype. Tip cells often do not proliferate, whereas stalk cells proliferate to contribute to the extension of the angiogenic sprouts (Jakobsson et al. 2010).

2.2.2 Lymphatic Vascular Physiology

2.2.2.1 Lymphatic Vessel Formation

Lymphatic vessels are differentiated from cardinal vein endothelium at early stages of embryo development (Tammela and Alitalo 2010). During the formation of primitive blood endothelium from angioblasts, Notch signaling dictates dorsal arterial cell fate differentiation. For Notch signaling inhibits COUP-TFII signaling that is crucial for cardinal vein cell fate differentiation, Notch-downregulated populations of the angioblasts express COUP-TFII (You et al. 2005) and SOX18 (Francois et al. 2008), which drive angioblast differentiation to the cardinal vein endothelial cells. In the mouse embryo, the cardinal vein is formed at around embryonic day 9.0 (E9.0), and the process of LEC differentiation from the cardinal vein endothelial cells occurs at E9.5 when the subpopulations of the cardinal vein endothelial cells begin to express Prospero homeobox protein 1 (PROX1), an LEC fate transcription factor (Oliver et al. 1993) (Figure 2.2a).

PROX1 expressed in the cardinal vein endothelial cells leads to expression of VEGFR receptor 3 (VEGFR3), Neuropilin 2 (NRP2), and Podoplanin, which are major regulators of lymphatic budding and lymphatic separation (Petrova et al. 2002). VEGFR3 is a key receptor for lymphangiogenesis, a new lymphatic vessel formation that can be driven by prolymphangiogenic growth factors, such as VEGF-C and VEGF-D (Achen et al. 1998). In addition to VEGFR3, NRP2 serves as a co-receptor of VEGF-C/D to facilitate prolymphangiogenic signal transduction through the VEGFR3 (Xu et al. 2010). While VEGFR3 and NRP2 promote lymphatic budding (Yang et al. 2012), podoplanin (PDPN) makes LECs separated

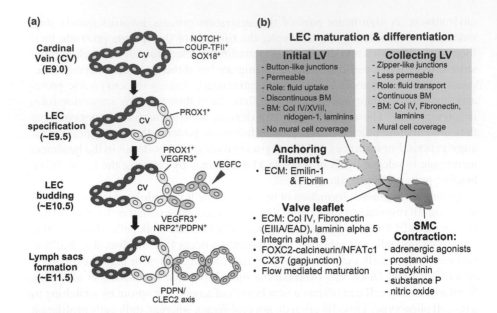

FIGURE 2.2 Lymphatic vessel formation and maturation. (a) Lymphatic vessels are derived from the CV that expresses COUP-TFII and SOX18. LEC specification occurs around at E9.5 by expressing PROX1, a lymphatic fate transcription factor, which leads to expression of VEGFR3, NRP2, and PDPN. VEGFR3-positive populations sprout LEC budding at E10.5, and the formed lymph sacs are separated from the vein via the platelet activation and blood coagulation at the blood/lymphatic interface through the PDPN-CLEC2 axis. (b) Primitive lymph sacs are further matured into initial LVs or collecting LVs. These initial and collecting LVs have distinct roles: interstitial fluid drainage and intraluminal fluid transport, respectively. This can be achieved by different cell–cell junction structure, permeability, ECM, BM, and mural cell engagement. Collecting LVs form intraluminal valves mediated by different gene expression. BM, basement membrane; CV, cardinal vein; ECM, extracellular matrix; HA, hyaluronan; LV, lymphatic vessel; NRP2, Neuropilin-2; PDPN, Podoplanin; VEGF, vascular endothelial growth factor.

from the cardinal vein endothelium (Herzog et al. 2013). Once lymphatic budding sprouts out from the vein endothelium, blood in the veins makes a contact with PDPN in LECs, which initiates blood coagulation by activation of CLEC2 (C-type lectin-like receptor 2) in platelets through the lymphatic PDPN interaction to CLEC2 at the conjunction area of blood and lymphatic vessels (Uhrin et al. 2010) (Figure 2.2a).

2.2.2.2 Lymphatic Vessel Morphogenesis and Maturation

After the primitive lymphatic vessels are formed, they experience maturation processes, resulting in two distinct types of lymphatic vessels in the lymphatic vascular network for efficient lymphatic function (Schulte-Merker, Sabine, and Petrova 2011). These two different lymphatic vessels are (1) initial lymphatic vessels and (2) collecting lymphatic vessels (Figure 2.2b).

The initial lymphatic vessels (or lymphatic capillaries) are small lymphatics (diameters less than 70 μm) that primarily drain interstitial fluid (Casley-Smith 1980). The initial lymphatic vessels are blind-ended and permeable by exhibiting discontinuous basement membrane, oak leaf-shaped LECs and shuffled cell–cell junctions ("button-like junctions") that can only open junctions under elevated interstitial fluid pressure and close the junctions in the pressure equilibrium (Pepper and Skobe 2003). This specialized cell–cell junction functions as an inlet valve (a primary valve) to enable one-way fluid uptake without reverse leakage. This inlet valve is only observed in LECs, because LECs are not fully attached to the extracellular matrix (ECM), but tethered to the surrounding ECM via highly elastic anchoring filaments that are composed of emilin-1 and fibrillin, which is not observed in blood endothelium (Pepper and Skobe 2003). The anchoring filaments maintain multicellular LEC architectures in spite of the discontinuous basement membrane and control the junctions under different interstitial fluid pressure (Leak 1970). The initial lymphatic vessels are not covered by mural cells, such as pericytes and SMCs to maintain loosened/flexible junctions (Figure 2.2b).

Once the initial lymphatic vessels drain the interstitial fluid, the fluid is referred to as "lymph". The lymph then goes through the bigger collecting lymphatic vessels (diameters, 100–600 μm) and regional lymph nodes to filter lymph debris, and merges to lymph trunks, bigger lymphatics, like thoracic ducts, and then finally goes back to the blood circulation through the junction of the lymph trunks with subclavian veins (Aspelund et al. 2016). Thus, the purpose of the collecting lymphatic vessels is transporting the intraluminal lymph fluid.

The collecting lymphatic vessels are much less permeable than the initial lymphatic vessels and carry the intraluminal lymph fluid to regional lymph nodes and lymph trunks without leakage. The collecting lymphatic vessels have well-defined basement membrane of collagen, laminin, and fibronectin, and they display tighter and more linear LEC junctions ("zipper-like junctions") compared to the initial lymphatic vessels with button-like junctions (Sweet et al. 2015). Mural cells, such as lymphatic SMCs, cover the collecting lymphatic vessels, potentiating the vessel contraction as an intrinsic lymphatic pump for propelling lymph fluid without the assistance of heartbeat that engines blood circulation. The SMC contraction is dependent on many factors, including adrenergic agonists, prostanoids, bradykinin, substance P, and nitric oxide (NO) (Trujillo et al. 2017, von der Weid and Zawieja 2004). The collecting lymphatic vessels are divided into valve segments ("lymphangions"), exhibiting intraluminal lymphatic valves to prevent lymph backflow (van Helden 2014) (Figure 2.2b).

At molecular levels, FOXC2, a member of the Forkhead/winged-helix family of transcription factors is one of the main players in the collecting lymphatic vessel maturation (Norrmen et al. 2009). In human, FOXC2 mutation causes lymphedema-distichiasis, a lymphatic stagnation featured by increased lymph backflow (Mellor et al. 2011). In mouse models, FOXC2 deficiency also showed dysregulated lymphatic valve formation and impaired mural cell coverage in the collecting lymphatic vessels during the maturation processes (Noon et al. 2006). Interestingly, the inactivation of the FOXC2 results in dilated lymphatic capillaries and ectopic coverage of mural cells on the lymphatic capillaries (de Mooij et al. 2009), suggesting that FOXC2 is critical for differential maturation of the initial and collecting lymphatic vessels. In lymphatic valve cells, FOXC2-calcineurin/NFATc1 signaling is activated and the loss of calcineurin at development results in lymphatic valve defects (Sabine and Petrova 2014). Besides, the gap junction protein CX37 is also known to be essential for the assembly of lymphatic valves (Kanady et al. 2015) (Figure 2.2b).

2.3 BLOOD AND LYMPHATIC VASCULAR PATHOLOGY

2.3.1 BLOOD VESSELS IN DISEASES

The blood vasculature is integral to the development and homeostasis of organism. Not only the vasculature forms a functional network to deliver nutrients to metabolically active tissues and remove waste products, the vascular cells also actively participate in maintaining the biological and physiological functions of the organs. Damaged or injured blood vessels can lead to substantial pathological diseases to the organism. For example, cardiovascular diseases, a class of diseases that involve the heart and the blood vessels, claim a significant number of deaths globally. Aortic aneurysms, pulmonary hypertension, supravalvular aortic stenosis, and atherosclerosis are among the most common cardiovascular diseases (Mazurek et al. 2017).

2.3.1.1 Aortic Aneurysms and Pulmonary Hypertension

Aortic aneurysms are defined as dilation of artery more than 50% of its diameter. When occurred near the renal arteries and abdominal arteries, aortic aneurysms lead to arterial ruptures with a mortality of 80%–90%. Pathological changes in aortic aneurysm include thinning of the arterial wall due to loss of SMCs and ECM remodeling (Rowe et al. 2000). Pulmonary hypertension is also another devastating disease involving the blood vessels, which claims approximately half of the patient demise within 3 years from initial diagnosis. Pulmonary hypertension has many causes including cardiac, parenchymal, lung, thromboembolic, infectious, and autoimmune diseases, genetic mutations, and idiopathic pulmonary arterial hypertension; but ultimately, patients with pulmonary hypertension exhibit pruning of small vessels, excessive accumulation of SMCs in proximal pulmonary vessels, and reduced compliance of pulmonary arterial vasculature (Simonneau et al. 2013).

2.3.1.2 Supravalvular Aortic Stenosis and Atherosclerosis

Supravalvular aortic stenosis and atherosclerosis share a common pathological change in the blood vessels. Both diseases have narrow arteries, which leads to arterial obstruction. Supravalvular aortic stenosis is a congenital disease caused by mutations that lead to the loss of function of elastin gene in patients (Curran et al. 1993). Atherosclerosis describes a narrowing of arteries by plagues, causing risks to blood flow, and it is also a usual cause of heart failure and strokes. The initial event of atherosclerosis is the accumulation of lipoproteins in the subendothelial space of the arteries and activation of endothelial cells. This triggers the circulating monocytes to adhere to the activated endothelial cells to extravasate into the vessel wall and differentiate into tissue macrophages. These macrophages then further accumulate more lipoproteins, while the SMCs are activated to secrete and deposit excessive amount of ECM proteins. This cascade of events leads to the buildup of plagues in the vessel walls and luminal space of the arteries (Libby, Ridker, and Hansson 2011).

2.3.1.3 Angiogenesis in Cancer Development

Angiogenesis is a highly coordinated process to generate new vessels. As a result, dysregulation of angiogenesis, whether angiogenesis is excessive or insufficient, can be devastating in pathological conditions. For example, angiogenesis in cancer is an excessive growth of new vessels. As the tumor continues to overgrow, tumor cells secrete several pro-angiogenic factors to recruit endothelial cells to form vessels towards nutrients- and oxygen-depleted tumor areas to enable the tumor to continue its expansion. Furthermore, the newly recruited vessels to the tumors then serve as an escaping route for the tumor cells to metastasize and colonize distinct organs (Carmeliet 2003, Lee, Song, and Chen 2016).

Thus far, therapeutic approaches to halt angiogenesis in tumors remain inefficient as the tumors continue to develop alternative pathways to trigger angiogenesis even when certain angiogenic pathways were blocked. However, anti-angiogenic therapies have shed light on a concept of vessel normalization to improve the efficacy of drugs (Jain 2005). In physiologic angiogenesis, both pro- and anti-angiogenic stimuli are tightly regulated to ensure a healthy angiogenesis. In contrast, in a tumor growth, endothelial cells are constantly exposed to an excessive amount of pro-angiogenic factors produced by tumor cells. As a result, these tumor-associated vessels are abnormal, leaky, and lack proper perivascular cell coverage, which render them inefficient to deliver blood flow and chemotherapeutic drugs. Thus, normalization of the vessels to improve the function and restore the blood flow of tumor vessels might beneficial to the treatment of cancer (Jain 2005).

Here, so far, we have only discussed a few of the physiological and pathological conditions where angiogenesis contributes. We encourage the readers to inquire a more detailed list of angiogenesis-related diseases elsewhere (Carmeliet 2003). For example, bone fracture fails to heal when angiogenesis inhibitors are used. Serious complication of rheumatoid arthritis can lead to vasculitis where the blood vessels are inflamed and sometimes become narrow to prevent adequate blood flow. In the eye, diabetic retinopathy is a complication in diabetic patients. Blood vessels are also inflamed in diabetic retinopathy and can leak fluid into the eye. At the advanced stage, angiogenesis occurs to cause excessive outgrowth of abnormal blood vessels in the eye. If not properly treated, diabetic retinopathy can lead to permanent vision loss (Carmeliet 2003).

2.3.2 Lymphatic Vascular Pathology

2.3.2.1 Lymphedema

Lymphedema, the excess accumulation of interstitial fluid resulting from impaired fluid drainage into the lymphatic vasculature, affects more than 150 million individuals worldwide. There are two types of lymphedema: (1) primary and (2) secondary lymphedema (Doller 2013). Primary lymphedema is caused by inherent genetic alterations in patients, which accounts for approximately 10% incidence among all the lymphedema cases worldwide. In human, Milroy's syndrome, lymphedema at the lower limbs presenting at birth or shortly after, has been described with mutations in VEGFR3 (Butler et al. 2007); lymphedema-distichiasis is known to involve FOXC2 mutations (Fang et al. 2000); and hypotrichosis–lymphedema–telangiectasia is caused by mutations in the transcription factor SOX18 (Irrthum et al. 2003). In mouse models, deficiencies or mutations in (1) lymphatics-related extracellular proteins (Angiopoietin-2, VEGF-C), (2) intracellular/membrane proteins (Ephrin B2, Integrin α9, NRP2, PDPN, SLP-76/SYK, SPRED1/2, VEGFR3), and (3) nuclear proteins (FOXC2, PROX1, SOX18) showed lymphatic dysfunction causing lymphedema-like phenotypes.

Secondary lymphedema, a major form of lymphedema (~90% by etiology), is an acquired formation of lymphedema that can be caused by lymphatic filariasis (a parasitic disease), lymph node dissection, tumor excision, trauma, infection, inflammation, fibrosis, obesity, and vascular anomaly (Rockson 2014). There is still no clinically available drug treatment, and the standard nonoperative care (e.g., limb elevation, compression garment, decongestive therapy) is largely palliative. Other operative managements (e.g., lymphaticovenular anastomosis, vascularized lymph node transfer) are highly invasive and involve complications.

2.3.2.2 Immune Dysfunction

The lymphatic system including lymphatic vessels and lymphoid organs modulates host immunity (Rockson 2013). Lymphatic vessels carry immune cells and lymphoid organs prepare, adapt and mature the immune cells. There are two types of lymphoid organs: (1) primary and (2) secondary lymphoid organs. The primary lymphoid organs are bone marrow and thymus. Bone marrow generates red and white blood cells, and the white blood cells include T/B lymphocytes and natural killer cells. Thymus differentiates T prolymphocytes to T lymphocytes. The secondary lymphoid organs include spleen, lymph nodes, bronchus-associated lymphatic tissue, mucosa-associated lymphatic tissue (MALT), and tonsils. Spleen is the largest lymphatic organ in the human body which filters old erythrocytes and thrombocytes through the fenestrated endothelium so that only intact erythrocytes migrate back to the blood stream, but old erythrocytes or thrombocytes are filtered and broken by spleen-residing macrophages. Lymph nodes filter cell debris in lymph fluid, but more importantly they store T/B lymphocytes and activate them via the proper antigen presentation. T/B lymphocytes are grouped in lymph follicles ("B lymphocyte zone") and in parafollicular tissues ("T lymphocyte zone"). Lymph nodes and lymphatic vessels cooperatively modulate adaptive immunity by draining antigen presenting cells, like dendritic cells (DCs), and delivering the antigen presenting cells to lymph nodes where T/B lymphocytes reside. Under infection, initial lymphatics rapidly uptake DCs, as they uptake interstitial fluid in an elevated fluid pressure, and then transport the DCs to draining lymph nodes, allowing T/B cells to be activated and propagate immune reactions (e.g., B-cell–mediated antibody production, and T-cell activation for direct killing the pathogens) in the body. Bronchus-associated lymphatic tissue and MALT are additional aggregate follicles to facilitate lymphocyte maturation and activation in different organs, such as the digestive and respiratory tracts. Tonsils have on their surface specialized antigen capture cells, called M cells, that allow for the uptake of antigens produced by pathogens. These M cells then alert the underlying B and T cells in the tonsil that a pathogen is present, and an immune response is stimulated.

When focused on the lymphatic vessel function and host immunity, the majority of lymphedema patients suffer from frequent skin infections, known as "cellulitis", owing to the impaired adaptive immunity. As the lymphedematous initial lymphatic vessels fail to drain interstitial fluid and the affected collecting lymphatic vessels fail to transport the fluid, the affected initial or collecting lymphatic vessels might also fail to capture and deliver the antigen presenting cells to the lymph nodes, so the T/B lymphocytes might not be activated in an appropriate manner.

2.3.2.3 Tumor Metastasis and Immunity

Lymphatic vessels are one of the routes of tumor dissemination in many types of cancers (Karkkainen, Makinen, and Alitalo 2002). The cancer cells can directly invade lymphatic vessels or express VEGF-C to induce tumor lymphangiogenesis in the tumor microenvironment. Genetic knock-out/knock-down or inhibitor-mediated blockage of VEGF-C, VEGFR3, and NRP2 ameliorated metastatic diseases in many types of tumors (Wang et al. 2016). In the clinic, lymph node dissection is a common process for tumor patients with local metastasis, evidencing that tumor lymphatics are the way of tumor cell exiting from the primary tumor sites. There have been studies describing certain cytokines or growth factors or adhesion signals as metastatic cues to promote tumor cell invading the lymphatic systems (Lee, Pandey, and Popel 2015). It is known that lymphatics express CXCL12 or CCL21 chemokines, and these chemokines recruit CXCR4- or CCR7-expressing tumor cells (Murphy 2001). It has been described that IL-6 expressed in cancer cells influence lymphatics to secrete CCL5 and recruit CCR5-expressing cancer cells to the lymphatic system (Lee et al. 2014). Hyaluronan (HA) expressed in cancer cells mediates cancer cell adhesion to the lymphatic endothelium by binding to LYVE-1.

Some studies described that after anti-angiogenic therapies, primary tumor burden is reduced, but metastasis is elevated via tumor invasion to lymphatics that are not fully inhibited by the current anti-angiogenic therapies. Similarly, hypoxic condition followed by inhibition of blood vessel growths boosted tumor spreading through the lymphatics. Further, it is described that tumor lymphatics directly support tumor growth by expressing tumor cell proliferating factors and inducing angiogenesis (Lee, Pandey, and Popel 2014). From these reasons, there have been trials to target tumor lymphangiogenesis and defeat lymphatic tumor invasion. Although no FDA-approved drugs are currently available to target lymphatic vessels, some are in preclinical development. They are antibodies, small molecules, and peptides mostly targeting VEGF-C/D, VEGFR3, and NRP2 (Lee, Pandey, and Popel 2015). For example, somatotropin and collagen IV mimetic anti-lymphangiogenic peptides were developed and tried to stop lymphatic vessel growth in the tumors and organs to slow down metastasis (Lee et al. 2011).

Tumor immunotherapies have been emerging, and a few strategies are already used in the clinics for blood cancers. Tumor lymphatics largely influence tumor immunity. Lymphatics-induced immune modification and tumor immune tolerance have been reported (Swartz and Lund 2012). Although the lymph nodes are abundant with T/B lymphocytes and tumor antigen-specific immune reactions normally occur in the lymph nodes, tumor cells invaded lymph nodes even establish and maintain tumor growing niche without normal lymph node function against tumor cells. This suggests that tumor cells might modify the lymph node environment, thus impairing host immunity. One proposed mechanism is that CCL21 expressed by lymphatics recruits CCR7-positive naive T cells into the lymph nodes and tumor stroma where they are educated to be less immune reactive. Similarly, CCR7-positive DCs are maintained in their immature state in the primary tumors and the tumor draining lymph nodes. Further, the immature DCs promote tumor-associated regulatory T cell activity to suppress cytotoxic T lymphocytes. TGFβ secreted in tumors also inhibits natural killer cell functions. TGFβ promotes tumor-associated regulatory T lymphocytes, causing tumor immune tolerance (Swartz and Lund 2012). Further tumor lymphatics expressed programmed death-ligand 1 (PD-L1) that can bind to PD-1 on T lymphocytes and induce an inhibitory signal that reduces proliferation of antigen-specific T lymphocytes (Dieterich et al. 2017). While several studies have described immunosuppressive roles of tumor lymphatics, one recent study revealed that tumor lymphatics can serve as a route of cytotoxic T-cell entrance to the tumor, synergistically boosting anti-tumor immunity (Fankhauser et al. 2017). Lymphatics-mediated tumor immunity is still controversial and needs further investigations.

2.4 *IN VITRO* BIOMIMETIC MICROSYSTEMS

2.4.1 MODELS FOR BLOOD VASCULATURE

The formation of blood vessels is a complex process that involves multiple steps. Thus, to understand blood vessel formation, both in vivo and in vitro models are essential and complementary to one another to unravel both the biophysical and biochemical stimuli that regulate blood vessel development. Here, we mainly discuss the advances of the field in developing in vitro models from two-dimensional (2D) to three-dimensional (3D) platforms to gain mechanistic insights in the formation of blood vessels through vasculogenesis and angiogenesis.

2.4.1.1 2D Wound Healing and Tube Formation Assays

One of the simplest in vitro assays for vessel formation is a simple 2D wound healing assay where endothelial cells are plated on tissue culture dish as a monolayer. The monolayer is disrupted by a sharp object to remove an area of endothelial cells to create a "wound" area. Over time, the area is gradually occupied by migrating endothelial cells (Liang, Park, and Guan 2007). Another common assay that captures some aspects of blood vessel formation is called tube formation assay, in which endothelial cells are plated on top of Matrigel and form a network of endothelial cells. This assay is known to capture some aspects of endothelial network formation during vasculogenesis (DeCicco-Skinner et al. 2014) (Figure 2.3a). However, these assays still lack the 3D process of vasculogenesis observed in vivo.

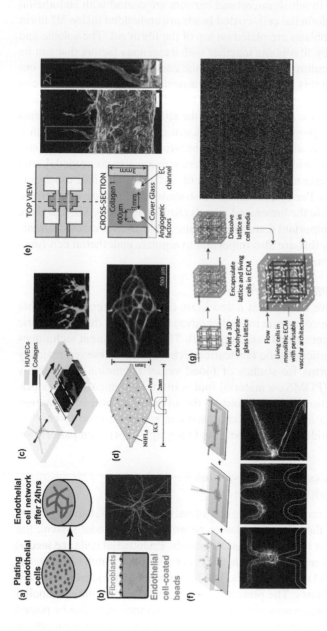

FIGURE 2.3 Biomimetic models for the studies of blood vasculature. (a) Endothelial cell tube formation assay to mimic the vasculogenic process. (b) Fibrin bead assay to mimic angiogenic invasion in 3D fibrin gel. (c) Microfluidic platform to mimic vascular walls and angiogenic invasion. (d) 3D vascular network formed through endothelial cell self-assembly in the presence of human lung fibroblasts. (e) Model of 3D biomimetic blood vessel and angiogenesis: endothelial cell channel is embedded inside hydrogel and sprout in response to angiogenic stimuli. (f) Soft-lithography to pattern complex vascular architecture. (g) 3D printing technology to generate a network of complex and multi-layer vascular channels. Scale bars: (b,i): 200 μm, (e) 100 μm (50 μm in inset), (f) 500 μm, and (g) 1 mm. (Illustration in panel b,i was adapted with permission from Eglinger et al. (2017, #164) under terms of the Creative Common Attribution 4.0 International License; panel c was adapted with permission from Song et al. (2012, #162); panel d was adapted with permission from Hsu et al. (2013); panel e was adapted with permission from Nguyen et al. (2013); panel f was adapted with permission from Baker et al. (2013); and panel g was adapted with permission from Miller et al. (2012).) 3D, three-dimensional.

2.4.1.2   3D Fibrin Bead Sprouting and Tubulogenesis Assays

To address the shortcoming of tube formation assay and wound healing assays, fibrin bead assay was developed in which microbead carriers are coated with endothelial cells (Figure 2.3b). The endothelial cell–coated beads are embedded inside 3D fibrin gel where human lung fibroblasts are plated on top of the fibrin gel. The soluble and insoluble factors secreted by fibroblasts together with exogenous factors that can be supplemented into the co-culture medium enable the endothelial cells to sprout out from the fibrin beads. With this system, some key aspects of angiogenic sprouts can be assessed such as sprout length, sprout density, branching, and lumen formation. Using this system, studies have also reported that the stiffness of the ECM proteins also regulates angiogenic sprouting in vitro (Nakatsu and Hughes 2008). With a similar approach, others have also developed a tubulogenesis assay. In this system, endothelial cells are seeded inside 3D collagen I. Pericytes are also co-seeded. Using this system, studies have reported the importance of different cytokines to support lumenized vessel formation. Pericyte co-culture also shows the increase in basement membrane deposition outside the lumenized vessels (Koh et al. 2008). Though these assays can capture many important features of 3D angiogenesis and vasculogenesis, they both lack shear forces that are also important to regulate endothelial cell function and homeostasis in vivo.

2.4.1.3   Microfluidic Angiogenesis Assay

Microfluidic platforms have emerged as a new class of model systems that enable perfusion of medium in the co-culture to further advance and refine the mechanistic understandings of angiogenesis and vasculogenesis. There are different techniques and model systems using microfluidic fabrication technologies. However, majority of the microfluidic platforms for studies of blood vessel formation rely heavily on polydimethylsiloxane (PDMS), a material that is inert, biocompatible, gas permeable, and can be functionalized to bind to the hydrogel. Another advantage of PDMS is that it is also a flexible material and can easily be used to cast off microchannels or microfluidic devices from soft-lithograph–fabricated patterns on silicon wafers.

2.4.1.3.1   Self-Assembly of Vascular Endothelial
Cells in Microfluidic Platforms

Even though there are different microfluidic platforms, they can be classified into two main categories to model the microvasculature network. In one model system, the microvasculature network is formed via vasculogenesis. Endothelial cells are seeded into a rectangular hydrogel region (with either fibrin or collagen), which is then sandwiched between two parallel microchannels. The microchannels are seeded with cell culture medium and fibroblasts. The presence of fibroblasts enables the endothelial cells to form a network of microvasculature. The microchannels can also be ported to a pump to provide shear forces into the system. The network of endothelial cells is fully lumenized as shown by perfusion of fluorescent microbeads (Song et al. 2018) (Figure 2.3c). Other groups also introduced some variants of the gel regions from a rectangular to a diamond shape (Hsu et al. 2013) (Figure 2.3d). With these systems, authors also incorporated tumor cells into the perfused microvasculature and studied the efficacy of different drugs, as tumor cells were interacting with the microvasculature (Phan et al. 2017).

2.4.1.3.2　Patterning Vascular Vessel Walls in Microfluidic Devices

In another system, instead of utilizing the vasculogenic properties of endothelial cells to self-assemble into microvasculature when cultured with fibrin and appropriate fibroblasts, several groups have utilized different technologies to mold and pattern vascular mimicry. In one approach, soft-lithography is used to generate patterns of rectangular microchannels on silicon wafers. PDMS is then cast off to form the silicon wafer master to obtain hollow rectangular microchannels. Endothelial cells are seeded inside the microchannel to form a monolayer of endothelium. However, due to the limitation of soft-lithography, although endothelial cells are lining the rectangular channel, only one of the surfaces of the rectangular channel is adjacent to a hydrogel, which constrains the endothelial cells to emanate into the hydrogel from one surface of the rectangular channels. With these analogs of vessel wall, studies have been used to elucidate the role of shear forces and pressure gradient to mediate the angiogenic process of endothelial cells (Song and Munn 2011; https://www.ncbi. nlm.nih.gov/pmc/articles/PMC3490212/).

2.4.1.3.3　Generation of Cylindrical Blood Vessel Analog

Others also reported an advanced improvement of generating biomimetic blood vessels with proper vessel geometry. In these systems, biomimetic blood vessels are created by threading acupuncture needles inside a collagen gel. Once the collagen gel is polymerized, acupuncture needles are retracted, leaving hollow cylindrical channels completely submerged inside the 3D collagen gel. Endothelial cells are then seeded into the hollow cylindrical channels to form biomimetic blood vessels. Acupuncture needle diameters ranged from 100 to 400 μm can be used to create biomimetic vessels of different calibers. Endothelial cells are free to emanate in all directions from the biomimetic blood vessels. However, when an additional channel is placed in parallel with the biomimetic blood vessel to introduce a gradient of angiogenic factors, endothelial cells are triggered to sprout towards the angiogenic gradient and gradually develop into perfusable neovessels connecting the two parallel channels (Nguyen et al. 2013) (Figure 2.3e). Using these biomimetic blood vessels, studies also reported the noncanonical Notch signaling pathway to regulate adherent junctions and vascular barrier function, and the importance of perivascular cells to regular vascular permeability via RhoA and Rac1 (Polacheck et al. 2017, Alimperti et al. 2017). Vessel branching is also reported to be regulated by Cdc42 using this system (Nguyen et al. 2017).

2.4.1.3.4　Patterning of Complex Vascular Network

Though individual biomimetic blood vessel has emerged as a valuable tool to study blood vessel properties, the blood circulation is a network of multiple blood vessels that are connected and branched out with defined geometry. Others have used different manufacturing approaches to build a network of vasculature.

Though they are different approaches, they share some common features as these methods often require a sacrificial material such as gelatin or biocompatible carbohydrate glass. In one method, pattern of a more complex vascular network is generated by using soft-lithography. Sacrificial gelatin is then injected into the patterns encased in a PDMS gasket. After a hydrogel is cast over the gelatin vascular network, and gelatin is dissolved, endothelial cells are seeded inside to form a complex network of vasculature (Baker et al. 2013) (Figure 2.3f). An advanced technology in 3D printing together with sacrificial material has enabled the generation of complex 3D vasculature network suspended in 3D environment. Miller et al. reported using a customized 3D printer to extrude carbohydrate glass to generate complex structure of sacrificial blood vessel channels. These channels were seeded with endothelial cells once the sacrificial carbohydrate channels were dissolved (Miller et al. 2012) (Figure 2.3g).

2.4.2 MODELS FOR LYMPHATIC VESSELS

Compared to blood vessels, biomimetic in vitro models of lymphatic vessels are much less investigated, in part because the history of lymphatic research is much shorter than that of blood vascular research owing to recent discoveries of lymphatic endothelial markers, such as LYVE-1, PDPN, and VEGFR3 in the 1990s–2000s. We describe some examples of biomimetic models of lymphatics and discuss factors to be considered in building up engineered lymphatics. We note that some traditional models of blood vessels such as 2D wound healing, tube formation, and fibrin bead assays (described in Section 2.4.1) have been similarly applied for modeling lymphatics in many of literatures, so that we did not include those examples here.

2.4.2.1 Examples of Biomimetics for Lymphatic Vascular Research

There have been some in vitro models for studying lymphatic vessel morphogenesis, which include vasculogenic lymphatic network formation and lymphangiogenic vessel sprouting. Lymphatic vasculogenesis is a network formation of the pre-existing LECs without external gradients of morphogens, mediated by autonomous cell migration and cell–cell junction formation and lumen formation (Figure 2.4a). Lymphatic vasculogenic models (Gibot et al. 2016) are often based on hydrogels (e.g., collagen, fibrin, HA) mixed with LECs supported by third cell types, such as lung/dermal fibroblasts or adipose-derived stromal cells, or by growth factor cocktails of VEGF-C and VEGF-A. The models showed lymphatic vascular network (Figure 2.4b) and lymphatic specific overlapping junctions in vitro (Figure 2.4c). Lymphatic sprouting is featured by directed lymphatic cell budding from the existing lymph endothelial clusters or lymphatic vessels via the gradients of morphogens (Figure 2.4d). Some studies using the gradients of lymphatic growth factors showed lymphatic sprouting in vitro in the fibrin/collagen-based hydrogel, testing out different growth factors and inhibitors with or without interstitial flow (Kim, Chung, and Jeon 2016).

Primary function of lymphatics is drainage of fluid and cells. There have been studies using different models (e.g., transwells, spheroid-beads, microfluidic, multichamber systems) of 2D or 3D LECs under interstitial flow, which showed

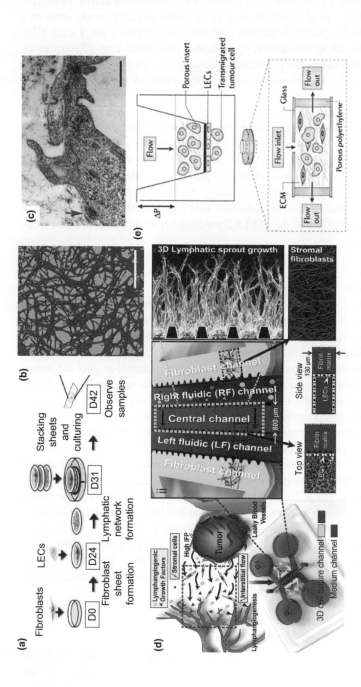

FIGURE 2.4 Biomimetic lymphatic vessel models. (a) A vasculogenesis model for 3D lymphatic network formation in vitro, which is supported by fibroblasts. (b) A representative image of the lymphatic network. Scale bar 500 μm. (c) A representative TEM image showing overlapping LECs, button-like junctions (green arrows), anchoring filaments (red arrow). Scale bar 500 nm. (d) A lymphatic sprouting model for 3D lymphangiogenesis under lymphangiogenic growth factors and interstitial fluid, mimicking tumor stroma. (e) In vitro models for introducing interstitial flow to LECs in 2D (upper panel) and 3D (lower panel). (Illustration in panels a, b, and c was adapted with permission from Gibot et al. (2016); panel d was adapted with permission from Kim, Chung, and Jeon (2016); and panel e was adapted with permission from Swartz and Lund (2012)). 3D, three-dimensional; LEC, lymphatic endothelial cell; TEM, transmission electron microscopy.

differential lymph cell morphology, sprouting, mass (e.g., fluid, solute), or cell (e.g., immune cells, cancer cells) transport via cell–cell openings or active transcytosis (Figure 2.4d) (Shields et al. 2007, Bonvin et al. 2010). A recent study showed a robust model of trans-lymphatic migration of diverse cell types, CD4 T cell, CD8 T cells, macrophages, DCs, and cancer cells under interstitial flow (Xiong et al. 2017). A 3D model of rudimentary lymphatic vessels was employed and flow was introduced, revealing that flow downregulates Notch signal in LECs and promotes lymphatic sprouting in vitro and in vivo (Choi et al. 2017).

Synthetic or engineering approach has been used for studying mechanical properties in lymphatic vessel phenotypes. There was an in vitro model that can introduce mechanical stretch to show remodeling of lymphatic muscle cells obtained from the hindlimb of sheep (Hooks et al. 2019). There was an in vitro model that can introduce mechanical stretch to show altered expression of inflammatory cytokines and fibrotic markers in LECs (Wang et al. 2017). An in vitro study using synthetic matrix for differential matrix stiffness showed that LEC migration is promoted in softer matrix by upregulation of the GATA2-mediated lymphangiogenic transcriptome (Frye et al. 2018).

2.4.2.2   Mechanical Stimuli Considered for Biomimetic Lymphatic Vessels

Biomimetic microdevice for lymphatics are still under-investigated. There are modes of mechanical stimuli that need to be considered in the biomimetic lymphatic models. LECs in nature respond to several types of mechanical stimuli: (1) luminal flow-induced shear stress, (2) interstitial flow pressure, and (3) ECM stiffness. (1) Luminal flow-induced shear stress is defined as the force that acts tangential to cells. Collecting lymphatic vessels experience unidirectional intraluminal flow inducing shear stress; thus, the flow-induced shear stress needs to be considered for understanding the morphogenesis, valve genesis, permeability, and mural cell interactions in the collecting lymphatics, compared to initial lymphatics that are less experiencing this type of shear. (2) Pressure is defined as the force applied to a given area. LECs in initial lymphatic vessels exposed to interstitial fluid pressure, which influences cell shape of LECs, cell–cell junction, degree of active transcytosis, tyrosine phosphorylation of VEGFR3, and LEC proliferation. (3) Stiffness is the ability of a cell to resist deformation. It is defined by the modulus of elasticity E and is mathematically defined as the force applied per unit area divided by the resultant strain. Lymphatic vessels exist almost in every organs and tissues, including both soft tissues (brain and adipose tissue) and stiff tissues (bone or tumors); thus, the consideration of diverse stiffness would be necessary to understand lymphatic structure and function in an organ-specific manner.

2.4.2.3   Extracellular Matrix Considered for Biomimetics of Lymphatic Vessels

The ECM is a complex meshwork of proteins and sugars, providing structural scaffolds and adhesive matrices for cells to form tissue-level, multicellular architectures for proper functions. In designing on-chip models, appropriate 3D structure is considered with what types of ECM can be applied to the structure for realistic lymphatic vessel structure and function is critical. The collecting lymphatic vessels have a continuous basement membrane composed of collagen IV, fibronectin, and laminins, which may be critical for collecting lymphatic zipper-like junctions and SMC coverage (Lutter and Makinen 2014). Luminal valves exist in the collecting lymphatic vessels, and the valve endothelial cells express high levels of integrin α9, a receptor of fibronectin–EIIIA/EDA spliced isoform (Altiok et al. 2015). Similarly, fibronectin–EIIIA/EDA is found in the ECM around lymphatic valves. Other matrix components of lymphatic valves include laminin-α5, whose function is still unknown. The smaller lymphatic capillaries lack SMCs and luminal valves, and have a discontinuous basement membrane containing gaps and button-like junctions. Expression of several laminins, collagens IV and XVIII, and nidogen-1 around initial lymphatic vessels was reported (Lutter et al. 2012). Since LECs largely express LYVE-1 in the initial lymphatics, its ligand HA is postulated as a functional ECM component.

2.5 CONCLUSIONS AND OUTLOOK

As discussed, there have been advances in in vitro biomimetic models of blood and lymphatic vessels. Given structural and functional interdependency of blood and lymphatic vasculatures, models of combined blood and lymphatic vessels in one system would be important to recapitulate overall dynamics of cell and mass transport between two vessels in homeostasis and diseases. Important challenge in future studies will also be to model organ-specific blood and lymphatic endothelium onchip using appropriately sourced organ-derived endothelial cells and organ-specific architectures. Given the emergence of the induced pluripotent stem cells (iPSCs) in the field of regenerative medicine, stem cell technologies for sourcing normal and disease blood/LECs from individuals would also be beneficial for improving personalized medicine. Regarding the role of stromal cells like mural cells in vessel contractility, permeability, and drainage, co-culturing of those stromal cells with endothelial cells would be important.

REFERENCES

Achen, M. G., M. Jeltsch, E. Kukk, T. Makinen, A. Vitali, A. F. Wilks, K. Alitalo, and S. A. Stacker. 1998. Vascular endothelial growth factor D (VEGF-D) is a ligand for the tyrosine kinases VEGF receptor 2 (Flk1) and VEGF receptor 3 (Flt4). *Proc Natl Acad Sci U S A* 95 (2):548–53.

Alimperti, S., T. Mirabella, V. Bajaj, W. Polacheck, D. M. Pirone, J. Duffield, J. Eyckmans, R. K. Assoian, and C. S. Chen. 2017. Three-dimensional biomimetic vascular model reveals a RhoA, Rac1, and N-cadherin balance in mural cell-endothelial cell-regulated barrier function. *Proc Natl Acad Sci U S A* 114 (33):8758–63.

Altiok, E., T. Ecoiffier, R. Sessa, D. Yuen, S. Grimaldo, C. Tran, D. Li, M. Rosner, N. Lee, T. Uede, and L. Chen. 2015. Integrin alpha-9 mediates lymphatic valve formation in corneal lymphangiogenesis. *Invest Ophthalmol Vis Sci* 56 (11):6313–9.

Aspelund, A., M. R. Robciuc, S. Karaman, T. Makinen, and K. Alitalo. 2016. Lymphatic system in cardiovascular medicine. *Circ Res* 118 (3):515–30.

Baker, B. M., B. Trappmann, S. C. Stapleton, E. Toro, and C. S. Chen. 2013. Microfluidics embedded within extracellular matrix to define vascular architectures and pattern diffusive gradients. *Lab Chip* 13 (16):3246–52.

Bautch, V. L., and K. M. Caron. 2015. Blood and lymphatic vessel formation. *Cold Spring Harb Perspect Biol* 7 (3):a008268.

Beenken, A., and M. Mohammadi. 2009. The FGF family: Biology, pathophysiology and therapy. *Nat Rev Drug Discov* 8 (3):235–53.

Bonvin, C., J. Overney, A. C. Shieh, J. B. Dixon, and M. A. Swartz. 2010. A multichamber fluidic device for 3D cultures under interstitial flow with live imaging: Development, characterization, and applications. *Biotechnol Bioeng* 105 (5):982–91.

Butler, M. G., S. L. Dagenais, S. G. Rockson, and T. W. Glover. 2007. A novel VEGFR3 mutation causes Milroy disease. *Am J Med Genet A* 143A (11):1212–7.

Carmeliet, P. 2003. Angiogenesis in health and disease. *Nat Med* 9 (6):653–60.

Carmeliet, P., V. Ferreira, G. Breier, S. Pollefeyt, L. Kieckens, M. Gertsenstein, M. Fahrig, A. Vandenhoeck, K. Harpal, C. Eberhardt, C. Declercq, J. Pawling, L. Moons, D. Collen, W. Risau, and A. Nagy. 1996. Abnormal blood vessel development and lethality in embryos lacking a single VEGF allele. *Nature* 380 (6573):435–9.

Casley-Smith, J. R. 1980. The fine structure and functioning of tissue channels and lymphatics. *Lymphology* 13 (4):177–83.

Choi, D., E. Park, E. Jung, Y. J. Seong, J. Yoo, E. Lee, M. Hong, S. Lee, H. Ishida, J. Burford, J. Peti-Peterdi, R. H. Adams, S. Srikanth, Y. Gwack, C. S. Chen, H. J. Vogel, C. J. Koh, A. K. Wong, and Y. K. Hong. 2017. Laminar flow downregulates Notch activity to promote lymphatic sprouting. *J Clin Invest* 127 (4):1225–40.

Corada, M., D. Nyqvist, F. Orsenigo, A. Caprini, C. Giampietro, M. M. Taketo, M. L. Iruela-Arispe, R. H. Adams, and E. Dejana. 2010. The Wnt/beta-catenin pathway modulates vascular remodeling and specification by upregulating Dll4/Notch signaling. *Dev Cell* 18 (6):938–49.

Curran, M. E., D. L. Atkinson, A. K. Ewart, C. A. Morris, M. F. Leppert, and M. T. Keating. 1993. The elastin gene is disrupted by a translocation associated with supravalvular aortic stenosis. *Cell* 73 (1):159–68.

de Mooij, Y. M., N. M. van den Akker, M. N. Bekker, M. M. Bartelings, L. J. Wisse, J. M. van Vugt, and A. C. Gittenberger-de Groot. 2009. Abnormal Shh and FOXC2 expression correlates with aberrant lymphatic development in human fetuses with increased nuchal translucency. *Prenat Diagn* 29 (9):840–6.

DeCicco-Skinner, K. L., G. H. Henry, C. Cataisson, T. Tabib, J. C. Gwilliam, N. J. Watson, E. M. Bullwinkle, L. Falkenburg, R. C. O'Neill, A. Morin, and J. S. Wiest. 2014. Endothelial cell tube formation assay for the in vitro study of angiogenesis. *J Vis Exp* (91):e51312.

Dieterich, L. C., K. Ikenberg, T. Cetintas, K. Kapaklikaya, C. Hutmacher, and M. Detmar. 2017. Tumor-associated lymphatic vessels upregulate PDL1 to inhibit T-cell activation. *Front Immunol* 8:66.

Doller, W. 2013. Lymphedema: Anatomy, physiology and pathophysiology of lymphedema, definition and classification of lymphedema and lymphatic vascular malformations. *Wien Med Wochenschr* 163 (7–8):155–61.

Fagiani, E., and G. Christofori. 2013. Angiopoietins in angiogenesis. *Cancer Lett* 328 (1):18–26.

Fang, J., S. L. Dagenais, R. P. Erickson, M. F. Arlt, M. W. Glynn, J. L. Gorski, L. H. Seaver, and T. W. Glover. 2000. Mutations in FOXC2 (MFH-1), a forkhead family transcription factor, are responsible for the hereditary lymphedema-distichiasis syndrome. *Am J Hum Genet* 67 (6):1382–8.

Fankhauser, M., M. A. S. Broggi, L. Potin, N. Bordry, L. Jeanbart, A. W. Lund, E. Da Costa, S. Hauert, M. Rincon-Restrepo, C. Tremblay, E. Cabello, K. Homicsko, O. Michielin, D. Hanahan, D. E. Speiser, and M. A. Swartz. 2017. Tumor lymphangiogenesis promotes T cell infiltration and potentiates immunotherapy in melanoma. *Sci Transl Med* 9 (407).

Francois, M., A. Caprini, B. Hosking, F. Orsenigo, D. Wilhelm, C. Browne, K. Paavonen, T. Karnezis, R. Shayan, M. Downes, T. Davidson, D. Tutt, K. S. Cheah, S. A. Stacker, G. E. Muscat, M. G. Achen, E. Dejana, and P. Koopman. 2008. Sox18 induces development of the lymphatic vasculature in mice. *Nature* 456 (7222):643–7.

Frye, M., A. Taddei, C. Dierkes, I. Martinez-Corral, M. Fielden, H. Ortsater, J. Kazenwadel, D. P. Calado, P. Ostergaard, M. Salminen, L. He, N. L. Harvey, F. Kiefer, and T. Makinen. 2018. Matrix stiffness controls lymphatic vessel formation through regulation of a GATA2-dependent transcriptional program. *Nat Commun* 9 (1):1511.

Furuyama, T., K. Kitayama, Y. Shimoda, M. Ogawa, K. Sone, K. Yoshida-Araki, H. Hisatsune, S. Nishikawa, K. Nakayama, K. Ikeda, N. Motoyama, and N. Mori. 2004. Abnormal angiogenesis in Foxo1 (Fkhr)-deficient mice. *J Biol Chem* 279 (33):34741–9.

Gibot, L., T. Galbraith, B. Kloos, S. Das, D. A. Lacroix, F. A. Auger, and M. Skobe. 2016. Cell-based approach for 3D reconstruction of lymphatic capillaries in vitro reveals distinct functions of HGF and VEGF-C in lymphangiogenesis. *Biomaterials* 78:129–39.

Herzog, B. H., J. Fu, S. J. Wilson, P. R. Hess, A. Sen, J. M. McDaniel, Y. Pan, M. Sheng, T. Yago, R. Silasi-Mansat, S. McGee, F. May, B. Nieswandt, A. J. Morris, F. Lupu, S. R. Coughlin, R. P. McEver, H. Chen, M. L. Kahn, and L. Xia. 2013. Podoplanin maintains high endothelial venule integrity by interacting with platelet CLEC-2. *Nature* 502 (7469):105–9.

Hooks, J. S. T., C. C. Clement, H. D. Nguyen, L. Santambrogio, and J. B. Dixon. 2019. In vitro model reveals a role for mechanical stretch in the remodeling response of lymphatic muscle cells. *Microcirculation* 26 (1):e12512.

Hsu, Y. H., M. L. Moya, C. C. Hughes, S. C. George, and A. P. Lee. 2013. A microfluidic platform for generating large-scale nearly identical human microphysiological vascularized tissue arrays. *Lab Chip* 13 (15):2990–8.

Irrthum, A., K. Devriendt, D. Chitayat, G. Matthijs, C. Glade, P. M. Steijlen, J. P. Fryns, M. A. Van Steensel, and M. Vikkula. 2003. Mutations in the transcription factor gene SOX18 underlie recessive and dominant forms of hypotrichosis-lymphedema-telangiectasia. *Am J Hum Genet* 72 (6):1470–8.

Jain, R. K. 2005. Normalization of tumor vasculature: An emerging concept in antiangiogenic therapy. *Science* 307 (5706):58–62.

Jakobsson, L., C. A. Franco, K. Bentley, R. T. Collins, B. Ponsioen, I. M. Aspalter, I. Rosewell, M. Busse, G. Thurston, A. Medvinsky, S. Schulte-Merker, and H. Gerhardt. 2010. Endothelial cells dynamically compete for the tip cell position during angiogenic sprouting. *Nat Cell Biol* 12 (10):943–53.

Kanady, J. D., S. J. Munger, M. H. Witte, and A. M. Simon. 2015. Combining Foxc2 and Connexin37 deletions in mice leads to severe defects in lymphatic vascular growth and remodeling. *Dev Biol* 405 (1):33–46.

Karkkainen, M. J., T. Makinen, and K. Alitalo. 2002. Lymphatic endothelium: A new frontier of metastasis research. *Nat Cell Biol* 4 (1):E2–5.

Kim, S., M. Chung, and N. L. Jeon. 2016. Three-dimensional biomimetic model to reconstitute sprouting lymphangiogenesis in vitro. *Biomaterials* 78:115–28.

Koh, W., A. N. Stratman, A. Sacharidou, and G. E. Davis. 2008. In vitro three dimensional collagen matrix models of endothelial lumen formation during vasculogenesis and angiogenesis. *Methods Enzymol* 443:83–101.

Leak, L. V. 1970. Electron microscopic observations on lymphatic capillaries and the structural components of the connective tissue-lymph interface. *Microvasc Res* 2 (4):361–91.

Lee, E., E. J. Fertig, K. Jin, S. Sukumar, N. B. Pandey, and A. S. Popel. 2014. Breast cancer cells condition lymphatic endothelial cells within pre-metastatic niches to promote metastasis. *Nat Commun* 5:4715.

Lee, E., N. B. Pandey, and A. S. Popel. 2014. Lymphatic endothelial cells support tumor growth in breast cancer. *Sci Rep* 4:5853.

Lee, E., N. B. Pandey, and A. S. Popel. 2015. Crosstalk between cancer cells and blood endothelial and lymphatic endothelial cells in tumour and organ microenvironment. *Expert Rev Mol Med* 17:e3.

Lee, E., E. V. Rosca, N. B. Pandey, and A. S. Popel. 2011. Small peptides derived from somato-tropin domain-containing proteins inhibit blood and lymphatic endothelial cell prolifer-ation, migration, adhesion and tube formation. *Int J Biochem Cell Biol* 43 (12):1812–21.

Lee, E., H. G. Song, and C. S. Chen. 2016. Biomimetic on-a-chip platforms for studying cancer metastasis. *Curr Opin Chem Eng* 11:20–27.

Lee, M. E., D. H. Temizer, J. A. Clifford, and T. Quertermous. 1991. Cloning of the GATA-binding protein that regulates endothelin-1 gene expression in endothelial cells. *J Biol Chem* 266 (24):16188–92.

Liang, C. C., A. Y. Park, and J. L. Guan. 2007. In vitro scratch assay: A convenient and inexpensive method for analysis of cell migration in vitro. *Nat Protoc* 2 (2):329–33.

Libby, P., P. M. Ridker, and G. K. Hansson. 2011. Progress and challenges in translating the biology of atherosclerosis. *Nature* 473 (7347):317–25.

Lugus, J. J., Y. S. Chung, J. C. Mills, S. I. Kim, J. Grass, M. Kyba, J. M. Doherty, E. H. Bresnick, and K. Choi. 2007. GATA2 functions at multiple steps in hemangioblast development and differentiation. *Development* 134 (2):393–405.

Lutter, S., and T. Makinen. 2014. Regulation of lymphatic vasculature by extracellular matrix. *Adv Anat Embryol Cell Biol* 214:55–65. doi:10.1007/978-3-7091-1646-3_5.

Lutter, S., S. Xie, F. Tatin, and T. Makinen. 2012. Smooth muscle-endothelial cell communication activates Reelin signaling and regulates lymphatic vessel formation. *J Cell Biol* 197 (6):837–49.

Mazurek, R., J. M. Dave, R. R. Chandran, A. Misra, A. Q. Sheikh, and D. M. Greif. 2017. Vascular cells in blood vessel wall development and disease. *Adv Pharmacol* 78:323–50.

Mellor, R. H., N. Tate, A. W. Stanton, C. Hubert, T. Makinen, A. Smith, K. G. Burnand, S. Jeffery, J. R. Levick, and P. S. Mortimer. 2011. Mutations in FOXC2 in humans (lymphoedema distichiasis syndrome) cause lymphatic dysfunction on dependency. *J Vasc Res* 48 (5):397–407.

Miller, J. S., K. R. Stevens, M. T. Yang, B. M. Baker, D. H. Nguyen, D. M. Cohen, E. Toro, A. A. Chen, P. A. Galie, X. Yu, R. Chaturvedi, S. N. Bhatia, and C. S. Chen. 2012. Rapid casting of patterned vascular networks for perfusable engineered three-dimensional tissues. *Nat Mater* 11 (9):768–74.

Murphy, P. M. 2001. Chemokines and the molecular basis of cancer metastasis. *N Engl J Med* 345 (11):833–5.

Nakatsu, M. N., and C. C. Hughes. 2008. An optimized three-dimensional in vitro model for the analysis of angiogenesis. *Methods Enzymol* 443:65–82.

Nguyen, D. T., L. Gao, A. Wong, and C. S. Chen. 2017. Cdc42 regulates branching in angiogenic sprouting in vitro. *Microcirculation* 24 (5). doi:10.1111/micc.12372.

Nguyen, D. H., S. C. Stapleton, M. T. Yang, S. S. Cha, C. K. Choi, P. A. Galie, and C. S. Chen. 2013. Biomimetic model to reconstitute angiogenic sprouting morphogenesis in vitro. *Proc Natl Acad Sci U S A* 110 (17):6712–7.

Noon, A., R. J. Hunter, M. H. Witte, B. Kriederman, M. Bernas, M. Rennels, D. Percy, S. Enerback, and R. P. Erickson. 2006. Comparative lymphatic, ocular, and metabolic phenotypes of Foxc2 haploinsufficient and aP2-FOXC2 transgenic mice. *Lymphology* 39 (2):84–94.

Norrmen, C., K. I. Ivanov, J. Cheng, N. Zangger, M. Delorenzi, M. Jaquet, N. Miura, P. Puolakkainen, V. Horsley, J. Hu, H. G. Augustin, S. Yla-Herttuala, K. Alitalo, and T. V. Petrova. 2009. FOXC2 controls formation and maturation of lymphatic collecting vessels through cooperation with NFATc1. *J Cell Biol* 185 (3):439–57.

Oliver, G., B. Sosa-Pineda, S. Geisendorf, E. P. Spana, C. Q. Doe, and P. Gruss. 1993. Prox 1, a prospero-related homeobox gene expressed during mouse development. *Mech Dev* 44 (1):3–16.

Park, C., T. M. Kim, and A. B. Malik. 2013. Transcriptional regulation of endothelial cell and vascular development. *Circ Res* 112 (10):1380–400.

Pendeville, H., M. Winandy, I. Manfroid, O. Nivelles, P. Motte, V. Pasque, B. Peers, I. Struman, J. A. Martial, and M. L. Voz. 2008. Zebrafish Sox7 and Sox18 function together to control arterial-venous identity. *Dev Biol* 317 (2):405–16.

Pepper, M. S., and M. Skobe. 2003. Lymphatic endothelium: Morphological, molecular and functional properties. *J Cell Biol* 163 (2):209–13.

Petrova, T. V., T. Makinen, T. P. Makela, J. Saarela, I. Virtanen, R. E. Ferrell, D. N. Finegold, D. Kerjaschki, S. Yla-Herttuala, and K. Alitalo. 2002. Lymphatic endothelial reprogramming of vascular endothelial cells by the Prox-1 homeobox transcription factor. *EMBO J* 21 (17):4593–9.

Phan, D. T. T., X. Wang, B. M. Craver, A. Sobrino, D. Zhao, J. C. Chen, L. Y. N. Lee, S. C. George, A. P. Lee, and C. C. W. Hughes. 2017. A vascularized and perfused organ-on-a-chip platform for large-scale drug screening applications. *Lab Chip* 17 (3):511–20.

Phng, L. K., M. Potente, J. D. Leslie, J. Babbage, D. Nyqvist, I. Lobov, J. K. Ondr, S. Rao, R. A. Lang, G. Thurston, and H. Gerhardt. 2009. Nrarp coordinates endothelial Notch and Wnt signaling to control vessel density in angiogenesis. *Dev Cell* 16 (1):70–82.

Polacheck, W. J., M. L. Kutys, J. Yang, J. Eyckmans, Y. Wu, H. Vasavada, K. K. Hirschi, and C. S. Chen. 2017. A non-canonical Notch complex regulates adherens junctions and vascular barrier function. *Nature* 552 (7684):258–62.

Potente, M., and T. Makinen. 2017. Vascular heterogeneity and specialization in development and disease. *Nat Rev Mol Cell Biol* 18 (8):477–94.

Rockson, S. G. 2013. Lymphatics: Where the circulation meets the immune system. *Lymphat Res Biol* 11 (3):115.

Rockson, S. G. 2014. Acquired lymphedema: Abnormal fluid clearance engenders tissue remodeling. *Lymphat Res Biol* 12 (1):1.

Rowe, V. L., S. L. Stevens, T. T. Reddick, M. B. Freeman, R. Donnell, R. C. Carroll, and M. H. Goldman. 2000. Vascular smooth muscle cell apoptosis in aneurysmal, occlusive, and normal human aortas. *J Vasc Surg* 31 (3):567–76.

Sabine, A., and T. V. Petrova. 2014. Interplay of mechanotransduction, FOXC2, connexins, and calcineurin signaling in lymphatic valve formation. *Adv Anat Embryol Cell Biol* 214:67–80.

Schulte-Merker, S., A. Sabine, and T. V. Petrova. 2011. Lymphatic vascular morphogenesis in development, physiology, and disease. *J Cell Biol* 193 (4):607–18.

Shalaby, F., J. Rossant, T. P. Yamaguchi, M. Gertsenstein, X. F. Wu, M. L. Breitman, and A. C. Schuh. 1995. Failure of blood-island formation and vasculogenesis in Flk-1-deficient mice. *Nature* 376 (6535):62–6.

Shibuya, M. 2011. Vascular endothelial growth factor (VEGF) and its receptor (VEGFR) signaling in angiogenesis: A crucial target for anti- and pro-angiogenic therapies. *Genes Cancer* 2 (12):1097–105.

Shields, J. D., M. E. Fleury, C. Yong, A. A. Tomei, G. J. Randolph, and M. A. Swartz. 2007. Autologous chemotaxis as a mechanism of tumor cell homing to lymphatics via interstitial flow and autocrine CCR7 signaling. *Cancer Cell* 11 (6):526–38.

Simonneau, G., M. A. Gatzoulis, I. Adatia, D. Celermajer, C. Denton, A. Ghofrani, M. A. Gomez Sanchez, R. Krishna Kumar, M. Landzberg, R. F. Machado, H. Olschewski, I. M. Robbins, and R. Souza. 2013. Updated clinical classification of pulmonary hypertension. *J Am Coll Cardiol* 62 (25 Suppl):D34–41.

Song, J., A. Miermont, C. T. Lim, and R. D. Kamm. 2018. A 3D microvascular network model to study the impact of hypoxia on the extravasation potential of breast cell lines. *Sci Rep* 8 (1):17949.

Song, J. W., and L. L. Munn. 2011. Fluid forces control endothelial sprouting. *Proc Natl Acad Sci U S A* 108 (37):15342–7.

Stamenkovic, I. 2003. Extracellular matrix remodelling: The role of matrix metalloproteinases. *J Pathol* 200 (4):448–64.

Swartz, M. A., and A. W. Lund. 2012. Lymphatic and interstitial flow in the tumour microenvironment: Linking mechanobiology with immunity. *Nat Rev Cancer* 12 (3):210–9.

Sweet, D. T., J. M. Jimenez, J. Chang, P. R. Hess, P. Mericko-Ishizuka, J. Fu, L. Xia, P. F. Davies, and M. L. Kahn. 2015. Lymph flow regulates collecting lymphatic vessel maturation in vivo. *J Clin Invest* 125 (8):2995–3007.

Tammela, T., and K. Alitalo. 2010. Lymphangiogenesis: Molecular mechanisms and future promise. *Cell* 140 (4):460–76.

Trujillo, A. N., C. Katnik, J. Cuevas, B. J. Cha, T. E. Taylor-Clark, and J. W. Breslin. 2017. Modulation of mesenteric collecting lymphatic contractions by sigma1-receptor activation and nitric oxide production. *Am J Physiol Heart Circ Physiol* 313 (4):H839–53.

Uhrin, P., J. Zaujec, J. M. Breuss, D. Olcaydu, P. Chrenek, H. Stockinger, E. Fuertbauer, M. Moser, P. Haiko, R. Fassler, K. Alitalo, B. R. Binder, and D. Kerjaschki. 2010. Novel function for blood platelets and podoplanin in developmental separation of blood and lymphatic circulation. *Blood* 115 (19):3997–4005.

van Helden, D. F. 2014. The lymphangion: A not so 'primitive' heart. *J Physiol* 592 (24):5353–4.

von der Weid, P. Y., and D. C. Zawieja. 2004. Lymphatic smooth muscle: The motor unit of lymph drainage. *Int J Biochem Cell Biol* 36 (7):1147–53.

Wang, J., Y. Huang, J. Zhang, Y. Wei, S. Mahoud, A. M. Bakheet, L. Wang, S. Zhou, and J. Tang. 2016. Pathway-related molecules of VEGFC/D-VEGFR3/NRP2 axis in tumor lymphangiogenesis and lymphatic metastasis. *Clin Chim Acta* 461:165–71.

Wang, S., D. Nie, J. P. Rubin, and L. Kokai. 2017. Lymphatic endothelial cells under mechanical stress: Altered expression of inflammatory cytokines and fibrosis. *Lymphat Res Biol* 15 (2):130–5.

Xiong, Y., C. C. Brinkman, K. S. Famulski, E. F. Mongodin, C. J. Lord, K. L. Hippen, B. R. Blazar, and J. S. Bromberg. 2017. A robust in vitro model for trans-lymphatic endothelial migration. *Sci Rep* 7 (1):1633.

Xu, Y., L. Yuan, J. Mak, L. Pardanaud, M. Caunt, I. Kasman, B. Larrivee, R. Del Toro, S. Suchting, A. Medvinsky, J. Silva, J. Yang, J. L. Thomas, A. W. Koch, K. Alitalo, A. Eichmann, and A. Bagri. 2010. Neuropilin-2 mediates VEGF-C-induced lymphatic sprouting together with VEGFR3. *J Cell Biol* 188 (1):115–30.

Yang, Y., J. M. Garcia-Verdugo, M. Soriano-Navarro, R. S. Srinivasan, J. P. Scallan, M. K. Singh, J. A. Epstein, and G. Oliver. 2012. Lymphatic endothelial progenitors bud from the cardinal vein and intersomitic vessels in mammalian embryos. *Blood* 120 (11):2340–8.

You, L. R., F. J. Lin, C. T. Lee, F. J. DeMayo, M. J. Tsai, and S. Y. Tsai. 2005. Suppression of Notch signalling by the COUP-TFII transcription factor regulates vein identity. *Nature* 435 (7038):98–104.

Zhong, T. P., S. Childs, J. P. Leu, and M. C. Fishman. 2001. Gridlock signalling pathway fashions the first embryonic artery. *Nature* 414 (6860):216–20.

3 Multispecies Microbial Communities and Synthetic Microbial Ecosystems

James Q. Boedicker
University of Southern California

CONTENTS

3.1 MICROBIAL COMMUNITY INTERACTIONS

Within a microbial ecosystem, cells interact in a variety of ways. Some exchange involves direct contact between cells. Contact-dependent interactions sometimes involve specialized appendages, such as pili or nanowires, to create direct contact between adjacent cells (Konovalova and Søgaard-Andersen 2011, Dubey and Ben-Yehuda 2011, Whitney et al. 2017). For cells that participate in extracellular electron transport, nanowires are used to carry electrical current between cells or from a cell to a metal surface (Reguera et al. 2005, Gorby et al. 2006, Pirbadian et al. 2014). Many interactions are contact independent, mediated by the extracellular release of molecules, including proteins, genetic material, and small molecules. In addition to individual molecules, supramolecular complexes also mediated interactions between cells, including extracellular vesicles, phages, and phage-like particles (Tran and Boedicker 2017, Lang, Zhaxybayeva, and Beatty 2012, Shikuma et al. 2014). Diffusion transports these messages and molecules between cells, enabling cells to interact over distances of microns to millimeters (Darch et al. 2018, Silva, Chellamuthu, and Boedicker 2017a). A comprehensive understanding of all the interactions that occur within a community will be essential to reliably predict and design microbial ecosystems. The following sections review how the exchange of signals, metabolites, genetic material, and ions facilitates interactions within microbial ecosystems and influences community function.

3.1.1 SIGNAL EXCHANGE

An important way in which microbes communicate is through the release and detection small molecules. The ability of released molecules to direct cellular function was first discovered in *Pneumococcus* cultures (Tomasz 1965). Shortly later another mechanism of molecular exchange was found in *Allovibrio fischeri* (Nealson, Platt, and Hastings 1970). These mechanisms are now known as quorum sensing (Boedicker and Nealson 2015, Miller and Bassler 2001), as the release of these molecules regulates gene expression in response to changes in cell density.

More recently quorum sensing systems have been reported in over 200 species of bacteria (Rajput, Kaur, and Kumar 2016). The types of molecules thought to be used as molecular signals have expanded to include several types of molecules beyond acyl-homoserine lactones and peptides (Rajput, Kaur, and Kumar 2016, Williams et al. 2007). Many such signaling molecules have multiple chemical variants. Variant signals are chemically similar, with small changes in chemical structure. For example, bacteria produce multiple variants of *N*-acyl-homoserine lactones, as shown in Figure 3.1. Multiple variants of cyclic peptide signals used by Gram-positive bacteria are also known (Ansaldi et al. 2002, Ji, Beavis, and Novick 1997, Stefanic et al. 2012). It is known that the quorum sensing receptor proteins, such as LuxR, are able to recognize multiple variants of the same signal (McClean et al. 1997, Silva, Chellamuthu, and Boedicker 2017b). The binding of non-cognate signals to cell receptors results in crosstalk. Such crosstalk influence quorum sensing activation within bacterial communities (Silva, Chellamuthu, and Boedicker 2017b, Ji, Beavis, and Novick 1997, Ansaldi et al. 2002, Miller et al. 2018).

Signal crosstalk between species influences gene regulation in diverse microbial communities, see Figure 3.1. The consequences of such crosstalk, especially in communities producing multiple signal variants, may have unexpected consequences in community activity (Wu, Menn, and Wang 2014). At its core, crosstalk results from the binding of multiple signals to the same receptor (Li and Nair 2012). This binding has downstream consequences on gene regulation, and examples of both excitatory and inhibitory crosstalk have been reported (Silva, Chellamuthu, and Boedicker 2017a, Ansaldi et al. 2002). The regulation of quorum sensing involves positive feedback on the production of the signal exchange machinery (synthase and receptor proteins), resulting in a complex interdependence of activity within communities of microbes that exchange signals. There is still significant work to be done, in both experiments and on the theory side, to properly account for the consequences of signal crosstalk on communitywide gene regulation. Given the ubiquity of crosstalk and the mixture of signals produced by many communities, future studies should focus on quorum sensing as an emergent property of a network of cells.

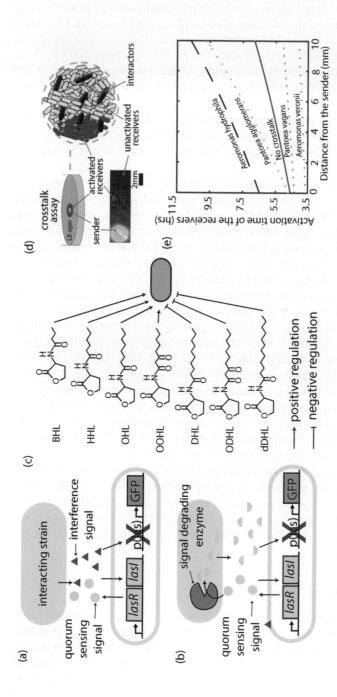

FIGURE 3.1 Signaling interactions within microbial communities. (a) In quorum sensing, signal production by a neighboring cell results in signal crosstalk, which potentially interferes with quorum sensing activation. (b) Neighboring cells can also cleave or chemically modify signals, another mechanism of signal interference. (c) Quorum sensing crosstalk depends on both the signal molecule and receptor protein, with different chemical variants of a signal resulting in activation or inhibition of quorum activation. Shown are crosstalk with the CviR receptor (orange) as measured in McClean et al. (1997). (d) Crosstalk between species has been measured using an assay that measures the influence of interactor cells on the propagation rate of a quorum sensing activation front. (e) Using the assay, the strength of crosstalk was measured for multiple quorum sensing strains isolated from a fresh water source (Silva, Chellamuthu, and Boedicker 2017b).

Crosstalk is only one type of signal interference that has been observed. Another mechanism through which one species can influence signal exchange in another species is through the chemical modification of released signals (Rampioni, Leoni, and Williams 2014). There are several enzymes known to interact with quorum sensing signals. For example the AiiA enzyme produced by *Bacillus* has been shown to cleave acyl-homoserine lactones, rendering them unable to activate quorum sensing (Chun et al. 2004, Dong et al. 2001). Other interactions which influence signal exchange are more indirect. Changes in the availability of nutrients or ion concentrations are known to modulate the response to quorum sensing signals (Boyle et al. 2015, Lee et al. 2013, Lazazzera 2000). These interactions within communities also need to be understood to accurately predict quorum sensing gene regulation within diverse cellular communities.

Several models have been developed to better understand signal exchange and quorum sensing within microbial communities. Signal exchange in one-species systems has been accurately described using coupled differential equations (Dockery and Keener 2001, Koerber, King, and Williams 2005, Pai et al. 2014, Williams et al. 2008). Some models abstract the system to focus only on signal concentrations (Pai et al. 2014), whereas others give more complete detail of synthase and receptor concentrations (Hunter, Vasquez, and Keener 2013). The influence of stochasticity on quorum sensing activation has also been analyzed (Koerber, King, and Williams 2005, Goryachev et al. 2005). While some modeling work has examined how the internal "circuitry" and feedback modulates quorum sensing activation (Dallidis and Karafyllidis 2014, Schaadt et al. 2013), even the integration of multiple signals by the same cell (Long et al. 2009, Mehta et al. 2009), little attention has been paid to quorum sensing within communities producing mulitple signal variants. Signaling interactions within a community may have surprising community-level responses not evident in one-species systems. Recently a neural newtork model was applied to examine the exchange of four signal variants within a community of *Staphylococcus aureus* (Yusufaly and Boedicker 2017). In the bacterial neural network model, each strain was represented by a node, see Figure 3.2. Nodes have two states, quorum sensing active and inactive. Connections between the nodes can be positive or negative, to variable strengths, representing signal crosstalk between strains. Analysis of the fully interconnected network revealed that inhibitory interactions between these strains constrained the number of community signaling states available and may have influenced the number of signal variants that evolved within the community.

3.1.2 METABOLIC INTERACTIONS

In addition to signal exchange, microbes also compete for resources and participate in cross-feeding (D'Souza et al. 2018). These interactions influence the growth rates of each species, which impacts the community composition and modulates activity. Whereas signaling interactions can be difficult to predict a priori, metabolic interactions, because the functions are highly conserved, can be accurately predicted from gene sequence alone (Durot, Bourguignon, and Schachter 2009). In signaling

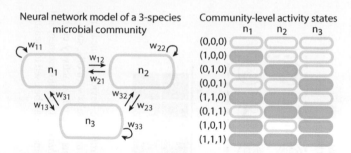

FIGURE 3.2 Signal interactions within a microbial community can be modeled as a neural network. Crosstalk between pairs of bacterial species are represented as interaction weights (w_{ij}) between species (n_i). Signal exchange between each species potentially activates quorum sensing in each species. In the diagram active cells are shown in green. Activity can also be represented as a binary string of 1's and 0's, indicating which species have activated quorum sensing. For a three-species community there are eight possible community-level quorum sensing activity state. The ratio of species and the strength and sign of each interaction weight determines the activity state of the community.

networks, although a given protein might be suspected to produce or bind to a specific type of signal, the exact signal molecules produced and the binding properties of a receptor cannot easily be predicted from sequence information alone. This advantage of metabolic networks has led to exploration of the metabolic capabilities and interactions within microbial communities. Metagenomics, in which the total, combined genetic material of a community is sequenced, has been used to infer the coupling of metabolic interactions and the overall flow of metabolites within communities (Durot, Bourguignon, and Schachter 2009, Faust and Raes 2012, Levy and Borenstein 2013). Such models can even be translated to interactions between individual species, as whole genome sequences are identified for the species involved (Morales and Holben 2011). Limits can be put on such models, such as flux-based constraints, to further refine the flow of nutrients through a community (Stolyar et al. 2007, Schilling et al. 1999). Such models have shown to accurately predict the metabolic capabilities of real communities and identify key metabolic linkages between species (McNally and Borenstein 2018).

In addition to inferring metabolic networks from sequencing results, experimental methods have been developed to quantify interaction strengths between species, see Figure 3.3. The generalized Lotka–Volterra model has emerged as a key conceptual model to understand and predict the consequences of interactions within microbial communities (Stein et al. 2013, Guo and Boedicker 2016, Venturelli et al. 2018). One noted success for this model was the identification of key interactions within the human microbiome involved in the development of disease. Work by Xavier et al. used time series of species abundance, as measured by sequencing of 16S rRNA, to parameterize a generalized Lotka–Volterra model for the species in the gastrointestinal tract of patients (Stein et al. 2013). By looking for correlations and anti-correlations in the fluctuations of each species, positive, negative, and

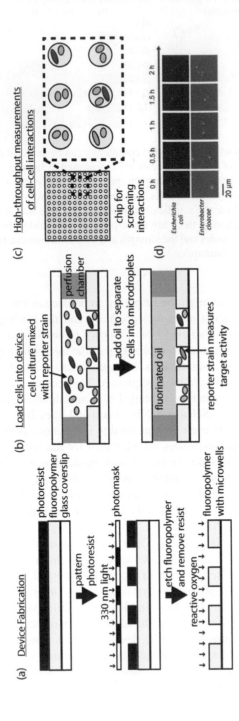

FIGURE 3.3 Microfluidic approach to quantify cell–cell interactions. (a) Example workflow for fabrication of a microfluidic device for measurement of cellular activity. (b) Microwells can be loaded with a mixture of cells and fluorescent reporter strains. Covering microwells with oils, such as fluorinated oil, creates multiple, independent microcultures. (c) Large arrays of microdroplets containing reporter strains for a target activity enable high-throughput measurements of interactions within microbial communities. (d) Images showing the growth of *Escherichia coli* (red fluorescence) and *Enterobacter cloacae* (green fluorescence) with in the same microdroplet (Guo, Silva, and Boedicker 2019). Such measurements can be used to measure the variability of interactions between bacterial species.

neutral interactions between species could be calculated for each pair of species in the network. Once the strength and direction of each interaction was defined, the model could be used to predict the future composition of the network and identify community compositions that favored the bloom of a pathogenic population, a major breakthrough in predicting how the composition of a microbial community determined key functions such as pathogenic potential.

Several studies have recently followed up on this idea to explore the capabilities of generalized Lotka–Volterra models to predict community dynamics. These follow-up studies took the approach of building the model from the bottom up, by measuring all the interactions within smaller communities and testing to see if the activity of larger community could be predicted (Venturelli et al. 2018, Guo and Boedicker 2016). One such study questioned the assumption that community dynamics were not influenced by higher-order interactions, i.e., not pairwise interactions (Guo and Boedicker 2016). By measuring interactions between all pairs, triples, and quadruples within a four-species network, it was found that a pairwise interaction network was indeed predictive, and that three- and four-species interactions were measurable but weak. The absence of high-order interactions has subsequently been supported by studies of other networks (Venturelli et al. 2018, Friedman, Higgins, and Gore 2017). One study did identify strong three-species interactions in a community of starch degrading microbes (Sanchez-Gorostiaga et al. 2018) and models have shown the potential importance of such high-order interactions (Bairey, Kelsic, and Kishony 2016). Other studies have cautioned that not all interactions follow a universal scaling with species composition (Momeni, Xie, and Shou 2017), and more work is warranted to determine when interactions require alternative mathematical representations.

Such studies are expanding to larger systems, thanks in part to the implementation of microfluidic techniques (Nagy et al. 2018, Guo, Silva, and Boedicker 2019). Building interaction models from the bottom up requires a large number of measurements of small subsets of a given community, with the number of measurements increasing approximately as the square of community size. Droplet-based techniques, which enable parallel cultivation and measurement of hundreds to millions of microbial combinations, offers the potential to measure pairwise interactions even within larger microbial networks (Park et al. 2011, Guo et al. 2012, Terekhov et al. 2017, Bai et al. 2013). One such method is shown in Figure 3.3. In these experiments (Guo, Silva, and Boedicker 2019), a mixed solution of cells was loaded into a microfluidic chip. As an immiscible oil layer was added to the device, aqueous droplets formed in the microwells that were randomly loaded with small numbers of cells. As a result of this random loading, in a single experiment multiple mixtures of cells could be assayed. These high-throughput techniques provide an overview of how the types and ratios of species combined influenced the activity of each individual cell and the microcommunity as a whole. One drawback of these measurements is the need to define or measure the species composition of each droplet, which was accomplished in simple systems by fluorescent labeling (Terekhov et al. 2017). This represents a potential drawback of such bottom-up approaches, especially for communities with difficult to culture microbes or microbes that are not amenable to genetic manipulation.

3.1.3 GENE EXCHANGE

Another important type of interaction that occur between cells within microbial communities is gene exchange. Compared to signaling and metabolic interactions, gene exchange occurs on a much slower timescale (Nazarian, Tran, and Boedicker 2018). Despite occurring over days to years, sequence comparison between genomes indicates that horizontal gene transfer strongly influences genomic composition of species. A study of microbes within the human body found thousands of horizontal gene transfer events within sequenced genomes of the human microbiome (Smillie et al. 2011). In fact, nearly 20% of the *E. coli* genome is thought to be the result of horizontal gene transfer (Lawrence and Ochman 1997).

Horizontal gene transfer, which occurs both within and between species of bacteria, occurs through a few molecular mechanisms: transduction, transformation, conjugation, and vesicle-mediated transfer (Nazarian, Tran, and Boedicker 2018, Ochman, Lawrence, and Groisman 2000). As shown in Figure 3.4, vesicle-mediated gene exchange enables gene exchange between multiple species of bacteria. For the results shown in Figure 3.4b,c, vesicles were harvested from cultures of donor strains via centrifugation. Some of these vesicles contained a plasmid with an antibiotic

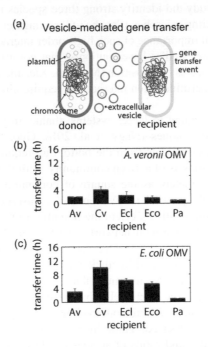

FIGURE 3.4 Gene exchange in extracellular vesicles has recently been shown to contribute to horizontal gene transfer between bacteria. (a) In vesicle-mediated gene exchange, donor cells produce membrane vesicles containing genetic material. Recipient cells uptake these vesicles, leading to gene transfer. (b and c) Vesicle-mediated gene exchange has been shown to facilitate gene transfer, here the exchange of a plasmid, between multiple pairs of recipient and donor species of bacteria (Tran and Boedicker 2017). Av, *Aeromonas veronii*; Ecl, *Enterobacter cloacae*; Eco, *Escherichia coli*; and Pa, *Pseudomonas aeruginosa*. See text for details.

resistance marker. These vesicles were added to a recipient culture, sensitive to the antibiotic, and the uptake of the plasmid was measured via gain of resistance in the recipient strain. As shown, vesicles from both *Aeromonas veronii* and *Escherichia coli* were able to facilitate gene transfer into five different species of Gram-negative bacteria, suggesting that vesicle-mediated gene transfer might be a ubiquitous and low barrier route of gene exchange within microbial populations.

In addition to these more universal exchange mechanisms, gene transfer agents have been reported. Gene transfer agents are virus-like particles that transfer genes similar to transduction (McDaniel et al. 2010, Lang, Zhaxybayeva, and Beatty 2012). Each mechanism transfers genes at a different rate and follows a unique set of exchange rules (Nazarian, Tran, and Boedicker 2018, Thomas and Nielsen 2005). These exchange rules determine which species are capable of gene exchange. For example, transduction is the transfer of genes within viruses. Viruses are capable of infecting only a limited number of strains, reducing the pairs of donor and recipient strains capable of gene exchange via transduction. Modeling efforts have examined horizontal gene transfer within bacterial populations (Levin, Stewart, and Rice 1979, Nielsen and Townsend 2004, Niehus et al. 2015, Mao and Lu 2016), and recently a model was published that incorporated all four forms of gene exchange within a diverse community of bacteria (Nazarian, Tran, and Boedicker 2018). Although the exchange rules were simplified, the model revealed unique roles for each mechanism of transfer. Calculations of exchange rates based on experimental measurements revealed the wide range of timescales over which gene transfer occurs, with transfer in vesicles being 16 orders of magnitude slower than conjugation (McDaniel et al. 2010, Nazarian, Tran, and Boedicker 2018).

Gene exchange is more than simply transferring a piece of DNA from one cell to another, maintenance and fixation of a DNA fragment are essential for such exchange to have long-term impact on cellular function (Thomas and Nielsen 2005). Bacteria have several defense mechanisms to limit the efficiency of horizontal gene transfer. Restriction-modification systems bind to specific sequences of DNA to identify and cleave foreign DNA (Oliveira, Touchon, and Rocha 2016). CRISPR interference, sometimes referred to as the bacterial immune system, also recognizes foreign DNA and targets it for cleavage (Marraffini and Sontheimer 2008, Bikard et al. 2012). Besides remaining intact in the new host, selection acts on the modified DNA fragment as another barrier to gene transfer. Gene which increase the fitness of the recipient strain are more likely to be maintained, leading to widespread genomic changes within the population. Fixation is influenced by many factors, including population size, relative fitness changes, and environmental conditions (Shapiro and Alm 2008, Thomas and Nielsen 2005, Bergstrom, Lipsitch, and Levin 2000). Together, both the physical transfer of a gene and fixation represent another important interaction that occurs within microbial communities, changing the composition and functional capabilities of such communities over time.

3.1.4 ION AND TOXIN EXCHANGE

Lastly, another important interaction which has been analyzed is ion exchange. It is well known that in many environments, metals such as iron are often a limiting resource (Schalk and Guillon 2013). Most microbial species have evolved several mechanisms to obtain iron from the environment (Schalk 2013, Poole and McKay 2003), and changes in iron availability strongly regulate gene expression (Kreamer et al. 2015, McHugh et al. 2003, Andrews, Robinson, and Rodríguez-Quiñones 2003). Interactions between species influence iron availability. For example in co-cultures, *Pseudomonas aeruginosa* harvests iron through lysis of *Staphylococcus aureus* (Mashburn et al. 2005). Many other ions also impact bacterial activity, including zinc, copper, manganese, molybdenum, and nickel (Dedyukhina and Eroshin 1991, Wackett, Dodge, and Ellis 2004). Recent studies have also focused on hydrogen ions as a mediator of species–species interactions (Ratzke and Gore 2018).

Many bacterial species release molecules that kill or inhibit the growth of neighboring cells. The exchange of toxic compounds is an important aspect of competition with neighbors for nutrients and space (Cornforth and Foster 2015, Kirkup and Riley 2004). For some of these compounds, whether or not they only function as antimicrobials has been questioned recently (Hibbing et al. 2010). For example phenazine exchange, although toxic at high concentrations, at lower concentrations the phenazine pyocyanin can act as a redox shuttle and regulate gene expression (Dietrich et al. 2008).

3.2 CONTROL OF MULTISPECIES BACTERIAL NETWORKS

There has been a push toward development of control strategies for diverse microbial communities (Brenner, You, and Arnold 2008). Such strategies would enable design of synthetic communities with desirable community-level traits. Although increased diversity within a community increases the complexity of interactions between cells, such communities may be able to utilize division of labor and incorporate strains with specialized functionality (Johns et al. 2016), such as exotic metabolic capabilities or the ability to reduce metals (Chellamuthu et al. 2019). Also, because most natural communities are already highly diverse, identifying control strategies for communities may lead to the ability to control or reprogram the activity of bacterial communities in the wild.

Several recent publications have examined the ability to control bacterial communities. The focus is typically on metabolic interactions, although toy systems with signal exchange have also been explored (Silva, Chellamuthu, and Boedicker 2017a, Brenner et al. 2007, Danino et al. 2010, Halleran and Murray 2018, Wu, Menn, and Wang 2014, Celik Ozgen et al. 2018, Kong et al. 2014). Cross-feeding of amino acids in communities of auxotrophs is one system that has been explored (Wintermute and Silver 2010, Mee et al. 2014). Such networks utilize a division of labor to more efficiently grow as a community (Pande et al. 2014). Other studies have focused on networks without known or engineered crosstalk. For example Venturelli et al. examined interactions within a 12-member community (Venturelli et al. 2018). Overall pairwise interactions controlled community behavior, with mostly negative interactions and several strong positive interactions. Together this network enabled stable coexistence of the community over multiple days. Modeling has shown that competitive interactions are important for community stability (Coyte, Schluter, and Foster 2015), and competitive interactions seem to dominate interactions of isolated communities (Foster and Bell 2012).

In many instances community dynamics appear to be stable and robust to perturbations. The composition of a closed ecosystem composed of the bacteria *E. coli* with *Tetrahymena thermophile* and *Chlamydomonas reinhardtii* was remarkably stable over several months (Frentz, Kuehn, and Leibler 2015). Long-term sequencing of the gut microbiome of two individuals also revealed a stable community composition over a year, with strong perturbations such as travel and illness causing only temporary fluctuations in community makeup (David et al. 2014). Recent reports point to very predictable compositions even in multispecies communities (Friedman, Higgins, and Gore 2017), suggesting that once interactions are known predicting and controlling the community composition may be straightforward. Another study shows that cross-feeding networks, even with large numbers of species, converge onto predictable compositions (Goldford et al. 2018). Although such findings demonstrate the potential to predict community dynamics, one wonders if these results also reveal the difficulty in driving communities to new, "user-defined" states. Communities resilient even to strong perturbations although predictable may not be controllable.

3.2.1 The Impact of Spatial Structure on Community Control

One aspect of bacterial communities that complicates accurate prediction of community behavior is spatial structure. In the lab, well-mixed systems can be used to remove spatial structure, simplifying analysis to more easily interrogate the relationship between cell–cell interactions and community-level behaviors. However the real world is seldom well-mixed, and bacteria inhabit intricate and complex spaces (Nunan et al. 2002, Tropini et al. 2017, Gutiérrez Castorena et al. 2016). There are many examples of microbes creating spatial structure, using directed motility and signal exchange to construct fruiting bodies or aggregation patterns (Sliusarenko, Zusman, and Oster 2007, Budrene and Berg 1991). More commonplace spatial structures are biofilms. Biofilms are surface attached communities of microbes (Davies et al. 1998). Because cells are stacked and relatively immotile over long periods of time, biofilms create gradients of nutrients, signaling molecules, and pH (Stewart and Franklin 2008). The structure of multispecies biofilms can also complicates predictive models, as each cell interacts within a unique neighborhood of cells.

The multitude of chemical microenvironments within biofilms makes prediction of activity a daunting task. Some community behaviors appear to have spatial requirements, such as balanced growth within a community of *Azotobacter vinelandii*, *Bacillus licheniformis*, and *Paenibacillus curdlanolyticus* (Kim et al. 2008), see Figure 3.5 for this and other examples of spatial structure impacting microbial activity. The stabilization of cell–cell interactions through imposed or emergent spatial structure has been shown in other systems (Momeni, Waite, and Shou 2013). It is known that mixed biofilms sometimes self-organize into predictable structures (Gloag et al. 2013, Hansen et al. 2007). It is also possible that when considering larger networks with billions of cells that the interactions more or less average out: the overall output of each species is predictable despite significant microscale heterogeneity of activity. Microscale heterogeneity does seem to be an important aspect

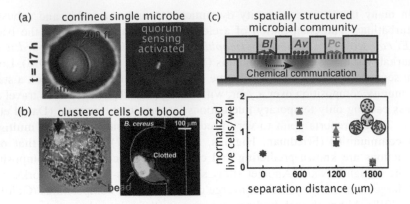

FIGURE 3.5 The activity of microbes and communities of microbes is influenced by spatial structural. (a) Upon confinement, even single bacterial cells are capable of activating high density behaviors such as quorum sensing (Boedicker, Vincent, and Ismagilov 2009). (b) The clustering of cells even influenced interactions between microbes and the environment. Clustered cells were able to induce coagulation in blood (Kastrup et al. 2008). (c) In a three-species community, competitive and cooperative interactions between species were balanced at intermediate spatial distributions. The overall growth of the community was measured as a function of spatial separation in a microfluidic device (Kim et al. 2008).

of predicting community function, and control strategies could take advantage of such heterogeneity. Assuming that the regulation of some bacterial activities has strict requirements, cell A must be near cell B not near cell C, this would potentially lead to activity hotspots within diverse spatially distributed communities. Changing the frequency of the occurrence of hotspots could have drastic community-wide influences on activity levels. Microfluidic technologies, now widespread in the analysis of cells and cellular networks, offers precise control of spatial structure and high-throughput measurements of network activity. Future work which incorporates microfluidics will likely help reveal the implications of spatial structures on community behavior and enable the validation of control strategies.

3.3 OUTLOOK

Designing and controlling multispecies microbial communities is an exciting and timely problem in microbiology and biological engineering. Advances in genetic engineering, community analysis, and microfluidics give us the tools needed to examine cellular networks across multiple scales. A focus on all of the interactions that occur within bacterial communities, including regulatory, metabolic, and evolutionary, should establish a stronger foundation on which to build predictive models of community behavior and strategies for control of complex cellular networks. Other disciplines, including physics and engineering, have expertise in complex dynamical systems, emergent behaviors, and communication networks, albeit largely from analysis of non-biological systems. Cross-disciplinary work, especially incorporating theoretical aspects of complex networks, should help in identifying universal rules that govern community behavior.

REFERENCES

Andrews, Simon C., Andrea K. Robinson, and Francisco Rodríguez-Quiñones. 2003. Bacterial iron homeostasis *FEMS Microbiology Reviews* 27 (2–3):215–237.

Ansaldi, Mireille, Darja Marolt, Tina Stebe, Ines Mandic-Mulec, and David Dubnau. 2002. Specific activation of the *Bacillus* quorum-sensing systems by isoprenylated pheromone variants *Molecular Microbiology* 44 (6):1561–1573.

Bai, Yunpeng, Santoshkumar N. Patil, Steven D. Bowden, Simon Poulter, Jie Pan, George P. C. Salmond, Martin Welch, Wilhelm T. S. Huck, and Chris Abell. 2013. Intra-species bacterial quorum sensing studied at single cell level in a double droplet trapping system *International Journal of Molecular Sciences* 14 (5):10570–10581.

Bairey, Eyal, Eric D. Kelsic, and Roy Kishony. 2016. High-order species interactions shape ecosystem diversity *Nature Communications* 7:12285–12285.

Bergstrom, Carl T., Marc Lipsitch, and Bruce R. Levin. 2000. Natural selection, infectious transfer and the existence conditions for bacterial plasmids *Genetics* 155 (4):1505.

Bikard, David, Asma Hatoum-Aslan, Daniel Mucida, and Luciano A. Marraffini. 2012. CRISPR interference can prevent natural transformation and virulence acquisition during in vivo bacterial infection *Cell Host & Microbe* 12 (2):177–186.

Boedicker, James, and Kenneth Nealson. 2015. Microbial communication via quorum sensing *IEEE Transactions on Molecular, Biological and Multi-Scale Communications* 1 (4):310–320.

Boedicker, James Q., Meghan E. Vincent, and Rustem F. Ismagilov. 2009. Microfluidic confinement of single cells of bacteria in small volumes initiates high-density behavior of quorum sensing and growth and reveals its variability *Angewandte Chemie-International Edition* 48 (32):5908–5911.

Boyle, Kerry E., Hilary Monaco, Dave van Ditmarsch, Maxime Deforet, and Joao B. Xavier. 2015. Integration of metabolic and quorum sensing signals governing the decision to cooperate in a bacterial social trait *PLoS Computational Biology* 11 (6):e1004279.

Brenner, Katie, Lingchomg You, and Frances H. Arnold. 2008. Engineering microbial consortia: A new frontier in synthetic biology *Trends in Biotechnology* 26 (9):483–439. doi:10.1016/j.tibtech.2008.05.004.

Brenner, Katie, David K. Karig, Ron Weiss, and Frances H. Arnold. 2007. Engineered bidirectional communication mediates a consensus in a microbial biofilm consortium *Proceedings of the National Academy of Sciences of the United States of America* 104 (44):17300–17304.

Budrene, Elena O., and Howard C. Berg. 1991. Complex patterns formed by motile cells of *Escherichia coli Nature* 349:630.

Celik Ozgen, Venhar, Wentao Kong, Andrew E. Blanchard, Feng Liu, and Ting Lu. 2018. Spatial interference scale as a determinant of microbial range expansion *Science Advances* 4 (11):eaau0695.

Chellamuthu, Prithiviraj, Frances Tran, Kalinga Pavan T. Silva, Marko S. Chavez, Mohamed Y. El-Naggar, and James Q. Boedicker. 2019. Engineering bacteria for biogenic synthesis of chalcogenide nanomaterials *Microbial Biotechnology* 12 (1):161–172.

Chun, Carlene K., Egon A. Ozer, Michael J. Welsh, Joseph Zabner, and Everett Peter Greenberg. 2004. Inactivation of a *Pseudomonas aeruginosa* quorum-sensing signal by human airway epithelia *Proceedings of the National Academy of Sciences of the United States of America* 101 (10):3587.

Cornforth, Daniel M., and Kevin R. Foster. 2015. Antibiotics and the art of bacterial war *Proceedings of the National Academy of Sciences* 112 (35):10827.

Coyte, Katharine Z., Jonas Schluter, and Kevin R. Foster. 2015. The ecology of the microbiome: Networks, competition, and stability *Science* 350 (6261):663.

D'Souza, Glen, Shraddha Shitut, Daniel Preussger, Ghada Yousif, Silvio Waschina, and Christian Kost. 2018. Ecology and evolution of metabolic cross-feeding interactions in bacteria *Natural Product Reports* 35 (5):455–488.

Dallidis, Stylianos E., and Ioannis G. Karafyllidis. 2014. Boolean network model of the *Pseudomonas aeruginosa* quorum sensing circuits *IEEE Transactions on NanoBioscience* 13 (3):343–349.

Danino, Tal, Octavio Mondragon-Palomino, Lev Tsimring, and Jeff Hasty. 2010. A synchronized quorum of genetic clocks *Nature* 463 (7279):326–330.

Darch, Sophie E., Olja Simoska, Mignon Fitzpatrick, Juan P. Barraza, Keith J. Stevenson, Roger T. Bonnecaze, Jason B. Shear, and Marvin Whiteley. 2018. Spatial determinants of quorum signaling in a *Pseudomonas aeruginosa* infection model *Proceedings of the National Academy of Sciences* 115 (18):4779.

David, Lawrence A., Arne C. Materna, Jonathan Friedman, Maria I. Campos-Baptista, Matthew C. Blackburn, Allison Perrotta, Susan E. Erdman, and Eric J. Alm. 2014. Host lifestyle affects human microbiota on daily timescales *Genome Biology* 15 (7):R89.

Davies, David G., Matthew R. Parsek, James P. Pearson, Barbara H. Iglewski, J. William Costerton, and E. Peter Greenberg. 1998. The involvement of cell-to-cell signals in the development of a bacterial biofilm *Science* 280 (5361):295–298.

Dedyukhina, Emiliya G., and Valery K. Eroshin. 1991. Essential metal ions in the control of microbial metabolism *Process Biochemistry* 26 (1):31–37.

Dietrich, Lars E. P., Tracy K. Teal, Alexa Price-Whelan, and Dianne K. Newman. 2008. Redox-active antibiotics control gene expression and community behavior in divergent bacteria *Science* 321 (5893):1203–1206.

Dockery, Jack D., and James P. Keener. 2001. A mathematical model for quorum sensing in *Pseudomonas aeruginosa Bulletin of Mathematical Biology* 63 (1):95–116.

Dong, Yi-Hu, Lian-Hui Wang, Jin-Ling Xu, Hai-Bao Zhang, Xi-Fen Zhang, and Lian-Hui Zhang. 2001. Quenching quorum-sensing-dependent bacterial infection by an *N*-acyl homoserine lactonase *Nature* 411:813.

Dubey, Gyanendra P., and Sigal Ben-Yehuda. 2011. Intercellular nanotubes mediate bacterial communication *Cell* 144 (4):590–600.

Durot, Maxime, Pierre-Yves Bourguignon, and Vincent Schachter. 2009. Genome-scale models of bacterial metabolism: Reconstruction and applications *FEMS Microbiology Reviews* 33 (1):164–190.

Faust, Karoline, and Jeroen Raes. 2012. Microbial interactions: From networks to models *Nature Reviews. Microbiology* 10 (8):538–550.

Foster, Kevin R., and Thomas Bell. 2012. Competition, not cooperation, dominates interactions among culturable microbial species *Current Biology* 22 (19):1845–1850.

Frentz, Zak, Seppe Kuehn, and Stanislas Leibler. 2015. Strongly deterministic population dynamics in closed microbial communities *Physical Review X* 5 (4):041014.

Friedman, Jonathan, Logan M. Higgins, and Jeff Gore. 2017. Community structure follows simple assembly rules in microbial microcosms *Nature Ecology & Evolution* 1:0109.

Gloag, Erin S., Lynne Turnbull, Alan Huang, Pascal Vallotton, Huabin Wang, Laura M. Nolan, Lisa Mililli, Cameron Hunt, Jing Lu, Sarah R. Osvath, Leigh G. Monahan, Rosalia Cavaliere, Ian G. Charles, Matt P. Wand, Michelle L. Gee, Ranganathan Prabhakar, and Cynthia B. Whitchurch. 2013. Self-organization of bacterial biofilms is facilitated by extracellular DNA *Proceedings of the National Academy of Sciences* 110 (28):11541.

Goldford, Joshua E., Nanxi Lu, Djordje Bajić, Sylvie Estrela, Mikhail Tikhonov, Alicia Sanchez-Gorostiaga, Daniel Segrè, Pankaj Mehta, Alvaro Sanchez. 2018. Emergent simplicity in microbial community assembly *Science* 361 (6401): 469–474. doi:10.1126/science.aat1168.

Gorby, Yuri A., Svetlana Yanina, Jeffrey S. McLean, Kevin M. Rosso, Dianne Moyles, Alice Dohnalkova, Terry J. Beveridge, In Seop Chang, Byung Hong Kim, Kyung Shik Kim, David E. Culley, Samantha B. Reed, Margie F. Romine, Daad A. Saffarini, Eric A. Hill, Liang Shi, Dwayne A. Elias, David W. Kennedy, Grigoriy Pinchuk, Kazuya Watanabe, Shuñichi Ishii, Bruce Logan, Kenneth H. Nealson, and Jim K. Fredrickson. 2006. Electrically conductive bacterial nanowires produced by *Shewanella oneidensis* strain MR-1 and other microorganisms *Proceedings of the National Academy of Sciences of the United States of America* 103 (30):11358–11363.

Goryachev, Andrew B., Da-Jun Toh, Keng Boon Wee, Travis Lee, Hai-Bao Zhang, and Lian-Hui Zhang. 2005. Transition to quorum sensing in an agrobacterium population: A stochastic model *PLoS Computational Biology* 1 (4):e37.

Guo, Mira T., Assaf Rotem, John A. Heyman, and David A. Weitz. 2012. Droplet microfluidics for high-throughput biological assays *Lab on a Chip* 12 (12):2146–2155.

Guo, Xiaokan, and James Q. Boedicker. 2016. The contribution of high-order metabolic interactions to the global activity of a four-species microbial community *PLoS Computational Biology* 12 (9):e1005079.

Guo, Xiaokan, Kalinga Pavan Thushara Silva, and James Q. Boedicker. 2019. Single-cell variability of growth interactions within a two-species bacterial community *Physical Biology* 16 (3):036001.

Gutiérrez Castorena, Edgar Vladimir, Ma del Carmen Gutiérrez-Castorena, Tania González Vargas, Lenom Cajuste Bontemps, Julián Delgadillo Martínez, Enrique Suástegui Méndez, and Carlos Alberto Ortiz Solorio. 2016. Micromapping of microbial hotspots and biofilms from different crops using digital image mosaics of soil thin sections *Geoderma* 279:11–21.

Halleran, Andrew D., and Richard M. Murray. 2018. Cell-free and in vivo characterization of Lux, Las, and Rpa quorum activation systems in *E. coli ACS Synthetic Biology* 7 (2):752–755.

Hansen, Susse Kirkelund, Paul B. Rainey, Janus A. J. Haagensen, and Søren Molin. 2007. Evolution of species interactions in a biofilm community *Nature* 445:533.

Hibbing, Michael E., Clay Fuqua, Matthew R. Parsek, and S. Brook Peterson. 2010. Bacterial competition: Surviving and thriving in the microbial jungle *Nature Reviews. Microbiology* 8 (1):15–25.

Hunter, Geoffrey A., Fernando Guevara Vasquez, and James P. Keener. 2013. A mathematical model and quantitative comparison of the small RNA circuit in the *Vibrio harveyi* and *Vibrio cholerae* quorum sensing systems *Physical Biology* 10 (4):046007.

Ji, Guangyong, Ronald Beavis, and Richard P. Novick. 1997. Bacterial interference caused by autoinducing peptide variants *Science* 276 (5321):2027.

Johns, Nathan I., Tomasz Blazejewski, Antonio L. C. Gomes, and Harris H. Wang. 2016. Principles for designing synthetic microbial communities *Current Opinion in Microbiology* 31:146–153.

Kastrup, Christian J., James Q. Boedicker, Andrei P. Pomerantsev, Mahtab Moayeri, Yao Bian, Rebecca R. Pompano, Timothy R. Kline, Patricia Sylvestre, Feng Shen, Stephen H. Leppla, Wei-Jen Tang, and Rustem F. Ismagilov. 2008. Spatial localization of bacteria controls coagulation of human blood by 'quorum acting' *Nature Chemical Biology* 4 (12):742–750.

Kim, Hyun Jung, James Q. Boedicker, Jang Wook Choi, and Rustem F. Ismagilov. 2008. Defined spatial structure stabilizes a synthetic multispecies bacterial community *Proceedings of the National Academy of Sciences of the United States of America* 105 (47):18188–18193.

Kirkup, Benjamin C., and Margaret A. Riley. 2004. Antibiotic-mediated antagonism leads to a bacterial game of rock-paper-scissors in vivo *Nature* 428 (6981):412–414.

Koerber, Adrian J., John R. King, and Paul Williams. 2005. Deterministic and stochastic modelling of endosome escape by *Staphylococcus aureus*: Quorum sensing by a single bacterium *Journal of Mathematical Biology* 50 (4):440–488.

Kong, Wentao, Venhar Celik, Chen Liao, Qiang Hua, and Ting Lu. 2014. Programming the group behaviors of bacterial communities with synthetic cellular communication *Bioresources and Bioprocessing* 1 (1):24.

Konovalova, Anna, and Lotte Søgaard-Andersen. 2011. Close encounters: Contact-dependent interactions in bacteria *Molecular Microbiology* 81 (2):297–301.

Kreamer, Naomi N., Rob Phillips, Dianne K. Newman, and James Q. Boedicker. 2015. Predicting the impact of promoter variability on regulatory outputs *Scientific Reports* 5:18238–18238.

Lang, Andrew S., Olga Zhaxybayeva, and J. Thomas Beatty. 2012. Gene transfer agents: Phage-like elements of genetic exchange *Nature Reviews. Microbiology* 10 (7):472–482.

Lawrence, Jeffrey G., and Howard Ochman. 1997. Amelioration of bacterial genomes: Rates of change and exchange *Journal of Molecular Evolution* 44 (4):383–397.

Lazazzera, Beth A. 2000. Quorum sensing and starvation: Signals for entry into stationary phase *Current Opinion in Microbiology* 3 (2):177–182.

Lee, Jasmine, Jien Wu, Yinyue Deng, Jing Wang, Chao Wang, Jianhe Wang, Changqing Chang, Yi-Hu Dong, Paul Williams, and Lian-Hui Zhang. 2013. A cell-cell communication signal integrates quorum sensing and stress response *Nature Chemical Biology* 9 (5):339–343.

Levin, Bruce R., Frank M. Stewart, and Virginia A. Rice. 1979. The kinetics of conjugative plasmid transmission: Fit of a simple mass action model *Plasmid* 2 (2):247–260.

Levy, Roie, and Elhanan Borenstein. 2013. Metabolic modeling of species interaction in the human microbiome elucidates community-level assembly rules *Proceedings of the National Academy of Sciences* 110 (31):12804.

Li, Zhi, and Satish K. Nair. 2012. Quorum sensing: How bacteria can coordinate activity and synchronize their response to external signals? *Protein Science* 21 (10):1403–1417.

Long, Tao, Kimberly C. Tu, Yufang F. Wang, Pankaj Mehta, N. Phuan Ong, Bonnie L. Bassler, and Ned S. Wingreen. 2009. Quantifying the integration of quorum-sensing signals with single-cell resolution *PLoS Biology* 7 (3):640–649.

Mao, Junwen, and Ting Lu. 2016. Population-dynamic modeling of bacterial horizontal gene transfer by natural transformation *Biophysical Journal* 110 (1):258–268.

Marraffini, Luciano A., and Erik J. Sontheimer. 2008. CRISPR interference limits horizontal gene transfer in staphylococci by targeting DNA *Science* 322 (5909):1843–1845.

Mashburn, Lauren M., Amy M. Jett, Darrin R. Akins, and Marvin Whiteley. 2005. *Staphylococcus aureus* serves as an iron source for *Pseudomonas aeruginosa* during in vivo coculture *Journal of Bacteriology* 187 (2):554.

McClean, Kay H., Michael K. Winson, Leigh Fish, Adrian Taylor, Siri Ram Chhabra, Miguel Camara, Mavis Daykin, John H. Lamb, Simon Swift, Barrie W. Bycroft, Gordon S.A.B. Stewart, and Paul Williams. 1997. Quorum sensing and *Chromobacterium violaceum*: Exploitation of violacein production and inhibition for the detection of *N*-acylhomoserine lactones *Microbiology* 143 (Pt 12):3703–3711.

McDaniel, Lauren D., Elizabeth Young, Jennifer Delaney, Fabian Ruhnau, Kim B. Ritchie, and John H. Paul. 2010. High frequency of horizontal gene transfer in the oceans *Science* 330 (6000):50.

McHugh, Jonathan P., Francisco Rodríguez-Quiñones, Hossein Abdul-Tehrani, Dimitri A. Svistunenko, Robert K. Poole, Chris E. Cooper, and Simon C. Andrews. 2003. Global iron-dependent gene regulation in *Escherichia coli*: A new mechanism for iron homeostasis *Journal of Biological Chemistry* 278(32):29478–29486.

McNally, Colin P., and Elhanan Borenstein. 2018. Metabolic model-based analysis of the emergence of bacterial cross-feeding via extensive gene loss *BMC Systems Biology* 12 (1):69–69.

Mee, Michael T., James J. Collins, George M. Church, and Harris H. Wang. 2014. Syntrophic exchange in synthetic microbial communities *Proceedings of the National Academy of Sciences of the United States of America* 111 (20):E2149–E2156.

Mehta, Pankaj, Sidhartha Goyal, Tao Long, Bonnie L. Bassler, and Ned S. Wingreen. 2009. Information processing and signal integration in bacterial quorum sensing *Molecular Systems Biology* 5:325–325.

Miller, Eric L., Morten Kjos, Monica I. Abrudan, Ian S. Roberts, Jan-Willem Veening, and Daniel E. Rozen. 2018. Eavesdropping and crosstalk between secreted quorum sensing peptide signals that regulate bacteriocin production in *Streptococcus pneumoniae The ISME Journal* 12 (10):2363–2375.

Miller, Melissa B., and Bonnie L. Bassler. 2001. Quorum sensing in bacteria *Annual Review of Microbiology* 55:165–199.

Momeni, Babak, Adam James Waite, and Wenying Shou. 2013. Spatial self-organization favors heterotypic cooperation over cheating *Elife* 2:e00960.

Momeni, Babak, Li Xie, and Wenying Shou. 2017. Lotka-Volterra pairwise modeling fails to capture diverse pairwise microbial interactions *Elife* 6. doi:10.7554/eLife.25051.

Morales, Sergio E., and William E. Holben. 2011. Linking bacterial identities and ecosystem processes: Can 'omic' analyses be more than the sum of their parts? *FEMS Microbiology Ecology* 75 (1):2–16.

Nagy, Krisztina, Agnes Abraham, Juan E. Keymer, and Peter Galajda. 2018. Application of microfluidics in experimental ecology: The importance of being spatial *Frontiers in Microbiology* 9:496.

Nazarian, Phillip, Frances Tran, and James Q. Boedicker. 2018. Modeling multispecies gene flow dynamics reveals the unique roles of different horizontal gene transfer mechanisms. *Frontiers in Microbiology* 9:2978–2978.

Nealson, Kenneth H., Terry Platt, and J. Woodland Hastings. 1970. Cellular control of the synthesis and activity of the bacterial luminescent system *Journal of Bacteriology* 104 (1):313–322.

Niehus, Rene, Sara Mitri, Alexander G. Fletcher, and Kevin R. Foster. 2015. Migration and horizontal gene transfer divide microbial genomes into multiple niches *Nature Communications* 6:8924.

Nielsen, Kaare M., and Jeffrey P. Townsend. 2004. Monitoring and modeling horizontal gene transfer *Nature Biotechnology* 22 (9):1110.

Nunan, Naoise, Kuan Wu, Iain M. Young, John W. Crawford, and Karl Ritz. 2002. In situ spatial patterns of soil bacterial populations, mapped at multiple scales, in an arable soil *Microbial Ecology* 44 (4):296–305.

Ochman, Howard, Jeffrey G. Lawrence, and Eduardo A. Groisman. 2000. Lateral gene transfer and the nature of bacterial innovation *Nature* 405 (6784):299–304.

Oliveira, Pedro H., Marie Touchon, and Eduardo P. C. Rocha. 2016. Regulation of genetic flux between bacteria by restriction–modification systems *Proceedings of the National Academy of Sciences* 113 (20):5658–5663.

Pai, Aanand, Jaydeep K. Srimani, Yu Tanouchi, and Lingchong You. 2014. Generic metric to quantify quorum sensing activation dynamics *ACS Synthetic Biology* 3 (4):220–227.

Pande, Samay, Holger Merker, Katrin Bohl, Michael Reichelt, Stefan Schuster, Luis F. de Figueiredo, Christoph Kaleta, and Christian Kost. 2014. Fitness and stability of obligate cross-feeding interactions that emerge upon gene loss in bacteria *ISME Journal* 8 (5):953–962.

Park, Jihyang, Alissa Kerner, Mark A. Burns, and Xiaoxia Nina Lin. 2011. Microdroplet-enabled highly parallel co-cultivation of microbial communities *PLoS One* 6 (2):e17019.

Pirbadian, Sahand, Sarah E. Barchinger, Kar M. Leung, Hye S. Byun, Yamini Jangir, Rachida A. Bouhenni, Samantha B. Reed, Margaret F. Romine, Daad A. Saffarini, Liag Shi, Yuri A. Gorby, John H. Golbeck, and Mohamed Y. El-Naggar. 2014. *Shewanella oneidensis*

MR-1 nanowires are outer membrane and periplasmic extensions of the extracellular
    electron transport components *Proceedings of the National Academy of Sciences of the
    United States of America* 111 (35):12883–12888.

Poole, Keith, and Geoffery A. McKay. 2003. Iron acquisition and its control in *Pseudomonas
    aeruginosa*: Many roads lead to Rome *Frontiers in Bioscience* 8 (1):661–686.

Rajput, Akanksha, Karambir Kaur, and Manoj Kumar. 2016. SigMol: Repertoire of
    quorum sensing signaling molecules in prokaryotes *Nucleic Acids Research*
    44 (D1):D634–D639.

Rampioni, Giordano, Livia Leoni, and Paul Williams. 2014. The art of antibacterial war-
    fare: Deception through interference with quorum sensing-mediated communication
    Bioorganic Chemistry 55:60–68.

Ratzke, Christoph, and Jeff Gore. 2018. Modifying and reacting to the environmental pH can
    drive bacterial interactions *PLoS Biology* 16 (3):e2004248.

Reguera, Gemma, Kevin D. McCarthy, Teena Mehta, Julie S. Nicoll, Mark T. Tuominen, and
    Derek R. Lovley. 2005. Extracellular electron transfer via microbial nanowires *Nature*
    435 (7045):1098–1101.

Sanchez-Gorostiaga, Alicia, Djordje Bajić, Melisa L. Osborne, Juan F. Poyatos, and Alvaro
    Sanchez. 2018. High-order interactions dominate the functional landscape of microbial
    consortia *bioRxiv*:333534. doi:10.1101/333534.

Schaadt, Nadine S., Anke Steinbach, Rolf W. Hartmann, and Volkhard Helms. 2013. Rule-
    based regulatory and metabolic model for quorum sensing in *P. aeruginosa* *BMC
    Systems Biology* 7:81.

Schalk, Isabelle J. 2013. Innovation and originality in the strategies developed by bacteria to
    get access to iron *ChemBioChem* 14 (3):293–294.

Schalk, Isabelle J., and Laurent Guillon. 2013. Fate of ferrisiderophores after import
    across bacterial outer membranes: Different iron release strategies are observed in
    the cytoplasm or periplasm depending on the siderophore pathways *Amino Acids*
    44 (5):1267–1277.

Schilling, Christophe H., Stefan Schuster, Bernhard O. Palsson, and Reinhart Heinrich. 1999.
    Metabolic pathway analysis: Basic concepts and scientific applications in the post-
    genomic era *Biotechnology Progress* 15 (3):296–303.

Shapiro, B. Jesse, and Eric J. Alm. 2008. Comparing patterns of natural selection across spe-
    cies using selective signatures *PLoS Genetics* 4 (2):e23.

Shikuma, Nicholas J., Martin Pilhofer, Gregor L. Weiss, Michael G. Hadfield, Grant J. Jensen,
    and Dianne K. Newman. 2014. Marine tubeworm metamorphosis induced by arrays of
    bacterial phage tail-like structures *Science* 343 (6170):529–533.

Silva, Kalinga Pavan, Prithivi Chellamuthu, and James Q. Boedicker. 2017a. Signal destruc-
    tion tunes the zone of activation in spatially distributed signaling networks *Biophysical
    Journal* 112 (5):1037–1044.

Silva, Kalinga Pavan T., Prithiviraj Chellamuthu, and James Q. Boedicker. 2017b.
    Quantifying the strength of quorum sensing crosstalk within microbial communities
    PLoS Computational Biology 13 (10):e1005809.

Sliusarenko, Oleksii, David R. Zusman, and George Oster. 2007. Aggregation during fruiting
    body formation in *Myxococcus xanthus* is driven by reducing cell movement *Journal
    of Bacteriology* 189 (2):611.

Smillie, Chris S., Mark B. Smith, Jonathan Friedman, Otto X. Cordero, Lawrence A. David,
    and Eric J. Alm. 2011. Ecology drives a global network of gene exchange connecting
    the human microbiome *Nature* 480 (7376):241–244.

Stefanic, Polonca, Francesca Decorosi, Carlo Viti, Janine Petito, Frederick M. Cohan, and
    Ines Mandic-Mulec. 2012. The quorum sensing diversity within and between ecotypes
    of *Bacillus subtilis* *Environmental Microbiology* 14 (6):1378–1389.

Stein, Richard R., Vanni Bucci, Nora C. Toussaint, Charlie G. Buffie, Gunnar Ratsch, Eric G. Pamer, Chris Sander, and Joao B. Xavier. 2013. Ecological modeling from time-series inference: Insight into dynamics and stability of intestinal microbiota *PLoS Computational Biology* 9 (12):e1003388.

Stewart, Philip S., and Michael J. Franklin. 2008. Physiological heterogeneity in biofilms *Nature Reviews. Microbiology* 6:199.

Stolyar, Sergey, Steve Van Dien, Kristina Linnea Hillesland, Nicolas Pinel, Thomas J. Lie, John A. Leigh, and David A. Stahl. 2007. Metabolic modeling of a mutualistic microbial community *Molecular Systems Biology* 3 (1):92.

Terekhov, Stanislav S., Ivan V. Smirnov, Anastasiya V. Stepanova, Tatyana V. Bobik, Yuliana A. Mokrushina, Natalia A. Ponomarenko, Alexey A. Belogurov, Maria P. Rubtsova, Olga V. Kartseva, Marina O. Gomzikova, Alexey A. Moskovtsev, Anton S. Bukatin, Michael V. Dubina, Elena S. Kostryukova, Vladislav V. Babenko, Maria T. Vakhitova, Alexander I. Manolov, Maja V. Malakhova, Maria A. Kornienko, Alexander V. Tyakht, Anna A. Vanyushkina, Elena N. Ilina, Patrick Masson, Alexander G. Gabibov, and Sidney Altman. 2017. Microfluidic droplet platform for ultrahigh-throughput single-cell screening of biodiversity *Proceedings of the National Academy of Sciences* 114 (10):2550.

Thomas, Christopher M., and Kaare M. Nielsen. 2005. Mechanisms of, and barriers to, horizontal gene transfer between bacteria *Nature Reviews. Microbiology* 3 (9):711–721.

Tomasz, Alexander. 1965. Control of the competent state in *Pneumococcus* by a hormone-like cell product: An example for a new type of regulatory mechanism in bacteria *Nature* 208:155.

Tran, Frances, and James Q. Boedicker. 2017. Genetic cargo and bacterial species set the rate of vesicle-mediated horizontal gene transfer *Scientific Reports* 7 (1):8813.

Tropini, Carolina, Kristen A. Earle, Kerwyn Casey Huang, and Justin L. Sonnenburg. 2017. The gut microbiome: Connecting spatial organization to function *Cell Host & Microbe* 21 (4):433–442.

Venturelli, Ophelia S., Alex V. Carr, Garth Fisher, Ryan H. Hsu, Rebecca Lau, Benjamin P. Bowen, Susan Hromada, Trent Northen, and Adam P. Arkin. 2018. Deciphering microbial interactions in synthetic human gut microbiome communities *Molecular Systems Biology* 14 (6):e8157.

Wackett, Lawrence P., Anthony G. Dodge, and Lynda B. M. Ellis. 2004. Microbial genomics and the periodic table *Applied and Environmental Microbiology* 70 (2):647.

Whitney, John C., S. Brook Peterson, Jungyun Kim, Manuel Pazos, Adrian J. Verster, Matthew C. Radey, Hemantha D. Kulasekara, Mary Q. Ching, Nathan P. Bullen, Diane Bryant, Young Ah Goo, Michael G. Surette, Elhanan Borenstein, Waldemar Vollmer, and Joseph D. Mougous. 2017. A broadly distributed toxin family mediates contact-dependent antagonism between gram-positive bacteria *Elife* 6. doi:10.7554/eLife.26938.

Williams, Joshua W, Xiaohui Cui, Andre Levchenko, and Ann M Stevens. 2008. Robust and sensitive control of a quorum-sensing circuit by two interlocked feedback loops *Molecular Systems Biology* 4 (1):234.

Williams, Paul, Klaus Winzer, Weng C. Chan, and Miguel Cámara. 2007. Look who's talking: Communication and quorum sensing in the bacterial world *Philosophical Transactions of the Royal Society of London. Series B, Biological Sciences* 362 (1483):1119–1134.

Wintermute, Edwin H., and Pamela A. Silver. 2010. Emergent cooperation in microbial metabolism *Molecular Systems Biology* 6:407.

Wu, Fuqing, David J. Menn, and Xiao Wang. 2014. Quorum-sensing crosstalk-driven synthetic circuits: From unimodality to trimodality *Chemistry & Biology* 21 (12):1629–1638.

Yusufaly, Tahir I., and James Q. Boedicker. 2017. Mapping quorum sensing onto neural networks to understand collective decision making in heterogeneous microbial communities *Physical Biology* 14 (4):046002.

Part II

Enabling Technologies for
Building a Biomimetic Model

4 Stem Cell Engineering

Yi Sun Choi, Kisuk Yang, Jin Kim,
and Seung-Woo Cho

Yonsei University

CONTENTS

4.1 INTRODUCTION

Stem cells are an invaluable cell source for regenerative medicine owing to their self-renewal ability and potential to differentiate into more specialized tissue cell lineages. Different types of stem cells can be categorized with respect to the time of isolation during the lifetime of an organism and the cells' differentiation capacity (Figure 4.1). Embryonic stem cells (ESCs) are derived from the inner mass of developing embryos. ESCs are pluripotent stem cells that can differentiate into three-germ layers. Adult stem cells (ASCs) are categorized as multipotent stem cells, since they can be isolated from various postnatal sources that include the bone marrow, adipose tissue, peripheral blood, skin, and nervous system. ASCs can differentiate into several lineages, such as mesenchymal and hematopoietic cells. However, their differentiation capacity is rather limited. The demand for efficient treatments of diseases and injuries has sharpened the focus on stem cells, given their potential as a source of replacement cells for tissue regeneration. Transplanted stem cells can repopulate the desired locations and initiate the production of diverse cell types with the appropriate functions. ASCs have been applied clinically to treat various diseases, and autologous and allogenic therapeutic formulations have been commercialized for human use in conditions that include cartilage injury (Wohn 2012), graft-versus-host disease (Locatelli et al. 2017), Crohn's disease, and type I diabetes. Clinical trials have been and are being undertaken for human ESC (hESC)-derived therapeutics. For example, the subretinal transplantation of hESC-derived retinal pigment epithelium showed a good clinical outcome in preventing the progression of macular degeneration and restoring vision (Schwartz et al. 2012). Despite these and other promising stem cell therapeutics, limitations exist for each type of stem cell. ASCs usually have limited lineage differentiation and proliferative capacity. Use of ESCs continues to be burdened by the ethical dilemma of the use of human embryos as a cell source, in addition to concerns of tumorigenicity due to uncontrolled differentiation and the problem of immunogenicity.

To overcome the abovementioned limitations of ESCs and ASCs, induced pluripotent stem cells (iPSCs) were established through pluripotent reprogramming. This approach can provide an ideal source of cells for autologous cell-based patient therapy and patient-specific disease modeling (Takahashi and Yamanaka 2006). iPSCs are reprogrammed to have the properties of undifferentiated stem cells. This is accomplished by the introduction of transcription factors, including

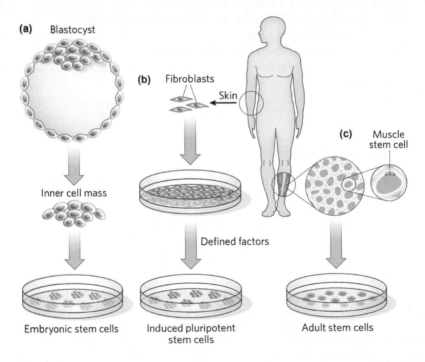

FIGURE 4.1 Stem cell types according to origin (Lutolf, Gilbert, and Blau 2009). (a) Embryonic stem cells, (b) Induced pluripotent stem cells, (c) Resident tissue-specific adult stem cells.

Oct4, Sox2, cMyc, and Klf4, into the somatic cells of patients. iPSCs are capable of differentiating into diverse three-germ layer cell types. As summarized in Figure 4.2, this innovative discovery has opened up a whole new avenue in regenerative medicine and disease modeling, since iPSCs possess the pluripotent property of ESCs, while avoiding the ethical and immunogenicity concerns. There have been numerous breakthroughs in iPSC research. iPSCs can now be created without disrupting healthy genes or using viral delivery systems (Okita et al. 2008, Woltjen et al. 2009). Indeed, patient-specific iPSC technology has contributed to functional stem cell therapeutics for various diseases and a wide variety of disease modeling (Robinton and Daley 2012).

Although stem cell technology has tremendous potential for therapeutic regenerative medicine, conventional methods for stem cell culture and transplantation continue to face a number of obstacles that hinder the desired clinical outcomes. One of the challenges is the difficulty in optimizing culture conditions to improve the proliferation, differentiation, and functional activity of stem cells. Conventional two-dimensional (2D) culture systems often do not provide highly effective stem

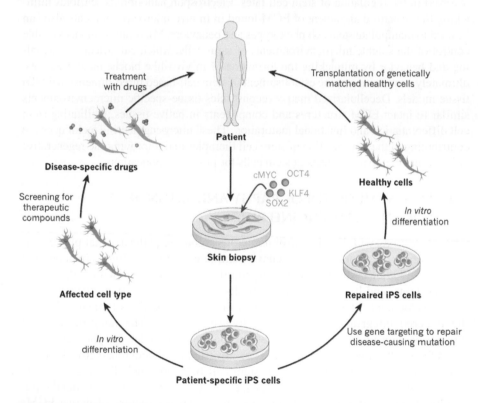

FIGURE 4.2 Biomedical applications of patient-specific iPSCs (Robinton and Daley 2012). iPSCs have the potential to model and treat various human diseases. Patient-specific iPSCs can be used for cell transplantation after repairing disease-causing genetic defects (right) or be differentiated into the affected cell type for in vitro disease modeling and therapeutic drug screening (left). iPSCs, induced pluripotent stem cells.

cell therapeutic formulations with sufficient viability and functional activity to facilitate their survival and engraftment, and the induction of therapeutic effects upon transplantation into diseased and defective tissues. This is because the culture systems cannot recapitulate the stem cell microenvironments. The consequences include inadequate stem cell proliferation, lineage-specific differentiation, and functionality (Becerra et al. 2011, Burdick and Vunjak-Novakovic 2008). In seeking a solution, attention has turned to microengineering techniques that closely mimic the biophysical, biochemical, and structural aspects of complex native tissues. This could circumvent the current limitations of stem cell technology and lead to the development of stem cell therapeutics that are clinically applicable.

This chapter reviews stem cell engineering strategies combined with various microengineering techniques that can recapitulate the microenvironments of complex tissues. The techniques include surface topography, electrospun fiber, microfluidic device, and decellularized matrix. The substrates and scaffolds with surface topography at the nano- or microscale can mimic topographical stimulation from the extracellular matrix (ECM) in vivo, which modulates mechanotransduction signaling cascades involved in the regulation of stem cell fates. Electrospun nanofibrous scaffolds mimicking the structural alignment of ECM found in in vivo microenvironments also can be used to manipulate stem cell phenotypes and behaviors. Microfluidic devices enable control of the soluble microenvironments of stem cells, which can affect cell signaling and behavior by mimicking the formation of in vivo-like biochemical gradients, ultimately contributing to the establishment of stem cell-based three-dimensional (3D) tissue models. Decellularized matrix reconstitutes tissue-specific microenvironments similar to intact ECM structures and components in native tissues, facilitating stem cell differentiation and functional maturation. These microengineering techniques are contributing to the optimization of stem cell therapies and improving the regenerative efficacy and therapeutic effects of stem cells for practical applications.

4.2 ECM–MIMETIC TOPOGRAPHY AND ALIGNMENT FOR STEM CELL ENGINEERING

Cell interactions with ECM control numerous signaling pathways that play essential roles in the regulation of cell behavior and fate (Gumbiner 1996, Giancotti and Ruoslahti 1999). ECMs provide binding motifs for cell adhesion with collagen and glycoproteins, and hence can guide cell elongation and migration during embryonic development. The cell adhesion provided by the ECMs is typically through the tethered integrin family of heterodimeric proteins. The integrin-mediated interaction with ECMs induces substantial changes in cellular morphology, which alters focal adhesion development and organization of filamentous-actin (F-actin). These changes control all aspects of stem cell behavior, including proliferation, differentiation, and function (Yang, Jung, et al. 2013). Accordingly, artificial topographical and structural cues that mimic the physical structure of native ECMs have been important in the development of biomaterials for the manipulation of stem cell behavior (Figure 4.3). This section discusses the surface topography and aligned fibrous structures of biomaterials to mimic the roles of ECMs to guide stem cell fate.

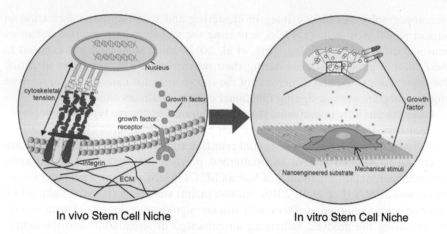

In vivo Stem Cell Niche   In vitro Stem Cell Niche

FIGURE 4.3 Artificial topographical cues to provide microenvironments mimicking in vivo stem cell niches (Zhang et al. 2015).

4.2.1 Micro- and Nanopatterned Surface Topography for Stem Cell Engineering

Artificial stem cell niches have been studied with the aim of increasing therapeutic efficacy of stem cells by manipulating the fate of these cells. Various factors control stem cell fate by generating biochemical and biophysical signals (Teo et al. 2010, Yang et al. 2012). Accumulating evidence indicates that biophysical cues, including topography, stiffness, and elasticity, can alter stem cell signaling and change cell behavior and function, similar to the effects produced by chemical induction (Seidlits, Lee, and Schmidt 2008). Especially, surface topographical cues of ECM can control stem cell proliferation and differentiation by altering intracellular signal transduction (Dalby et al. 2007, Chen et al. 2012). Further improvement of the efficacy of stem cell therapy requires engineered substrates and scaffolds with optimal surface topography to provide more favorable microenvironments that enhance stem cell proliferation or direct differentiation.

Fabrication of surface topography with precise control of geometry and rigidity can be done using diverse methods. These include soft lithography, photolithography, 3D printing, and electrospinning (Basnar and Willner 2009, Van Dorp et al. 2012). The fabricated substrates have been investigated for their effects on the stem cell behaviors with different parameters, including topographical size at the micro- or nanoscale, and shape (ridges, pillars, pits, or grooves) (McBeath et al. 2004, McBride and Knothe Tate 2008). Topographies with nano- and/or microscale cues promote and facilitate self-renewal, proliferation, and lineage-specific differentiation of stem cells. Specifically, nanoscale-patterned polymer surfaces, such as poly(ε-caprolactone) (PCL), polycarbonate, or polystyrene, support the long-term maintenance of the undifferentiated phenotype of mesenchymal stem cells (MSCs) (McMurray et al. 2011). In addition, nanopatterned substrates contribute to the maintenance of the self-renewal capacity of mouse ESCs (Jeon et al. 2012).

Nanotopography can induce integrin clustering and focal adhesion formation in human neural stem cells (hNSCs), activating the signaling pathway that enhances neuronal differentiation (Yang, Jung, et al. 2013). Since stem cells are exposed to their specific biophysical cues during their development, it is important to define the appropriate shape and dimension of the topography that can mimic the relevant biophysical cues when designing functional culture substrates and scaffolds.

Other studies have investigated stem cells cultured on diverse types of topographical patterns. One study demonstrated that nanopillar structures elicit specialized adhesion shapes that can provide spatial cues that alter directional signaling (Bucaro et al. 2012). Aligned groove nanopatterned polymer substrates can significantly enhance neuronal differentiation of human MSCs (hMSCs) (Yim, Pang, and Leong 2007) and hESCs (Lee et al. 2010). Human neural stem cells (hNSCs) cultured on grooved micropatterned surfaces will display significantly increased neurite outgrowth along the grooves, indicating an enhanced differentiation into the neuronal lineage (Béduer et al. 2012). Topographical cues generated by specific scales of submicron- or nanopatterns significantly promote osteogenic differentiation of hMSCs (Watari et al. 2012). To simultaneously provide microgroove and nanopore structural cues, a multiscale hierarchical patterned substrate was fabricated. hNSCs grown on the engineered substrate displayed a highly elongated cellular morphology and increased integrin clustering and focal adhesion development, leading to the significantly promoted differentiation into the neuronal lineage (Yang et al. 2014). These studies elucidated the synergistic topographical effects of aligned micro- and nanostructures on stem cell differentiation. The surface topography typically facilitates integrin clustering and focal adhesion assembly in stem cells, which induces cytoskeleton reorganization and alters cytoskeletal tension (Dalby, Gadegaard, and Oreffo 2014; Figure 4.4). The mechanical tension altered by topographical cues also affects nuclear mechanotransduction, which modulates specific cell signaling pathways through deformation of the nucleus (Pan et al. 2012). These changes influence stem cell proliferation and differentiation, and ultimately affect stem cell phenotypes and functions (Teo et al. 2010).

Surface topographies created for stem cell engineering by micro- and nanofabrication techniques have been prepared on many diverse biomaterial surfaces, including biodegradable and electroconductive materials. One study described that poly-L-lactic acid (PLLA) substrates with a grated nanostructure can induce osteogenic differentiation of mouse ESCs (Smith et al. 2009). Biodegradable poly(lactic-co-glycolic acid) (PLGA) polymer used as a nanotopography substrate can induce focal adhesion and increase the differentiation of hNSCs into the neuronal lineage (Yang, Park, Lee, et al. 2015). To increase ECM immobilization, mussel-inspired surface modification was done in the same study using a 3,4-dihydroxy-L-phenylalanine on the PLGA nanopatterned substrate. The modification synergistically enhanced neurogenesis of hNSCs (Yang, Park, Lee, et al. 2015). Nanotubes composed of titanium dioxide (TiO_2) that featured a spacing interval of 15–30 nm reportedly accelerated integrin clustering and focal adhesion formation, which in turn significantly promoted osteogenic differentiation of rat MSCs (Park et al. 2007). In another study, TiO_2 topographical substrates further improved neuronal differentiation of hNSCs through pulsed electrical stimulation (Yang et al. 2017).

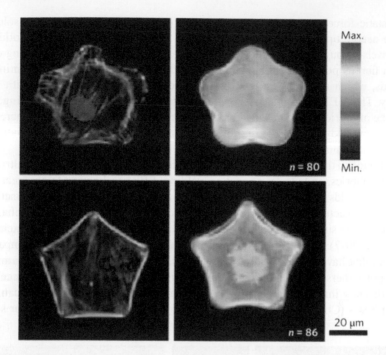

FIGURE 4.4 Myosin heat maps showing the alteration of cytoskeletal tension of MSCs on the different shapes of topographical surfaces (soft curves and sharp corners) (Dalby, Gadegaard, and Oreffo 2014). MSCs, mesenchymal stem cells.

The collective data support the importance of selecting the appropriate topography type and biomaterial that are optimal for each unique stem cell type to reconstitute ECM-mimetic topographical cues for stem cell manipulation. Identification of the optimal scales and shapes of surface micro- and/or nanostructures to induce significant effects of topography on proliferation and differentiation of stem cells is an ongoing and more recently intensifying field of research that could prove fundamentally valuable as a means of manipulating stem cell fate and function to potentiate stem cell therapy by creating advanced biomaterials that are responsive to their topographical microenvironments for intended applications.

4.2.2 Electrospun and Structurally Aligned Fibrous Scaffolds for Stem Cell Engineering

Electrospinning is another technique that has been used to mimic the complex structural cues of ECMs. The technique has been applied to fabricate functional biomaterial scaffolds for stem cell engineering. In electrospinning, polymer fibers with diameters in the range of submicron to nanometer are produced, generating a fiber mesh with high porosity and large surface-area-to-volume ratio (Pham, Sharma, and Mikos 2006). The set-up for electrospinning includes a syringe pump, high voltage source, and a collector. A high voltage is applied at the needle tip to generate an electrostatic force to overcome the surface tension between the polymer solution and the needle tip. This allows the ejection of a stream of nanofiber that solidifies as it travels through the air and reaches the grounded collector. The resulting electrospun fibrous polymers (EFPs) have emerged as a versatile source of biomimetic scaffolds, which are characterized by a network of stacked nanofibers with high porosity. These EFP-based scaffolds allow efficient cell penetration, signaling, and exchange of nutrients and gas, and have produced promising outcomes of increased cell adhesion and growth. The scaffolds may prove valuable in the differentiation of stem cells into multiple lineages.

In addition to the myriad non-biological applications of EFP, such as fluid filtration, protective fabrics, and electromagnetic shielding, EFP can be used to engineer biomimetic scaffolds, considering biocompatibility, provision of similar ECM functions, appropriate structural features like pore size and diameter, and sufficient mechanical strength for the support of cell adhesion, infiltration, growth, and nutrient exchange (Lins et al. 2017). EFP scaffolds can be fabricated from a variety of biocompatible polymers that have included poly(ethersulfone), PLGA, PLLA, and PCL. Laminin-coated poly(ethersulfone) fiber scaffolds with diverse diameters have been successful in significantly influencing the differentiation and proliferation of rat hippocampus-derived NSCs (Christopherson, Song, and Mao 2009; Figure 4.5). In another study,

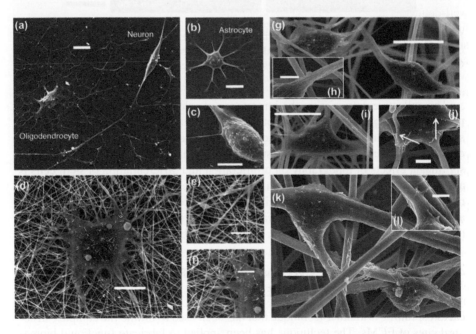

FIGURE 4.5 Scanning electron microscopy images of rat hippocampus-derived NSCs cultured on electrospun nanofiber meshes for 5 days. (a–c) Cells cultured on laminin-coated PES mesh. Cells cultured on fiber mesh with diameters of 283 nm (d–f), 749 nm (g–j), and 1,452 nm (k and l). Cells are highlighted in yellow in (e and f). Cell attachment to fibers is indicated by arrows in (j). Scale bars in (a, b, d, g, i, k), (c, e, f), and (h, j, l) are 10, 5, and 2 µm, respectively (Christopherson, Song, and Mao 2009). NSC, neural stem cell; PES, poly(ethersulfone).

PLGA-based bioactive EFP scaffolds with immobilized bone-forming peptides were developed using polydopamine coating. The scaffolds supported the compact distribution of collagen type I and spreading of hMSCs. Transplantation of hMSCs using the PLGA EFP scaffolds also led to the significant improvement of bone regeneration in a mouse calvarial bone defect model (Lee, Lee, et al. 2013). Using electrospun PLLA scaffolds, another research group demonstrated the differentiation of adipose-derived MSCs into insulin-producing cells in vivo (Fazili et al. 2016). In addition, others demonstrated that PCL-based electrospun scaffolds can enhance the neuronal differentiation of mouse ESCs and guide the neurite outgrowth of differentiated neuronal lineage cells (Xie et al. 2009). Natural materials, such as collagen, silk, fibrinogen, and chitosan, can also be used as the backbone of electrospun scaffolds that have the attributes of pronounced biocompatibility and biodegradability, and which only minimally trigger inflammation (Pham, Sharma, and Mikos 2006). Electrospun silk fibroin supports the attachment and proliferation of bone marrow-derived MSCs, keratinocytes, and fibroblasts (Jin et al. 2004).

Customization of EFP for functional biomaterial scaffolds can be achieved through tuning the experimental parameters and equipment designs for electrospinning. Highly aligned nanofibers can be generated from rotating cylindrical collectors, auxiliary electrodes, and collectors with sharpened edges. Aligned fibers could induce the development of cardiomyocytes with mature cytoskeleton structures of well-defined sarcomeres and intercalated disks (Zong et al. 2005). A developed multilayered alignment technique facilitated the formation of aligned scaffolds that significantly enhanced the expression of tenomodulin in adipose-derived hMSCs compared to non-aligned scaffolds (Orr et al. 2015). When prepared in combination with 3D stereolithographic printing, aligned EFP scaffolds increased NSC attachment and proliferation, and enabled directional control of neurite outgrowth (Lee, Nowicki, et al. 2017). Alternatively, by using sacrificial agents or by changing the collector design to porous meshes or patterned surfaces, further increases in the porosity of EFP could be realized, promoting the infiltration of cells. As one example, poly(ethylene oxide) was incorporated into a solution containing PCL, collagen, and hydroxyapatite during the electrospinning process. The poly(ethylene oxide) fibers readily dissolved in the post-fabrication washing step, which left a highly porous scaffold that supported greater infiltration of hMSCs (Phipps et al. 2012). Other authors used a collector covered with metal strips to generate porous 3D PCL scaffolds and demonstrated higher proliferation and viability of hMSCs as well as enhanced osteogenic differentiation (Rampichová et al. 2013).

The versatile nature of electrospinning allows the engineering of more complex forms of EFP scaffolds through layering, coaxial-electrospinning, and surface coating to initiate the targeted mimicry of specific cells or tissues of interest. Layered scaffolds can be formed by sequentially spinning different polymer solutions in a process where each layer can be tailored for a specific role in cellular hierarchy (Khorshidi et al. 2016). In one study, a five-layer scaffold with three PLA layers in the upper region and two PLA layers with embedded tricalcium phosphate nanoparticles in the lower region was created. The scaffold simultaneously promoted site-specific chondrogenic and osteogenic differentiation of adipose-derived hMSCs (Mellor et al. 2015). Co-spinning of polymers can be applied to improve the scaffold's biocompatibility and biodegradability, increase the rigidity of EFP, or achieve a better composition, like the co-spinning of collagen type I and III for a better mimicry of native collagen structure (Matthews et al. 2002). Other authors reported that co-spun graphene oxide and PLGA 3D scaffolds accelerated hMSC adhesion and proliferation compared to pure PLGA scaffolds (Luo et al. 2015). To provide biological cues for increasing the biological functionality, several bioactive molecules can be deposited onto the surface of EFP scaffolds. The molecules include collagen, hydroxyapatite, chitosan, or specific proteins. They enhance the growth and differentiation of various stem cells of interest. For example, osteoinductive scaffolds have been engineered from electrospun PCL (Lee, Jin, et al. 2017) and silk fibroin (Sawkins et al. 2013) modified with hydroxyapatite particulates using polydopamine adhesive coating. These osteoinductive EFP scaffolds possess improved osteoinductive capacity, which are shown to promote osteogenic differentiation and mineralization of MSCs, and facilitate bone regeneration to repair critical-sized calvarial defects (Lee, Jin, et al. 2017). Others combined multiple techniques to produce a scaffold with concentric layers with PLLA in the core to provide mechanical support, with gelatin containing embedded retinoic acid and purmorphamine on the shell. The surface gelatin layer can release instructive cues for NSCs to differentiate into motor neurons with significant increase in neurite length (Binan et al. 2014).

In summary, EFP is a highly versatile source for biomimetic scaffolds generally comprising classes of synthetic and natural materials for stem cell engineering. EFP scaffolds can be biologically and structurally modified to reproduce the complex niches necessary for various stem cells of interest. Current studies are focusing on developing EFP scaffolds that are more effective for functional stem cell therapeutics by integrating electrospinning with many other techniques to create biomimetic stem cell microenvironments.

4.3 MICROFLUIDIC DEVICES FOR STEM CELL-BASED 3D TISSUE MODELS

Stem cells respond sensitively to the coordinated interaction with other cells and surrounding microenvironments. Therefore, the major challenge for controlling stem cell survival, proliferation, and differentiation is preparing a finely tunable cell culture platform that facilitates complex and dynamic regulation of such stem cell niches. The environmental cues determining the stem cell fate include biochemical, biophysical, and mechanical stimulation. The specific concentrations of paracrine and autocrine signaling molecules, and the balance between endogenous and exogenous factors also determine the fate of stem cells to maintain their pluripotency or to facilitate their differentiation into specific lineages (Gupta et al. 2010). Since these stimulations need to be precisely controlled in terms of time and space, conventional cell culture techniques cannot recapitulate the critical aspects of in vivo microenvironment conditions. To overcome this hurdle, microfluidics that can provide appropriate spatiotemporal control of chemicals, oxygen, pH, signaling factors, and physical stimulations (e.g., flow shear stress, stretching) has been applied to stem cell technology (Figure 4.6). ECM-derived surface coatings or hydrogel platforms can be

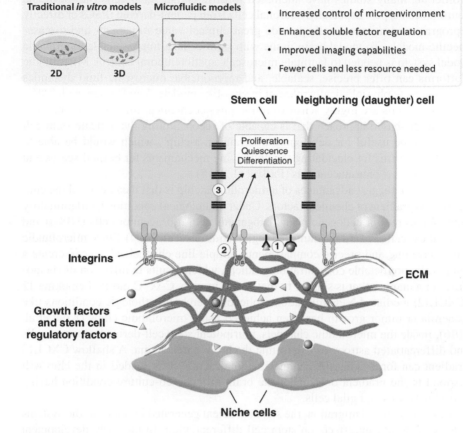

FIGURE 4.6 Advantages of microfluidic models over traditional in vitro models (Boussommier-Calleja et al. 2016). Microenvironment control through soluble and matrix-bound factor gradients directs the fate of stem cells (Volckaert and De Langhe 2014).

integrated into microfluidic systems to manipulate the behavior of stem cells. Since stem cells have force-coupled signaling pathways to sense the mechanical property of the niche, microfluidic systems enabling well-controlled dynamic flow can alter stem cell properties and fates (Discher, Mooney, and Zandstra 2009).

4.3.1 BRAIN-MIMETIC CHIP WITH STEM CELLS

The brain is the most sophisticated organ in the human body. Despite a long history of active research, the complex crosstalk between specific neural cell types has not been fully elucidated. Although a conventional cell culture system can provide fundamental information concerning cellular reactions in the central nervous system (CNS), studying the CNS has been particularly challenging because CNS functions are derived from complex interactions between different cell types, rather than a single cell type (Paşca 2018). Since fully differentiated neuronal cells rarely proliferate, many studies have attempted to differentiate stem cells into desirable neuronal cell types. Especially, neuronal cells from patient-derived iPSCs or directly reprogrammed neuronal cells provide great clinical value in patient- and disease-specific models. To model brain tissue with stem cells, a finely tunable biomimetic model system is needed to facilitate microscale spatiotemporal control. Microfluidic platforms can offer precise, scalable, and reproducible microscale fluid dynamics that can create physiologically relevant in vivo-like models (Van Duinen et al. 2015). Since stem cells are highly sensitive to the physicochemical microenvironments, a microfluidic-based approach that is capable of reconstituting the delicate stem cell niches can be useful for developing a "brain-on-a-chip", which would be able to provide platforms for elucidating the underlying mechanisms for brain diseases and discovering new pharmaceuticals (Park et al. 2015).

One of the biggest advantages of a microfluidic chip is that it can control the concentration gradient of chemical factors. Under pathological conditions, inflammatory chemokines recruit endogenous and exogenous neural progenitor cells (NPCs) and guide these cells toward the damaged region (Imitola et al. 2004). Since microfluidic chips have fine and precise compartments, simple line channels alone can create a temporally predictable concentration gradient, which results in diffusion of chemokines in a manner that is similar to that found in vivo. C–X–C motif chemokine 12 (CXCL12)-mediated NPC migration replicating cerebral pathologic conditions like ischemia or tumor growth has been achieved using microfluidic chips (Kilic et al. 2016). Inside the microfluidic channels, pluripotent stem cell-derived neuronal cells and differentiated astrocytes can mimic the brain parenchyma. A shallow CXCL12 gradient can form along the channels. Human fetal NPCs seeded in the chip will respond to the gradient only under the brain-mimetic co-culture condition having neural networks and glial cells.

Besides stem cell migration, the signal gradient generated in microfluidic systems can exert significant effects on stem cell differentiation. In the early development of the vertebrate nervous system, the gradient of diffusible paracrine factors plays important roles in stem cell differentiation and various functional developments. Cells recognize their position through the gradient of extracellular signaling molecules and determine their developmental fate. Signaling molecules like sonic hedgehog (Shh), fibroblast growth factor (FGF), and bone morphogenic protein (BMP) engage in an interplay to guide stem cell differentiation along the dorsoventral and anteroposterior axes (Figure 4.7; Park et al. 2009). Inductive signaling molecules regulate the development of the neural plate at the mid-hindbrain junction (FGF8) and floor plate (Shh). The cross-sectional image in Figure 4.7 illustrates the Shh and BMP gradients, which specify ventral neurons and dorsal neurons, respectively. By replenishing reagents in the microfluidic chip using an osmotic pump, different factors in two fluid streams can generate multiplex gradient via diffusion (Park et al. 2009). Thus, the concentration gradient of Shh, BMP, and FGF that is similar to that of dorsoventral and anteroposterior axes during the early development of the vertebrate nervous system can be formed on the gradient-generating microfluidic chip. Such a chip with multiplex factor gradients had been used with hESC-derived NPCs to stimulate an increased proliferation rate and the expression of Tuj in the region with a combination of Shh and FGF8 (Park et al. 2009). This phenomenon proves the

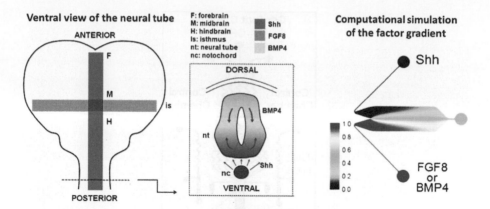

FIGURE 4.7 Signaling factor gradient in the early developmental stage of the vertebrate nervous system. The computational simulation of factor gradient is shown in the right panel (Park et al. 2009).

efficacy of the multigradient chip, since the individual treatment of each factor had minimal influence on the hESC-derived NPCs.

Another type of microfluidic chip was used to elucidate the response of stem cells to specific factor concentration. A microfluidic chamber connected to the gradient-generating part with parallel diffusive mixing channels was used to demonstrate the effect of growth factors on NSCs (Figure 4.8; Chung et al. 2005). NSCs can differentiate into major cell types of the nervous system, including neurons, astrocytes, and oligodendrocytes. In this device, NSCs could be continuously exposed to the steady-state gradient of the growth factor mixtures of FGF2, epidermal growth factor (EGF), and platelet-derived growth factor throughout the culture. The increased concentration of growth factors in the chips escalated the proliferation of NSCs but hindered the differentiation of NSCs into the astrocyte lineage.

2D laminar flow-based gradient-generating chips aim to directly identify the effect of exogenous factors by clearing away the cell-driven cytokines, whereas 3D-based microfluidic chips focus on implementing an in vivo-like microenvironment. Conventionally, 3D cell culture methods have been used to generate microenvironments similar to the cells' original niches to maintain their in vivo characteristics. However, problems with the approach include the small volume-to-surface ratio, which leads to hypoxia in the center of the culture system, and the difficulty in observation due to thick z-axis layers (Boussommier-Calleja et al. 2016). To overcome these limitations of current 3D culture systems, a microfluidic chip could incorporate a 3D ECM hydrogel. This could be a powerful tool for investigating cell–cell communication, cell–matrix interaction, and cellular response under reconstituted in vivo-like 3D conditions. To this end, a study demonstrated the effectiveness of a microfluidic chip design by comparing NSC culture in a 3D ECM microfluidic system with conventional 2D and 3D culture conditions (Han et al. 2012). In addition, the authors documented significantly increased levels of hypoxia-inducible factor 1-alpha, a transcription factor that controls signaling pathways

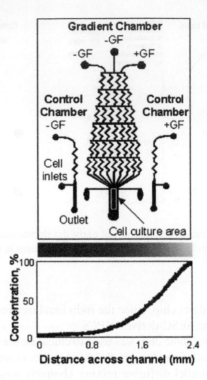

FIGURE 4.8 Illustration of the gradient-generating microfluidic device. The gradient profile of the cell culture chamber was visualized using fluorescent particles (Chung et al. 2005).

promoting stem cell self-renewal at low oxygen concentrations, under hypoxic condition in NSCs cultured in the 3D microfluidic chip, as compared to the conventional methods (Yang, Han, et al. 2013). These results indicate that microfluidic platforms can create in vivo-like hypoxic microenvironments that facilitate the stem cells to adapt to hypoxia and retain their capability of self-renewal. In another study, a brain-mimetic 3D microfluidic chip with NSCs was created and used to test the paracrine effect of glial cell-derived neurotrophic factor (GDNF)-overexpressing MSCs on NSC differentiation (Yang, Park, Han, et al. 2015). The GDNF-overexpressing MSCs facilitated the neuronal differentiation of NSCs, particularly toward dopaminergic neurons. The same study used hypoxic-ischemic stroke animal models to demonstrate that the transplanted GDNF-overexpressing MSCs improved neurobehavioral functional recovery by inducing endogenous NSC differentiation, which validated the potential of microfluidic platform for better prediction of in vivo paracrine actions (Yang, Park, Han, et al. 2015).

4.3.2 VASCULATURE-MIMETIC CHIP WITH STEM CELLS

Blood vessels comprise an integral part of tissues and are involved in most of the metastatic cascade. The implementation of vasculature is critical, especially in drug screening platforms, because the molecular structure of a drug can change during the transport through blood vessels (Trachsel and Neri 2006). Furthermore, in stem cell research, the differentiation of stem cells can be significantly affected by the vasculature network. In one study, a 3D microfluidic co-culture platform clearly showed the effects of brain endothelial cells on NSCs (Figure 4.9). The brain endothelial cells cultured in channels connected to 3D hydrogel channels with NSCs secreted paracrine effectors, such as vascular endothelial growth factor (VEGF), EGF, basic FGF (bFGF), brain-derived neurotrophic factor, and pigment epithelium-derived factor. This design facilitated the creation of a concentration gradient of the growth factor through the NSC-containing hydrogel. In this system, the differentiation of NSCs to astrocytes was more active in proximity to brain endothelial cells, whereas the neuronal differentiation was suppressed along the gradient. Overall, this microfluidic co-culture platform recapitulated the actual brain tissue where the astrocyte end-feet structure surrounds brain vessels. Computational simulations and experiments confirmed that the 3D microfluidic co-culture system was more effective than conventional culture methods in accumulating the paracrine factors from endothelial cells and clearly identifying their effects (Shin et al. 2014).

Unlike other organs, the brain has a unique selective barrier—the blood–brain barrier (BBB)—that restricts molecular exchange between blood and brain. This protects the CNS in real life. However, it has been a pharmacokinetic hurdle in the development of drugs that target the CNS. Using animal models to develop human brain-targeting drugs often fails to generate accurate drug testing results for two reasons. First, it is difficult to analyze the transfer and function of drug candidates inside the brain tissue of animals. Second, there is an inherently different drug response that reflects the species-specific physiological difference

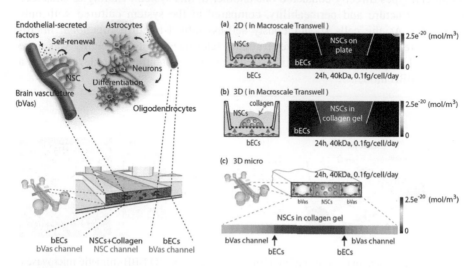

FIGURE 4.9 Illustration of a 3D microfluidic brain vasculature model. In brain tissue, endothelial cells secrete signaling factors that regulate NSC self-renewal and differentiation. The secreted factors diffuse and are diluted in conventional culture platforms (a and b), and accumulate and are concentrated in the 3D microfluidic model (c), as in actual brain tissue in vivo (Shin et al. 2014). 3D, three-dimensional; NSC, neural stem cell.

(Maoz et al. 2018). Thus, there is a pressing need to establish functional in vitro human BBB models. A BBB-on-a-chip system with the endothelial barrier combining the stem cell-derived CNS components was developed and shown to recapitulate the neurovascular unit of brain tissue in microfluidic system (Kilic et al. 2016, Maoz et al. 2018). The endothelial barrier and the brain compartment were reconstituted on the both sides of a porous membrane between an upper chamber representing the blood side and a lower chamber representing the brain side. BBB functions of this artificial system were also evaluated by measuring the permeability with transendothelial electrical resistance or the transfer rate of fluorescent particles. Several useful features of this artificial BBB chip, such as accurate control, clear visualization, and in vivo-like fluidic shear stress to the endothelial layer, which are critical to enhance the barrier integrity, allowed this system to represent some parts of BBB function (Maoz et al. 2018). The authors also demonstrated the metabolic interactions between the neurons and microvasculature of the brain by connecting microfluidic chips to give directional flow of the culture media and sequentially analyzed the media before and after the exposure to the brain part (Maoz et al. 2018). With this chip system, they proved that the direct utilization of vascular metabolites by neurons and astrocytes could increase the neuronal synthesis of glutamate and gamma-aminobutyric acid.

A brain chip model that improves the structural similarity of microvascular network of the brain has also been demonstrated by Campisi et al. (2018). The system comprised a 3D microfluidic system composed of iPSC-ECs, brain pericytes, and astrocytes with fibrin gel, in which 3D brain microvascular structures were generated in a vasculogenesis-like process (Campisi et al. 2018; Figure 4.10). Three different cell types directly contacted one another in this system, leading to enhanced vascular structure and permeability, compared to the system cultured with only iPSC-ECs. As the cells spontaneously self-assembled to form functional microvasculature within the system, the structural relevance and functionality was quite comparable to in vivo tissues. This system has great potential to be developed as a practical BBB model.

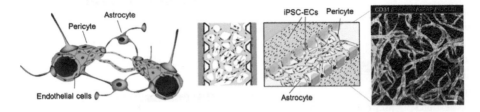

FIGURE 4.10 Schematic illustration of the actual BBB and 3D BBB-mimetic microvascular network model in a microfluidic system. The far right panel shows a confocal microscopy image of self-assembled microvasculature including iPSC-endothelial cells (CD31, green), pericytes (F-actin, red), and astrocytes (GFAP, magenta) (Campisi et al. 2018). BBB, blood–brain barrier; iPSC, induced pluripotent stem cell.

4.4 DECELLULARIZED MATRIX FOR STEM CELL ENGINEERING

Tissue regeneration using stem cells requires functional biomaterials that can provide appropriate stem cell niches for maximal therapeutic efficiency. Signaling induced by the cell–matrix interaction is one of the key factors in determining cellular behaviors and functions, thus making it crucial in stem cell niche formation (Watt and Huck 2013). Therefore, reconstitution of stem cell microenvironments based on the actual ECM components, including collagen, laminin, and glycoproteins, has garnered interest for stem cell engineering. Even though the combination of some ECM components or modification of polymer scaffolds with ECM has been used to generate native tissue-like environments, it remains a major challenge to accurately mimic the precise components, compositions, and factors in the actual tissue-specific niches for stem cells.

To provide the stem cells with tissue-specific microenvironments that have optimal biophysical and biochemical cues, along with structural integrity, decellularization of the native organs or tissues has progressed markedly over the past few decades. Through the decellularization process, antigen-presenting cellular components are completely removed from organs, tissues, or matrix-depositing cells, while the biochemical composition and microstructure of the ECM are preserved to reconstitute a tissue-specific microenvironment (Badylak, Freytes, and Gilbert 2009). Therefore, various stem cell behaviors, such as self-renewal, proliferation, and differentiation, and functions, can be improved and guided by such tissue-specific effects of a decellularized matrix. This kind of microenvironment engineering based on natural tissue-derived materials for stem cell manipulation holds great promise for improved regenerative medicine using stem cells.

4.4.1 DECELLULARIZED TISSUE MATRIX-DERIVED HYDROGELS FOR STEM CELL ENGINEERING

The 3D hydrogel platform has been extensively explored to support stem cell growth and differentiation. Both natural and synthetic materials have been used as 3D hydrogels to modulate stem cell behaviors. Natural hydrogels, such as hyaluronic acid, can maintain the pluripotency of hESCs (Gerecht et al. 2007). Hydrogels based on synthetic polymers including poly(ethylene glycol) (Hwang et al. 2006), and other derivatives have also been studied for MSC and ESC cultures (Li et al. 2006, Williams et al. 2003). However, it has become apparent that the ideal stem cell microenvironment requires more complex cell–ECM and cell–cell interactions for better therapeutic efficacy. Conventional natural and synthetic polymers are inadequate to recreate such sophisticated microenvironments. Decellularized tissue-derived ECM has emerged as an ideal material for creation of a 3D hydrogel to recapitulate the intricate structures and functions of actual tissues so that the stem cells can be provided with the natural microenvironment similar to their original tissue (Sellaro et al. 2009). In addition, a decellularized matrix can provide unique ECM compositions and structures specific to each type of tissue, which is optimal for improving the function and potency of each stem cell type (Uygun et al. 2010).

Decellularized matrix materials for hydrogel platforms are harvested from various organs and tissues, including skin (Wolf et al. 2012), small intestine (Okada et al. 2010), urinary bladder (Freytes et al. 2008), brain (Jin, Lee, et al. 2018), liver (Jin, Kim, et al. 2018, Lee, Shin, et al. 2013), heart (Duan et al. 2011), muscle (DeQuach et al. 2012), and bone (Sawkins et al. 2013). The decellularization process varies depending upon organ and tissue types. The aim remains the same—to remove as many cells as possible while leaving the structural ECM components and tethered signaling molecules intact for hydrogel formation. Decellularization of the native organs and tissues utilizes diverse decellularization agents, which include ionic (e.g., sodium dodecyl sulfate, sodium deoxycholate) or nonionic (e.g., Triton X-100) detergents and enzymatic treatments (e.g., trypsin, DNase, RNase) (Crapo, Gilbert, and Badylak 2011). Table 4.1 provides an overview of the protocols for tissue decellularization for 3D ECM hydrogel formation.

To induce hydrogel formation with decellularized tissue-derived ECM, decellularized tissues need to be solubilized with pepsin in an acidic condition (Young et al. 2011; Figure 4.11). After solubilization, the resultant ECM solution still contains numerous bioactive components, including growth factors and cell adhesion proteins, as well as various ECMs found in native tissues, such as collagens, proteoglycans, and glycosaminoglycans. The solubilized matrix can then undergo self-assembly to form the 3D hydrogel, as the decellularized matrix mainly consists of collagens (Figure 4.11). Polymerization kinetics that induce gelation are influenced by temperature and pH. Decellularized matrix-based hydrogels can reconstitute collagen nanofibrous network incorporating tissue-specific ECM components, which provides in vivo-like microenvironments for stem cells (Lee, Shin, et al. 2013, Jin, Kim, et al. 2018).

Decellularized matrix hydrogel promotes stem cell differentiation and also potentiates the therapeutic efficacy of stem cells for tissue regeneration. A study examined the adipogenic and chondrogenic differentiation of human adipose-derived stem cells cultured in decellularized matrix hydrogels (Pati et al. 2014). The decellularized matrix hydrogel maintained higher levels of cell viability and promoted differentiation into each desired lineage, compared to the collagen hydrogel. The encapsulated cells in the decellularized matrix hydrogel did not undergo stress-induced apoptosis and retained appropriate phenotypes due to cell–ECM interactions. Other authors evaluated the odontogenic differentiation of dental pulp stem cells on hydrogel derived from decellularized bone ECM in comparison to collagen (Paduano et al. 2016). Dental pulp stem cells cultured on decellularized matrix hydrogel displayed higher levels of mRNA expression of odontogenic genes without a negative effect on cell viability, proving once again that decellularized matrix hydrogel scaffolds provide an excellent environment for specific differentiation of stem cells. Others used hydrogels prepared from decellularized porcine spinal cord tissue to induce therapeutic effects of MSCs in spinal cord injury repair (Tukmachev et al. 2016). This decellularized matrix-based hydrogel stimulated neovascularization and axonal ingrowth into the lesion when injected into a spinal cord cavity. The decellularized matrix hydrogel showed advantages over conventional synthetic nondegradable materials in terms of the biological activities of modulating the immune response and stimulating vascularization and axonal growth.

TABLE 4.1
Decellularization Reagents Used to Produce ECM Hydrogels for Each Tissue Type

Tissue	Decellularization Reagents	Reference(s)
Human adipose	1% SDS or 2.5 mM sodium deoxycholate 2.4 mM sodium deoxycholate with 500 U lipase and colipase	Young et al. (2011)
Porcine brain	0.05% trypsin/EDTA, 3% Triton X-100, 3% SDS	Jin, Lee, et al. (2018)
Bovine bone	0.5 M HCl 1:1 Chloroform:methanol 0.05% trypsin/0.02% EDTA 1% (w/v) penicillin/streptomycin in PBS	Sawkins et al. (2013)
Porcine cartilage	10 mM Tris-HCl at pH 8 0.25% trypsin 1.5 M NaCl in 50 mM Tris-HCl at pH 7.6 50 U/mL DNase and 1 U/mL RNase in 10 mM Tris-HCl 1% Triton X-100 10 mM Tris-HCl 0.1% peracetic acid/4% ethanol	Pati et al. (2014)
Human tendon	0.1% EDTA	Farnebo et al. (2014)
Porcine heart	1% SDS, 0.5% penicillin/streptomycin	Wassenaar et al. (2016)
Porcine liver	0.02% trypsin and 0.05% EDTA 3% Triton X-100 4% sodium deoxycholic acid 0.1% peracetic acid	Sellaro et al. (2009)
Porcine lung	1× penicillin/streptomycin 0.1% Triton X-100 2% sodium deoxycholate DNase NaCl	Pouliot et al. (2016)
Porcine skeletal muscle	1% SDS	Pouliot et al. (2016)
Porcine dermis	0.25% trypsin 70% ethanol 3% H_2O_2 1% Triton X-100 in 0.26% EDTA/0.69% Tris 0.1% peracetic acid/4% ethanol	Wolf et al. (2012), Faulk et al. (2014)
Porcine bladder	0.1% peracetic acid/4% ethanol	Freytes et al. (2008)

EDTA, ethylenediaminetetraacetic acid; PBS, phosphate-buffered saline; SDS, sodium dodecyl sulfate.

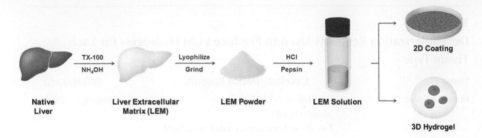

FIGURE 4.11 Decellularization and solubilization of native tissue to generate 3D ECM hydrogel platforms (Lee, Shin, et al. 2013). 3D, three-dimensional; ECM, extracellular matrix.

The authors mentioned that these enhanced properties of stem cells are likely due to the creation of native stem cell niches with intricate structures and cell–ECM interactions. The applications of decellularized matrix hydrogel, especially concerning the cell delivery vehicle, were demonstrated in other studies with lung-derived ECM hydrogel (Pouliot et al. 2016) and cardiac-derived ECM hydrogel (Jang et al. 2017). The first study developed a lung-derived ECM hydrogel as a 3D MSC culture platform as well as a carrier system for in vivo delivery of cells. The MSCs showed high viability and stemness in the decellularized hydrogel system, which were then applied to in vivo system to show a high retention rate. The second study also involved the preparation of a cardiac-derived ECM hydrogel to support the 3D culture of both cardiac progenitor cells and MSCs to increase their regenerative efficacy (Jang et al. 2017). The transplanted cells formed functional vessels in the ischemic environment of in vivo infarction model by amplifying the activity of host myocardial precursor cells within the cardiac-derived ECM hydrogel. These promising results confirm the high potential of decellularized matrix hydrogel as a cell delivery vehicle for stem cell therapy.

More recently, the decellularized matrix hydrogel has been applied for improving transcription factor-mediated direct reprogramming. Direct reprogramming of terminally differentiated somatic cells into other tissue lineage cells without passing through the stem cell state was achieved by transduction and forced expression of tissue-specific transcription factors (Sekiya and Suzuki 2011). However, conventional culture systems cannot provide optimal microenvironments for cells undergoing direct reprogramming. A study reported the development of 3D hydrogel culture platforms from decellularized liver and brain tissues (Jin, Lee, et al. 2018). The authors reported successful generation of vascularized liver organoids from reprogrammed hepatocytes using decellularized liver matrix hydrogel in the microfluidic system with a dynamic fluid flow (Jin, Kim, et al. 2018). This liver organoid platform proved useful for drug testing and also could be integrated with other internal organoids in multiple array device to provide multiorgan models. The same study also demonstrated that 3D hydrogel prepared from decellularized brain matrix facilitated the direct reprogramming of primary fibroblasts into therapeutic neurons via epigenetic modulation and mechanosensitive signaling pathways (Figure 4.12; Jin, Lee, et al. 2018).

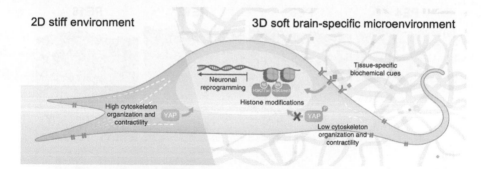

FIGURE 4.12   Potential mechanisms of enhanced direct neuronal reprogramming by decellularized brain matrix-derived 3D hydrogel (Jin, Lee, et al. 2018). 3D, three-dimensional.

4.4.2   Decellularized Matrix-Modified Scaffolds for Stem Cell Engineering

The decellularized matrix has gained great interest for scaffold modification as it can endow the scaffolds with naturally occurring tissue-specific ECM microenvironments. Modification of the polymer scaffolds with decellularized matrix can be achieved by decellularizing the cell-grown scaffolds after ECM-secreting cells, such as MSCs and fibroblasts, are cultured on the scaffolds for a certain period of time (Figure 4.13; Liao, Guo, Nelson, et al. 2010). The decellularization methods vary in small details. However, in many studies, the cell-cultured scaffolds were either decellularized through freeze-thaw procedures using liquid nitrogen (Sadr et al. 2012, Liao, Guo, Nelson, et al. 2010, Datta et al. 2005) or by brief incubation in a mixture of solution containing 0.5% Triton X-100 and 20 mM NH_4OH (Kang et al. 2012). Both approaches produce structurally intact ECM attached to the surface and inner walls of the scaffolds. Stem cells are reseeded onto the modified scaffolds and interact with the deposited ECMs. Numerous studies have documented the accelerated osteogenic differentiation of MSCs on decellularized matrix-modified scaffolds, including titanium fiber mesh scaffolds (Datta et al. 2005), beta-tricalcium phosphate scaffolds (Kang et al. 2012), polyester urethane foams (Sadr et al. 2012), and PCL microfiber scaffolds (Liao, Guo, Nelson, et al. 2010). Studies that examined the osteoinductive potential of decellularized matrix-modified scaffolds confirmed the significant enhancements of mineralization as well as expression of osteoblastic markers and osteogenic factors after recellularization with stem cells. Enhanced chondrogenic differentiation of MSCs using PCL microfiber scaffolds decorated with decellularized matrix has been described (Liao, Guo, Grande-Allen, et al. 2010). The decellularized matrix-modified scaffolds appeared to support the differentiation of stem cells to chondrocytes with upregulated expression of the chondrogenic markers aggrecan and collagen type II.

The collective results support the suggestion that ECM derived from matrix-secreting cells can create a native tissue-like microenvironment when deposited onto synthetic polymer scaffolds, and that these decellularized matrix-modified scaffolds can promote

PE4   PE8   PE12   PE16

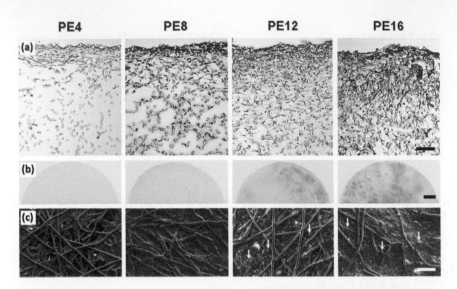

FIGURE 4.13   Cell culture on the synthetic polymer constructs modified with decellularized tissue-derived ECM. Histological stains to visualize the distribution of cells and ECM proteins are shown in (a) with the scale bar representing 100 µm. (b) Mineralized matrix revealed via X-ray images with the scale bar representing 1 mm. Scanning electron micrographs of the surface of the scaffold are shown in (c) with the scale bar representing 100 µm and the arrows indicating mineral nodules (Liao, Guo, Nelson, et al. 2010). ECM, extracellular matrix.

stem cell differentiation for osteogenesis and chondrogenesis. The incorporation of bioactive ECM signals into synthetic scaffolding platforms provides insight toward the development of ideal biomaterials for stem cell engineering and tissue regeneration.

4.5   CONCLUSIONS

This chapter provides an overview of the progress and advances of stem cell technology combined with microenvironment engineering. Although stem cell research has opened up whole new therapeutic avenues for incurable diseases and tissue injuries, current stem cell therapy is faced with obstacles that are preventing the translation of its experimental potential to clinical reality. Stem cell researchers should more actively consider employing engineering approaches to improve stem cell functionality for therapeutic purposes. Engineering stem cell microenvironments could usefully tackle the current dilemmas by providing optimal conditions for self-renewal, proliferation, differentiation, survival, and functionality of stem cells. Several microengineering techniques, including surface topography, electrospinning, microfluidic system, and tissue-specific matrix, have been explored to manipulate stem cell microenvironments. These studies will inform future breakthroughs that will overcome the problems of conventional stem cell therapy. With the continuing development of technologies and increasing knowledge in stem cell research, stem cell technology integrated with microengineering may be able to provide innovative regenerative medicine.

REFERENCES

Badylak, Stephen, Donald Freytes, and Thomas Gilbert. 2009. Extracellular matrix as a biological scaffold material: Structure and function. *Acta Biomaterialia* 5 (1):1–13.

Basnar, Bernhard, and Itamar Willner. 2009. Dip-pen-nanolithographic patterning of metallic, semiconductor, and metal oxide nanostructures on surfaces. *Small* 5 (1):28–44.

Becerra, José, Leonor Santos-Ruiz, José A Andrades, and Manuel Marí-Beffa. 2011. The stem cell niche should be a key issue for cell therapy in regenerative medicine. *Stem Cell Reviews* 7 (2):248–255.

Béduer, Amélie, Christophe Vieu, Florent Arnauduc, Jean-Christophe Sol, Isabelle Loubinoux, and Laurence Vaysse. 2012. Engineering of adult human neural stem cells differentiation through surface micropatterning. *Biomaterials* 33 (2):504–514.

Binan, Loïc, Charlène Tendey, Gregory De Crescenzo, Rouwayda El Ayoubi, Abdellah Ajji, and Mario Jolicoeur. 2014. Differentiation of neuronal stem cells into motor neurons using electrospun poly-L-lactic acid/gelatin scaffold. *Biomaterials* 35 (2):664–674.

Boussommier-Calleja, Alexandra, Ran Li, Michelle B Chen, Siew Cheng Wong, and Roger D Kamm. 2016. Microfluidics: A new tool for modeling cancer–immune interactions. *Trends in Cancer* 2 (1):6–19.

Bucaro, Michael A, Yolanda Vasquez, Benjamin D Hatton, and Joanna Aizenberg. 2012. Fine-tuning the degree of stem cell polarization and alignment on ordered arrays of high-aspect-ratio nanopillars. *ACS Nano* 6 (7):6222–6230.

Burdick, Jason A, and Gordana Vunjak-Novakovic. 2008. Engineered microenvironments for controlled stem cell differentiation. *Tissue Engineering Part A* 15 (2):205–219.

Campisi, Marco, Yoojin Shin, Tatsuya Osaki, Cynthia Hajal, Valeria Chiono, and Roger D Kamm. 2018. 3D self-organized microvascular model of the human blood-brain barrier with endothelial cells, pericytes and astrocytes. *Biomaterials* 180:117–129.

Chen, Weiqiang, Luis G Villa-Diaz, Yubing Sun, Shinuo Weng, Jin Koo Kim, Raymond HW Lam, Lin Han, Rong Fan, Paul H Krebsbach, and Jianping Fu. 2012. Nanotopography influences adhesion, spreading, and self-renewal of human embryonic stem cells. *ACS Nano* 6 (5):4094–4103.

Christopherson, Gregory T, Hongjun Song, and Hai-Quan Mao. 2009. The influence of fiber diameter of electrospun substrates on neural stem cell differentiation and proliferation. *Biomaterials* 30 (4):556–564.

Chung, Bong Geun, Lisa A Flanagan, Seog Woo Rhee, Philip H Schwartz, Abraham P Lee, Edwin S Monuki, and Noo Li Jeon. 2005. Human neural stem cell growth and differentiation in a gradient-generating microfluidic device. *Lab on a Chip* 5 (4):401–406.

Crapo, Peter M, Thomas W Gilbert, and Stephen F Badylak. 2011. An overview of tissue and whole organ decellularization processes. *Biomaterials* 32 (12):3233–3243.

Dalby, Matthew J, Nikolaj Gadegaard, and Richard OC Oreffo. 2014. Harnessing nanotopography and integrin–matrix interactions to influence stem cell fate. *Nature Materials* 13 (6):558.

Dalby, Matthew J, Nikolaj Gadegaard, Rahul Tare, Abhay Andar, Mathis O Riehle, Pawel Herzyk, Chris DW Wilkinson, and Richard OC Oreffo. 2007. The control of human mesenchymal cell differentiation using nanoscale symmetry and disorder. *Nature Materials* 6 (12):997.

Datta, Neha, Heidi L Holtorf, Vassilios I Sikavitsas, John A Jansen, and Antonios G Mikos. 2005. Effect of bone extracellular matrix synthesized in vitro on the osteoblastic differentiation of marrow stromal cells. *Biomaterials* 26 (9):971–977. doi:10.1016/j.biomaterials.2004.04.001.

DeQuach, Jessica A, Joy E Lin, Cynthia Cam, Diane Hu, Michael A Salvatore, Farah Sheikh, Karen L Christman. 2012. Injectable skeletal muscle matrix hydrogel promotes neovascularization and muscle cell infiltration in a hindlimb ischemia model. *European Cells and Materials* 23:400.

Discher, Dennis E, David J Mooney, and Peter W Zandstra. 2009. Growth factors, matrices, and forces combine and control stem cells. *Science* 324 (5935):1673–1677.

Duan, Yi, Zen Liu, John O'Neill, Leo Q Wan, Donald O Freytes, and Gordana Vunjak-Novakovic. 2011. Hybrid gel composed of native heart matrix and collagen induces cardiac differentiation of human embryonic stem cells without supplemental growth factors. *Journal of Cardiovascular Translational Research* 4 (5):605.

Farnebo, Simon, Colin Y Woon, Taliah Schmitt, Lydia-Marie Joubert, Maxwell Kim, Hung Pham, and James Chang. 2014. Design and characterization of an injectable tendon hydrogel: A novel scaffold for guided tissue regeneration in the musculoskeletal system. *Journal of Tissue Engineering Part A* 20 (9–10):1550–1561.

Faulk, Denver M, Ricardo Londono, Matthew T Wolf, Christian A Ranallo, Christopher A Carruthers, Justin D Wildemann, Christopher L Dearth, and Stephen F Badylak. 2014. ECM hydrogel coating mitigates the chronic inflammatory response to polypropylene mesh. *Biomaterials* 35 (30):8585–8595.

Fazili, Afsaneh, Soghra Gholami, Bagher Minaie Zangi, Ehsan Seyedjafari, and Mahdi Gholami. 2016. In vivo differentiation of mesenchymal stem cells into insulin producing cells on electrospun poly-L-lactide acid scaffolds coated with *Matricaria chamomilla* L. oil. *Cell* 18 (3):310.

Freytes, Donald O, Jeffrey Martin, Sachin S Velankar, Annie S Lee, and Stephen F Badylak. 2008. Preparation and rheological characterization of a gel form of the porcine urinary bladder matrix. *Journal of Biomaterials* 29 (11):1630–1637.

Gerecht, Sharon, Jason A Burdick, Lino S Ferreira, Seth A Townsend, Robert Langer, and Gordana Vunjak-Novakovic. 2007. Hyaluronic acid hydrogel for controlled self-renewal and differentiation of human embryonic stem cells. *Proceedings of the National Academy of Sciences* 104 (27):11298–11303.

Giancotti, Filippo G, and Erkki Ruoslahti. 1999. Integrin signaling. *Science* 285 (5430):1028–1033.

Gumbiner, Barry M. 1996. Cell adhesion: The molecular basis of tissue architecture and morphogenesis. *Cell* 84 (3):345–357.

Gupta, Kshitiz, Deok-Ho Kim, David Ellison, Christopher Smith, Arnab Kundu, Jessica Tuan, Kahp-Yang Suh, and Andre Levchenko. 2010. Lab-on-a-chip devices as an emerging platform for stem cell biology. *Lab on a Chip* 10 (16):2019–2031.

Han, Sewoon, Kisuk Yang, Yoojin Shin, Jung Seung Lee, Roger D Kamm, Seok Chung, and Seung-Woo Cho. 2012. Three-dimensional extracellular matrix-mediated neural stem cell differentiation in a microfluidic device. *Lab on a Chip* 12 (13):2305–2308.

Hwang, Nathaniel S, Myoung Sook Kim, Somponnat Sampattavanich, Jin Hyen Baek, Zijun Zhang, and Jennifer Elisseeff. 2006. Effects of three-dimensional culture and growth factors on the chondrogenic differentiation of murine embryonic stem cells. *Stem Cells* 24 (2):284–291.

Imitola, Jaime, Khadir Raddassi, Kook In Park, Franz-Josef Mueller, Marta Nieto, Yang D Teng, Dan Frenkel, Jianxue Li, Richard L Sidman, and Christopher A Walsh. 2004. Directed migration of neural stem cells to sites of CNS injury by the stromal cell-derived factor 1α/CXC chemokine receptor 4 pathway. *Proceedings of the National Academy of Sciences* 101 (52):18117–18122.

Jang, Jinah, Hun-Jun Park, Seok-Won Kim, Heejin Kim, Ju Young Park, Soo Jin Na, Hyeon Ji Kim, Moon Nyeo Park, Seung Hyun Choi, and Sun Hwa Park. 2017. 3D printed complex tissue construct using stem cell-laden decellularized extracellular matrix bioinks for cardiac repair. *Biomaterials* 112:264–274.

Jeon, Kilsoo, Hyun-Jik Oh, Hyejin Lim, Jung-Hyun Kim, Duk Hyun Lee, Eung-Ryoung Lee, Bae Ho Park, and Ssang-Goo Cho. 2012. Self-renewal of embryonic stem cells through culture on nanopattern polydimethylsiloxane substrate. *Biomaterials* 33 (21):5206–5220.

Jin, Hyoung-Joon, Jingsong Chen, Vassilis Karageorgiou, Gregory H Altman, and David L Kaplan. 2004. Human bone marrow stromal cell responses on electrospun silk fibroin mats. *Biomaterials* 25 (6):1039–1047.

Jin, Yoonhee, Jin Kim, Jung Seung Lee, Sungjin Min, Suran Kim, Da-Hee Ahn, Yun-Gon Kim, and Seung-Woo Cho. 2018. Vascularized liver organoids generated using induced hepatic tissue and dynamic liver-specific microenvironment as a drug testing platform. *Advanced Functional Materials* 28 (37):1801954.

Jin, Yoonhee, Jung Seung Lee, Jin Kim, Sungjin Min, Soohyun Wi, Ji Hea Yu, Gyeong-Eon Chang, Ann-Na Cho, Yeeun Choi, and Da-Hee Ahn. 2018. Three-dimensional brain-like microenvironments facilitate the direct reprogramming of fibroblasts into therapeutic neurons. *Nature Biomedical Engineering* 2 (7):522.

Kang, Yunqing, Sungwoo Kim, Julius Bishop, Ali Khademhosseini, and Yunzhi Yang. 2012. The osteogenic differentiation of human bone marrow MSCs on HUVEC-derived ECM and β-TCP scaffold. *Biomaterials* 33 (29):6998–7007.

Khorshidi, Sajedeh, Atefeh Solouk, Hamid Mirzadeh, Saeedeh Mazinani, Jose M Lagaron, Shahriar Sharifi, and Seeram Ramakrishna. 2016. A review of key challenges of electrospun scaffolds for tissue-engineering applications. *Journal of Tissue Engineering Regenerative Medicine* 10 (9):715–738.

Kilic, Onur, David Pamies, Emily Lavell, Paula Schiapparelli, Yun Feng, Thomas Hartung, Anna Bal-Price, Helena T Hogberg, Alfredo Quinones-Hinojosa, and Hugo Guerrero-Cazares. 2016. Brain-on-a-chip model enables analysis of human neuronal differentiation and chemotaxis. *Lab on a Chip* 16 (21):4152–4162.

Lee, Jong Seung, Yoonhee Jin, Hyun-Ji Park, Kisuk Yang, Min Suk Lee, Hee Seok Yang, and Seung-Woo Cho. 2017. In situ bone tissue engineering with an endogenous stem cell mobilizer and osteoinductive nanofibrous polymeric scaffolds. *Biotechnology Journal* 12 (12):1700062.

Lee, Man Ryul, Keon Woo Kwon, Hosup Jung, Hong Nam Kim, Kahp Y Suh, Keesung Kim, and Kye-Seong Kim. 2010. Direct differentiation of human embryonic stem cells into selective neurons on nanoscale ridge/groove pattern arrays. *Biomaterials* 31 (15):4360–4366.

Lee, Se-Jun, Margaret Nowicki, Brent Harris, and Lijie Grace Zhang. 2017. Fabrication of a highly aligned neural scaffold via a table top stereolithography 3D printing and electro-spinning. *Tissue Engineering Part A* 23 (11–12):491–502.

Lee, Young Jun, Ji-Hye Lee, Hyeong-Jin Cho, Hyung Keun Kim, Taek Rim Yoon, and Heungsoo Shin. 2013. Electrospun fibers immobilized with bone forming peptide-1 derived from BMP7 for guided bone regeneration. *Biomaterials* 34 (21):5059–5069.

Lee, Jung Seung, Jisoo Shin, Hae-Min Park, Yun-Gon Kim, Byung-Gee Kim, Jong-Won Oh, and Seung-Woo Cho. 2013. Liver extracellular matrix providing dual functions of two-dimensional substrate coating and three-dimensional injectable hydrogel platform for liver tissue engineering. *Biomacromolecules* 15 (1):206–218.

Li, Qiang, Jun Wang, Shilpa Shahani, Danny DN Sun, Blanka Sharma, Jennifer H Elisseeff, and Kam W Leong. 2006. Biodegradable and photocrosslinkable polyphosphoester hydrogel. *Biomaterials* 27 (7):1027–1034.

Liao, Jiehong, Xuan Guo, K Jane Grande-Allen, F Kurtis Kasper, and Antonios G Mikos. 2010. Bioactive polymer/extracellular matrix scaffolds fabricated with a flow perfusion bioreactor for cartilage tissue engineering. *Biomaterials* 31 (34):8911–8920.

Liao, Jiehong, Xuan Guo, Dan Nelson, F Kurtis Kasper, and Antonios G Mikos. 2010. Modulation of osteogenic properties of biodegradable polymer/extracellular matrix scaffolds generated with a flow perfusion bioreactor. *Acta Biomaterialia* 6 (7):2386–2393.

Lins, Luanda C, Florence Wianny, Sebastien Livi, Colette Dehay, Jannick Duchet-Rumeau, and Jean-François Gérard. 2017. Effect of polyvinylidene fluoride electrospun fiber orientation on neural stem cell differentiation. *Journal of Biomedical Materials Research Part B: Applied Biomaterials* 105 (8):2376–2393.

Locatelli, Franco, Mattia Algeri, Valentina Trevisan, and Alice Bertaina. 2017. Remestemcel-L for the treatment of graft versus host disease. *Expert Review of Clinical Immunology* 13 (1):43–56.

Luo, Yu, He Shen, Yongxiang Fang, Yuhua Cao, Jie Huang, Mengxin Zhang, Jianwu Dai, Xiangyang Shi, and Zhijun Zhang. 2015. Enhanced proliferation and osteogenic differentiation of mesenchymal stem cells on graphene oxide-incorporated electrospun poly (lactic-co-glycolic acid) nanofibrous mats. *ACS Applied Materials Interfaces* 7 (11):6331–6339.

Lutolf, Matthias P, Penney M Gilbert, and Helen M Blau. 2009. Designing materials to direct stem-cell fate. *Nature* 462 (7272):433.

Maoz, Ben M, Anna Herland, Edward A FitzGerald, Thomas Grevesse, Charles Vidoudez, Alan R Pacheco, Sean P Sheehy, Tae-Eun Park, Stephanie Dauth, and Robert Mannix. 2018. A linked organ-on-chip model of the human neurovascular unit reveals the metabolic coupling of endothelial and neuronal cells. *Nature Biotechnology* 36 (9):865.

Matthews, Jamil A, Gary E Wnek, David G Simpson, and Gary L Bowlin. 2002. Electrospinning of collagen nanofibers. *Biomacromolecules* 3 (2):232–238.

McBeath, Rowena, Dana M Pirone, Celeste M Nelson, Kiran Bhadriraju, and Christopher S Chen. 2004. Cell shape, cytoskeletal tension, and RhoA regulate stem cell lineage commitment. *Developmental Cell* 6 (4):483–495.

McBride, Sara H, and Melissa L Knothe Tate. 2008. Modulation of stem cell shape and fate A: The role of density and seeding protocol on nucleus shape and gene expression. *Tissue Engineering Part A* 14 (9):1561–1572.

McMurray, Rebecca J, Nikolaj Gadegaard, P Monica Tsimbouri, Karl V Burgess, Laura E McNamara, Rahul Tare, Kate Murawski, Emmajayne Kingham, Richard OC Oreffo, and Matthew J Dalby. 2011. Nanoscale surfaces for the long-term maintenance of mesenchymal stem cell phenotype and multipotency. *Nature Materials* 10 (8):637.

Mellor, Liliana F, Mahsa Mohiti-Asli, John Williams, Arthi Kannan, Morgan R Dent, Farshid Guilak, and Elizabeth G Loboa. 2015. Extracellular calcium modulates chondrogenic and osteogenic differentiation of human adipose-derived stem cells: A novel approach for osteochondral tissue engineering using a single stem cell source. *Tissue Engineering Part A* 21 (17–18):2323–2333.

Okada, Masaho, Thomas R Payne, Hideki Oshima, Nobuo Momoi, Kimimasa Tobita, and Johnny Huard. 2010. Differential efficacy of gels derived from small intestinal submucosa as an injectable biomaterial for myocardial infarct repair. *Biomaterials* 31 (30):7678–7683.

Okita, Keisuke, Masato Nakagawa, Hong Hyenjong, Tomoko Ichisaka, and Shinya Yamanaka. 2008. Generation of mouse induced pluripotent stem cells without viral vectors. *Science* 322 (5903):949–953.

Orr, Steven B, Abby Chainani, Kirk J Hippensteel, Alysha Kishan, Christopher Gilchrist, N William Garrigues, David S Ruch, Farshid Guilak, and Dianne Little. 2015. Aligned multilayered electrospun scaffolds for rotator cuff tendon tissue engineering. *Acta Biomaterialia* 24:117–126.

Paduano, Francesco, Massimo Marrelli, Lisa J White, Kevin M Shakesheff, and Marco Tatullo. 2016. Odontogenic differentiation of human dental pulp stem cells on hydrogel scaffolds derived from decellularized bone extracellular matrix and collagen type I. *PLoS One* 11 (2):e0148225.

Pan, Zhen, Ce Yan, Rong Peng, Yingchun Zhao, Yao He, and Jiandong Ding. 2012. Control of cell nucleus shapes via micropillar patterns. *Biomaterials* 33 (6):1730–1735.

Park, Jung, Sebastian Bauer, Klaus von der Mark, and Patrik Schmuki. 2007. Nanosize and vitality: TiO_2 nanotube diameter directs cell fate. *Nano Letters* 7 (6):1686–1691.

Park, Joong Yull, Suel-Kee Kim, Dong-Hun Woo, Eun-Joong Lee, Jong-Hoon Kim, and Sang-Hoon Lee. 2009. Differentiation of neural progenitor cells in a microfluidic chip-generated cytokine gradient. *Stem Cells* 27 (11):2646–2654.

Park, DoYeun, Jaeho Lim, Joong Yull Park, and Sang-Hoon Lee. 2015. Concise review: Stem cell microenvironment on a chip: Current technologies for tissue engineering and stem cell biology. *Stem Cells Translational Medicine* 4 (11):1352–1368.

Paşca, Sergiu P. 2018. The rise of three-dimensional human brain cultures. *Nature* 553 (7689):437.

Pati, Falguni, Jinah Jang, Dong-Heon Ha, Sung Won Kim, Jong-Won Rhie, Jin-Hyung Shim, Deok-Ho Kim, and Dong-Woo Cho. 2014. Printing three-dimensional tissue analogues with decellularized extracellular matrix bioink. *Journal of Nature Communications* 5:3935.

Pham, Quynh P, Upma Sharma, and Antonios G Mikos. 2006. Electrospinning of polymeric nanofibers for tissue engineering applications: A review. *Tissue Engineering* 12 (5):1197–1211.

Phipps, Matthew C, William C Clem, Jessica M Grunda, Gregory A Clines, and Susan L Bellis. 2012. Increasing the pore sizes of bone-mimetic electrospun scaffolds comprised of polycaprolactone, collagen I and hydroxyapatite to enhance cell infiltration. *Biomaterials* 33 (2):524–534.

Pouliot, Robert A, Patrick A Link, Nabil S Mikhaiel, Matthew B Schneck, Michael S Valentine, Franck J Kamga Gninzeko, Joseph A Herbert, Masahiro Sakagami, and Rebecca L Heise. 2016. Development and characterization of a naturally derived lung extracellular matrix hydrogel. *Journal of Biomedical Materials Research Part A* 104 (8):1922–1935.

Rampichová, Michala, Jiri Chvojka, Matej Buzgo, Eva Prosecká, Petr Mikeš, Lucie Vysloužilová, Daniel Tvrdik, Petra Kochová, Tomáš Gregor, and David Lukáš. 2013. Elastic three-dimensional poly (ε-caprolactone) nanofibre scaffold enhances migration, proliferation and osteogenic differentiation of mesenchymal stem cells. *Cell Proliferation* 46 (1):23–37.

Robinton, Daisy A, and George Q Daley. 2012. The promise of induced pluripotent stem cells in research and therapy. *Nature* 481 (7381):295.

Sadr, Nasser, Benjamin E Pippenger, Arnaud Scherberich, David Wendt, Sara Mantero, Ivan Martin, and Adam Papadimitropoulos. 2012. Enhancing the biological performance of synthetic polymeric materials by decoration with engineered, decellularized extracellular matrix. *Biomaterials* 33 (20):5085–5093.

Sawkins, Michael J, William Bowen, Pam Dhadda, Hareklea Markides, Laura E Sidney, Adam James Taylor, Felicity RAJ Rose, Stephen F Badylak, Kevin M Shakesheff, and Lisa J White. 2013. Hydrogels derived from demineralized and decellularized bone extracellular matrix. *Acta Biomaterialia* 9 (8):7865–7873.

Schwartz, Steven D, Jean-Pierre Hubschman, Gad Heilwell, Valentina Franco-Cardenas, Carolyn K Pan, Rosaleen M Ostrick, Edmund Mickunas, Roger Gay, Irina Klimanskaya, and Robert Lanza. 2012. Embryonic stem cell trials for macular degeneration: A preliminary report. *The Lancet* 379 (9817):713–720.

Seidlits, Stephanie K, Jae Y Lee, and Christine E Schmidt. 2008. Nanostructured scaffolds for neural applications. *Nanomedicine* 3(2):183–199. doi:10.2217/17435889.3.2.183.

Sekiya, Sayaka, and Atsushi Suzuki. 2011. Direct conversion of mouse fibroblasts to hepatocyte-like cells by defined factors. *Nature* 475 (7356):390.

Sellaro, Tiffany L, Aarati Ranade, Denver M Faulk, George P McCabe, Kenneth Dorko, Stephen F Badylak, and Stephen C Strom. 2009. Maintenance of human hepatocyte function in vitro by liver-derived extracellular matrix gels. *Tissue Engineering Part A* 16 (3):1075–1082.

Shin, Yoojin, Kisuk Yang, Sewoon Han, Hyun-Ji Park, Yun Seok Heo, Seung-Woo Cho, and Seok Chung. 2014. Reconstituting vascular microenvironment of neural stem cell niche in three-dimensional extracellular matrix. *Advanced Healthcare Materials* 3 (9):1457–1464.

Smith, Laura A, Xiaohua Liu, Jiang Hu, Peng Wang, and Peter X Ma. 2009. Enhancing osteogenic differentiation of mouse embryonic stem cells by nanofibers. *Tissue Engineering Part A* 15 (7):1855–1864.

Takahashi, Kazutoshi, and Shinya Yamanaka. 2006. Induction of pluripotent stem cells from mouse embryonic and adult fibroblast cultures by defined factors. *Cell* 126 (4):663–676.

Teo, Benjamin Kim Kiat, Soneela Ankam, Lesley Y Chan, and Evelyn KF Yim. 2010. Nanotopography/mechanical induction of stem-cell differentiation. *Methods in Cell Biology* 98:241–294.

Trachsel, Eveline, and Dario Neri. 2006. Antibodies for angiogenesis inhibition, vascular targeting and endothelial cell transcytosis. *Advanced Drug Delivery Reviews* 58 (5–6):735–754.

Tukmachev, Dmitry, Serhiy Forostyak, Zuzana Koci, Kristyna Zaviskova, Irena Vackova, Karel Vyborny, Ioanna Sandvig, Axel Sandvig, Christopher J Medberry, and Stephen F Badylak. 2016. Injectable extracellular matrix hydrogels as scaffolds for spinal cord injury repair. *Tissue Engineering Part A* 22 (3–4):306–317.

Uygun, Basak E, Alejandro Soto-Gutierrez, Hiroshi Yagi, Maria-Louisa Izamis, Maria A Guzzardi, Carley Shulman, Jack Milwid, Naoya Kobayashi, Arno Tilles, and Francois Berthiaume. 2010. Organ reengineering through development of a transplantable recellularized liver graft using decellularized liver matrix. *Nature Medicine* 16 (7):814.

Van Dorp, Willem F, Xiaoyan Zhang, Ben L Feringa, Thomas W Hansen, Jakob B Wagner, and Jeff Th M De Hosson. 2012. Molecule-by-molecule writing using a focused electron beam. *ACS Nano* 6 (11):10076–10081.

Van Duinen, Vincent, Sebastiaan J Trietsch, Jos Joore, Paul Vulto, and Thomas Hankemeier. 2015. Microfluidic 3D cell culture: From tools to tissue models. *Current Opinion in Biotechnology* 35:118–126.

Volckaert, Thomas, and Stijn De Langhe. 2014. Lung epithelial stem cells and their niches: Fgf10 takes center stage. *Fibrogenesis & Tissue Repair* 7 (1):8.

Wassenaar, Jean W, Rebecca L Braden, Kent G Osborn, and Karen L Christman. 2016. Modulating in vivo degradation rate of injectable extracellular matrix hydrogels. *Journal of Materials Chemistry B* 4 (16):2794–2802.

Watari, Shinya, Kei Hayashi, Joshua A Wood, Paul Russell, Paul F Nealey, Christopher J Murphy, and Damian C Genetos. 2012. Modulation of osteogenic differentiation in hMSCs cells by submicron topographically-patterned ridges and grooves. *Biomaterials* 33 (1):128–136.

Watt, Fiona M, and Wilhelm T Huck. 2013. Role of the extracellular matrix in regulating stem cell fate. *Nature Reviews. Molecular Cell Biology* 14 (8):467–473. doi:10.1038/nrm3620.

Williams, Christopher G, Tae Kyun Kim, Anya Taboas, Athar Malik, Paul Manson, and Jennifer Elisseeff. 2003. In vitro chondrogenesis of bone marrow-derived mesenchymal stem cells in a photopolymerizing hydrogel. *Tissue Engineering* 9 (4):679–688.

Wohn, D Yvette. 2012. Korea okays stem cell therapies despite limited peer-reviewed data. *Nature Medicine* 18(3):329. doi:10.1038/nm0312-329a.

Wolf, Matthew T, Kerry A Daly, Ellen P Brennan-Pierce, Scott A Johnson, Christopher A Carruthers, Antonio D'Amore, Shailesh P Nagarkar, Sachin S Velankar, and Stephen F Badylak. 2012. A hydrogel derived from decellularized dermal extracellular matrix. *Biomaterials* 33 (29):7028–7038.

Woltjen, Knut, Iacovos P Michael, Paria Mohseni, Ridham Desai, Maria Mileikovsky, Riikka Hämäläinen, Rebecca Cowling, Wei Wang, Pentao Liu, and Marina Gertsenstein. 2009. piggyBac transposition reprograms fibroblasts to induced pluripotent stem cells. *Nature* 458 (7239):766.

Xie, Jingwei, Stephanie M Willerth, Xiaoran Li, Matthew R Macewan, Allison Rader, Shelly E Sakiyama-Elbert, and Younan Xia. 2009. The differentiation of embryonic stem cells seeded on electrospun nanofibers into neural lineages. *Biomaterials* 30 (3):354–362.

Yang, Kisuk, Sewoon Han, Yoojin Shin, Eunkyung Ko, Jin Kim, Kook In Park, Seok Chung, and Seung-Woo Cho. 2013. A microfluidic array for quantitative analysis of human neural stem cell self-renewal and differentiation in three-dimensional hypoxic microenvironment. *Biomaterials* 34 (28):6607–6614.

Yang, Kisuk, Kyuhwan Jung, Eunkyung Ko, Jin Kim, Kook In Park, Jinseok Kim, and Seung-Woo Cho. 2013. Nanotopographical manipulation of focal adhesion formation for enhanced differentiation of human neural stem cells. *ACS Applied Materials Interfaces* 5 (21):10529–10540.

Yang, Kisuk, Hyunjung Jung, Hak-Rae Lee, Jong Seung Lee, Su Ran Kim, Ki Yeong Song, Eunji Cheong, Joona Bang, Sung Gap Im, and Seung-Woo Cho. 2014. Multiscale, hierarchically patterned topography for directing human neural stem cells into functional neurons. *ACS Nano* 8 (8):7809–7822.

Yang, Yong, Karina Kulangara, Ruby TS Lam, Rena Dharmawan, and Kam W Leong. 2012. Effects of topographical and mechanical property alterations induced by oxygen plasma modification on stem cell behavior. *ACS Nano* 6 (10):8591–8598.

Yang, Kisuk, Hyun-Ji Park, Sewoon Han, Joan Lee, Eunkyung Ko, Jin Kim, Jong Seung Lee, Ji Hea Yu, Ki Yeong Song, and Eunji Cheong. 2015. Recapitulation of in vivo-like paracrine signals of human mesenchymal stem cells for functional neuronal differentiation of human neural stem cells in a 3D microfluidic system. *Biomaterials* 63:177–188.

Yang, Kisuk, Esther Park, Jong Seung Lee, Il-Sun Kim, Kwonho Hong, Kook In Park, Seung-Woo Cho, and Hee Seok Yang. 2015. Biodegradable nanotopography combined with neurotrophic signals enhances contact guidance and neuronal differentiation of human neural stem cells. *Macromolecular Bioscience* 15 (10):1348–1356.

Yang, Kisuk, Seung Jung Yu, Jong Seung Lee, Hak-Rae Lee, Gyeong-Eon Chang, Jungmok Seo, Taeyoon Lee, Eunji Cheong, Sung Gap Im, and Seung-Woo Cho. 2017. Electroconductive nanoscale topography for enhanced neuronal differentiation and electrophysiological maturation of human neural stem cells. *Nanoscale* 9 (47):18737–18752.

Yim, Evelyn KF, Stella W Pang, and Kam W Leong. 2007. Synthetic nanostructures inducing differentiation of human mesenchymal stem cells into neuronal lineage. *Experimental Cell Research* 313 (9):1820–1829.

Young, D Adam, Dina O Ibrahim, Diane Hu, and Karen L Christman. 2011. Injectable hydrogel scaffold from decellularized human lipoaspirate. *Acta Biomaterialia* 7 (3):1040–1049.

Zhang, Yan, Andrew Gordon, Weiyi Qian, and Weiqiang Chen. 2015. Engineering nanoscale stem cell niche: Direct stem cell behavior at cell–matrix interface. *Advanced Healthcare Materials* 4 (13):1900–1914.

Zong, Xinhua, Harold Bien, Chiung-Yin Chung, Lihong Yin, Dufei Fang, Benjamin S Hsiao, Benjamin Chu, and Emilia Entcheva. 2005. Electrospun fine-textured scaffolds for heart tissue constructs. *Biomaterials* 26 (26):5330–5338.

5 Organoid Technology for Basic Science and Biomedical Research

Szu-Hsien (Sam) Wu, Jihoon Kim, and Bon-Kyoung Koo
IMBA

CONTENTS

5.1   INTRODUCTION

Complex multicellular organisms are comprised of specialized tissues and organs that coordinate their functions in order to maintain normal physiology and prevent pathology, morbidity, and death. A major focus of study within the fields of biological and biomedical research is on the cellular and subcellular mechanisms underlying the processes that can form or destroy the body. However, the internal organs and tissues of mammals are often inaccessible to experimental manipulation, posing a major challenge for the study of developmental processes (i.e., morphogenesis and patterning), homeostatic turnover, and disease progression in these tissues.

Instead of observing these processes directly, researchers have relied on a range of model systems to investigate the principles underlying various cellular and molecular processes, with commonly used systems including animal models (e.g., mouse, rat, zebrafish, drosophila) and cell culture models. Animal models have taught us a great deal about the basic principles of developmental processes, adult stem cell (AdSC) regulation and disease progression, but such models involve laborious, time-consuming work, and costly husbandry. In addition, the relevance of animal models to human physiology is sometimes questionable. Two-dimensional (2D) cell culture systems, such as immortalized human cell lines, have also been used extensively in various areas of research ranging from basic cell biology to cell signalling pathways, differentiation, and toxicology. Although 2D cell culture is much simpler to perform and to a certain extent will retain the physiological functions of the tissue, much of the three-dimensional (3D) organization and many of the signaling resulting from cell–cell interactions in the in vivo tissue microenvironment are lost (Shamir and Ewald 2014; Kamb 2005).

Recently developed 3D organoid culture systems have been developed to overcome some of the aforementioned drawbacks. Organoid culture systems are robust model systems which recapitulate in vivo organ physiology and are amenable to physical, chemical, and/or genetic manipulation, and so are a more optimal experimental system. Historically, the term organoid has been used loosely to describe systems such as explant cultures maintained in Matrigel (laminin-rich extracellular matrix extracted from Engelbreth–Holm–Swarm sarcoma, a mouse tumor) (Kleinman and Martin 2005; Hughes 2010), such as mammary gland cultures, where the branching of mammary epithelial cells could be observed (Streuli, Bailey, and Bissell 1991). Whilst such approaches did not result in the formation of long-term, stable cell cultures, nevertheless these early experiments highlighted the importance of 3D cultures (Shamir and Ewald 2014; Hagios et al. 1998).

The current definition of an organoid culture system is (Lancaster and Knoblich 2014; Clevers 2016) a long-term, 3D culture system derived from stem or progenitor cells that are able to self-organize and develop 3D structures which recapitulate the composition of the tissue of origin. In this chapter, we will discuss the generation of various organoid systems and highlight applications and studies that demonstrate the wide-ranging utility of organoids.

5.2   GENERATION OF ORGANOIDS

Organoids can be generated from two alternative sources (Figure 5.1), starting either from pluripotent stem cells (PSCs) or AdSCs. The two approaches are fundamentally different in terms of the starting cell population and the number of cell types present in the final culture, and thus have been used to address different research questions. In this section, we will discuss the generation of both types of organoid culture (Figure 5.2).

5.2.1   PSC-Derived Organoids

In principal, PSCs can give rise to all cell types of the body if provided with the appropriate conditions. Many research labs have generated various differentiated cell types from PSCs using knowledge gained from studies of embryonic development. However, most protocols were established using 2D PSC cultures, which do not fully recapitulate developmental processes such as spatial patterning and

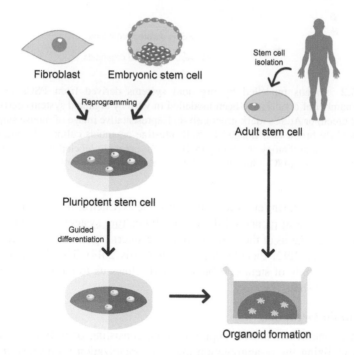

FIGURE 5.1   Schematic representation of organoid establishment. Organoids can be established with cells from two alternative sources: PSCs and AdSCs. For the former, PSCs are first established via reprogramming of somatic cells such as terminally differentiated fibroblasts or derived from embryos at blastocyst stage. Subsequently, PSCs are differentiated towards particular lineages to generate organoids. AdSC-derived organoid establishment requires isolation of resident adult stem cells which are embedded in ECM-containing gel. Both methods utilize media containing a cocktail of growth factors which recapitulate the endogenous stem cell niche. AdSC, adult stem cell; ECM, extracellular matrix; PSC, pluripotent stem cell.

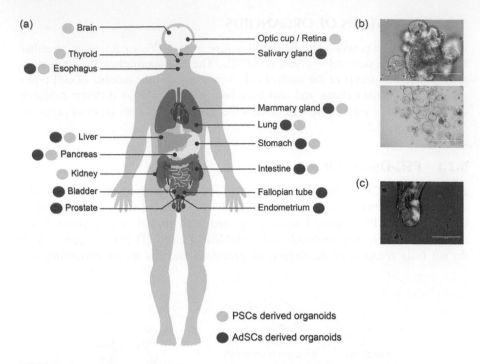

FIGURE 5.2 Organs modelled by organoid systems derived from PSCs or AdSCs. (a) A large number of organs have been modelled in vitro by organoid systems derived from PSCs (light green) or AdSCs (dark green). (b–c) Representative image of mouse small intestinal organoid (b) bright field image of small intestine organoids cultured in budding (top panel) and cystic morphology (bottom panel) and (c) organoid labeled with green and red fluorescence proteins. AdSC, adult stem cell; PSC, pluripotent stem cell.

cell–cell interaction during embryonic morphogenesis (Chen et al. 2014). The research group of Yoshiki Sasai pioneered the use of 3D culture systems to generate structures mimicking regions of the brain, retina, and anterior pituitary/adenohypophysis (Eiraku and Sasai 2012; Piccolo 2014; De Robertis 2014). This has opened new avenues for the study of stem cell biology and allows us to model developmental processes "in a dish."

5.2.1.1 Brain Organoids

The brain is one of the most complex organs, consisting of billions of neuronal networks. Utilizing the re-aggregation and self-organization properties of embryonic stem cells in 3D aggregates called embryoid bodies (EBs), instrumental work from the group of Yoshiki Sasai has defined ways to generate specific regions of the brain in vitro. For example, using chemically defined, serum-free, EB-like aggregate culture (SFEBq), the group generated cortical cells from both mouse and human PSCs (Eiraku et al. 2008). When cells in SFEBq culture were re-aggregated in low-adhesion plates, they organized into 3D cortical-like structures, such as the formation of a layer of neuroepithelium with apical-basal polarity.

One of the defined cortical regions generated *via* SFEBq culture was the adenohypophysis, or anterior pituitary gland, which contains hormone-secreting endocrine cells. Sasai et al. (Suga et al. 2011) sought in particular to recreate the crosstalk of the forming adenohypophysis with the surrounding tissues in 3D cultures starting from ES cells. Using cell aggregates and with the addition of a hedgehog signaling agonist, the ES cells formed a thickened placodal epithelium which invaginated to form a hollow vesicle resembling Rathke's pouch (a precursor structure of the adenohypophysis). Multiple endocrine lineages could be generated from the ES cell-derived adenohypophysis. Using a similar approach, the group could also recapitulate the development of the human adenohypophysis (Ozone et al. 2016).

The cerebellum contains several neural cell types, such as Purkinje cells and interneurons, which together confer motor function. During development, the cerebellum derives from the isthmic organizer located at the midbrain–hindbrain boundary. Treatment of 3D aggregated mouse ESCs (mESCs) with FGF2 and insulin led to the formation of an isthmic organizer-like tissue, and further treatment of this structure with an inhibitor of sonic hedgehog (Shh) signaling resulted in the generation of Purkinje cell progenitors (Muguruma et al. 2010). In a follow-up study, the authors identified two growth factors, SDF1 and FGF9, which drive the formation from hESCs of polarized neuroepithelium that resembles cerebellum of the first trimester (Muguruma et al. 2015).

The hippocampus develops from an intermediate structure called the medial pallium in the dorsomedial telencephalon. Starting with SFEBq human ESC cultures, stimulation with Wnt and Bone Morphogenetic Protein (BMP) agonists led to the formation of choroid plexus-like tissue, which corresponds to the most dorsomedial part of the telencephalon. A shortened period of Wnt and BMP stimulation led to the formation of medial pallium-like tissue, from which functional granule neurons and pyramidal neurons could be derived (Kadoshima et al. 2013).

Inspired by the 3D culture system pioneered by the group of Yoshiki Sasai, cerebral organoids containing cells representing different brain regions have also been generated (Lancaster et al. 2013). The approach taken involved the re-aggregation of human PSCs into free-floating EBs before allowing spontaneous neural differentiation of the EBs without the addition of exogenous growth factors. The EBs were then transferred to Matrigel droplet cultures and subsequently to a spinning bioreactor. As a result, millimeter-scale cerebral organoids formed, containing neuroepithelium with regional patterns, including the forebrain, midbrain–hindbrain boundary, choroid plexus, and hippocampus (Lancaster et al. 2013). More recently, an improved protocol incorporated bioengineered microfilaments in hPSC-EBs to achieve guided self-organization of cortical structures prior to embedment in Matrigel droplet. Pulse stimulation of Wnt signaling and supplement of ECM and vitamin A resulted in robust generation of polarized cortical plate (Lancaster et al. 2017). Detailed gene expression analyses of cerebral organoids using single-cell RNA-seq (scRNA-seq) have revealed the close resemblance between the cells in cerebral organoids and their corresponding counterparts in vivo (Camp et al. 2015; Quadrato et al. 2017). By fusing hPSC-derived ventral and dorsal forebrain organoids, migration of interneurons across organoids could also be observed in real time (Bagley et al. 2017). Therefore, cerebral organoids can be used to study multiple aspects of human cortical development.

5.2.1.2 Retinal Organoids

The retina is the neural region of the eye which responds to stimulation by light. During development, the optic vesicle evaginates from the pseudostratified neuroepithelium and the distal part invaginates to form the optic cup. To explore the dynamics of cellular interaction and behavior, Sasai et al. (Eiraku et al. 2011) utilized the SFEBq mESC culture with Matrigel in order to generate a self-organizing optic cup. The resulting organoids expressed markers found in the developing retina and recapitulated in vivo development (Eiraku et al. 2011). Using a similar approach with some further optimization, optic cup organoids could also be generated from human PSCs. Interestingly, the hPSC-derived optic cup organoids grew to a larger size and formed multi-layered tissue containing rod and cone cells, thus recapitulating species-specific features of development (Nakano et al. 2012).

5.2.1.3 Intestinal/Gastric/Esophageal Organoids

During development, the definitive endoderm gives rise to the gut tube, which can be subdivided into foregut, midgut, and hindgut. In later development, the foregut gives rise to the oral cavity, pharynx, respiratory tract, stomach, pancreas, and liver, whereas the midgut gives rise to the small intestine and ascending colon, whilst the hindgut generates the bulk of the colon (large intestine) and rectum. Therefore, attempting to derive particular part(s) of the gastrointestinal tract from PSCs requires knowledge of the factors involved in gut tube specification during development, including factors driving: (1) differentiation of PSCs towards definitive endoderm; (2) specification into distinct regions of gut tube; (3) patterning within regions of the gut tube; and (4) expansion of epithelia with differentiated cell types (Wells and Spence 2014).

The generation of PSC-derived intestinal organoids involves the stepwise induction of specification. Firstly, the differentiation of definitive endoderm requires the addition of Activin A (a member of the TGF-β family) to a 2D monolayer culture of hPSCs (Spence et al. 2011). Later, both Wnt3a and FGF4 were needed to specify hindgut and small intestinal fate in the definitive endoderm. During hindgut specification, the flat sheet of cells spontaneously budded off from the plastic plate and condensed to form hindgut spheroids. When the spheroids are embedded in Matrigel and treated with a well-established "mini-gut" growth factor cocktail (Sato et al. 2009), the culture transitions in a fashion that resembles fetal small intestinal development (Spence et al. 2011). A follow-up study identified BMP activation in hPSC-derived hindgut spheroids as crucial to promote colonic differentiation. The resulting hPSC-derived colonic organoids expressed SATB2 (a marker of posterior large intestine) with the expression of corresponding HOX genes (Múnera et al. 2017).

Gastric organoids can be generated from hPSCs in a similar fashion. The stomach derives from the posterior foregut and is divided into two parts in humans, fundic and antral, which correspond to the proximal and distal parts. Treatment of definitive endoderm with Wnt3a, FGF4 and a BMP inhibitor (Noggin) led to the formation of spheroids with foregut identity. These foregut spheroids were exposed to Noggin, retinoic acid (RA) and epidermal growth factor (EGF) leading to the specification of antral (proximal) gastric fate, showing morphogenesis of glandular structures (McCracken et al. 2014). A follow-up study identified that Wnt/β-Catenin signaling is important for the fundic fate specification (McCracken et al. 2017). Continuous stimulation of Wnt signaling in hPSC-derived posterior foregut spheroids led to the formation of gastric fundic organoids devoid of antral or intestinal fate (McCracken et al. 2017).

The esophagus is a stratified squamous epithelium surrounded by muscles and enteric nerves. Two independent studies have sought to generate human esophageal organoids from human PSCs (Trisno et al. 2018; Zhang et al. 2018). Their protocol includes definitive endoderm induction (with Wnt3a, FGF4 and Noggin), specification into anterior foregut (low Wnt+low RA or Noggin followed by Wnt inhibition), dorsal esophageal differentiation (BMP or TGF-β inhibition) and maturation (FGF10+EGF+Noggin). The resulting human esophageal organoids contained esophageal progenitor cells, which could be used to form mature esophageal-derived organoids.

5.2.1.4 Liver Organoids

The liver is composed of two epithelial cell types, hepatocytes and biliary ducts, which originate from the endodermal foregut. During liver development, a structure known as the liver bud is formed. The formation of the liver bud involves the delamination and condensation of hepatic progenitors from the foregut endoderm which is followed by rapid vascularization. Takebe et al. (2013, 2014) sought to recapitulate these complex events by mixing human PSC-derived hepatic endodermal cells (specified via treatment with activin-A followed by treatment with bFGF and BMP4), mesenchymal cells and endothelial cells. 3D aggregates were formed when the cell mixture was plated at high density on a layer of Matrigel and these liver bud-like aggregates were found to contain a well-connected network of blood vessels. These aggregates connected with the host vasculature within 48 h upon transplantation. Remarkably, human metabolites were detected in the blood of the xenografted mice. Next, in-depth scRNA-seq analysis revealed significant similarity between the 3D liver-bud cultures and human fetal liver. Moreover, the authors also identified the importance of VEGF-A/KDR signaling for successful outgrowth in vivo upon transplantation (Camp et al. 2017).

5.2.1.5 Lung and Thyroid Organoids

Both lung and thyroid develop from the ventral foregut and derive from a common Nkx2.1+ progenitor population. These Nkx2.1+ progenitors have been generated in vitro via the inhibition of TGF-β and BMP signaling in mESC-derived definitive endoderm cells, and the progenitors retained the potential to give rise to both lung and thyroid cells (Longmire et al. 2012). Rossant et al. (Wong et al. 2012) generated cystic fibrosis transmembrane conductance regulator (CFTR)-expressing proximal airway cells in human iPSC-derived definitive endoderm via induction with Shh and FGF2, to direct foregut specification, and subsequent maturation into lung progenitors through treatment with FGF10, FGF7 and a low dose of BMP4 in an air–liquid interface culture. Huang et al. (2014) improved upon this method by sequentially inhibiting the BMP, TGF-β, and Wnt signaling pathways to increase the efficiency of foregut endoderm specification from human PSCs. Subsequent induction of Wnt signaling and treatment with FGF10, KGF, BMP4, and RA was used to direct ventral fate specification and so obtain lung/airway progenitors. Various mature epithelial cell types were then obtained via treatment with Wnt, FGF, c-AMP, and glucocorticoids. Dye et al. (2015) achieved the same endpoint through a different approach, by treating hPSC-derived definitive endoderm with BMP/TGF-β inhibitors and FGF4/ Wnt activators to direct the culture towards a foregut endoderm fate. Moreover, additional activation of the Hedgehog pathway in the definitive endoderm culture stimulated ventral specification toward the lung lineage. Floating spheroids were then embedded in Matrigel and exposed to FGF10 in order to form mature lung organoids. The resulting human lung organoids resembled the upper airway epithelium, with a similar composition of basal cells and ciliated cells surrounded by smooth muscle and myofibroblasts.

Thyroid organoids could be generated from the sorted Nkx2.1-Pax8 double positive progenitors *via* embedding the cells in Matrigel and treating them with thyroid stimulating hormone (Kurmann et al. 2015). Upon in vivo transplantation under the kidney capsules of hypothyroid mice, thyroid organoids maintained their structures and functional markers, and elevated plasma T4 and T3 hormone levels.

5.2.1.6 Kidney Organoids

The kidney is comprised of more than 20 cell types, making up nephrons with a dense network of vasculature and interstitial compartments. Several protocols have been reported to generate nephron progenitor cells and kidney organoids from hPSCs (Morizane and Bonventre 2017). Of particular note, Takasato et al. (2015) used sequential induction with Wnt and FGF9 in 2D-cultured hPSCs to specify intermediate mesoderm before the cells were then pelleted and transferred onto a Transwell for 3D culture. A short pulse of Wnt stimulation was found to increase the number of nephrons in the culture. In another study, Morizane et al. (2015) first generated posterior intermediate mesoderm by using Wnt stimulation followed by Activin treatment. FGF9 was then added to generate metanephric mesenchyme, which formed pretubular aggregates with additional Wnt stimulation and upon transfer to low-attachment plates. Removal of the Wnt agonist caused the culture to develop further to a renal vesicle-like state, and removal of FGF9 allowed its subsequent maturation. The resulting kidney organoids contained nephrons with defined glomeruli containing Bowman's capsule-like structures made of podocytes, and both proximal and distal tubules connected by the loop of Henle (Takasato et al. 2015; Morizane et al. 2015).

5.2.1.7 Inner Ear Organoids

The inner ear is important for sensory functions such as detecting movement, gravity, and sound. The developmental origin of the inner ear is the nonneural and preplacodal ectoderm. Koehler et al. (2013) utilized the SFEBq culture system to direct mESCs towards a nonneural ectodermal fate by providing BMP activation, TGF-β inhibition and 2% Matrigel in the media. To induce placodal ectoderm formation, BMP inhibition and FGF2 activation were applied to the culture and the resulting aggregates resembled the developing otic placode. The aggregates underwent spontaneous differentiation in serum-free suspension culture, allowing the formation of vesicles from the placodal epithelium. Hair cells formed in this culture had functional mechanosensitive properties and formed synapses with sensory neurons, thus representing a useful platform for studying inner ear development and disease.

5.2.1.8 Mammary Gland Organoids

The mammary gland is of nonneural ectodermal origin and is where milk is produced to feed offspring. Qu et al. (2017) sought to generate mammary gland organoids via the EB approach, where hiPSC-EBs were guided towards a nonneural fate with medium commonly used for the culture of mammary cancer cell lines. 10-day-old EBs were transferred to a floating Matrigel-collagen I mixed-gel culture to which had been added pTHrP, a hormone that is important for normal mammary development. Additional treatment with hydrocortisone, insulin, FGF10, and HGF for 20 days increased the yield of mammary cells and resulted in the formation of alveolar mammary-like structures. The resulting mammary organoids could produce milk when cultured in lactogenic medium containing prolactin, hydrocortisone and insulin (Hatsumi et al. 2006). The semi-solid, floating, mixed-gel cultures of mammary gland organoids would therefore be a physiologically relevant model for the study of mammary development.

5.2.2 ADULT STEM CELL-DERIVED ORGANOIDS

The establishment of AdSC-derived organoids relies on the self-renewal capacity and self-organization properties of tissue-resident stem cells. While PSC-derived organoids are formed by activating in PSCs the signaling pathways active during development of the tissue of interest, AdSC-derived organoids are formed by providing growth factors supplied by the tissue's niche environment to maintain AdSC characteristics. Most approaches involve isolation of the tissue-resident AdSC population (if known) and culture of the AdSCs in a laminin-rich matrix such as Matrigel or basement membrane extract (BME). The culture medium contains a cocktail of growth factors which activates the in vivo signaling pathways responsible for AdSC maintenance including Wnt, BMP, EGF, Notch, and Shh, amongst others. Defining the appropriate medium composition for AdSC maintenance is a major bottleneck in the establishment of AdSC-derived organoids. Here we discuss the establishment of organoids derived from different organs.

5.2.2.1 Intestinal Organoids

The adult intestine is composed of repeating units of protruding villi, with invaginated structures called the crypts of Lieberkühn nestled in between the protrusions. The villi contain differentiated cells such as enterocytes which carry out absorptive functions, goblet cells that secrete mucus into the lumen of the intestinal epithelium and enteroendocrine cells that secrete hormones (Barker, Bartfeld, and Clevers 2010). The intestinal crypts contain multipotent, self-renewing stem cells, or crypt base columnar cells, marked by the expression of Lgr5 (Barker et al. 2007) and that give rise to daughter cells which make up the entire intestinal epithelium. The base of the crypt is also the location of another differentiated cell type—Paneth cells.

These cells are long-lived (generally 2–3 months) and in addition to producing antibacterial peptides, they also play a major role in regulating stem cell behavior by supplying growth factors (Clevers 2013). The intestinal stem cells (ISCs) are governed by Wnt, Notch, EGF, and BMP signaling pathways, and the relevant growth factors as supplied by their niche.

Mindful of the features of ISCs, such as their multipotency and massive proliferation capacity, Sato et al. (2009) were the first to establish 3D mouse intestinal organoid cultures. The authors isolated ISCs or entire crypt compartments and embedded them in Matrigel. In addition to the Matrigel support, the authors supplied culture medium containing R-spondin-1 (a Wnt amplifier and ligand of Lgr4/5/6 and Rnf43/Znrf3), EGF, and Noggin (a BMP inhibitor) to mimic the in vivo microenvironment in a dish. Small intestinal organoids were found to grow without exogenous Wnt3a (as Paneth cells secrete enough Wnt ligand in culture), whereas colonic organoids require additional Wnt in culture since the colonic epithelium possesses a limited capacity for endogenous Wnt production. These "mini-guts" are polarized, with a basal domain in direct contact with the Matrigel and an apical domain of enterocyte brush borders forming an enclosed lumen. Crypt-like budding structures project outward from the lumen, and Lgr5+ stem cells and Paneth cells can be found at the tip of the protrusions. Remarkably, the organoids retain the ratio of various intestinal epithelial cell types and their anatomical cross-section as found in vivo (Sato et al. 2009; Grün et al. 2015). The organoids can also be passaged over time, and remain genetically and phenotypically stable.

While mouse intestinal organoids can grow in Matrigel with a supply of R-spondin-1, EGF, and Noggin, human small intestinal organoids and colonic organoids require additional factors including Wnt3a, nicotinamide, and inhibitors of ALK and p38 (involved in TGF-β signaling) for long-term maintenance (Jung et al. 2011; Sato et al. 2011).

5.2.2.2 Stomach Organoids

The stomach contains two main anatomical parts: the corpus and antrum in humans, and the forestomach (rostral), corpus, and antrum (caudal) in mice. Unlike the protruding villus structures in the small intestine, the stomach contains invaginated glands in the corpus and antral parts. The gastric glands are divided into four anatomical segments (from top to bottom of the gland): the gastric pit, isthmus, neck, and base regions (Kim and Shivdasani 2016). Each part of the stomach gland contains different cell types which carry out specific functions, such as mucus-secreting pit cells, rapidly dividing cells in the isthmus region, mucus-secreting neck cells and terminally differentiated enzyme-secreting chief cells. In between these cells, hormone-secreting enteroendocrine cells are also found throughout the gland and parietal cells, which secrete hydrochloric acid, are abundant within the corpus glands.

Similar to the small intestine, a proportion of chief cells at the bottom of the gland are marked by the expression of the membrane receptors Lgr5 and Troy, which are both downstream Wnt target genes (Barker et al. 2010; Stange et al. 2013; Leushacke et al. 2017). Lgr5+ gastric stem cells were first identified in the pyloric antral part of the stomach, a region of the stomach where stem cells are known to divide rapidly. In contrast, chief cells in the corpus that express Lgr5 and Troy divide slowly.

Upon injury to the stomach, Lgr5+/Troy+ chief cells become more proliferative and can efficiently give rise to the different cell types of the gastric epithelium, hence they are termed "reserve quiescent stem cells" (Barker et al. 2010; Stange et al. 2013). Isolated glands or FACS-isolated single cells expressing the stem cell markers Lgr5 and Troy efficiently give rise to budding gastric organoids when cultured in Matrigel supplemented with medium containing EGF, R-spondin-1, Wnt3a, Noggin, FGF10, and Gastrin (Stange et al. 2013).

5.2.2.3 Esophageal Organoids

Esophageal organoids could be generated by enzymatically dissociating the esophageal mucosa and embedding single cells in droplets of Matrigel, and maintained in medium containing Wnt3a, R-spondin-2, Noggin, Gastrin, EGF, and TGF-β inhibitors (DeWard, Cramer, and Lagasse 2014). Such esophageal organoids recapitulate the architecture of the native tissue, as confirmed by antibody staining. Proliferative basal cells are found on the outermost ring of the organoids, with squamous epithelium attached to the basal layer and a keratinized layer in the lumen.

5.2.2.4 Liver/Pancreas Organoids

The adult liver is a bulky tissue that contains two epithelial cell types—hepatocytes and ductal cells. The proliferation rate of these cells is extremely low during homeostasis, but upon injury the cells become rapidly dividing and so confer upon the organ its high regenerative capacity (Huch, Boj, and Clevers 2013; Miyajima, Tanaka, and Itoh 2014). The expression of Lgr5, which is a marker of AdSCs in the stomach, small intestine, and colon, is barely detectable in the healthy liver. Nevertheless, upon acute injury, Lgr5+ cells emerge around the bile duct and divide to replenish the lost tissue (Huch, Dorrell, et al. 2013). Moreover, organoids can be generated by the sorting and seeding of single Lgr5+ cells from the injured liver into Matrigel with medium containing Wnt3a, R-spondin-1, EGF, Noggin, Gastrin, FGF10, and HGF. Wnt3a and Noggin are dispensable for long-term maintenance of the culture after the first week of organoid establishment. The resulting organoids consist of cells expressing both hepatocyte and ductal lineage markers, indicating the bipotency of Lgr5+ cells, although further gene expression profiling revealed a biased fate commitment towards the ductal lineage, and mature hepatocyte markers were only weakly expressed. Inhibition of either Notch or TGF-β signaling resulted in the upregulation of mature liver and hepatocyte markers (Huch, Dorrell, et al. 2013). To assess the capacity of the Lgr5+ cell-derived liver organoids to perform physiological functions in vivo, organoids were transplanted into fumarylacetoacetate hydrolase (FAH)-deficient mice, which require constant administration of NTBC to avoid death due to liver failure (Azuma et al. 2007). Strikingly, Lgr5+ cell-derived liver organoids which had undergone pre-treatment with hepatocyte differentiation medium could prolong the survival of FAH-deficient mice after the withdrawal of NTBC (Huch, Dorrell, et al. 2013). The same protocol can be applied to derive human liver organoids, with the addition of forskolin (a cyclic AMP pathway agonist) and inhibitors of TGF-β signaling to the growth factors required for formation of murine liver organoids in order to allow the long-term maintenance of the bipotent state of Lgr5+ cells (Huch et al. 2015). Cells remained genetically stable over time in culture, as assessed by whole genome sequencing. Mature hepatocytes were formed from the organoids upon removal of R-spondin-1 from the culture medium, and were functional upon transplantation in mice.

Comparison of the pancreas to the liver reveals many areas of similarity. Similar to the liver, under conditions of homeostasis the pancreas is also a slowly proliferating organ, with the Wnt signaling pathway downregulated in mature exocrine/acinar, endocrine and ductal cells. Ligation of the pancreatic duct damaged the tissue, and it was found that this led to the upregulation of Lgr5 expression in ductal cells, from which pancreatic organoid cultures could be derived starting from single cells (Xu et al. 2008). Lgr5+ pancreatic ductal cells embedded in Matrigel could be cultured as organoids by using a growth factor cocktail similar to that for liver organoids (EGF, R-spondin 1, Noggin, Gastrin, FGF10). The multipotent state of the pancreatic organoids could be sustained long-term and organoids could be differentiated into mature functional duct, exocrine/acinar, and endocrine cells in vitro and upon transplantation in vivo (Huch, Bonfanti, et al. 2013). Human pancreatic organoids could be established in a similar manner after enzymatic dissociation into single cells and the addition of a TGF-β inhibitor into the culture medium (Boj et al. 2015).

Although the expression of the stem cell marker Lgr5 is elevated only upon injury, both liver and pancreatic organoids could be established with or without injury from the duct cells of the respective tissues. For mouse ductal organoid establishment, the bulk organ is first minced into small fragments, followed by enzymatic dissociation. After dissociation, ducts usually become visible within the tissue suspension and can be manually picked for the generation of ductal organoids (Broutier et al. 2016). Human biopsies are usually dissociated into single cells to enable culture due to the low amount of source material. The resulting organoids consist of bi- or multipotent ductal cells expressing Lgr5 and can be differentiated into mature functional cells as described above. Most strikingly, the plasticity retained in ductal organoid cultures makes them a potential source for tissue replacement therapy, exemplified by the extended survival displayed upon transplantation of organoids into the damaged livers of FAH-deficient mice (Dorrell et al. 2014). Intriguingly, the identities of ductal cells in liver and pancreas were found to be similar, with overlapping surface markers, further highlighting the similarities in the induced stem cell compartments in these two tissues.

5.2.2.5   Gallbladder Organoids

The gallbladder stores and concentrates the bile that is produced by the liver. After feeding, the gallbladder contracts and releases bile into the small intestine, aiding the processes of digestion and absorption (Housset et al. 2016). Historically, there has been no method available which has allowed the culture and propagation of cholangiocytes from the gallbladder, meaning that cell replacement therapy in the form of the supply of healthy cholangiocytes to patients suffering from common bile duct disorders (i.e., biliary atresia or ischemic strictures) was not possible. Sampaziotis et al. sought to develop a method to culture extrahepatic cholangiocytes using a 3D culture system. In order to generate organoids, the authors seeded mechanically dissociated human tissue in droplets of Matrigel and cultured the tissue in medium containing EGF, R-spondin-1 and DKK-1 (Dickkopf-related protein 1, a Wnt antagonist). Whilst it may be counterintuitive to have both a Wnt potentiator (R-spondin 1) and antagonist (DKK-1) in the same medium, and in stark contrast to liver or pancreatic organoid culture systems, gene expression analysis suggested that noncanonical Wnt signaling (such as the planar cell polarity pathway) and Rho-kinase activity are upregulated in extrahepatic cholangiocyte organoids (ECOs) cultured in the presence of both R-spondin and DKK-1. Cells in ECOs maintain the expression of ductal markers (CK7 and CK19). As a proof-of-concept for their clinical potential, ECOs cultured in a bioengineered bile duct scaffold could be surgically implanted into the midportion of the native common bile duct in mice, and so could replace the duct of the host mouse. Notably, and dissimilar to the previously described ductal organoids, the differentiation potential of ECOs was restricted to the ductal lineage, and mature hepatocytes or pancreatic lineages could not be generated (Sampaziotis et al. 2017).

5.2.2.6 Prostate Organoids

The prostate is a male gland consisting mainly of luminal and basal cells. Karthaus et al. generated murine and human prostate organoids by embedding dissociated cells into Matrigel and culturing them in the minigut medium (containing R-spondin-1, Noggin, and EGF), with the additional of FGF10, FGF2, PGE2, nicotinamide, and a p38 inhibitor (SB202190) for human prostate organoid culture. The resulting cystic organoid structure consists of outer and inner layers which represent the basal and luminal components found in vivo. The authors also investigated the growth factor requirements for organoids established from different parts of the organ (anterior, dorsolateral, or ventral prostate). Extrinsic Noggin and R-spondin enhanced organoid establishment efficiency but were dispensable for the passaging and maintenance of both murine and human prostate organoid cultures. In murine organoid cultures, EGF was required for both establishment and passaging, whereas TGF-β inhibition was only required for the maintenance of anterior- and ventral prostate-derived organoids. In human prostate organoid cultures, most other components were essential for both the establishment and long-term maintenance of the culture. Both luminal and basal cells could give rise to organoids starting from single cells in both murine and human systems, and prostate organoids remained genetically stable after months of cultures (Karthaus et al. 2014).

5.2.2.7 Fallopian Tube and Endometrial Organoids

The female reproductive system has a complicated organisation tightly regulated by cyclic hormonal changes. The fallopian tube epithelium is a tubular structure formed by a simple columnar epithelium containing secretory and ciliated cells. Since the cyclic hormonal changes have a profound effect on the maintenance of the fallopian tube epithelium, it is important to understand the regulatory mechanisms underlying fallopian tube stem cell self-renewal and maintenance. In 2015, Kessler et al. reported a 3D organoid culture system which supports the long-term, stable growth of human fallopian tissue. The authors first cultured isolated epithelial cells in a 2D system before transferring these cells into a 3D Matrigel matrix in culture medium supplemented with a cocktail of growth factors including regulators of Wnt, Notch, EGF, FGF10, and TGF-β signaling. The organoid cultures, derived from single fallopian tube epithelial cells, can develop both secretory and ciliated cells and so maintain plasticity.

Over the lifetime of an adult female, the endometrial layer extensively and dynamically changes over 400 times in synchronization with the hormonal cycle. The tight regulation of endometrial stem and progenitor cells is known to be a crucial factor for reproductive capability (Gargett, Chan, and Schwab 2008). Two separate studies, Turco et al. (2017) and Boretto et al. (2017), have established a 3D endometrial organoid culture system recapitulating the functions of human and murine endometrium in vitro. The system successfully maintained genetically stable and functional endometrium with the capacity to respond to hormonal stimulation for a long period. The authors obtained glands containing stem and progenitor cells from endometrial or decidual tissue and embedded single cells into Matrigel in culture medium supplemented with A83-01 (a TGF-β inhibitor), FGF10, HGF, EGF, Noggin, R-spondin-1, and nicotinamide. Importantly, this organoid system recapitulates in vivo molecular signatures as confirmed by global gene expression profiling analysis comparing established organoids to the original tissue sample and to cultured stromal cells established from the sample biopsy. Moreover, the organoids could respond to ovarian hormones, showing phenotypic responses similar to the endometrium.

5.2.2.8   Mammary Gland Organoids

The mammary gland is actively remodelled throughout the menstrual cycle and during pregnancy, which indicates the existence of stem/progenitor cells within the tissue. Moreover, reconstitution of the functional mammary gland in the context of a mouse transplantation model supported the existence of a bipotent mammary gland stem cell. In 2015, Linnemann et al. reported an organoid culture system supporting human mammary gland organoid formation within collagen gels in mammary epithelial cell growth medium supplemented with Rho-kinase inhibitor and forskolin. The terminal ductal tubular unit was recapitulated in the system and was composed of a population of basal and luminal cells. However, the signaling pathways required for the establishment and maintenance of mammary gland organoid cultures are not fully understood. A report from Plaks et al. (2013) suggests the importance of the Wnt signaling pathway for mammary gland organogenesis. A study by Jamieson et al. (2017) supports this by reporting advanced culture conditions which generate mammary organoids recapitulating in vivo features of mammary gland architecture and function.

5.2.2.9   Taste Bud Organoids

Taste buds are the main sensory organ perceiving taste located in the tongue epithelium. Ren et al. have described an Lgr5- or Lgr6-expressing taste bud stem/progenitor population that can generate taste buds ex vivo and which mimics functional taste buds in a mouse model in vivo. In the study, the authors isolated Lgr5+ or Lgr6+ stem and progenitor cells and embedded them into Matrigel in the same culture medium as used for the establishment of intestinal organoids. In the established organoid culture, the authors could detect actively cycling cells in addition to mature, differentiated taste bud cells, marked by the expression of marker genes such as gustducin, carbonic anhydrase 4, taste receptor type 1 member 3, nucleoside triphosphate diphosphohydrolase-2, or cytokeratin K8 (Ren et al. 2014).

5.2.2.10   Salivary Gland Organoids

The salivary glands are the source of saliva, which initiates digestion and maintains homeostasis in the oral cavity. In a report (Nanduri et al. 2014) and a follow-up study (Maimets et al. 2016), the authors demonstrate the existence of a murine salivary gland stem cell population and describe the conditions required to culture this population in vitro. The authors could expand the salivary gland stem cell population by embedding the cells into Matrigel and culturing in the presence of FGF and EGF, but the authors did not demonstrate the long-term maintenance capacity of this culture system. In their follow-up study, the authors identified the importance of Wnt pathway activation for the long-term maintenance of salivary gland organoids. Importantly, salivary gland organoids can be used to successfully repair damaged salivary glands without tumor formation or dysplastic changes, hinting at the immense therapeutic potential of the platform in general.

5.2.2.11   Lung Organoids

The lung is the organ where gas exchange takes place and consists of trachea, bronchi, bronchioles, and alveoli. Due to its position at the interface of the body with the external environment, the lung epithelium is continuously exposed to environmental stimuli. The first report of the development of murine bronchiolar lung organoid culture was from Rock et al. (2009). The authors established bronchiolar lung organoids originating from basal cells embedded into Matrigel and cultured in the presence of EGF. The authors demonstrated that single basal cells isolated from the trachea could form tracheospheres, which could be continuously passaged. However, the tracheospheres were limited in their cellular composition, and did not contain mature Clara+, neuroendocrine or mucus-producing cells. In the follow-up study, the authors used the system to screen for factors involved in regulating the decision of stem cells to differentiate into ciliated cells or secretory cells.

Another type of murine lung organoid has been generated from a different region of the airway, the alveoli. Alveoli are the place where the actual gas-exchange takes place and consist of two types of cells: alveolar type 1 and 2 (AT1 and AT2, respectively). Desai, Brownfield, and Krasnow (2014) demonstrated both AT1 and AT2 derive from the same progenitors, but the rare population of AT2 cells also function as stem cells to provide AT1 cells for alveolar renewal. The authors showed that the EGFR pathway regulates the stem cell potential of AT2 cells in vitro and that oncogenic KRAS causes dysregulation in vivo. In addition to these studies, Lee et al. (2014) have recently established a co-culture system of bronchioalveolar stem cell-derived lung organoids and endothelial cells and have used the system to demonstrate the multipotency of bronchioalveolar stem cells and their differentiation into both bronchiolar and alveolar cells in vitro.

5.3   APPLICATIONS OF ORGANOIDS

In this section, we will highlight some of the recent studies which have developed and expanded the use of organoids in biomedical research. We have divided studies into the broad areas of infectious disease/host–pathogen interactions, hereditary/genetic disorders, genome engineering, oncology and biobanking (Figure 5.3).

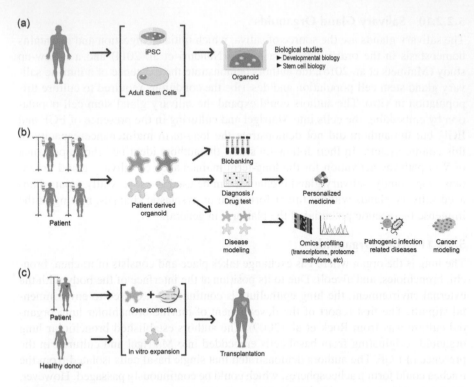

FIGURE 5.3 Applications of organoid systems. Organoid culture systems can be used: (a) to address basic biological questions such as understanding stem cell characteristics and their interaction with neighboring cells and stem cell development (Huch and Koo 2015), (b) as a platform for the study of disease processes e.g. novel cancer drivers (Drost and Clevers 2018) and disease related gene mutation identification, pathogen or virus culture platform (Heo et al. 2018), (c) to generate biologically matched material for transplantation, for example, CFTR gene correction of cystic fibrosis patient organoids (Schwank et al. 2013) and establishment of liver organoid for the transplantation (Huch et al. 2015). CFTR, cystic fibrosis transmembrane conductance regulator.

5.3.1 INFECTIOUS DISEASE STUDIES

3D organoid systems not only contain organ-specific cell types but also recapitulate some of the microenvironmental features which make it a desirable system for studying host–pathogen interactions, especially for pathogens that specifically infect humans.

One very recent outbreak of infectious disease which was widely reported was the outbreak of Zika virus (ZIKV), predominantly in South America, which is transmitted by mosquitoes or sexual intercourse, with the greatest threat posed to pregnant women. The most severe and obvious pathology are the congenital malformations, such as microcephaly, in newborns. Many research labs have taken advantage of the cerebral organoid system, which mimics human fetal brain development, to directly visualize and dissect host cellular responses following the onset of ZIKV infection. Organoids infected with ZIKV were smaller due to the massive loss of neural progenitors through programmed cell death (autophagy) and neuronal development was greatly impaired (Garcez et al. 2016). One study found that ZIKV infection of cerebral organoids led to the upregulation of the innate immune receptor, Toll-like-Receptor 3 (TLR3), suggesting that TLR3 may mediate aspects of pathology downstream of ZIKV infection. Subsequent inhibition of TLR3 rescued cells in the developing cerebral organoids from ZIKV/TLR3-mediated apoptosis (Dang et al. 2016).

Helicobacter pylori (*H. pylori*) is a bacterial pathogen which propagates in proximity to the epithelial lining of the stomach and is associated with human diseases such as gastric ulcers and gastric cancer. By injecting live *H. pylori* into the lumen of gastric organoids, it has been possible for the first time to study the host–pathogen interactions associated with this bacterium without the use of animals (McCracken et al. 2014; Bartfeld et al. 2015; Bartfeld 2016), an approach which has also been used to study *Clostridium difficile* (Leslie et al. 2015). More recently, Heo et al. injected oocysts from *Cryptosporidium parvum* (a common cause of diarrhea) into human AdSC-derived lung and small intestinal organoids, allowing *C. parvum* to propagate through its full life cycle. Temporal gene expression analyses of the infected organoids and the parasite itself revealed genes associated with infectivity and propagation during the *C. parvum* life cycle (Heo et al. 2018).

It is also possible to infect human organoids with common viruses such as enterovirus (Drummond et al. 2017), norovirus (Ramani, Atmar, and Estes 2014; Ettayebi et al. 2016), and rotavirus (Saxena et al. 2016; Yin et al. 2015) in order to study the molecular pathways involved in infection.

5.3.2 HEREDITARY DISEASE STUDIES

Due to the self-organizing capacity of organoids, which leads to the formation of different cell types into structures similar to the architecture found in vivo, these cultures have represented an attractive method for modeling organ-specific genetic disorders. Organoids have been used to elucidate the developmental timing of the onset of phenotypic changes, the cell types that are affected, physical alterations such as changes in morphology, size, patterning, or polarity, and potential targets for therapeutic intervention.

Cerebral organoids have been used to study one of the most common neurodevelopmental disorders, microcephaly, where newborns are either missing particular parts of the brain or have insufficient growth across all regions, resulting in a significantly smaller brain. Lancaster et al. (2013) generated iPSCs from a patient with compound heterozygous truncating mutations in *CDK5RAP2* resulting in loss of the protein, which had previously been reported to be the cause of microcephaly and reduced stature (Bond et al. 2005). Remarkably, doubling time and iPSC colony morphology appeared normal; however, cerebral organoids formed from the patient iPSCs displayed several marked differences when compared with control organoids. Patient organoids had a smaller neuroepithelium, with a smaller neural progenitor (radial glial cell) population and abnormal spindle orientation. Re-introducing *CDK5RAP2* via electroporation of an expression vector into the developing cerebral organoids could rescue these phenotypes, suggesting that loss of *CDK5RAP2* causes premature neural differentiation which leads to microcephaly (Lancaster et al. 2013).

Cystic fibrosis (CF) is caused by various mutations in the CFTR, with the most commonly occurring mutation worldwide being deletion of phenylalanine at position 508. In most cases, even with medical intervention, patients have a reduced quality of life and a short life expectancy of less than 40 years (Chin, Earlam, and Aaron 2015). Moreover, a significant portion of patients do not respond to the available drugs because of the variability in the mutations causing CF (Ratjen and Döring 2003). Dekkers et al. (2013) used human AdSC-derived intestinal organoids as a platform to develop functional assays for CFTR activity. In brief, treatment with forskolin induces swelling in normal organoids *via* water influx into the lumen; functional CFTR mediates the influx of water and therefore organoid swelling becomes a readout of CFTR functionality. For organoids with the common mutant version CFTR^{F508del}, forskolin-mediated organoid swelling is compromised. The authors further demonstrated that the functionality of CFTR^{F508del} could be restored through exposure to lower temperatures (27°C) as well as treatment with compounds such as VX-809 and VX-770 (Dekkers et al. 2013). This highlights the potential of organoid cultures to screen for therapeutic interventions that will be efficacious for individual patients.

Alpha 1-antitrypsin (A1AT) is a secreted protein encoded by the *SERPINA1* gene, produced by hepatocytes and transported to the lungs in the circulation in order to protect the lungs from the proteolytic activity of elastase produced by neutrophils. A1AT deficiency leads to respiratory complications, whilst an accumulation of misfolded A1AT in hepatocytes causes cirrhosis and increases the risk of pulmonary emphysema and hepatic disease (Fairbanks and Tavill 2008; Lawless et al. 2008). The most common disease-causing mutation detected is a single copy of the autosomal co-dominant Z allele (Glu342Lys) of *SERPINA1* (ZZ mutant). Huch et al. established liver organoids from patients with the ZZ mutant genotype, and whilst expansion in the bipotent state appeared normal in terms of organoid morphology, growth rate and functionality, upon differentiation into hepatocytes, misfolded A1AT aggregates rapidly accumulated in the patient-derived organoids (Huch et al. 2015).

5.3.3 GENOME ENGINEERING IN ORGANOIDS

CRISPR/Cas9 technology utilizes a bacterial adaptive defense mechanism made up of a complex of a dual RNA hybrid containing single-stranded CRISPR RNA (crRNA) and transactivating crRNA (tracrRNA) together with the Cas9 endonuclease. The crRNA contains 20 nucleotides complementary to a target sequence and together with the tracrRNA directs Cas9 endonuclease to target site. Then, Cas9 generates double-strand breaks on target DNA adjacent to the PAM (protospacer adjacent motif) site, which allows self- and nonself-determination (Jiang and Doudna 2017). Using knowledge gained from basic scientific research into microbial defense mechanisms, engineering of the dual-tracrRNA:crRNA hybrid into a single molecule (termed single-strand guide RNA or gRNA) directing Cas9 to cleave double-stranded DNA (Jinek et al. 2012) has allowed gene targeting in various model systems through simply designing the 20 nucleotides of the gRNA sequence to match the desired genomic locus.

As described above, one of the remarkable features of 3D organotypic cultures derived from AdSCs is their long-term genomic stability through culture and passaging, which makes them an attractive system for the study of genetic disease and a versatile platform for genetic engineering, such as gene knock-out or targeted knock-in. Several protocols, including liposomal transfection (Schwank et al. 2013), electroporation (Fujii et al. 2015), or retro- or lenti-viral transduction (Koo et al. 2012), have been developed to deliver genetic elements into organoids efficiently and so allow their genetic manipulation. AdSC-organoids grow in Matrigel with their basal domain forming the outer layer, making them inaccessible to the above delivery methods in this state. Prior to delivery of genetic elements, organoids are usually mechanically or enzymatically dissociated into smaller fragments or into single cells in order to expose the apical luminal domain. These methods have allowed the delivery of conditional genetic elements, such as vectors containing a tamoxifen/ 4-OHT (e.g., Cre-ERT) or Dox-inducible (Tet-On/Off) system, as well as more standard vectors into organoids. More recently, an optimized Apobec-Cas9 base-editor (with the addition of a flag tag and nuclear localization signals at the N-terminus of the Apobec sequence) was shown to have efficient base editing activity in organoids. Through the introduction of APC^{Q1405X} and $PIK3ca^{E545K}$ mutations, targeted editing allowed organoids to continue rapidly expanding in the presence of a MEK inhibitor and in the absence of R-spondin, the continued growth being indicative of efficient editing activities (Zafra et al. 2018). In general, when selecting organoid clones containing the desired manipulation(s), flow-cytometry for single-cell reporter sorting, antibiotic selection, growth factor withdrawal or single organoid picking have all been reported to specifically isolate the desired clones. In summary, organoid technology is entirely amenable to combination with various genetic engineering technologies, just as with 2D cell lines, making them a versatile ex vivo system.

Following the first successful demonstration of CRISPR/Cas9 gene editing in an organoid system, where the $CFTR^{F508del}$ mutation was corrected in CF patient-derived intestinal organoids (Schwank et al. 2013), many studies have applied organoid technology hand-in-hand with CRISPR/Cas9 technology to various areas of biomedical research. More recently, organoids were used to study the origin of specific mutational signatures in human colonic cancers. Two genes predisposing to cancer, *MLH1* (involved in mismatch repair) and *NTHL1* (involved in base excision repair), were individually knocked-out in human colon organoids, and clonal and subclonal organoids were sequenced by whole genome sequencing. As a result, it was found that the mutational landscapes of *MLH1*-deficient organoids resemble the mutation profiles observed in mismatch repair-deficient colorectal cancers, and *NTHL1*-deficient organoids displayed features of mutation signature 30 (Drost et al. 2017).

Brain tumors remain one of the most devastating cancers in both adolescents and adults. Historically, there have been no suitable models for investigating the onset of cortical tumorigenesis. The group of Jürgen Knoblich developed so-called neoplastic cerebral organoids, which have allowed for the screening of various clinically prevalent mutations to identify combinations that would lead to glioblastoma-like features (Bian et al. 2018). Therefore, by combining the cerebral organoid platform with genome engineering, 3D fluorescent imaging, gene expression analysis, and drug/chemical treatment techniques, the system has served as a powerful and versatile platform to investigate the molecular mechanisms underlying brain tumor progression. By combining CRISPR/Cas9 and organoid technologies, one could make use of the publicly available sequencing data for different human cancers to carry out combinatorial mutation studies or customized, positive-selection genetic screens to study those cancers.

5.3.4 ORGANOIDS FOR STUDIES IN ONCOLOGY

Cancer remains a major global health problem despite decades of biomedical research. The translation of scientific knowledge into effective clinical treatment relies, at least in part, on testing in a model which accurately recapitulates tissue physiology. Indeed, choosing the right model systems for studying cancer biology has been an ongoing challenge, as many drug treatments are effective in models but not in patients (Caponigro and Sellers 2011; Drost and Clevers 2018).

Organoids can be generated from normal tissues with growth factors that act as a surrogate AdSC niche; organoids can also be generated from tumor tissues in the same fashion. Remarkably, tumor-derived organoids (tumoroids) recapitulate genomic landscape of tumor tissues over time. So far, patient-derived tumoroids have been derived from various tissues, which have been reviewed elsewhere (Drost and Clevers 2018). Since biopsies of cancer tissues often consist of a mixture of normal and cancerous cells, and as normal organoids usually outgrow tumoroids in culture, the isolation of a pure culture of cancer organoids is crucial for downstream analyses and studies. As an example of this, most colorectal cancers have abnormal Wnt activation due to mutations in the negative regulators (i.e., *APC, RNF43/ZNRF3, AXIN1,* etc.), therefore medium lacking R-spondin is often used to enrich for colorectal cancer organoids and under these suboptimal conditions, cancer clones can outgrow normal cells (van de Wetering et al. 2015). The same approach can be used to select for clones carrying individual mutations or a combination of mutations in driver genes in signaling pathways such as the EGF, p53 or TGFβ/BMP pathways.

Patient-derived cancer organoids recapitulate the histological morphology, intratumor heterogeneity and drug sensitivity profile of the original tumor. Of particular, one study tested anti-cancer drugs in a library of cancer organoids derived from patients, most of whom had metastatic colorectal cancer. By comparing the drug responses in cancer organoids to those in the patients, the authors found that the organoids were very highly predictive for therapeutic efficacy (Vlachogiannis et al. 2018).

5.4   CONCLUSION AND OUTLOOK

In this chapter, we have summarized the various methods used to generate organoids either from PSCs or AdSCs. The development of most organoid systems has only been possible because of studies in model organisms which have contributed to our understanding of developmental biology, thus highlighting the importance of basic research for the advancement of biomedical applications. We have also highlighted several applications of organoid technology, which more accurately recapitulates in vivo physiology when compared with 2D cell lines.

It is clear that the organoid system successfully opened a new era of biology with great potential to extend our knowledge of stem cell and developmental biology in a dish and further possibilities of translational researches. However, there are still many difficulties remains to overcome. For instance, the organoid generation protocol is very different between protocols, so the characteristics of the established organoid are not identical to each other. Also, maturation process of PSC-derived organoid is still complicated and difficult to standardize. Moreover, the organoid system still has difficulties in studying diseases which require other types of cell or tissues such as immune cells. Besides, CRISPR/Cas9 mediated genome-wide screening on the organoid system is still technically challenging. To overcome such obstacles, researchers investigating various possibilities such as developing co-culture conditions with other types of cells, improving delivering methods for CRISPR/Cas9 mediated genome editing, enhancing maturation protocols, etc. Further work and more complex functional studies are still needed to better understand the processes of vascularization and interaction with immune cells, as well as further translational research in areas such as transplantation therapeutics.

REFERENCES

Azuma, Hisaya, Nicole Paulk, Aarati Ranade, Craig Dorrell, Muhsen Al-Dhalimy, Ewa Ellis, Stephen Strom, Mark A. Kay, Milton Finegold, and Markus Grompe. 2007. Robust expansion of human hepatocytes in Fah–/–/Rag2–/–/Il2rg–/– mice. *Nature Biotechnology* 25(8): 903–10.

Bagley, Joshua A., Daniel Reumann, Shan Bian, Julie Lévi-Strauss, and Juergen A. Knoblich. 2017. Fused cerebral organoids model interactions between brain regions. *Nature Methods* 14(7): 743–51.

Barker, Nick, Sina Bartfeld, and Hans Clevers. 2010. Tissue-resident adult stem cell populations of rapidly self-renewing organs. *Cell Stem Cell* 7(6): 656–70.

Barker, Nick, Johan H. van Es, Jeroen Kuipers, Pekka Kujala, Maaike Van Den Born, Miranda Cozijnsen, Andrea Haegebarth, et al. 2007. Identification of stem cells in small intestine and colon by marker gene Lgr5. *Nature* 449(7165): 1003–7.

Barker, Nick, Meritxell Huch, Pekka Kujala, Marc van de Wetering, Hugo J. Snippert, Johan H. van Es, Toshiro Sato, et al. 2010. Lgr5+ve stem cells drive self-renewal in the stomach and build long-lived gastric units in vitro. *Cell Stem Cell* 6(1): 25–36.

Bartfeld, Sina. 2016. Modeling infectious diseases and host-microbe interactions in gastrointestinal organoids. *Developmental Biology* 420(2): 262–70.

Bartfeld, Sina, Tülay Bayram, Marc van de Wetering, Meritxell Huch, Harry Begthel, Pekka Kujala, Robert Vries, Peter J. Peters, and Hans Clevers. 2015. In vitro expansion of human gastric epithelial stem cells and their responses to bacterial infection. *Gastroenterology* 148(1): 126–36.e6.

Bian, Shan, Marko Repic, Zhenming Guo, Anoop Kavirayani, Thomas Burkard, Joshua A. Bagley, Christian Krauditsch, and Jürgen A. Knoblich. 2018. Genetically engineered cerebral organoids model brain tumor formation. *Nature Methods* 15(8): 631–39.

Boj, Sylvia F., Chang-Il Hwang, Lindsey A. Baker, Iok In Christine Chio, Dannielle D. Engle, Vincenzo Corbo, Myrthe Jager, et al. 2015. Organoid models of human and mouse ductal pancreatic cancer. *Cell* 160(1–2): 324–38.

Bond, Jacquelyn, Emma Roberts, Kelly Springell, Sophia Lizarraga, Sheila Scott, Julie Higgins, Daniel J. Hampshire, et al. 2005. A centrosomal mechanism involving CDK5RAP2 and CENPJ controls brain size. *Nature Genetics* 37(4): 353–55.

Boretto, Matteo, Benoit Cox, Manuel Noben, Nikolai Hendriks, Amelie Fassbender, Heleen Roose, Frédéric Amant, et al. 2017. Development of organoids from mouse and human endometrium showing endometrial epithelium physiology and long-term expandability. *Development* 144(10): 1775–86.

Broutier, Laura, Amanda Andersson-Rolf, Christopher J. Hindley, Sylvia F. Boj, Hans Clevers, Bon-Kyoung Koo, and Meritxell Huch. 2016. Culture and establishment of self-renewing human and mouse adult liver and pancreas 3D organoids and their genetic manipulation. *Nature Protocols* 11(9): 1724–43.

Camp, J. Gray, Farhath Badsha, Marta Florio, Sabina Kanton, Tobias Gerber, Michaela Wilsch-Bräuninger, Eric Lewitus, et al. 2015. Human cerebral organoids recapitulate gene expression programs of fetal neocortex development. *Proceedings of the National Academy of Sciences* 112(51): 201520760.

Camp, J. Gray, Keisuke Sekine, Tobias Gerber, Henry Loeffler-Wirth, Hans Binder, Malgorzata Gac, Sabina Kanton, et al. 2017. Multilineage communication regulates human liver bud development from pluripotency. *Nature* 546(7659): 533–38.

Caponigro, Giordano, and William R. Sellers. 2011. Advances in the preclinical testing of cancer therapeutic hypotheses. *Nature Reviews Drug Discovery* 10(3): 179–87.

Chen, Kevin G., Barbara S. Mallon, Ronald D.G. McKay, and Pamela G. Robey. 2014. Human pluripotent stem cell culture: Considerations for maintenance, expansion, and therapeutics. *Cell Stem Cell* 14(1): 13–26.

Chin, Melanie, Karen Earlam, and Shawn D. Aaron. 2015. Survival in cystic fibrosis: Trends, clinical factors, and prediction models. *Pediatric Allergy, Immunology, and Pulmonology* 28(4): 244–49.

Clevers, Hans. 2013. The intestinal crypt, a prototype stem cell compartment. *Cell* 154(2): 274–84.

Clevers, Hans. 2016. Modeling development and disease with organoids. *Cell* 165(7): 1586–97.

Dang, Jason, Shashi Kant Tiwari, Gianluigi Lichinchi, Yue Qin, Veena S. Patil, Alexey M. Eroshkin, and Tariq M. Rana. 2016. Zika virus depletes neural progenitors in human cerebral organoids through activation of the innate immune receptor TLR3. *Cell Stem Cell* 19(2): 258–65.

Dekkers, Johanna F., Caroline L. Wiegerinck, Hugo R. de Jonge, Inez Bronsveld, Hettie M. Janssens, Karin M. de Winter-de Groot, Arianne M. Brandsma, et al. 2013. A functional CFTR assay using primary cystic fibrosis intestinal organoids. *Nature Medicine* 19(7): 939–45.

Desai, Tushar J., Douglas G. Brownfield, and Mark A. Krasnow. 2014. Alveolar progenitor and stem cells in lung development, renewal and cancer. *Nature* 507(7491): 190–94.

DeWard, Aaron D., Julie Cramer, and Eric Lagasse. 2014. Cellular heterogeneity in the mouse esophagus implicates the presence of a nonquiescent epithelial stem cell population. *Cell Reports* 9(2): 701–11.

Dorrell, Craig, Branden Tarlow, Yuhan Wang, Pamela S. Canaday, Annelise Haft, Jonathan Schug, Philip R. Streeter, et al. 2014. The organoid-initiating cells in mouse pancreas and liver are phenotypically and functionally similar. *Stem Cell Research* 13(2): 275–83.

Drost, Jarno, Ruben van Boxtel, Francis Blokzijl, Tomohiro Mizutani, Nobuo Sasaki, Valentina Sasselli, Joep de Ligt, et al. 2017. Use of CRISPR-modified human stem cell organoids to study the origin of mutational signatures in cancer. *Science* 358(6360): 234–38.

Drost, Jarno, and Hans Clevers. 2018. Organoids in cancer research. *Nature Reviews Cancer* 18(7): 407–18.

Drummond, Coyne G., Alexa M. Bolock, Congrong Ma, Cliff J. Luke, Misty Good, and Carolyn B. Coyne. 2017. Enteroviruses infect human enteroids and induce antiviral signaling in a cell lineage-specific manner. *Proceedings of the National Academy of Sciences* 114(7): 1672–77.

Dye, Briana R., David R. Hill, Michael A.H. Ferguson, Yu-Hwai Tsai, Melinda S. Nagy, Rachel Dyal, James M. Wells, et al. 2015. In vitro generation of human pluripotent stem cell derived lung organoids. *ELife* 4(March): 1–25. doi:10.7554/eLife.05098.

Eiraku, Mototsugu, and Yoshiki Sasai. 2012. Self-formation of layered neural structures in three-dimensional culture of ES cells. *Current Opinion in Neurobiology* 22(5): 768–77.

Eiraku, Mototsugu, Nozomu Takata, Hiroki Ishibashi, Masako Kawada, Eriko Sakakura, Satoru Okuda, Kiyotoshi Sekiguchi, Taiji Adachi, and Yoshiki Sasai. 2011. Self-organizing optic-cup morphogenesis in three-dimensional culture. *Nature* 472(7341): 51–6.

Eiraku, Mototsugu, Kiichi Watanabe, Mami Matsuo-Takasaki, Masako Kawada, Shigenobu Yonemura, Michiru Matsumura, Takafumi Wataya, Ayaka Nishiyama, Keiko Muguruma, and Yoshiki Sasai. 2008. Self-organized formation of polarized cortical tissues from ESCs and its active manipulation by extrinsic signals. *Cell Stem Cell* 3(5): 519–32.

Ettayebi, Khalil, Sue E. Crawford, Kosuke Murakami, James R. Broughman, Umesh Karandikar, Victoria R. Tenge, Frederick H. Neill, et al. 2016. Replication of human noroviruses in stem cell-derived human enteroids. *Science* 353(6306): 1387–93.

Fairbanks, Kyrsten D., and Anthony S. Tavill. 2008. Liver disease in alpha 1-antitrypsin deficiency: A review. *The American Journal of Gastroenterology* 103(8): 2136–41.

Fujii, Masayuki, Mami Matano, Kosaku Nanki, and Toshiro Sato. 2015. Efficient genetic engineering of human intestinal organoids using electroporation. *Nature Protocols* 10(10): 1474–85.

Garcez, Patricia P., Erick Correia Loiola, Rodrigo Madeiro da Costa, Luiza M. Higa, Pablo Trindade, Rodrigo Delvecchio, Juliana Minardi Nascimento, Rodrigo Brindeiro, Amilcar Tanuri, and Stevens K. Rehen. 2016. Zika virus impairs growth in human neurospheres and brain organoids. *Science* 352(6287): 816–18.

Gargett, Caroline E., Rachel W.S. Chan, and Kjiana E. Schwab. 2008. Hormone and growth factor signaling in endometrial renewal: Role of stem/progenitor cells. *Molecular and Cellular Endocrinology* 288(1–2): 22–9.

Grün, Dominic, Anna Lyubimova, Lennart Kester, Kay Wiebrands, Onur Basak, Nobuo Sasaki, Hans Clevers, and Alexander van Oudenaarden. 2015. Single-cell messenger RNA sequencing reveals rare intestinal cell types. *Nature* 525(7568): 251–55.

Hagios, Carmen, André Lochter, and Mina J. Bissell. 1998. Tissue architecture: The ultimate regulator of epithelial function? *Philosophical Transactions of the Royal Society of London. Series B: Biological Sciences* 353(1370): 857–70.

Hatsumi, Toshinobu, and Yutaka Yamamuro. 2006. Downregulation of estrogen receptor gene expression by exogenous 17β-estradiol in the mammary glands of lactating mice. *Experimental Biology and Medicine* 231(3): 311–16.

Heo, Inha, Devanjali Dutta, Deborah A. Schaefer, Nino Iakobachvili, Benedetta Artegiani, Norman Sachs, Kim E. Boonekamp, et al. 2018. Modelling cryptosporidium infection in human small intestinal and lung organoids. *Nature Microbiology* 3(7): 814–23.

Housset, Chantal, Yves Chrétien, Dominique Debray, and Nicolas Chignard. 2016. Functions of the gallbladder. In Ronald Terjung (Ed.), *Comprehensive Physiology*, 6: 1549–77 Hoboken, NJ: John Wiley & Sons, Inc.

Huang, Sarah X.L., Mohammad Naimul Islam, John O'Neill, Zheng Hu, Yong-Guang Yang, Ya-Wen Chen, Melanie Mumau, et al. 2014. Efficient generation of lung and airway epithelial cells from human pluripotent stem cells. *Nature Biotechnology* 32(1): 84–91.

Huch, Meritxell, Sylvia F. Boj, and Hans Clevers. 2013. Lgr5 +liver stem cells, hepatic organoids and regenerative medicine. *Regenerative Medicine* 8(4): 385–87.

Huch, Meritxell, Paola Bonfanti, Sylvia F. Boj, Toshiro Sato, Cindy J.M. Loomans, Marc van de Wetering, Mozhdeh Sojoodi, et al. 2013. Unlimited in vitro expansion of adult bi-potent pancreas progenitors through the Lgr5/R-spondin axis. *The EMBO Journal* 32(20): 2708–21.

Huch, Meritxell, Craig Dorrell, Sylvia F. Boj, Johan H. van Es, Vivian S.W. Li, Marc van de Wetering, Toshiro Sato, et al. 2013. In vitro expansion of single Lgr5+ liver stem cells induced by Wnt-driven regeneration. *Nature* 494(7436): 247–50.

Huch, Meritxell, Helmuth Gehart, Ruben van Boxtel, Karien Hamer, Francis Blokzijl, Monique M.A. Verstegen, Ewa Ellis, et al. 2015. Long-term culture of genome-stable bipotent stem cells from adult human liver. *Cell* 160(1–2): 299–312.

Huch, Meritxell, and Bon-Kyoung Koo. 2015. Modeling mouse and human development using organoid cultures. *Development*. doi:10.1242/dev.118570.

Hughes, Chris S., Lynne M. Postovit, and Gilles A. Lajoie. 2010. Matrigel: A complex protein mixture required for optimal growth of cell culture. *Proteomics* 10(9): 1886–90.

Jamieson, Paul R., Johanna F. Dekkers, Anne C. Rios, Nai Yang Fu, Geoffrey J. Lindeman, and Jane E. Visvader. 2017. Derivation of a robust mouse mammary organoid system for studying tissue dynamics. *Development* 144(6): 1065–71.

Jiang, Fuguo, and Jennifer A. Doudna. 2017. CRISPR–Cas9 structures and mechanisms. *Annual Review of Biophysics* 46(1): 505–29.

Jinek, Martin, Krzysztof Chylinski, Ines Fonfara, Michael Hauer, Jennifer A. Doudna, and Emmanuelle Charpentier. 2012. A programmable dual-RNA-guided DNA endonuclease in adaptive bacterial immunity. *Science* 337(6096): 816–21.

Jung, Peter, Toshiro Sato, Anna Merlos-Suárez, Francisco M. Barriga, Mar Iglesias, David Rossell, Herbert Auer, et al. 2011. Isolation and in vitro expansion of human colonic stem cells. *Nature Medicine* 17(10): 1225–27.

Kadoshima, Taisuke, Hideya Sakaguchi, Tokushige Nakano, Mika Soen, Satoshi Ando, Mototsugu Eiraku, and Yoshiki Sasai. 2013. Self-organization of axial polarity, inside-out layer pattern, and species-specific progenitor dynamics in human ES cell-derived neocortex. *Proceedings of the National Academy of Sciences* 110(50): 20284–89.

Kamb, Alexander. 2005. What's wrong with our cancer models? *Nature Reviews Drug Discovery* 4(2): 161–65.

Karthaus, Wouter R., Phillip J. Iaquinta, Jarno Drost, Ana Gracanin, Ruben van Boxtel, John Wongvipat, Catherine M. Dowling, et al. 2014. Identification of multipotent luminal progenitor cells in human prostate organoid cultures. *Cell* 159(1): 163–75.

Kessler, Mirjana, Karen Hoffmann, Volker Brinkmann, Oliver Thieck, Susan Jackisch, Benjamin Toelle, Hilmar Berger, et al. 2015. The Notch and Wnt pathways regulate stemness and differentiation in human fallopian tube organoids. *Nature Communications* 6(1): 8989.

Kim, Tae-Hee, and Ramesh A. Shivdasani. 2016. Stomach development, stem cells and disease. *Development* 143(4): 554–65.

Kleinman, Hynda K., and George R. Martin. 2005. Matrigel: Basement membrane matrix with biological activity. *Seminars in Cancer Biology* 15(5): 378–86.

Koehler, Karl R., Andrew M. Mikosz, Andrei I. Molosh, Dharmeshkumar Patel, and Eri Hashino. 2013. Generation of inner ear sensory epithelia from pluripotent stem cells in 3D culture. *Nature* 500(7461): 217–21.

Koo, Bon-Kyoung, Daniel E. Stange, Toshiro Sato, Wouter Karthaus, Henner F. Farin, Meritxell Huch, Johan H. van Es, and Hans Clevers. 2012. Controlled gene expression in primary Lgr5 organoid cultures. *Nature Methods* 9(1): 81–3.

Kurmann, Anita A., Maria Serra, Finn Hawkins, Scott A. Rankin, Munemasa Mori, Inna Astapova, Soumya Ullas, et al. 2015. Regeneration of thyroid function by transplantation of differentiated pluripotent stem cells. *Cell Stem Cell* 17(5): 527–42.

Lancaster, Madeline A., Nina S. Corsini, Simone Wolfinger, E. Hilary Gustafson, Alex W. Phillips, Thomas R. Burkard, Tomoki Otani, Frederick J. Livesey, and Juergen A. Knoblich. 2017. Guided self-organization and cortical plate formation in human brain organoids. *Nature Biotechnology* 35(7): 659–66.

Lancaster, Madeline A., and Juergen A. Knoblich. 2014. Organogenesis in a dish: Modeling development and disease using organoid technologies. *Science* 345(6194): 1247125.

Lancaster, Madeline A., Magdalena Renner, Carol-Anne Martin, Daniel Wenzel, Louise S. Bicknell, Matthew E. Hurles, Tessa Homfray, Josef M. Penninger, Andrew P. Jackson, and Juergen A. Knoblich. 2013. Cerebral organoids model human brain development and microcephaly. *Nature* 501(7467): 373–79.

Lawless, Matthew W., Arun K. Mankan, Steven G. Gray, and Suzanne Norris. 2008. Endoplasmic reticulum stress—A double edged sword for Z alpha-1 antitrypsin deficiency hepatoxicity. *The International Journal of Biochemistry & Cell Biology* 40(8): 1403–14.

Lee, Joo-Hyeon, Dong Ha Bhang, Alexander Beede, Tian Lian Huang, Barry R. Stripp, Kenneth D. Bloch, Amy J. Wagers, Yu-Hua Tseng, Sandra Ryeom, and Carla F. Kim. 2014. Lung stem cell differentiation in mice directed by endothelial cells via a BMP4-NFATc1-thrombospondin-1 axis. *Cell* 156(3): 440–55.

Leslie, Jhansi L., Sha Huang, Judith S. Opp, Melinda S. Nagy, Masayuki Kobayashi, Vincent B. Young, and Jason R. Spence. 2015. Persistence and toxin production by *Clostridium difficile* within human intestinal organoids result in disruption of epithelial paracellular barrier function. *Infection and Immunity* 83(1): 138–45.

Leushacke, Marc, Si Hui Tan, Angeline Wong, Yada Swathi, Amin Hajamohideen, Liang Thing Tan, Jasmine Goh, et al. 2017. Lgr5-expressing chief cells drive epithelial regeneration and cancer in the oxyntic stomach. *Nature Cell Biology* 19(7): 774–86.

Linnemann, Jelena R., Haruko Miura, Lisa K. Meixner, Martin Irmler, Uwe J. Kloos, Benjamin Hirschi, Harald S. Bartsch, et al. 2015. Quantification of regenerative potential in primary human mammary epithelial cells. *Development* 142(18): 3239–51.

Longmire, Tyler A., Laertis Ikonomou, Finn Hawkins, Constantina Christodoulou, Yuxia Cao, Jyh Chang Jean, Letty W. Kwok, et al. 2012. Efficient derivation of purified lung and thyroid progenitors from embryonic stem cells. *Cell Stem Cell* 10(4): 398–411.

Maimets, Martti, Cecilia Rocchi, Reinier Bron, Sarah Pringle, Jeroen Kuipers, Ben N.G. Giepmans, Robert G.J. Vries, et al. 2016. Long-term in vitro expansion of salivary gland stem cells driven by Wnt signals. *Stem Cell Reports* 6(1): 150–62.

McCracken, Kyle W., Eitaro Aihara, Baptiste Martin, Calyn M. Crawford, Taylor Broda, Julie Treguier, Xinghao Zhang, John M. Shannon, Marshall H. Montrose, and James M. Wells. 2017. Wnt/β-catenin promotes gastric fundus specification in mice and humans. *Nature* 541(7636): 182–87.

McCracken, Kyle W., Emily M. Catá, Calyn M. Crawford, Katie L. Sinagoga, Michael Schumacher, Briana E. Rockich, Yu-Hwai Tsai, et al. 2014. Modelling human development and disease in pluripotent stem-cell-derived gastric organoids. *Nature* 516(7531): 400–4.

Miyajima, Atsushi, Minoru Tanaka, and Tohru Itoh. 2014. Stem/progenitor cells in liver development, homeostasis, regeneration, and reprogramming. *Cell Stem Cell* 14(5): 561–74.

Morizane, Ryuji, and Joseph V. Bonventre. 2017. Kidney organoids: A translational journey. *Trends in Molecular Medicine* 23(3): 246–63.

Morizane, Ryuji, Albert Q. Lam, Benjamin S. Freedman, Seiji Kishi, M. Todd Valerius, and Joseph V. Bonventre. 2015. Nephron organoids derived from human pluripotent stem cells model kidney development and injury. *Nature Biotechnology* 33(11): 1193–200.

Muguruma, Keiko, Ayaka Nishiyama, Hideshi Kawakami, Kouichi Hashimoto, and Yoshiki Sasai. 2015. Self-organization of polarized cerebellar tissue in 3D culture of human pluripotent stem cells. *Cell Reports* 10(4): 537–50.

Muguruma, Keiko, Ayaka Nishiyama, Yuichi Ono, Hiroyuki Miyawaki, Eri Mizuhara, Seiji Hori, Akira Kakizuka, et al. 2010. Ontogeny-recapitulating generation and tissue integration of ES cell–derived Purkinje cells. *Nature Neuroscience* 13(10): 1171–80.

Múnera, Jorge O., Nambirajan Sundaram, Scott A. Rankin, David Hill, Carey Watson, Maxime Mahe, Jefferson E. Vallance, et al. 2017. Differentiation of human pluripotent stem cells into colonic organoids via transient activation of BMP signaling. *Cell Stem Cell* 21(1): 51–64.e6.

Nakano, Tokushige, Satoshi Ando, Nozomu Takata, Masako Kawada, Keiko Muguruma, Kiyotoshi Sekiguchi, Koichi Saito, Shigenobu Yonemura, Mototsugu Eiraku, and Yoshiki Sasai. 2012. Self-formation of optic cups and storable stratified neural retina from human ESCs. *Cell Stem Cell* 10(6): 771–85.

Nanduri, Lalitha S.Y., Mirjam Baanstra, Hette Faber, Cecilia Rocchi, Erik Zwart, Gerald de Haan, Ronald van Os, and Robert P. Coppes. 2014. Purification and ex vivo expansion of fully functional salivary gland stem cells. *Stem Cell Reports* 3(6): 957–64.

Ozone, Chikafumi, Hidetaka Suga, Mototsugu Eiraku, Taisuke Kadoshima, Shigenobu Yonemura, Nozomu Takata, Yutaka Oiso, Takashi Tsuji, and Yoshiki Sasai. 2016. Functional anterior pituitary generated in self-organizing culture of human embryonic stem cells. *Nature Communications* 7(1): 10351.

Piccolo, Stefano. 2014. Yoshiki Sasai: Stem cell Sensei. *Development* 141(19): 3613–14.

Plaks, Vicki, Audrey Brenot, Devon A. Lawson, Jelena R. Linnemann, Eline C. Van Kappel, Karren C. Wong, Frederic de Sauvage, Ophir D. Klein, and Zena Werb. 2013. Lgr5-expressing cells are sufficient and necessary for postnatal mammary gland organogenesis. *Cell Reports* 3(1): 70–8.

Qu, Ying, Bingchen Han, Bowen Gao, Shikha Bose, Yiping Gong, Kolja Wawrowsky, Armando E. Giuliano, Dhruv Sareen, and Xiaojiang Cui. 2017. Differentiation of human induced pluripotent stem cells to mammary-like organoids. *Stem Cell Reports* 8(2): 205–15.

Quadrato, Giorgia, Tuan Nguyen, Evan Z. Macosko, John L. Sherwood, Sung Min Yang, Daniel R. Berger, Natalie Maria, et al. 2017. Cell diversity and network dynamics in photosensitive human brain organoids. *Nature* 545(7652): 48–53.

Ramani, Sasirekha, Robert L. Atmar, and Mary K. Estes. 2014. Epidemiology of human noroviruses and updates on vaccine development. *Current Opinion in Gastroenterology* 30(1): 25–33.

Ratjen, Felix, and Gerd Döring. 2003. Cystic fibrosis. *The Lancet* 361(9358): 681–89.

Ren, Wenwen, Brian C. Lewandowski, Jaime Watson, Eitaro Aihara, Ken Iwatsuki, Alexander A. Bachmanov, Robert F. Margolskee, and Peihua Jiang. 2014. Single Lgr5- or Lgr6-expressing taste stem/progenitor cells generate taste bud cells ex vivo. *Proceedings of the National Academy of Sciences* 111(46): 16401–6.

De Robertis, Edward M. 2014. Yoshiki Sasai 1962–2014. *Cell* 158(6): 1233–35.

Rock, Jason R., Mark W. Onaitis, Emma L. Rawlins, Yun Lu, Cheryl P. Clark, Yan Xue, Scott H. Randell, and Brigid L.M. Hogan. 2009. Basal cells as stem cells of the mouse trachea and human airway epithelium. *Proceedings of the National Academy of Sciences* 106(31): 12771–5.

Sampaziotis, Fotios, Alexander W. Justin, Olivia C. Tysoe, Stephen Sawiak, Edmund M. Godfrey, Sara S. Upponi, Richard L. Gieseck, et al. 2017. Reconstruction of the mouse extrahepatic biliary tree using primary human extrahepatic cholangiocyte organoids. *Nature Medicine* 23(8): 954–63.

Sato, Toshiro, Daniel E. Stange, Marc Ferrante, Robert G.J. Vries, Johan H. van Es, Stieneke van den Brink, Winan J. van Houdt, et al. 2011. Long-term expansion of epithelial organoids from human colon, adenoma, adenocarcinoma, and Barrett's epithelium. *Gastroenterology* 141(5): 1762–72.

Sato, Toshiro, Robert G. Vries, Hugo J. Snippert, Marc van de Wetering, Nick Barker, Daniel E. Stange, Johan H. van Es, et al. 2009. Single Lgr5 stem cells build crypt-villus structures in vitro without a mesenchymal niche. *Nature* 459(7244): 262–5.

Saxena, Kapil, Sarah E. Blutt, Khalil Ettayebi, Xi-Lei Zeng, James R. Broughman, Sue E. Crawford, Umesh C. Karandikar, et al. 2016. Human intestinal enteroids: A new model to study human rotavirus infection, host restriction, and pathophysiology. *Journal of Virology* 90(1): 43–56.

Schwank, Gerald, Bon-Kyoung Koo, Valentina Sasselli, Johanna F. Dekkers, Inha Heo, Turan Demircan, Nobuo Sasaki, et al. 2013. Functional repair of CFTR by CRISPR/Cas9 in intestinal stem cell organoids of cystic fibrosis patients. *Cell Stem Cell* 13(6): 653–8.

Shamir, Eliah R., and Andrew J. Ewald. 2014. Three-dimensional organotypic culture: Experimental models of mammalian biology and disease. *Nature Reviews Molecular Cell Biology* 15(10): 647–64.

Spence, Jason R., Christopher N. Mayhew, Scott A. Rankin, Matthew F. Kuhar, Jefferson E. Vallance, Kathryn Tolle, Elizabeth E. Hoskins, et al. 2011. Directed differentiation of human pluripotent stem cells into intestinal tissue in vitro. *Nature* 470(7332): 105–9.

Stange, Daniel E., Bon-Kyoung Koo, Meritxell Huch, Greg Sibbel, Onur Basak, Anna Lyubimova, Pekka Kujala, et al. 2013. Differentiated Troy+ chief cells act as reserve stem cells to generate all lineages of the stomach epithelium. *Cell* 155(2): 357–68.

Streuli, Charles H., Nina Bailey, and Mina J. Bissell. 1991. Control of mammary epithelial differentiation: Basement membrane induces tissue-specific gene expression in the absence of cell-cell interaction and morphological polarity. *The Journal of Cell Biology* 115(5): 1383–95.

Suga, Hidetaka, Taisuke Kadoshima, Maki Minaguchi, Masatoshi Ohgushi, Mika Soen, Tokushige Nakano, Nozomu Takata, et al. 2011. Self-formation of functional adenohypophysis in three-dimensional culture. *Nature* 480(7375): 57–62.

Takasato, Minoru, Pei X. Er, Han S. Chiu, Barbara Maier, Gregory J. Baillie, Charles Ferguson, Robert G. Parton, et al. 2015. Kidney organoids from human IPS cells contain multiple lineages and model human nephrogenesis. *Nature* 526(7574): 564–8.

Takebe, Takanori, Keisuke Sekine, Masahiro Enomura, Hiroyuki Koike, Masaki Kimura, Takunori Ogaeri, Ran-Ran Zhang, et al. 2013. Vascularized and functional human liver from an IPSC-derived organ bud transplant. *Nature* 499(7459): 481–4.

Takebe, Takanori, Ran-Ran Zhang, Hiroyuki Koike, Masaki Kimura, Emi Yoshizawa, Masahiro Enomura, Naoto Koike, Keisuke Sekine, and Hideki Taniguchi. 2014. Generation of a vascularized and functional human liver from an IPSC-derived organ bud transplant. *Nature Protocols* 9(2): 396–409.

Trisno, Stephen L., Katherine E.D. Philo, Kyle W. McCracken, Emily M. Catá, Sonya Ruiz-Torres, Scott A. Rankin, Lu Han, et al. 2018. Esophageal organoids from human pluripotent stem cells delineate Sox2 functions during esophageal specification. *Cell Stem Cell* 23(4): 501–15.e7.

Turco, Margherita Y., Lucy Gardner, Jasmine Hughes, Tereza Cindrova-Davies, Maria J. Gomez, Lydia Farrell, Michael Hollinshead, et al. 2017. Long-term, hormone-responsive organoid cultures of human endometrium in a chemically defined medium. *Nature Cell Biology* 19(5): 568–77.

Vlachogiannis, Georgios, Somaieh Hedayat, Alexandra Vatsiou, Yann Jamin, Javier Fernández-Mateos, Khurum Khan, Andrea Lampis, et al. 2018. Patient-derived organoids model treatment response of metastatic gastrointestinal cancers. *Science* 359(6378): 920–6.

Wells, James M., and Jason R. Spence. 2014. How to make an intestine. *Development* 141(4): 752–60.

van de Wetering, Marc, Hayley E. Francies, Joshua M. Francis, Gergana Bounova, Francesco
    Iorio, Apollo Pronk, Winan van Houdt, et al. 2015. Prospective derivation of a living
    organoid biobank of colorectal cancer patients. *Cell* 161(4): 933–45.

Wong, Amy P., Christine E. Bear, Stephanie Chin, Peter Pasceri, Tadeo O. Thompson, Ling-
    Jun Huan, Felix Ratjen, James Ellis, and Janet Rossant. 2012. Directed differentia-
    tion of human pluripotent stem cells into mature airway epithelia expressing functional
    CFTR protein. *Nature Biotechnology* 30(9): 876–82.

Xu, Xiaobo, Joke D'Hoker, Geert Stangé, Stefan Bonné, Nico De Leu, Xiangwei Xiao, Mark
    Van De Casteele, et al. 2008. β cells can be generated from endogenous progenitors in
    injured adult mouse pancreas. *Cell* 132(2): 197–207.

Yin, Yuebang, Marcel Bijvelds, Wen Dang, Lei Xu, Annemiek A. van der Eijk, Karen
    Knipping, Nesrin Tuysuz, et al. 2015. Modeling rotavirus infection and antiviral ther-
    apy using primary intestinal organoids. *Antiviral Research* 123(November): 120–31.

Zafra, Maria Paz, Emma M. Schatoff, Alyna Katti, Miguel Foronda, Marco Breinig, Anabel
    Y. Schweitzer, Amber Simon, et al. 2018. Optimized base editors enable efficient edit-
    ing in cells, organoids and mice. *Nature Biotechnology* 36(9): 888–93.

Zhang, Yongchun, Ying Yang, Ming Jiang, Sarah Xuelian Huang, Wanwei Zhang, Denise Al
    Alam, Soula Danopoulos, et al. 2018. 3D modeling of esophageal development using
    human PSC-derived basal progenitors reveals a critical role for notch signaling. *Cell
    Stem Cell* 23(4): 516–29.e5.

6 Design, Fabrication, and Microflow Control Techniques for Organ-on-a-Chip Devices

Zachary Estlack
University of Utah

Ali Khodayari Bavil
Texas Tech University

Jungkyu Kim
University of Utah

CONTENTS

6.1   INTRODUCTION

Recently, organ-on-a-chip (organ chip) devices using microfabrication techniques have been proposed to replace animal models and to provide more systematic drug tests. These chips use various types of cells seeded in microfluidic chips to recapitulate physiologically relevant microenvironments for pharmacological and/or cellular response studies (Ronaldson-Bouchard and Vunjak-Novakovic 2018). Organ chips utilizing the excellent properties of microfluidics allow for precise control of parameters in three-dimensional models to closely match the microenvironment seen in the human body as well as an increase in the controllability of an experiment when compared to animal models.

The design of organ chips focuses on the number of channels, cell types, support structure, feeding, and flow control (Sosa-Hernandez et al. 2018). The support for cells in the microfluidic environment is dependent upon the experimental requirements and the chip design. Typically, glass, polycarbonate (PC), polydimethylsiloxanes (PDMS), or hydrogel is used as support in organ chips. The combination of design and fabrication leads to a variety of chip designs, from single channel chips with fibronectin treated glass to support cells and promote attachment, to active multichannel chips with pneumatically actuated PDMS membranes that apply strain to co-cultured cells at in vivo relevant intervals (Ingber 2018).

On-chip cell culture is more complex than off-chip culture due to the limited volume and normally restricted access to the cell layer. Conventionally, static or perfusion culture models are considered when designing both ex vivo experiments and organ chips. However, static models are unable to recapitulate the flow response of body systems (Sontheimer-Phelps, Hassell, and Ingber 2019). Considering the impact of dynamic flow profiles on cell proliferation and signaling, a deep understanding of flow patterns in the body is needed for creating physiologically relevant organ chips. Various flow patterns in the human body can be observed from cardiovascular (heart and blood circulation), respiratory (dynamics of gas exchange), renal (urination process), and gastrointestinal systems and can be classified as perfusion, pulsatile, or continuous flows. To create physiologically relevant flow patterns, hydrostatic pressure, syringe pumps, and peristaltic pumps are commonly used (Byun et al. 2014). For complex flow profiles, on-chip micropumps are often used to create desired flow patterns.

In this chapter, we first introduce the basic notions of organ chip design and fabrication. This section covers the key biological and design considerations that are required to accurately recapitulate an in vivo environment on-chip. Next, there is an in-depth look at the flow control utilized in organ chips. From in vivo flow conditions, various flow patterns in microfluidic platforms will be presented and discussed to provide a selection guideline of flow control techniques to support organ chips. Finally, examples of recently developed organ-on-a-chip devices are listed with a brief description of their design, fabrication, and flow control.

6.2   DESIGN AND FABRICATION OF AN ORGAN-ON-A-CHIP

The design of an organ chip starts by identifying the target in vivo system, organ, or structure that is to be mimicked. Considering anatomical and physiological parameters, the cell type(s), the materials necessary for each component, the shape of the channels, and cell support can be determined. Once initial design parameters are selected for the organ, a fabrication procedure with possible materials can be developed and culture parameters such as extracellular matrix (ECM), media, and feeding method, and physiochemical manipulation can be selected to form an accurate in vitro tissue model.

6.2.1 Chip Materials, Fabrication, and Surface Modification

The common bulk materials for fabricating microfluidic chips are PDMS, polymethylmethacrylate (PMMA), and glass. Among these materials, PDMS is most commonly used due to its clarity, gas permeability, flexible material properties, and easy and inexpensive fabrication using soft lithography techniques (Bein et al. 2018, Quiros-Solano et al. 2018, Huh et al. 2012, Hirama et al. 2018, Chan et al. 2011). In the case that organ chips require high rigidity, high-resolution imaging, or intensive surface chemistry, PMMA or glass is used instead (Guckenberger et al. 2015, Okagbare et al. 2010, Hirama et al. 2018). However, microfabrication using PMMA and glass requires hot-embossing, imprinting, and wet etching steps with more complicated bonding and packaging procedures (Dolník, Liu, and Jovanovich 2000, Seiler, Harrison, and Manz 1993).

The support material selection is highly dependent upon the design of the organ chip and by extension is based on both the organ system being studied as well as the requirements of the experiment to be performed. Simply, an organ chip can be either single or multilayer. For single layered chips, glass is commonly used for the support layer, Figure 6.1a. To fabricate a single layer chip, a PDMS microchannel is cured and a PDMS–glass bond is then formed through oxygen plasma activation (Bhattacharya et al. 2005). Once bonded the microchannels will be formed by PDMS on three sides and glass as the bottom surface. Single layered organ chips have been used to create a variety of organ chips, especially when reconstructing the cardiovascular system (Agarwal et al. 2013, Lee et al. 2018, Dodson et al. 2015). For example, a heart valve-on-a-chip used glass as the primary support for its porcine aortic valvular endothelial cells (Lee et al. 2018). This chip utilized normal

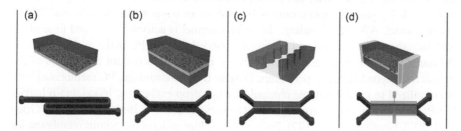

FIGURE 6.1 Example cross sections and shape for the four primary types of organ chips. (a) Single channel that provides feeding and flow to the cells for experiments. Cell support is normally provided by glass or bulk PDMS. (b) Multilayer chip with cells seeded on one or both sides of rigid porous support. (c) Multichannel chip that feeds cells from different sides of ECM hydrogel. (d) Active multichannel device that allows for strain application to the seeded cells. ECM, extracellular matrix; PDMS, polydimethylsiloxane.

soft lithography techniques to fabricate a bifurcating channel in PDMS (Xia and Whitesides 1998). The bifurcating channel allowed for testing a range of shear stress values with a single flow setup and the glass support allowed for easy imaging of the morphological changes to the cells based on the differences in applied shear. The simplicity of the single-channeled organ chip enables parallelization to improve overall throughput for drug screenings (Chen et al. 2012, Toh et al. 2009, Ye et al. 2007). In these chips, the single-channeled design can simply be repeated to create multiple identical channels to separate seeded cells and allow for independent testing. With this high-throughput chip design in mind, a liver cancer-on-a-chip showed the power of parallelization as its design gives researchers a way to monitor the impact of different concentrations and mixtures of drugs on liver carcinoma cells (Ye et al. 2007). This chip starts as a single channel with a glass wafer backing and extends outwards radially, bifurcating multiple times to dilute and mix the applied drugs at the same time before passing over cell culture channels. The bifurcation drastically decreases the time and increases the information gained from experiments with a relatively simple organ chip design.

However, due to the impermeability of glass, it cannot be used for cell support in multilayered organ chips. A multilayer (two or more channels stacked vertically) or multichannel (two or more separated horizontal channels) chip allows for more complex experiments, e.g., permeability of a cell layer or the interaction of cells, with the consequence of more complex design and fabrication. As shown in Figure 6.1b–d, various membranes are normally used to support or separate cells and the decision on the style of membrane comes from the requirements of the target organ. Thermoplastics (mainly PC, polystyrene, polyethylene terephthalate, and Cyclic Olefin Copolymer), PDMS, and ECM support like collagen vitrigel (CV) or hydrogel can be used to create multichannel chips. For multilayer chips, the pore size of the membrane is chosen based on the size of the cells being used and the fluidic resistance required for accurate transport of applied chemicals. From thermoplastics, PC is normally used due to its availability as it is used in macroscale filter applications. However, PC–PDMS bonding requires additional surface activation steps beyond what is required for PDMS/PDMS or PDMS/Glass bonding (Sunkara et al. 2011). In brief, the membrane is incubated in a 5% (3-Aminopropyl) triethoxysilane solution after initial oxygen plasma activation. After drying the membrane with N_2 gas, this membrane is bonded to an oxygen plasma-activated PDMS microchannel. After this bonding, this microchannel is activated a second time along with the other microchannel and the two are bonded with the PC in the middle. Because PC membranes are commercially available, it is difficult to control parameters of the membrane beyond available pore sizes. Despite this disadvantage, PC membranes have been utilized to create multiple channels for eye drop evaluation, blood–brain barrier (BBB) permeation, and spermatogenesis (Komeya et al. 2017, Bennet et al. 2018, Griep et al. 2013, Booth and Kim 2012). Each of these chips utilizes the porous membrane for its permeability (Komeya et al. 2017), or to study the permeability of a cell layer or layers (Bennet et al. 2018, Booth and Kim 2012, Griep et al. 2013). For example, Bennet et al. (2018) fabricated a corneal epithelium-on-a-chip for permeation study. This chip was created because the eye is a complex structure and it is difficult to study with traditional methods. The cornea is the multilayered structure on the anterior part of the eye that is a primary site of exposure to bacteria and toxins. Due to its complex structure, the common animal models used to study ocular treatments and diseases have character-istic differences to the human cornea. For instance, the most commonly used animal model, the rabbit, has different number of layers of corneal epithelial cells than seen in humans (Estlack et al. 2017). These characteristic differences are present in ex vivo experiments as well. While most engineered experiments are done with more control in layers of cells and easier access to permeation data, they are normally done at static or continuous flow conditions that do not match the pulsing flow that is seen in vivo due to blinking. The cornea chip was fabricated with an upper donor channel and a lower receiver channel that surrounded a PC membrane that provided support for the cells. Chips with one to five layers of cells were all used as a part of the permeation study under three different flow conditions: static, continuous, and simulated blinking gener-ated by a syringe pump. Results of the permeation study were compared to determine the effect of the number of cell layers as well as the flow conditions. The primary con-cern for this chip was permeation, which is unable to be measured with a simple glass backed chip and the PC membrane also allowed for control of some of the mechanical properties of the membrane which is important for corneal epithelial cell proliferation (LeBleu, Macdonald, and Kalluri 2007).

In contrast to rigid PC or thermoplastic membranes, PDMS can be used as a flex-ible supporting membrane. These are normally fabricated in-house which allows for a high level of control of the properties and structure of the membrane. A common fabrication method involves spin-coating a thin layer of PDMS onto a salinized wafer with an array of posts (Huh et al. 2010, Xia and Whitesides 1998, Quiros-Solano et al. 2018). Organ chips that aim to replicate the lung or the gut commonly use this type of support as it allows for controlled mechanical strain application on the cells (Kim et al. 2012, Shin and Kim 2018, Huh et al. 2010). Huh et al. (2010) created a "breathing" lung on a chip designed to study the impact of strain on the alveolar–capillary interface and this system's response to bacteria and inflammation. It utilized a thin, flexible, porous PDMS membrane to provide structure to both human alveolar epithelial cells and human pulmonary microvascular endothelial cells that were co-cultured on opposite sides of the PDMS membrane. The design of the microfluidic chip included pneumatic side channels that could be put under negative pressure using a vacuum. This was to recapitulate the stresses experienced by these cells in vivo dur-ing normal respiration and thus more closely replicate the in vivo environment. Once this lung chip was successfully cultured, Huh et al. used it to study how ultrafine silica nanoparticles introduced to the system effected ICAM-1, a pulmonary inflammatory response marker, expression in cases with and without strain and were able to show a statistical difference in expression when the system was put under strain.

6.2.2 EXTRACELLULAR MATRIX

To culture cells in organ chips, attachment and proliferation of the cells should be enhanced by coating the surface of microchannels with an ECM to mediate cell adhesion through focal adhesions (Chen et al. 2003). ECMs can be a variety of pro-teins which are compatible with the chosen cell type(s) (Howe et al. 1998). Common coatings for organ chips include collagen, fibronectin, and hydrogels and should be chosen based on the interactions with both the cells and the supporting material

(Huh et al. 2010, Bennet et al. 2018). For instance, collagen can be used in a way that causes it to form into gelatin. This gelatin can be utilized to form a three-dimensional environment, and Puleo et al. (2009) utilized collagen in a microfluidic environment to create corneal tissue cultures. In this work, the researchers utilized a vitrification process with the collagen support to control the material properties of the collagen. The collagen was used to support cells and was also etched away to allow for direct co-culture of stromal and epithelial cells. To fabricate the CV, an equal volume of a 0.5% solution of acid solubilized type I collagen and culture medium is mixed at 4°C and then poured into a culture dish. The dish is incubated for 2 h to complete gelation and then is dried for at least 2 days. The now rigid material is rehydrated with PBS before seeding cells (Takezawa et al. 2004). The chip was verified using a standard transepithelial permeability assay and showed viable barrier properties on the chip. The gut chip model developed by Li et al. (2013) also utilized collagen gelatin to incorporate the three-dimensional aspects of the in vivo gut environment. The chip exhibited decreased transepithelial electrical resistance, a measure of permeability, and cell marker expression when compared to a two-dimensional culture model, showing the benefit of the three-dimensional gelatin support and ECM.

Fibronectin can also be used for the ECM and is important for a variety of in vivo properties especially cell adhesion (Pankov and Yamada 2002). When added to an organ chip, fibronectin utilizes protein adsorption to bind to the support structure and form a coating (Meadows and Walker 2005). The protein presents many possible binding sites for cells and more binding sites can be presented even after initial cell attachment for a variety of organ chips (Huh et al. 2010, Bennet et al. 2018, Lee et al. 2018). Like collagen, the material properties of fibronectin can be modified using UV crosslinking to control membrane stiffness to match in vivo properties. This is important for cells like the corneal epithelium where the stiffness of the supporting membrane has been shown to impact the growth of the cells (Liliensiek et al. 2006).

Hydrogels can be fabricated to create an ECM-imbedded three-dimensional cell culture environment similar to collagen gelatin (Seo, Jung, and Kim 2018). A general procedure to fabricate an ECM hydrogel composite is given (Massensini et al. 2015). Briefly, a porcine urinary bladder is prepared by delamination and decellularization before being rinsed with PBS and deionized water and solubilized with pepsin. The mixture is then diluted to the desired ECM concentration and forms, over time, into a hydrogel based on ambient temperature and concentration. This hydrogel can be fabricated with material properties close to in vivo tissue and has been used for repairing structurally defected tissues (Seo, Jung, and Kim 2018, Faulk et al. 2014, Kim et al. 2018). The ability to mimic in vivo tissue material properties also makes hydrogel a promising material for use in organ chips as both a support structure and an ECM. In multichannel chips, hydrogels are used to provide an area to observe the interactions of different types of cells, especially interactions between cancer cells and healthy cells (Zervantonakis et al. 2012, Jeon et al. 2015, Lee et al. 2017). For instance, a microfluidic platform was designed for real-time imaging of the three-dimensional simulated tumor interaction in the vascular interface and measurement of endothelial barrier (Zervantonakis et al. 2012). The chip incorporated PDMS channels with an interior hydrogel where normal (human microvascular endothelial and umbilical vein endothelial) cells and cancerous (human breast carcinoma) cells were seeded. The seeding process was done after the formation of the hydrogel. Tumor and endothelial cells were flown in on opposite sides of the hydrogel and the endothelial cells bound to the surface of the hydrogel while the tumor cells invade the hydrogel in response to growth factor gradients. The three-dimensional matrix created by the hydrogel allowed for increased signaling between the tumor and endothelial cells more closely matching what is seen in vivo.

6.2.3 PHYSICOCHEMICAL ENVIRONMENT CONTROL

Once organ chips form the target tissue properly, the chips can be used under various physicochemical environments. Controlling the physicochemical environment is important during both culture and physiologically relevant experiments. For instance, the lung and gut systems are almost always under physical stress or strain in vivo. The previously discussed PDMS membrane allows for dynamic motion during on-chip culture. The general procedure for creating a chip with active functionality is to bond both layers of the chip to the PDMS membrane with channels on both sides of the cell culture channel. An etchant is run through these channels to remove the piece of the membrane that remains there. The two outer channels then become pneumatic channels that, when vacuum is applied, stretches the inner membrane, applying strain to the cells (Huh et al. 2010, Kim et al. 2012, Bein et al. 2018). The breathing lung-on-a-chip found that the application of strain to the cell layer promoted alignment of the endothelial cells to obtain physiologically relevant responses. The stretching of the cell layers increased the permeability of the layer by increasing the transcellular pore sizes and this allowed for immune cells to pass through the endothelium and membrane and attack bacteria that had attached to the epithelial cell layer (Huh et al. 2010).

Furthermore, it has been shown that both conventional two-dimensional cell cultures and transwell-based assays are unable to provide information about multicellular interactions between healthy cells and cancer cells under the flow and mechanical stimuli (Sontheimer-Phelps, Hassell, and Ingber 2019). Hassell et al. (2017) utilized a "breathing" lung chip to study the effect of microenvironment on cancer growth and influence in the alveolus. Using co-culture methods for nonsmall cell lung cancer the researchers were able to show that the cancer cells proliferated much less in the presence of normal breathing strain than when grown in a no strain environment. The researchers believe that the breathing motions of the chip limited the expansion and multilayer growth of the cancer cells that is seen without the in vivo motion. Thus, if the motion of a lung decreases for some reason, the growth of the present cancer would increase drastically. In vivo, the growth of the cancer leads to the loss of normal lung motion creating a positive feedback loop increasing the proliferation of the cancer cells.

A primary method of chemical environment control is to recapitulate mass transport phenomena using diffusion and convection. In the body, diffusion is one of the most important transportation mechanisms. Atoms, ions, molecules, or suspended colloids can diffuse due to differences in the concentrations of two species and continues until an equilibrium is found. Diffusion flux, which is the amount of specious passing through an area in the unit of time, can be calculated with Fick's first law of diffusion $J = -D_{diff} \, \partial c/\partial x$, where D_{diff} and $\partial c/\partial x$ are diffusion coefficient and gradient of concentration, respectively. Osmosis is a good example of diffusion in which molecules of a solvent passes through a permeable membrane toward the higher solute concentration on the other side and is seen in vivo with the digestive system. This passive motion is dependent on the pressure on both sides of the membrane. Diffusion can be further controlled through fluid flow, creating a convection–diffusion environment. Due to the normally slow rate of diffusion, in these cases, the convective flow is the dominant factor in the location or distribution of a species.

In some in vivo tissues such as the pulmonary alveolus, corneal epithelial, and epidermis there exists an air–liquid interface. To recapitulate this environment systematically in a microchip, chip design and fabrication procedures should accommodate the formation of this interface. For example, Hiemstra et al. (2018) studied the off-chip effects of the formation of this interface as a possible method for more accurate epithelial lung cell experiments. The air–liquid interface, in this case, allows for the production of in vivo physical barriers. A chip-based liver–intestine and liver–skin culture by Maschmeyer et al. (2015) shows the importance and effects of the air–liquid interface in an organ chip environment. In this chip, the interface was formed by controlling the height that the cells were seeded and cultured at when compared to the amount of media added to the organ chip. Based on the type of environment required cells were cultured just at the surface of the media using culture inserts or submerged in the media. The successful formation of this interface during long-term culture (around 14 days) allows for a wider range of useful experiments to be performed. The air–liquid interface is important for lung, eye, skin, chips as well as some other specific cases (Maschmeyer et al. 2015, Bennet et al. 2018, Hiemstra et al. 2018). It has been shown to be an important component of those in vivo systems and should be included in organ chips targeting those systems. The inclusion of the interface can also allow for more realistic experimental methods as it represents the organ's interface to outside stimulus. A lung chip or eye chip can be exposed to an airborne toxin or contaminant in a representative way of the in vivo situation.

In summary, the design and fabrication of organ chips starts from understanding target in vivo systems. Based on the requirements of the systems, the chip layout, bulk chip material, cell support material, and physiochemical considerations are developed. Chip layout can be a simple single channel system with a coated support or a complex multichannel system for testing permeation of an analyte. PDMS is used most commonly as the bulk chip material due to its ease of use and cost but chips can be fabricated from glass or PMMA if specific surface or material characteristics are required. Cells can be supported by an impermeable material like glass for experiments focusing on fluid shear or porous membranes made of a thermoplastic or PDMS for permeation-based experiments. The combination of design, material choice, and support contribute to the physiochemical environment of the chip. The environment can also be controlled through strain application, diffusion control, and air–liquid interface formation. Careful considerations of these parameters are required to design and fabricate a physiologically relevant organ chip.

6.3   FLOW CONTROL FOR ORGAN CHIPS

Flow control in organ chips should be carefully considered, not only for initial cell seeding, but also experimental design for specific cell response, drug-delivery, and permeation studies. To characterize flow patterns, the Reynolds number (Re) is used as it gives the ratio of inertial forces to viscous forces. This number defines if a flow is laminar (ordered) or turbulent (disordered) and the transition between the two occurs at Re between 1,200 and 2,500. Even though flow in the body is usually considered laminar, turbulent flow can occur with normal cardiac function (e.g., ascending aorta). However, due to the size of the microchannels commonly used in organ chips, it is difficult to develop truly turbulent flow on-chip. Instead, researchers focus on flow or shear stress profile mimicry to create physiologically relevant microenvironments. Thus, information on shear stress profiles and flow rates are influential when designing an organ chip as they give information on the target nondimensional values and shear stresses that are necessary to recapitulate an environment. Furthermore, the mechanical properties of body organs should be properly scaled down or adopted into the organ chips according to the particular purpose of the study.

Typically, the flow patterns of in vivo systems can be classified as perfusion, pulsatile, or continuous flows. To create these flow patterns, hydrostatic pressure, syringe pumps, and peristaltic pumps have been with organ chips. With appropriate flow control, the study of pharmacokinetic/pharmacodynamic properties of drugs in an organ chip is more reliable under flow than compared to static conditions in terms of adsorption, distribution, metabolism, elimination, and toxicity properties (Bhatia and Ingber 2014). In contrast, static conditions are commonly used during cultures on plastic tissue culture plates, in cultures separated by semi-permeable membranes, or in static Trans-well culture systems (Wilmer et al. 2016). Despite this simplicity, these culture systems have difficulties inducing relevant physiological responses, e.g., directional elongation or generation of biochemical modulators, in cells when compared to the dynamic flow circumstances seen in vivo. For instance, it is observed that epithelial cells under dynamic flow conditions demonstrate more in vivo behavior, such as cytoskeletal reorganization, than in static conditions and valvular endothelial cell alignment and elongation under physiological flow has been demonstrated with the heart valve-on-a-chip platform, a development which has not been seen in static culture (Lee et al. 2018, Jang and Suh 2010). These cellular responses are mainly due to the shear forces on the cells in dynamic conditions which can be controlled through design parameters, height (h) and width (w), and flow rate (Q) based on the Hagen-Poiseuille equation

$$\tau = 6Q\mu/h^2w \qquad (6.1)$$

where μ is the dynamic viscosity. Another issue with static conditions occurs during drug delivery-based experiments. Sedimentation of nanoparticles under static condition inhibits homogenous delivery and can lessen the usefulness of results (Bhise et al. 2014). Various flow patterns are required for organ chip development, depending on the stage of the chip preparation. In general, the process of seeding cells inside a microfluidic chip is similar to normal subculture procedure (Hassell et al. 2017, Huh et al. 2010, Benam et al. 2017). However, caution is needed to ensure little to no flow occurs on the chip to allow for settling of cells onto the support structure. Usually 12–24 h after initial seeding, any unattached cells are removed to maintain overall cell viability and avoid apoptosis (Kobuszewska et al. 2017, Kimura et al. 2015). After the target cells are seeded inside the microfluidic chip, the application of media is important to maintain cell proliferation (Bhatia and Ingber 2014). Depending on the target organs and the purpose of the experiment, flow can be either simple gravity fed, perfusion flow, continuous flow, or in vivo like flow patterns. The selection of flow conditions is dependent upon the types of organs and experimental goals. For instance, if shear and flow pattern are not critical parameters, organ chips will utilize simple static or perfusion flow conditions to ensure simplicity. In most cases, organ chips utilize external equipment to manipulate and control the fluid flow. However, for complex flow profiles, on-chip micropumps can be used to recapitulate in vivo flow conditions. Both are discussed in the following sections as well as summarized in Table 6.1 below.

6.3.1 DECAYING FLOW PATTERN IN ORGAN CHIPS

The simplest flow utilized in organ chips is hydrostatic, or gravity-driven flow. In hydrostatic flow (Marimuthu and Kim 2013, Sung, Kam, and Shuler 2010, Goral et al. 2013, Gnyawali et al. 2017, Komeya et al. 2017), the pressure difference between the inlet and outlet reservoirs drives flow in a microchannel. Since the pressure

TABLE 6.1
Organ Chip Flow Pattern Generation

Flow Pattern	Method	External Power	Recirculating Capability	Integration	References
Decaying	Hydrostatic pressure	No	No	On-chip	Marimuthu and Kim (2013), Sung, Kam, and Shuler (2010), Goral et al. (2013), Gnyawali et al. (2017), Komeya et al. (2017), Esch et al. (2016)
	Capillary-driven	No	No	On-chip	Khodayari Bavil and Kim (2018), Meyvantsson et al. (2008), Kim, Paczesny, et al. (2013)
Steady	Gravity-driven	No	No	Off-chip	Zhu et al. (2004), Wang et al. (2018)
	Syringe-pump	Yes	No	Off-chip	Park et al. (2010), Bennet et al. (2018)
	Osmosis-driven	No	No	Off-chip	Byun et al. (2014), Ju et al. (2009), Park et al. (2009)
Pulsatile	Syringe-pump	Yes	No	Off-chip	Bennet et al. (2018), Estlack et al. (2017)
	Peristaltic Pump	Yes	Yes	Off-chip	Ma et al. (2012), Zhang, Chen, and Huang (2015)
	Pneumatic actuators	Yes	Yes	On/off-chip	Kim, Jensen, et al. (2013), Linshiz et al. (2016)

difference is a time-dependent decaying function, the flow rate decays to reach a steady-state value. The flow rate from this method can be easily understood and controlled with Bernoulli's equation. Consider a medium reservoir with an equivalent diameter of d_1 and outlet channel diameter of d_2. The necessary height of fluid in the reservoir for a flow velocity, v, can be calculated by:

$$h = \left(V^2/2g\right) + \lambda/d_2 \times V^2/2g \qquad (6.2)$$

where λ is the channel friction coefficient and theoretically found to be $\lambda = 64/Re$ (Komeya et al. 2017). By knowing $dh/dt = Q/A = v \times (d_2/d_1)^2$, a ratio of d_1 to d_2 can be chosen to get a desired outlet flow rate. When high precision long-term flow is needed, a hydrodynamic resistance circuit can be added to the organ chip design. Komeya et al. (2017) used a resistance circuit for long-term pumpless microfluidic cell culture on a two-chambered PDMS chip separated by a PC membrane. Media was flown from the upper channel, through a resistance circuit, before passing over the cells seeded in the lower channel. Due to the large reservoir and the resistance circuit comprised of a high fluidic resistance microchannel, media flow can be continued at an acceptable rate for long periods of time. The long-term media application of this chip allowed for the study of spermatogenesis in the rat teste tissue that was studied in this experiment over a long time period without the need for intervention, and this approach can be adapted in other devices due to its simple design.

A possible option for future on-chip controlled decaying flow is capillary driven flow. Capillary pumps are one of the popular passive pumping techniques used in microfluidics (Khodayari Bavil and Kim 2018, Meyvantsson et al. 2008, Kim, Paczesny, et al. 2013). Capillary-driven pumps rely on surface tension and usually generate low flow rates in the scale of micro/nanoliter per minute that decay with time. In order to get more precise and controlled flow in capillary-driven pumps, different designs of microstructures are used (Zimmermann et al. 2007). Recently, a capillary-driven pump has been introduced for sample liquid perfusion with a constant flow rate independent of sample surface energy by controlling the fluidic resistance and integrating an absorbent pad (Guo, Hansson, and van der Wijngaart 2018). Although they are limited and have not been implemented in an organ chip yet, these capillary-based micropumps can potentially be utilized to simplify an organ chip by avoiding the need for an external flow control system.

6.3.2 CONTINUOUS FLOW GENERATORS

The simplest way to generate continuous flow is by utilizing hydrostatic pressure using two horizontal channels at different height levels (Zhu et al. 2004). Gravity-driven flow can also approximate continuous flow by incorporating multiple channels or a siphon system. A unique method was recently developed for a continuous flow gravity-driven passive pump by coupling with a siphon-based autofill function (Wang et al. 2018). In this configuration, a storage container on top of the chip is connected to an inlet reservoir by low and high siphons to maintain constant hydrostatic pressure on the inlet. This enhanced hydrostatic pressure-driven flow device can control the overall flow rate accurately and enable selective perfusions to feed multiple cell channels simultaneously.

While hydrostatic flows are used for simple media infusion, they are limited beyond creating decaying flow patterns and low ranges of flow rate without introducing complex additions to the system. Instead, for a steady flow driven by an external source, a syringe pump is commonly used in typical microfluidic setups due to its simplicity, ease of control, and a wide range of possible flow rates. Due to their simple operation, syringe pumps have been employed in a variety of organ-on-a-chip applications as the primary flow control technique for either perfusion, generation of appropriate shear stress, or mass transport (Bennet et al. 2018, Huh et al. 2010, Lee et al. 2018, Ye et al. 2007). Some syringe pumps allow for backward movement to create negative pressure and act as a vacuum pump, enabling long-term application by refilling the inlet reservoir with fresh medium (Jeon et al. 2015). However, syringe pumps do have some flow rate fluctuations making them disadvantageous for precise flow control (Gnyawali et al. 2017). This fluctuation can originate from the mechanical parts of the pump, step motor, and syringe plunger which restricts their applications in precise microfluidic structures that are sensitive to flow pattern and cytometry platforms.

Another technique to generate a continuous flow pattern is the electroosmotic pump, i.e., the utilization of a concentration gradient between two solutes through a permeable membrane (Glawdel and Ren 2009, Park et al. 2007). The flow rate for osmosis-driven flow can be controlled by regulating the osmotic pressures through the solute concentrations and contact area. Using an electroosmotic pump, flow rates in the range of nanoliter per minute can be achieved, however, its use is limited to simple microfluidics without recirculation capabilities. Osmosis-driven flow also needs accurate design considerations to overcome electroporation, electrolysis, and Joule-heating issues.

6.3.3 IN VIVO FLOW PROFILE MIMICRY FOR ORGAN CHIPS

When working to replicate the cardiovascular system, diaphragm actuation, intestine deformation, or the blinking of the eye, it is important to mimic in vivo flow patterns for physiologically relevant results (Estlack et al. 2017, Lee et al. 2018, Bein et al. 2018, Nawroth et al. 2018). In these cases, programmable syringe pumps, peristaltic pumps, and on-chip micropumps are employed for organ chips.

A programmable syringe pump can go beyond simple infusion and create customized pumping methods. Bennet et al. used a programmable syringe pump with the previously discussed cornea-on-a-chip to test the permeation difference under continuous and pulsatile flow. The pulsatile flow was generated through a program that started and stopped the pump 14 times a minute to simulate the in vivo tear flow condition. The results from static, continuous flow, and simulated tear flow show that in vivo-based flow has an effect on the overall permeation of the tested drugs (Bennet et al. 2018). However, due to the limited bandwidth of mechanical motion, it is impossible to generate high frequency or complex flow patterns with a programmable syringe pump.

A peristaltic pump is commonly used for generating a simple pulsatile flow by squeezing a flexible tube sequentially as shown in Figure 6.2c. In a peristaltic pump, the confined liquid inside the flexible tube is driven by rotating a roller on a rotor. With

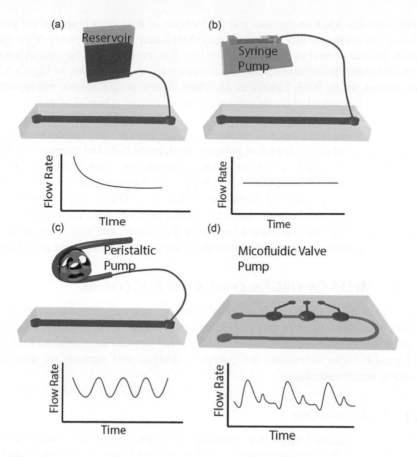

FIGURE 6.2 Flow control mechanisms in the organ on a chip. (a) Decaying flow for cell culture purposes. (b) Continues flow for cell culture, mimicking body flow, and drug injection. (c) Pulsatile flow for more precise medium injection during cell culture and simplified mimicry of cardiac and blinking flow patterns. (d) Microfluidic valve pump able to produce in vivo mimicking flow profiles and is embedded inside the chip.

proper tube maintenance, these valveless pumps can be used for long-term injection and the flow direction can be easily controlled by switching the direction of the rotor (Ma et al. 2012, Zhang, Chen, and Huang 2015). These pumps can be used as open-loop or closed-loop systems, requiring the refilling of the inlet reservoir or the recirculation of confined medium, respectively. Open-loop fluidic routs are used in chips that need fresh medium for cells, and, in contrast, a closed-loop scheme is usually used to mimic body recirculation to test a drug's effects on the microenvironment. An open-loop flow control system utilizing a peristaltic pump has been used to provide fresh medium from a reservoir for 72~96 h to investigate the effect of perfusion conditions on the proliferation and morphology of the cardiac cells (rat cardiomyoblasts-H9C2) (Kobuszewska et al. 2017), and a closed-loop flow control system was used with a heart-on-a-chip designed to study the effects of geometry and microenvironmental conditions on the same cardiomyoblasts. However, a peristaltic pumping system generates fixed magnitudes and frequencies of flow rates based on the given hardware. To overcome this limitation, microfabricated micropumps are often used for more precise control to generate in vivo like flow profiles. Among various actuation mechanisms, pneumatic actuators are commonly used as shown in Figure 6.2d (Kim, Jensen, et al. 2013, Linshiz et al. 2016). These programmable micropumps allow for precise control of the flow through the sequential actuation of microvalves. Although these micropumps need multilayer fabrication steps with off-chip hardware such as solenoid valves and pressure/vacuum sources, they generate accurate and well-controlled customized flow patterns. Both magnitude and frequency of flow can be easily controlled based on the closing pressure on a thin PDMS membrane and actuation interval time, respectively. This is well demonstrated in the previously discussed heart valve-on-a-chip developed by Lee et al. (2018), where microfluidic valves are used as proxy heart channels to produce pulsatile flow. The resulting flow could be controlled based on desired "heartbeat frequency" and magnitude allowing it to accurately mimic a variety of in vivo shear stress profiles seen by heart valves.

6.4 EXAMPLES OF ORGAN-ON-A-CHIP PLATFORMS

In addition to the examples of organ chips that have been discussed in the previous chapters, the following chips give insight into the capabilities of organ chip systems and some different design aspects required for some unique environments. Table 6.2 and Figure 6.3 also summarize the materials, designs, and methods for these and previously mentioned chips.

6.4.1 LUNG-ON-A-CHIP

The lung is a common target for organ-on-a-chip designs as the human lung is exposed to many hazards and toxins as well as repeated stresses. A lung-on-a-chip device allows for accurate studying of how these hazards and stresses can impact the human vascular system. The respiratory system serves several vital functions of CO_2/O_2 exchanges, controlling the blood pH, body heat exchange, balancing of the body fluid, and voice production. Lung expansion and contraction happens by downward/upward movement of the diaphragm and elevation/depression of the ribs. This movement provides pulsatile airflow in the lungs and causes O_2/CO_2 exchange in the alveolus. By ventilating the fresh air into the alveoli, oxygen diffuse into the pulmonary and in the opposite direction, carbon dioxide diffuses into the alveoli due to the difference in the partial pressure of the gas in the fluid–gas interfacial phase. This causes strain on the alveolar membrane with a maximum of 10% linear elongation (Stucki et al. 2015). By circulating this oxygen-enriched blood in the body, oxygen diffuses from tissue capillaries $\left(P_{O_2} = 40 \text{ mmHg}\right)$ into the circumference tissue cells $\left(P_{O_2} = 23 \text{ mmHg}\right)$. Utilizing the in vivo constraints and conditions, a "breathing" lung-on-a-chip was developed as described previously (Huh et al. 2010), and a general design and protocol of a lung-on-a-chip were created by Benam et al. (2017). The protocol covers a PDMS device that is inexpensive to produce and can model the inflammatory responses of the lung as well as guidelines for an experiment. The device created utilizes a polyethylene terephthalate (PET) membrane coated with a

TABLE 6.2
Presented Organ-on-a-Chip Device Summary

Chip	Flow	Bulk Material	Cell Support	ECM	Cell Type	Feeding	Reference
Heart valve	Non-uniform pulsatile	PDMS	Glass	Fibronectin	Primary porcine	Perfusion	Lee et al. (2018)
Corneal epithelium	Static, continuous, & pulse	PDMS	PC membrane	Fibronectin	Immortalized human	Static	Bennet et al. (2018)
Breathing lung	Continuous	PDMS	PDMS membrane	Fibronectin or collagen	Immortalized human	Perfusion	Huh et al. (2010)
Cornea	–	PDMS	CV membrane	Collagen	Primary rabbit	Perfusion	Puleo et al. (2009)
Blood brain barrier	Continuous	PDMS	PC membrane	Collagen	Immortalized human	Perfusion	Griep et al. (2013)
Renal	Continuous	PDMS	Polyester membrane	Fibronectin	Primary rat	Perfusion	Jang and Suh (2010)
Long term culture	Decaying	PDMS	PC membrane	–	Primary rat	Perfusion	Komeya et al. (2017)
Heart-on-a-chip	Pulsatile	PDMS	Glass	–	Primary rat	Perfusion	Kobuszewska et al. (2017)
Gut chip	Continuous	PDMS	PDMS membrane	Collagen	Immortalized human	Perfusion	Kim et al. (2012)
Gut chip	Continuous	PDMS	PDMS membrane	Collagen	Immortalized human	Perfusion	Shin and Kim (2018)
Heart	–	PDMS	Glass	–	Muscular thin films	–	Agarwal et al. (2013)
Retina	–	PDMS	Glass	Agar Gel	Excised rat retina	–	Dodson et al. (2015)
Lung	Continuous	PDMS	PET membrane	Collagen	Primary human	Static	Benam et al. (2017)
Breast cancer chip	–	PDMS	Fibrin gel	Fibrin gel	Immortalized and primary human	–	Jeon et al. (2015)
Cancer-on-a-chip	Continuous	PDMS	PDMS membrane	Fibronectin	–	Perfusion	Zhang et al. (2017)
Eye-cavity	–	PMMA	PMMA	Fibronectin	Immortalized rat	Perfusion	Chan et al. (2015)

ECM, extracellular matrix; PC, polycarbonate; PDMS, polydimethylsiloxane.

collagen ECM for support of primary human lung cells. Fabrication involves pouring PDMS onto three-dimensional printed molds to create the upper and lower channels and using a laser cutter to cut access ports into the PET membrane as well as extra holes to aid in bonding. During experiments, the device is under continuous flow provided by a syringe pump or similar device.

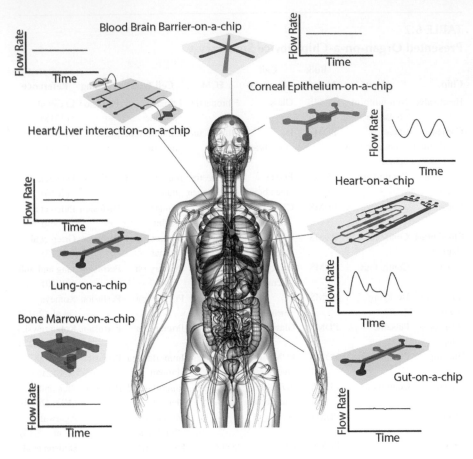

FIGURE 6.3 Examples of organ chips throughout the body (Torisawa et al. 2014, Huh et al. 2010, Bennet et al. 2018, Zhang et al. 2017, Shin and Kim 2018, Lee et al. 2018, Griep et al. 2013).

6.4.2 GUT-ON-A-CHIP

The intestine was one of the first bodily systems considered for organ-on-a-chip development. Its importance as an infection site and ease of recapitulation drove the development (Kim et al. 2012). Kim et al. developed an early gut-on-a-chip device in an attempt to replace animal testing. The device utilized a collagen-coated flexible and porous PDMS membrane, similar to the lung-on-a-chip discussed above, to mimic the physiologic stresses in the environment with low flow rates. The chip was fabricated using general PDMS fabrication techniques to create the upper layer, lower layer, and the membrane. They were then bonded using oxygen plasma treatment to create vacuum chambers (both sides) and culture chambers (split by the membrane). To isolate the external vacuum channels from the culture chambers, the membrane was bonded to the top layer using a support block and then torn to open the pneumatic channels. The researchers were able to show that under continuous flow, 40~50 μL/min based on physiologic shear stress (0.02 dyne/cm²), the immortalized human cells cultured in this environment developed folds comparable to the structure of the in vivo intestinal villi and is also able to support a microbial environment akin to that seen in vivo. Shin and Kim (2018) more recently fabricated a gut-on-a-chip to study the inflammatory response under physiological flow and motions. The chip was fabricated the same as the chip above, with upper and lower channels that were bonded to a PDMS membrane that was used for support of the cells and coated with type I collagen as the ECM for seeding of immortalized human cells. The elasticity of PDMS was utilized to allow for strain application from an external system to simulate in vivo motions. This chip was able to show that epithelial barrier dysfunction in the gut leads to the onset of inflammation and the results suggest that a person with a hyperpermeable intestinal epithelium may be more vulnerable to infection.

6.4.3 HEART-ON-A-CHIP

The heart has been targeted for organ-on-a-chip development due to its importance in the human body and the wide range of conditions that occur there. For instance, Agarwal et al. (2013) fabricated a high throughput hear-on-a-chip to study the stresses felt by tissue during contraction. The chip was designed to be totally autoclavable, made up of PDMS and glass, and was used to study the effects of isoproterenol at various doses on cardiac contractility. To fabricate the chip, an aluminum chamber with a PC cover was machined to support the PDMS coated chip. During testing, each chip would be secured with screws in the support. The microfluidic aspect of the device allows it to incorporate expensive or rare cells and treatment options at lower cost and higher applicability than more standard testing methods.

6.4.4 RETINA-ON-A-CHIP

In contrast to the anterior focused eye chip devices previously mentioned, some organ-on-a-chip devices have focused on the posterior segment of the eye. For example, Dodson et al. (2015) created a retina-on-a-chip, utilizing excised rat retinas in a PDMS microfluidic platform. Because the researchers used excised rat retina instead of culturing new tissue their support and feeding systems were different than what has previously been discussed. In this work, the researchers were not culturing cells from scratch but instead used formed structures and focused on keeping them alive on a glass and agar gel substrate. The design of the chip allowed for access to specific parts of the retina for testing with a variety of reagents and permits real-time imaging of biological processes. The chip was fabricated as three separate components, the PDMS microchannels, the tubing support layer, and the media cylinder which were brought together using standard bonding techniques. In their experiments the researchers were able to use lipopolysaccharides to show the ability to target specific tissue and that the chip can be used for real-time manipulation of cells as well as an ex vivo drug assay.

6.4.5 BONE MARROW-ON-A-CHIP

The bone marrow environment is, like the other in vivo systems described, extremely complex and difficult to recapitulate using traditional methods (Sacchetti et al. 2007). To combat these difficulties Torisawa et al. (2014) created a bone marrow-on-a-chip to study in vivo drug responses. The chip utilized a multistep fabrication process that used collagen in a PDMS cylinder implanted into a mouse to create accurate bone that can then be inserted into the final on-chip device. This design essentially utilized the mouse for both the bulk support and feeding of the culture environment. Once removed from the mouse, a traditional organ-on-a-chip device is used for feeding and testing. This chip shows promise as a living bone model for future experiments involving the bone marrow microenvironment.

6.4.6 CANCER-ON-A-CHIP

The main purpose of cancer-on-a-chip devices, as previously discussed, is a better understanding of cancer behavior while interacting with other tissues and determining the efficacy of cancer drugs (Zervantonakis et al. 2012, Ye et al. 2007). However, cancer metastasis, the spread of cancer cells from home organ (seed) to the secondary organ tissues (soil), has also been demonstrated by organ chip platforms (Jeon et al. 2015). Similar to the cancer chips explained previously, PDMS microchannels separated by hydrogel are bonded to glass and endothelial cells, mesenchymal stem cells, and osteoblast-differentiated cells are seeded into the hydrogel. These cells were cultured under static medium in the microfluidic channel, but a syringe pump was used for flow during experiments. From this study, extravasation of seeded human breast cancer cells was observed on the human microvascularized bone using a microfluidic chip that included muscle and bone mimicking units (Jeon et al. 2015). Using the proper design and relevant flows, these platforms demonstrate the effectiveness of cancer chips on the study of anticancer therapies by real-time imaging of drug metabolism.

6.4.7 BLOOD–BRAIN BARRIER-ON-A-CHIP

The BBB is an extremely selective barrier for the human nervous system and makes drug treatment to the area difficult. A model of the BBB was needed to provide insight into its role in disease and options for treatments through the barrier. Griep et al. (2013) fabricated an early BBB chip using platinum electrodes and human immortalized brain endothelial cells. The permeability of this chip was measured using trans-endothelial electrical resistance (TEER) and compared to conventional culture methods. Static cultured chips showed slightly higher TEER than off-chip measurements, but when constant shear stress was applied a sharp increase in TEER was seen matching the in vivo BBB function. Booth and Kim (2012) more recently created a more complicated microfluidic model of the BBB that was shown to closely match the properties seen in vivo. The chip was designed to include key properties identified by the researchers: endothelial cells with tight junction expression, astrocyte co-culture, shear stress, selective permeability, and high electrical resistance.

Both immortalized mouse endothelial brain cell (bEnd.3) and immortalized mouse astrocyte (C8-D1A) were seeded onto the fibronectin-coated PC membrane and allowed to form a co-culture environment under in vivo fluid shear provided by a peristaltic pump. The chip successfully incorporated the key properties discussed by the researchers and was verified by studying the tight junction expression as well as the glial fibrillary acidic protein (GFAP) expression, an astrocyte-specific marker, and TEER levels. The BBB chip was cultured both with and without the C8-D1A cells and the recorded TEER values were much higher for the co-cultured chips. In addition, the cell lines used in the chip were carefully selected based on previous verification of in vivo relevance, b.End3, or work done by the researchers themselves, C8-D1A (Omidi et al. 2003). The C8-D1A cell line on-chip expressed GFAP, a marker specific to astrocytes, and the cell line displayed astrocytic morphology.

6.5   CONCLUSION

The development of different types of microfluidic culture environments has led to the development of organ-on-a-chip devices. These chips have benefits over traditional animal and culture models because they are easily controllable, more closely match in vivo conditions, and lack the ethicality concerns that can be attributed to animal models. The fabrication and design techniques presented in this chapter show the basics for creating a microenvironment that can recapitulate the human body in both structure and flow. The selection of chip material (PDMS, PMMA, Glass), support structure (glass, PC, PDMS, ECM hydrogel, CV), cell type (human or animal and immortalized or primary), and other details are paramount in creating a useful and accurate organ chip. To obtain relevant experimental results, flow control is also extremely important. Static and perfusion flow are important for seeding and culturing cells and providing reasonable flow rates during experiments allows for results that are similar to in vivo cases. In some cases, complex flow profiles are required to study cellular responses.

In order to advance the organ chip field, researchers need to target three primary areas of improvement. First, while already much better than most conventional experiments, the in vivo accuracy of organ chips can be increased. Designs incorporating realistic scaling and shapes of channels and wells is lacking in most organ chips and true three-dimensional culture on-chip is still under development. Second, organ chips need to focus on ease of fabrication and ease of use. This will allow for the commercialization of organ chips as they move from proofs of concept to lab apparatus used in, for example, drug testing. While simple single layer chips have recently been parallelized, this concept needs to be expanded to multichannel and active organ chips to increase usability. Last, a larger focus can be made to incorporate in vivo flow patterns. We discuss a variety of chips that use flow patterns to study the effect of flow rates or shear stress on cells but as devices and experiments become more complex, recapitulating in vivo blood flow will become more important to the accuracy of the chip. A universal flow profiler generator equipped with programming functions would be beneficial to advance organ chip development.

Last, to improve the value of the organ chip development further, many current research efforts are moving toward the development of multiorgan systems.

As organ-on-a-chip devices grow in scope and use they will start to incorporate more complex structures and functions, possibly even combining multiple chips to form a full "respiratory system-on-a-chip" or "circulatory system-on-a-chip" (Ronaldson-Bouchard and Vunjak-Novakovic 2018). By combining multiple organs to create a "body-on-a-chip," in vivo interactions between different organs or systems can be studied to simulate and evaluate the movement and processing of chemicals and signals throughout the body (Wang et al. 2017). Recently, Skardal et al. (2017) successfully incorporated simple liver, heart, and lung chips for drug testing. Each individual system is relatively simple, the lung chip does not incorporate breathing motions and the system does not replicate in vivo flow. However, this body chip shows the promise of future multiorgan chip devices that incorporate designs similar to the currently developed complex single organ chips.

REFERENCES

Agarwal, Ashutosh, Josue Adrian Goss, Alexander Cho, Megan Laura McCain, and Kevin Kit Parker. 2013. Microfluidic heart on a chip for higher throughput pharmacological studies. *Lab on a Chip* 13 (18):3599–608.

Bein, Amir, Woojung Shin, Sasan Jalili-Firoozinezhad, Min Hee Park, Alexandra Sontheimer-Phelps, Alessio Tovaglieri, Angeliki Chalkiadaki, Hyun Jung Kim, and Donald E. Ingber. 2018. Microfluidic organ-on-a-chip models of human intestine. *Cellular and Molecular Gastroenterology and Hepatology* 5 (4):659–68.

Benam, Kambez H., Marc Mazur, Youngjae Choe, Thomas C. Ferrante, Richard Novak, and Donald E. Ingber. 2017. Human lung small airway-on-a-chip protocol. In *3D Cell Culture: Methods and Protocols*, edited by Zuzana Koledova, 345–65. New York: Springer.

Bennet, Devasier, Zachary Estlack, Ted Reid, and Jungkyu Kim. 2018. A microengineered human corneal epithelium-on-a-chip for eye drops mass transport evaluation. *Lab on a Chip* 18 (11):1539–51.

Bhatia, Sangeeta N., and Donald E. Ingber. 2014. Microfluidic organs-on-chips. *Nature Biotechnology* 32 (8):760–72.

Bhattacharya, Shantanu, Arindom Datta, Jordan M. Berg, and Shubhra Gangopadhyay. 2005. Studies on surface wettability of poly(dimethyl) siloxane (PDMS) and glass under oxygen-plasma treatment and correlation with bond strength. *Journal of Microelectromechanical Systems* 14 (3):590–97.

Bhise, Nupura S., João Ribas, Vijayan Manoharan, Yu Shrike Zhang, Alessandro Polini, Solange Massa, Mehmet R. Dokmeci, and Ali Khademhosseini. 2014. Organ-on-a-chip platforms for studying drug delivery systems. *Journal of Controlled Release* 190:82–93.

Booth, Ross, and Hanseup Kim. 2012. Characterization of a microfluidic in vitro model of the blood-brain barrier (muBBB). *Lab on a Chip* 12 (10):1784–92.

Byun, Chang Kyu, Kamel Abi-Samra, Yoon-Kyoung Cho, and Shuichi Takayama. 2014. Pumps for microfluidic cell culture. *Electrophoresis* 35 (2–3):245–57.

Chan, Yau Kei, Chiu On Ng, Paul C. Knox, Michael J. Garvey, Rachel L. Williams, and David Wong. 2011. Emulsification of silicone oil and eye movements. *Investigative Ophthalmology and Visual Science* 52 (13):9721–7.

Chan, Yau Kei, Kwun Hei Samuel Sy, Chun Yu Wong, Ping Kwan Man, David Wong, and Ho Cheung Shum. 2015. In vitro modeling of emulsification of silicone oil as intraocular tamponade using microengineered eye-on-a-chip. *Investigative Ophthalmology and Visual Science* 56 (5):3314–9.

Chen, Christopher S., Jose L. Alonso, Emanuele Ostuni, George M. Whitesides, and Donald E. Ingber. 2003. Cell shape provides global control of focal adhesion assembly. *Biochemical and Biophysical Research Communications* 307 (2):355–61.

Chen, Qiushui, Jing Wu, Yandong Zhang, and Jin-Ming Lin. 2012. Qualitative and quantitative analysis of tumor cell metabolism via stable isotope labeling assisted microfluidic chip electrospray ionization mass spectrometry. *Analytical Chemistry* 84 (3):1695–701.

Dodson, Kirsten H., Franklin D. Echevarria, Deyu Li, Rebecca M. Sappington, and Jon F. Edd. 2015. Retina-on-a-chip: A microfluidic platform for point access signaling studies. *Biomedical Microdevices* 17 (6):114.

Dolník, Vladislav, Shaorong Liu, and Stevan Jovanovich. 2000. Capillary electrophoresis on microchip. *Electrophoresis* 21 (1):41–54.

Esch, Mandy B., Hidetaka Ueno, Dawn R. Applegate, and Michael L. Shuler. 2016. Modular, pumpless body-on-a-chip platform for the co-culture of GI tract epithelium and 3D primary liver tissue. *Lab on a Chip* 16 (14):2719–29.

Estlack, Zachary, Devasier Bennet, Ted Reid, and Jungkyu Kim. 2017. Microengineered biomimetic ocular models for ophthalmological drug development. *Lab on a Chip* 17 (9):1539–51.

Faulk, Denver M., Ricardo Londono, Matthew T. Wolf, Christian A. Ranallo, Christopher A. Carruthers, Justin D. Wildemann, Christopher L. Dearth, and Stephen F. Badylak. 2014. ECM hydrogel coating mitigates the chronic inflammatory response to polypropylene mesh. *Biomaterials* 35 (30):8585–95.

Glawdel, Tomasz, and Carolyn L. Ren. 2009. Electro-osmotic flow control for living cell analysis in microfluidic PDMS chips. *Mechanics Research Communications* 36 (1):75–81.

Gnyawali, Vaskar, Mohammadali Saremi, Michael C. Kolios, and Scott S.H. Tsai. 2017. Stable microfluidic flow focusing using hydrostatics. *Biomicrofluidics* 11 (3):034104.

Goral, Vasiliy N., Chunfeng Zhou, Fang Lai, and Po Ki Yuen. 2013. A continuous perfusion microplate for cell culture. *Lab on a Chip* 13 (6):1039–43.

Griep, Lonneke M., Floor Wolbers, Bjorn de Wagenaar, Paul M. ter Braak, Babette B. Weksler, Ignacio A. Romero, Pierre Olivier Couraud, et al. 2013. BBB on chip: Microfluidic platform to mechanically and biochemically modulate blood-brain barrier function. *Biomedical Microdevices* 15 (1):145–50.

Guckenberger, David J., Theodorus E. de Groot, Alwin M.D. Wan, David J. Beebe, and Edmond W.K. Young. 2015. Micromilling: A method for ultra-rapid prototyping of plastic microfluidic devices. *Lab on a Chip* 15 (11):2364–78.

Guo, Weijin, Jonas Hansson, and Wouter van der Wijngaart. 2018. Capillary pumping independent of the liquid surface energy and viscosity. *Microsystems & Nanoengineering* 4 (1). doi:10.1038/s41378-018-0002-9.

Hassell, Bryan A., Girija Goyal, Esak Lee, Alexandra Sontheimer-Phelps, Oren Levy, Christopher S. Chen, and Donald E. Ingber. 2017. Human organ chip models recapitulate orthotopic lung cancer growth, therapeutic responses, and tumor dormancy in vitro. *Cell Reports* 21 (2):508–16.

Hiemstra, Pieter S., Gwendolynn Grootaers, Anne M. van der Does, Cyrille A.M. Krul, and Ingeborg M. Kooter. 2018. Human lung epithelial cell cultures for analysis of inhaled toxicants: Lessons learned and future directions. *Toxicology in Vitro* 47:137–46.

Hirama, Hirotada, Taku Satoh, Shinji Sugiura, Kazumi Shin, Reiko Onuki-Nagasaki, Toshiyuki Kanamori, and Tomoya Inoue. 2018. Glass-based organ-on-a-chip device for restricting small molecular absorption. *Journal of Bioscience and Bioengineering* 127 (5):641–6.

Howe, Alan, Andrew E. Aplin, Suresh K. Alahari, and Rudolph L. Juliano. 1998. Integrin signaling and cell growth control. *Current Opinion in Cell Biology* 10 (2):220–31.

Huh, Dongeun, Benjamin D. Matthews, Akiko Mammoto, Martín Montoya-Zavala, Hong Yuan Hsin, and Donald E. Ingber. 2010. Reconstituting organ-level lung functions on a chip. *Science* 328 (5986):1662–8.

Huh, Dongeun, Yu-Suke Torisawa, Geraldine A. Hamilton, Hyun Jung Kim, and Donald E. Ingber. 2012. Microengineered physiological biomimicry: Organs-on-chips. *Lab on a Chip* 12 (12):2156–64.

Ingber, Donald E. 2018. Developmentally inspired human 'organs on chips'. *Development* 145 (16). doi:10.1242/dev.156125.

Jang, Kyung-Jin, and Kahp-Yang Suh. 2010. A multi-layer microfluidic device for efficient culture and analysis of renal tubular cells. *Lab on a Chip* 10 (1):36–42.

Jeon, Jessie S., Simone Bersini, Mara Gilardi, Gabriele Dubini, Joseph L. Charest, Matteo Moretti, and Roger D. Kamm. 2015. Human 3D vascularized organotypic microfluidic assays to study breast cancer cell extravasation. *Proceedings of the National Academy of Sciences of the United States of America* 112 (1):214–9.

Ju, Jongil, Jung-Moon Ko, Hyeon-Cheol Cha, Joong Yull Park, Chang-Hwan Im, and Sang-Hoon Lee. 2009. An electrofusion chip with a cell delivery system driven by surface tension. *Journal of Micromechanics and Microengineering* 19 (1). doi:10.1088/0960-1317/19/1/015004.

Khodayari Bavil, Ali, and Jungkyu Kim. 2018. A capillary flow-driven microfluidic system for microparticle-labeled immunoassays. *Analyst* 143 (14):3335–42.

Kim, Hyun Jung, Dongeun Huh, Geraldine Hamilton, and Donald E. Ingber. 2012. Human gut-on-a-chip inhabited by microbial flora that experiences intestinal peristalsis-like motions and flow. *Lab on a Chip* 12 (12):2165–74.

Kim, Jungkyu, Erik C. Jensen, Amanda M. Stockton, and Richard A. Mathies. 2013. Universal microfluidic automaton for autonomous sample processing: Application to the Mars Organic Analyzer. *Analytical Chemistry* 85 (16):7682–8.

Kim, Su-Hwan, Sang-Hyuk Lee, Ju-Eun Lee, Sung Jun Park, Kyungmin Kim, In Seon Kim, Yoon Sik Lee, Nathaniel S. Hwang, and Byung-Gee Kim. 2018. Tissue adhesive, rapid forming, and sprayable ECM hydrogel via recombinant tyrosinase crosslinking. *Biomaterials* 178:401–12.

Kim, Sung-Jin, Sophie Paczesny, Shuichi Takayama, and Katsuo Kurabayashi. 2013. Preprogrammed capillarity to passively control system-level sequential and parallel microfluidic flows. *Lab on a Chip* 13 (11):2091–8.

Kimura, Hiroshi, Takashi Ikeda, Hidenari Nakayama, Yasuyuki Sakai, and Teruo Fujii. 2015. An on-chip small intestine-liver model for pharmacokinetic studies. *Journal of Laboratory Automation* 20 (3):265–73.

Kobuszewska, Anna, Ewelina Tomecka, Kamil Zukowski, Elzbieta Jastrzebska, Michal Chudy, Artur Dybko, Philippe Renaud, and Zbigniew Brzozka. 2017. Heart-on-a-chip: An investigation of the influence of static and perfusion conditions on cardiac (H9C2) cell proliferation, morphology, and alignment. *SLAS Technology* 22 (5):536–46.

Komeya, Mitsuru, Kazuaki Hayashi, Hiroko Nakamura, Hiroyuki Yamanaka, Hiroyuki Sanjo, Kazuaki Kojima, Takuya Sato, et al. 2017. Pumpless microfluidic system driven by hydrostatic pressure induces and maintains mouse spermatogenesis in vitro. *Scientific Reports* 7 (1):15459.

LeBleu, Valerie S., Brian Macdonald, and Raghu Kalluri. 2007. Structure and function of basement membranes. *Experimental Biology and Medicine* 232 (9):1121–9.

Lee, Joohyung, Zachary Estlack, Himali Somaweera, Xinmei Wang, Carla M.R. Lacerda, and Jungkyu Kim. 2018. A microfluidic cardiac flow profile generator for studying the effect of shear stress on valvular endothelial cells. *Lab on a Chip* 18 (19):2946–54.

Lee, Seung Hwan, Kyu Young Shim, Bumsang Kim, and Jong Hawn Sung. 2017. Hydrogel-based three-dimensional cell culture for organ-on-a-chip applications. *Biotechnology Progress* 33 (3):580–9.

Li, Na, Dandan Wang, Zhigang Sui, Xiaoyi Qi, Liyun Ji, Xiuli Wang, and Ling Yang. 2013. Development of an improved three-dimensional in vitro intestinal mucosa model for drug absorption evaluation. *Tissue Engineering Part C: Methods* 19 (9):708–19.

Liliensiek, Sara J., Sean Campbell, Paul F. Nealey, and Christopher J. Murphy. 2006. The scale of substratum topographic features modulates proliferation of corneal epithelial cells and corneal fibroblasts. *Journal of Biomedical Materials Research Part A* 79 (1):185–92.

Linshiz, Gregory, Erik Jensen, Nina Stawski, Changhao Bi, Nick Elsbree, Hong Jiao, Jungkyu Kim, et al. 2016. End-to-end automated microfluidic platform for synthetic biology: From design to functional analysis. *Journal of Biological Engineering* 10:3.

Ma, Liang, Jeremy Barker, Changchun Zhou, Wei Li, Jing Zhang, Bioyang Lin, Gregory Foltz, Jenni Kublbeck, and Paavo Honkakoski. 2012. Towards personalized medicine with a three-dimensional micro-scale perfusion-based two-chamber tissue model system. *Biomaterials* 33 (17):4353–61.

Marimuthu, Mohana, and Sanghyo Kim. 2013. Pumpless steady-flow microfluidic chip for cell culture. *Analytical Biochemistry* 437 (2):161–3.

Maschmeyer, Ilka, Tobias Hasenberg, Annika Jaenicke, Marcus Lindner, Alexandra Katharina Lorenz, Julie Zech, Leif-Alexander Garbe, et al. 2015. Chip-based human liver-intestine and liver-skin co-cultures: A first step toward systemic repeated dose substance testing in vitro. *European Journal of Pharmaceutics and Biopharmaceutics* 95 (Pt A):77–87.

Massensini, Andre R., Harmanvir Ghuman, Lindsey T. Saldin, Christopher J. Medberry, Timothy J. Keane, Francesca J. Nicholls, Sachin S. Velankar, Stephen F. Badylak, and Michel Modo. 2015. Concentration-dependent rheological properties of ECM hydrogel for intracerebral delivery to a stroke cavity. *Acta Biomaterialia* 27:116–30.

Meadows, Pamela Y., and Gilbert C. Walker. 2005. Force microscopy studies of fibronectin adsorption and subsequent cellular adhesion to substrates with well-defined surface chemistries. *Langmuir* 21 (9):4096–107.

Meyvantsson, Ivar, Jay W. Warrick, Steven Hayes, Allyson Skoien, and David J. Beebe. 2008. Automated cell culture in high density tubeless microfluidic device arrays. *Lab on a Chip* 8 (5):717–24.

Nawroth, Janna C., Riccardo Barrile, David Conegliano, Sander van Riet, Pieter S. Hiemstra, and Remi Villenave. 2018. Stem cell-based Lung-on-Chips: The best of both worlds? *Advanced Drug Delivery Reviews* 140:12–32.

Okagbare, Paul I., Jason M. Emory, Proyag Datta, Jost Goettert, and Steven A. Soper. 2010. Fabrication of a cyclic olefin copolymer planar waveguide embedded in a multi-channel poly(methyl methacrylate) fluidic chip for evanescence excitation. *Lab on a Chip* 10 (1):66–73.

Omidi, Yadollah, Lee Campbell, Jaleh Barar, David Connell, Saeed Akhtar, and Mark Gumbleton. 2003. Evaluation of the immortalised mouse brain capillary endothelial cell line, b.End3, as an in vitro blood–brain barrier model for drug uptake and transport studies. *Brain Research* 990 (1–2):95–112.

Pankov, Roumen, and Kenneth M. Yamada. 2002. Fibronectin at a glance. *Journal of Cell Science* 115 (20):3861–3.

Park, Joong Yull, Chang Mo Hwang, Soon Hyuck Lee, and Sang-Hoon Lee. 2007. Gradient generation by an osmotic pump and the behavior of human mesenchymal stem cells under the fetal bovine serum concentration gradient. *Lab on a Chip* 7 (12):1673–80.

Park, Joong Yull, Suel-Kee Kim, Dong-Hun Woo, Een-Joong Lee, Jong-Hoon Kim, and Sang-Hoon Lee. 2009. Differentiation of neural progenitor cells in a microfluidic chip-generated cytokine gradient. *Stem Cells* 27 (11):2646–54.

Park, Joong Yull, Mina Morgan, Aaron N. Sachs, Julia Samorezov, Ryan Teller, Ye Shen, Kenneth J. Pienta, and Shuichi Takayama. 2010. Single cell trapping in larger microwells capable of supporting cell spreading and proliferation. *Microfluidics and Nanofluidics* 8 (2):263–8.

Puleo, Christopher M., Winnette McIntosh Ambrose, Toshiaki Takezawa, Jennifer Elisseeff, and Tza-Huei Wang. 2009. Integration and application of vitrified collagen in multilayered microfluidic devices for corneal microtissue culture. *Lab on a Chip* 9 (22):3221–7.

Quiros-Solano, William F., Nikolas Gaio, Cinzia Silvestri, Yusuf B. Arik, Oscar M.J.A. Stassen, Ad D. van der Meer, Carlijn V.C. Bouten, et al. 2018. A novel method to transfer porous PDMS membranes for high throughput Organ-on-Chip and Lab-on-Chip assembly. In *2018 IEEE Micro Electro Mechanical Systems (MEMS)*. Belfast.

Ronaldson-Bouchard, Kacey, and Gordana Vunjak-Novakovic. 2018. Organs-on-a-Chip: A fast track for engineered human tissues in drug development. *Cell Stem Cell* 22 (3):310–24.

Sacchetti, Benedetto, Alessia Funari, Stefano Michienzi, Silvia Di Cesare, Stefania Piersanti, Isabella Saggio, Enrico Tagliafico, et al. 2007. Self-renewing osteoprogenitors in bone marrow sinusoids can organize a hematopoietic microenvironment. *Cell* 131 (2):324–36.

Seiler, Kurt, D. Jed Harrison, and A. Manz. 1993. Planar glass chips for capillary electrophoresis: Repetitive sample injection, quantitation, and separation efficiency. *Analytical Chemistry* 65 (10):1481–8.

Seo, Yoojin, Youngmee Jung, and Soo Hyun Kim. 2018. Decellularized heart ECM hydrogel using supercritical carbon dioxide for improved angiogenesis. *Acta Biomaterialia* 67:270–81.

Shin, Woojung, and Hyun Jung Kim. 2018. Intestinal barrier dysfunction orchestrates the onset of inflammatory host-microbiome cross-talk in a human gut inflammation-on-a-chip. *Proceedings of the National Academy of Sciences of the United States of America* 115 (45):E10539–47.

Skardal, Aleksander, Sean V. Murphy, Mahesh Devarasetty, Ivy Mead, Hyun-Wook Kang, Young-Joon Seol, Yu Shrike Zhang, et al. 2017. Multi-tissue interactions in an integrated three-tissue organ-on-a-chip platform. *Scientific Reports* 7 (1):8837.

Sontheimer-Phelps, Alexandra, Bryan A. Hassell, and Donald E. Ingber. 2019. Modelling cancer in microfluidic human organs-on-chips. *Nature Reviews Cancer* 19, 65–81.

Sosa-Hernandez, Juan Eduardo, Angel M. Villalba-Rodriguez, Kenya D. Romero-Castillo, Mauricio A. Aguilar-Aguila-Isaias, Isaac E. Garcia-Reyes, Arturo Hernandez-Antonio, Ishtiaq Ahmed, et al. 2018. Organs-on-a-chip module: A review from the development and applications perspective. *Micromachines (Basel)* 9 (10):536.

Stucki, Andreas O., Janick D. Stucki, Sean R.R. Hall, Marcel Felder, Yves Mermoud, Raplh A. Schmid, Thomas Geiser, and Olivier T. Guenat. 2015. A lung-on-a-chip array with an integrated bio-inspired respiration mechanism. *Lab on a Chip* 15 (5):1302–10.

Sung, Jong Hwan, Carrie Kam, and Michael L. Shuler. 2010. A microfluidic device for a pharmacokinetic-pharmacodynamic (PK-PD) model on a chip. *Lab on a Chip* 10 (4):446–55.

Sunkara, Vijaya, Dong-Kyu Park, Hyundoo Hwang, Rattikan Chantiwas, Steven A. Soper, and Yoon-Kyoung Cho. 2011. Simple room temperature bonding of thermoplastics and poly(dimethylsiloxane). *Lab on a Chip* 11 (5):962–5.

Takezawa, Toshiaki, Katsuyuki Ozaki, Aya Nitani, Chiyuki Takabayashi, and Tadashi Shimo-Oka. 2004. Collagen vitrigel: A novel scaffold that can facilitate a three-dimensional culture for reconstructing organoids. *Cell Transplantation* 13 (4):463–74.

Toh, Yi-Chin, Teck Chuan Lim, Dean Tai, Guangfa Xiao, Danny van Noort, and Hanry Yu. 2009. A microfluidic 3D hepatocyte chip for drug toxicity testing. *Lab on a Chip* 9 (14):2026–35.

Torisawa, Yu-suke, Catherine S. Spina, Tadanori Mammoto, Akiko Mammoto, James C. Weaver, Tracy Tat, James J. Collins, and Donald E. Ingber. 2014. Bone marrow-on-a-chip replicates hematopoietic niche physiology in vitro. *Nature Methods* 11 (6):663–9.

Wang, Ying I., Carlota Oleaga, Christopher J. Long, Mandy B. Esch, Christopher W. McAleer, Paula G. Miller, James J. Hickman, and Michael L. Shuler. 2017. Self-contained, low-cost body-on-a-chip systems for drug development. *Experimental Biology and Medicine (Maywood)* 242 (17):1701–13.

Wang, Xiaolin, Da Zhao, Duc T.T. Phan, Jingguan Liu, Xiang Chen, Bin Yang, Christopher C.W. Hughes, Weijia Zhang, and Abraham P. Lee. 2018. A hydrostatic pressure-driven passive micropump enhanced with siphon-based autofill function. *Lab on a Chip* 18 (15):2167–77.

Wilmer, Martijn J., Chee Ping Ng, Henriëtte Lanz, Paul Vulto, Laura Suter-Dick, and Rosalinde Masereeuw. 2016. Kidney-on-a-chip technology for drug-induced nephrotoxicity screening. *Trends in Biotechnology* 34 (2):156–70. doi:10.1016/j.tibtech.2015.11.001.

Xia, Younan, and George M. Whitesides. 1998. Soft lithography. *Annual Review of Materials Science* 28:153–84.

Ye, Nannan, Jianhua Qin, Weiwei Shi, Xin Liu, and Bingcheng Lin. 2007. Cell-based high content screening using an integrated microfluidic device. *Lab on a Chip* 7 (12):1696–704.

Zervantonakis, Ioannis K., Shannon K. Hughes-Alford, Joseph L. Charest, John S. Condeelis, Frank B. Gertler, and Roger D. Kamm. 2012. Three-dimensional microfluidic model for tumor cell intravasation and endothelial barrier function. *Proceedings of the National Academy of Sciences of the United States of America* 109 (34):13515–20.

Zhang, Yu Shrike, Julio Aleman, Su Ryon Shin, Tugba Kilic, Duckjin Kim, Seyed Ali Mousavi Shaegh, Solange Massa, et al. 2017. Multisensor-integrated organs-on-chips platform for automated and continual in situ monitoring of organoid behaviors. *Proceedings of the National Academy of Sciences of the United States of America* 114 (12):E2293–302.

Zhang, Xiannian, Zitian Chen, and Yanyi Huang. 2015. A valve-less microfluidic peristaltic pumping method. *Biomicrofluidics* 9 (1):014118.

Zhu, Xioyue, Leonard Yi Chu, Bor-han Chueh, Minguwu Shen, Bhaskar Hazarika, Nandita Phadke, and Shuichi Takayama. 2004. Arrays of horizontally-oriented mini-reservoirs generate steady microfluidic flows for continuous perfusion cell culture and gradient generation. *Analyst* 129 (11):1026–31.

Zimmermann, Martin, Heinz Schmid, Patrick Hunziker, and Emmanuel Delamarche. 2007. Capillary pumps for autonomous capillary systems. *Lab on a Chip* 7 (1):119–25.

7 Microfluidic Techniques for High-Throughput Cell Analysis

Dongwei Chen, Juanli Yun,
Yuxin Qiao, and Jian Wang
Chinese Academy of Sciences

Ran Hu, Beiyu Hu, and Wenbin Du
Chinese Academy of Sciences
University of the Chinese Academy of Sciences

CONTENTS

7.1 MICROFLUIDIC DEVICES FOR CHEMOTAXIS ASSAYS

Chemotaxis refers to the ability of motile organisms to respond to concentration gradients of a chemoeffector, either chemoattractant or chemorepellent. Chemotaxis is a basic behavior of cells to adapt to environmental changes (Eisenbach 2004). Chemotaxis enables microorganisms or cells to migrate toward food sources, or to escape from unfavorable chemicals. Therefore, studies of chemotaxis are of great importance in bioremediation of contaminated environment, controlling pathogenic infection, and understanding of human biology, development, and diseases.

To study the chemotactic ability of bacterial or eukaryotic cells, several traditional chemotaxis assays have been developed, such as swarm plate assays (Adler 1966) and capillary assays (Adler and Dahl 1967). However, these methods are mostly qualitative methods, are time-consuming, and have low efficiency. Alternative techniques, including microfluidic devices, have recently been developed to quantitatively study chemotaxis (Ahmed et al. 2010, Wu et al. 2013). Overall, the microfluidic chemotaxis devices (μFCDs) typically include a concentration gradient region, a source region, a sink region, and inlets and outlets of the channels. The microfluidic gradient generators can be roughly divided into flow-based and diffusion-based methods according to the methods of chemical gradient generation. Flow-based methods generate a gradient of chemicals by convective transport of laminar flow. The initial concentration of chemical molecular determines the overall characteristics of the concentration gradient. Diffusion-based methods build a concentration gradient in the convection-free region by diffusion propagation of chemical molecules. Notably, as the microfluidic channels are comparable in size to mammalian or bacterial cells, the possibility is raised of studying chemotaxis at the level of individual cells quantitatively, and facilitating streamlined chemotactic cell capture and manipulation.

7.1.1 FLOW-BASED CHEMOTAXIS

The laminar flow at the microscale is one of the most important features of fluid physics relative to the macroscopic scale (Squires and Quake 2005). T-shaped or Y-shaped channels can generate turbulent diffusion between two-phase laminar flow, which is a simple and effective for create gradients (Ahmed et al. 2010, Wolfram et al. 2016). Jeon et al. (2000) first introduced the microchannel network gradient generators. The low Reynolds number of microfluidic characteristics allows two aqueous streams to merge and flow in parallel. The diffusion interface between parallel flows allows formation of a concentration gradient perpendicular to the flow direction.

Mao et al. (2003) first developed a parallel-flow device to study the chemotactic behavior of bacteria (Figure 7.1a). The device was fabricated from polydimethylsiloxane (PDMS) and contains three lateral inlets for the chemo-effector, *Escherichia coli* cells and buffer introduction. Solutions flow from left to right along the main channel, and solutions exit the device through sector-shaped outlets on the right side. Diffusion between chemoeffector and buffer streams generates a stable concentration gradient across the microchannel perpendicular to the direction of flow. The bacterial cells encounter the gradient diffusion interface, and chemotaxis occurs and is accessed by the distribution of bacteria in each of 22 outlets. This device provides high sensitivity to detect responses with concentration of L-aspartate as low as 3.2 nM, nearly three orders of magnitude lower than traditional capillary assays.

Englert et al. (2009) introduced a μFCD using the microchannel network resembles a "Christmas tree" shape, in which a splitting and mixing process is conducted repeatedly in an array of bifurcated microchannels (Figure 7.1b). Different concentrations of the solution are initially introduced from the two inlets, are each divided in two, and are then mixed in a serpentine channel repeatedly. At the end of the

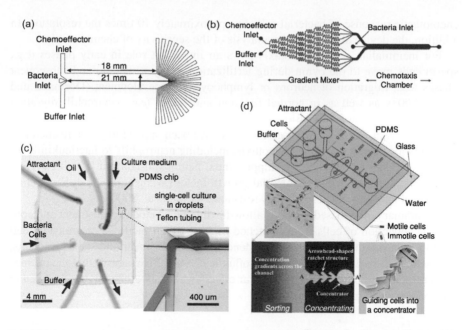

FIGURE 7.1 Flow-based microfluidic devices for microbial chemotaxis assays. (a) Parallel-flow device developed by Mao et al. (2003). (Reprinted by permission from Mao et al, 2003. *PNAS*, 100(9):5449–5454. Copyright (2003) National Academy of Sciences, USA.) (b) Microfluidic chemotaxis device developed by Englert et al. (2009), which consists a micro-mixer array and a chemotaxis chamber. (Adapted from Springer Nature Customer Service Centre GmbH: Humana Press, *Chemotaxis* by Tian Jin and Dale Hereld © 2009.) (c) An integrated microfluidic SlipChip device for chemotactic sorting and cultivation. (Reprinted from Dong et al. 2016. *Scientific Reports*, 6:24192 licensed under CC BY 4.0.) (d) Microfluidic device developed by Kim et al. (2011) consists of a Y-shaped microchannel with arrowhead-shaped ratchet structures to trap and accumulate chemotactic cells. (Reproduced from Kim et al., 2011. *Analyst*, 136:3238–3243. Copyright (2011) Royal Society of Chemistry.)

pre-mixed network, the fluid streams converge into a wide channel to form a gradient across the channel. An arbitrarily shaped concentration profile can be achieved for bacterial chemotaxis in stable, competing gradients by adjusting the relative flow rates of inlets.

Single-cell isolation and enumeration of chemotactic cells in droplets is demonstrated in a microfluidic device developed by Dong et al. (2016) (Figure 7.1c). The device contains a concentration gradient generator and a droplet single encapsulating unit for subsequent assays. The device can be used for continuous sorting of chemotactic cells as long as there is flow through the channel followed by high-throughput single-cell cultivation and enumeration in droplets. Other researchers (Kim et al. 2011) have developed this microfluidic flow-based chemotaxis assays by integrating an arrowhead-shaped ratchet "concentrator" into the main channel of the device (Figure 7.1d). In the microfluidic device, bacterial cells are sorted by chemotactic response and accumulated in the concentrator array, consequently amplifying the chemotactic response of bacterial cells by approximately 10 times the resolution. In addition, the device also allows the analysis of the sensitivity of chemoreceptors.

For mammalian cells, chemotaxis plays an important role in early phases (e.g., sperm cells move to the ovum during fertilization and gastrulation) and subsequent phases (e.g., migration of neurons or lymphocytes) of development (Dormann and Weijer 2003), as well as in normal function and health (e.g., neutrophil migration during inflammation and invasion and metastasis of cancer cells) (Franz et al. 2002, de Oliveira et al. 2016). Flow-based methods have been applied to generate chemical gradients for mammalian cell chemotaxis, including neutrophils to interleukin-8 (Li Jeon et al. 2002, Lin et al. 2004), hippocampal neurons to laminin (Dertinger et al. 2002), breast cancer cells to epidermal growth factor (Wang et al. 2004, Saadi et al. 2006), and lung cancer chemotherapy resistance (Siyan et al. 2009).

In summary, flow-based methods allow flexible flow velocity adjustment and continuous collection of cells from branched outlets, which facilitate chemotactic cell sorting from mixtures. Despite the tendency to generate linear and stable gradients, the major drawback of flow-based microfluidic chemotaxis assays is that the chemotaxis response might be masked by shear effects when bacterial cells are exposed in the concentration gradient flow field.

7.1.2 Diffusion-Based Chemotaxis

The free diffusion of molecules between solutions is another way to generate a concentration gradient. Chemical molecules spontaneously diffuse from a region of higher concentration to a region of lower concentration, and a gradient is generated in the diffusion region connecting higher and lower concentration regions. Unlike traditional methods, microfluidic methods utilize microchannels with high fluidic resistance (Atencia et al. 2009, Shen et al. 2014), porous membranes (Diao et al. 2006, Nagy et al. 2015), or hydrogels (Cheng et al. 2007, Si et al. 2012) to provide physical barriers against fluid convection, thereby separating the gradient generation region from the high- and low-concentration chemical sources. Since chemical molecular transport in the absence of flow is driven by diffusion, the concentration gradient will reach a steady-state when the transmission of influx and efflux reaches equilibrium. There is a balance between the time it takes to form a stable gradient and the fluidic resistance used. When using high-resistance material, the molecular diffusion is slow, and it takes a long time to reach a steady-state gradient, but it will be stable for a longer time. Otherwise, when a low-resistance material is used, the time to reach the steady-state gradient is short, but stability time of the gradient will also be shorter.

Diao et al. (2006) developed a three-channel microfluidic chip using nitrocellulose membrane as barrier to generate a linear concentration gradient. Three parallel microchannels were cut out by CO_2 laser in the nitrocellulose membrane and sandwiched between the top polycarbonate manifold and the bottom glass slide (Figure 7.2a). Diffusion through the porous nitrocellulose membrane from the source channel to the sink channel could establish a steady linear concentration gradient in the center channel, in which microbial cells moved only by chemotactic response. This device can afford a static chemical concentration gradient and is easy to operate without additional experimental complexities. However, the use of nitrocellulose membranes has

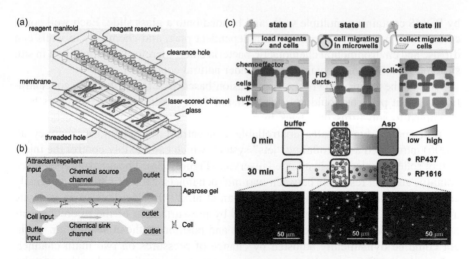

FIGURE 7.2 Diffusion-based microfluidic chemotaxis devices for microbial cells. (a) A three-channel microfluidic chip developed by Diao et al. (2006). (Reprinted from Diao et al. 2006. *Lab on a Chip*, 6:381–388. Copyright (2006) Royal Society of Chemistry.) (b) A hydrogel-based linear gradient generator of Cheng et al. (2007). (Reprinted from Cheng et al. 2007. *Lab on a Chip*, 7:763–769. Copyright (2007) Royal Society of Chemistry.) (c) Schematic illustration and operation of a microfluidic SlipChip device based on FID between microwells. SlipChip allows instrument-free cell loading, chemotaxis assays, and collection of chemotactic cells. (Reprinted from Shen et al. 2014. *Lab on a Chip*, 14:3074–3080. Copyright (2014) Royal Society of Chemistry.) FID, free interface diffusion.

some limitations, including long gradient establishing time and short gradient maintenance time. Derived from the three-channel design by Diao et al. (Cheng et al. 2007) built an agarose-based microfluidic device to generate a steady and long-term linear chemical concentration gradient (Figure 7.2b). The agarose gel channels are obtained by pouring hot agarose solution onto the silicon master and cooling to patterned gel. Agarose provides an effective diffusion matrix and is biocompatible for cells. The chemotaxis of *E. coli* and differentiated HL-60 cells was monitored using this device.

Shen et al. (2014) introduced a microfluidic SlipChip device for microbial chemotaxis study and collection of chemotactic bacteria (Figure 7.2c). The device consists of two glass plates with reconfigurable microchannels. The connection states of the microchannels can be switched by a simple slipping operation to establish a dynamic diffusion gradient based on free interface diffusion, and to bring about the chemotactic migration of microbial cells in the SlipChip. Chemotaxis index was introduced to assess the chemotactic ability by monitoring and counting the number of bacterial cells migrated into the microwells. In addition, the migrated cells could be collected by pipetting for further studies off the chip. The SlipChip-based chemotaxis has been applied in chemotactic screening of imidazolinone-degrading bacteria from contaminated soil (Chen et al. 2019), as well as identification of chemoeffectors for plant growth-promoting rhizobacteria (Feng et al. 2018, 2019).

A microfluidics-based in situ chemotaxis assay was developed by Lambert et al. (2017) to study the behaviors of marine microorganisms. This device was fabricated by PDMS containing multiple wells and bonded onto a glass slide. Each well could connect to external environment via independent port, and chemicals in wells can diffuse and environmental chemotactic bacteria enter; therefore, it supplies an in situ system to examine microbial behavior under natural conditions.

One of the major drawbacks of diffusion-based µFCDs is that the concentration gradient patterns could not be alternated dynamically. Some researchers have attempted to solve this drawback by actively controlling diffusion. To establish a spatiotemporal dynamically controllable concentration gradient, Keenan et al. (2006) introduced a microfluidic "jets" system, which can flexibly control the injection of soluble molecules in an open reservoir (Figure 7.3a). The microfluidic device contains an open reservoir between two microchannels, each of which is internally connected by an array of channels (Keenan et al. 2010). The device can generate a stable, reproducible gradient in minutes by pressurization of the microchannels to eject solutions filled in them. The slope and position of the concentration gradient could be dynamically adjusted by change of pressures on two main channels independently. This microjet device was used to study gradient-induced neutrophil desensitization (Keenan et al. 2010).

To realize spatiotemporal control of diffusive chemical gradient, a microfluidic device, known as a microfluidic palette, was introduced by Atencia et al. (2009) (Figure 7.3b). The chemical delivery channels are separated from diffusion chamber in different layers; therefore, the overlapping chemical gradient can be generated in the circular chamber without any convection. The chemical gradient can either be held constant for long term by pressure balance of three convection units or

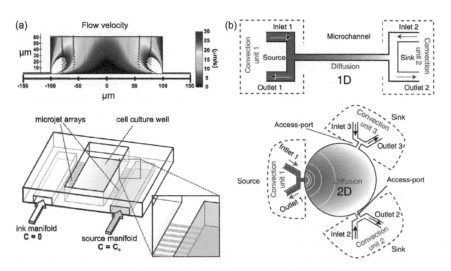

FIGURE 7.3 Dynamically controlled diffusion-based chemotaxis devices for microbial cells. (a) Schematic of the micro-jet device and its fluid dynamic simulation. (Modified from Nirveek Bhattacharjee et al. 2010. *Integrative Biology*, 2(11–12):669–679. Copyright © 2010, Oxford University Press.) (b) The microfluidic palette capable of generating overlapping and spatiotemporal controlled gradients. (Reprinted from Atencia et al. 2009. *Lab on a Chip*, 9:2707–2714. Copyright (2009) Royal Society of Chemistry.)

dynamically changed by switching the pump and introducing the reagent from each inlet. The chemotactic response of *Pseudomonas aeruginosa* to glucose over both space and time was studied in this device.

In vivo, mammalian cells grow in three-dimensional (3D) microenvironments within the support structure called the extracellular matrix (Abbott 2003). Recent research has revealed important differences in cellular behaviors in 2D and 3D cultures (Abbott 2003), including chemotaxis. Therefore, it is highly desirable to develop microfluidic devices that incorporate 3D extracellular matrices, such as collagen gel and agarose gel, to provide physiologically relevant microenvironments for cell migration. A number of diffusion-based microfluidic devices were developed for 3D chemotaxis studies of mammalian cells (Kim and Wu 2012).

Saadi et al. (2007) presented a "Ladder Chamber", which uses a two-compartment diffusion system to generate steady-state gradients (Figure 7.4a). The chamber consists

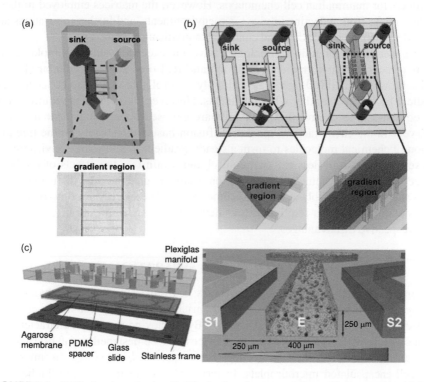

FIGURE 7.4 Diffusion-based microfluidic devices for mammalian cell chemotaxis. (a) The Ladder Chamber designed by Saadi et al. (2007) to generate concentration gradients in microgrooves. (Adapted from Springer Nature Customer Service Center GmbH: Springer Nature, *Biomedical Microdevices* by Saadi et al. © 2007.) (b) Schematic of microfluidic device developed by Mosadegh et al. (2007) to produce spatiotemporal gradients. (Reprinted from Mosadegh et al. 2007. *Langmuir*, 23(22):10910–10912. Copyright (2007) American Chemical Society.) (c) Agarose-based 3D μFCD presented by Haessler et al. (2009). (Reprinted from Springer Nature Customer Service Center GmbH: Springer Nature, *Biomedical Microdevices* by Haessler et al. © 2009.) 3D, three-dimensional; μFCD, microfluidic chemotaxis device.

of two main channels and an array of microchannels that bridge them. Concentration gradients are obtained by diffusion of soluble molecules through collagen gel filled in the microchannels, which provides a 3D environment for cells. Neutrophil migration under chemoattractant (interleukin-8) gradients was observed in the chamber. Mosadegh et al. (2007) further evolved the ladder chamber by changing the geometries of gradient-generating region (Figure 7.4b). Aizel et al. (2017) developed a tunable microfluidic system for chemotaxis studies in a 3D collagen matrix. This microsystem can generate either homogeneous gradients by pure diffusion or spatially evolving gradients by convection diffusion. A stable gradient can be maintained over days for long-duration chemotaxis experiments. Cancer cells migrating along chemokine gradients was observed in this device. Agarose gels are also used as the matrix for mammalian cell chemotaxis in a 3D μFCD (Figure 7.4). In summary, by introducing biomatrices into microfluidic devices to mimic extracellular matrix, these studies which provides more physiologically compatible microenvironments than two-dimensional diffusion chambers for mammalian cell chemotaxis. However, the matrices employed in these studies are far from reflecting in vivo microenvironments, and further innovations are needed to enable more flexible and controllable gradient generation.

In general, microfluidic technologies have far exceeded other technologies in terms of concentration gradient regulation and real-time study of cellular chemotaxis. Flow-based μFCDs generate a highly controllable chemical concentration gradient by laminar mixing within channels. However, these devices require external equipments to control flow, and cells are exposed under fluid shear forces to intervene in the chemotactic response. Diffusion-based μFCDs rely on the free diffusion of chemical molecules to form a steady gradient in a flow-free environment. These devices require less external control and minimizes the effects of fluids on the cells. However, chemical concentration gradients generated by diffusion-based μFCDs cannot be flexibly adjusted to some extent. Therefore, both flow-based and diffusion-based μFCDs require further development and innovation to overcome their respective disadvantages.

7.2 MICROFLUIDIC SORTING AND SEPARATION OF CELLS

Continuous sorting of micron-sized particles has great significance in biomedical engineering, since it allows separation and enrichment of target cells from complex matrices or mixtures for downstream studies. By employing the unique characteristics of microscale phenomena, microfluidic devices can be designed and fabricated for rapid and accurate separation and sorting of microparticles, including cells, microbes, and cell-encapsulated microdroplets. In terms of driving mode, microfluidic sorting of microparticles can be divided into passive and active methods (Gossett et al. 2010). Passive sorting methods separate different sizes of microparticles mainly by designing the geometric dimensions of the microchannels. Dean flow (Seo et al. 2007, Bhagat et al. 2008), pinched flow fractionation (PFF) (Yamada et al. 2004), multi-orifice flow fractionation (Park and Jung 2009), deterministic lateral displacement (Huang 2004), and microscale filtration (Ji et al. 2008) belong to the passive sorting methods. The active sorting technology controls the microparticles under an external field, which results in high efficiency and specificity, including dielectrophoresis

(DEP) (Markx et al. 1994, Agresti et al. 2010, Tao et al. 2016), surface acoustic wave (SAW) (Wixforth et al. 2004, Franke et al. 2009), magnetic force (Adams et al. 2008), and optical manipulation (Applegate et al. 2004).

7.2.1 PASSIVE SORTING TECHNIQUES

To sort the cells/droplets/particles for separation and analysis, passive sorting techniques have been established without rely on external fields. Among these techniques, Dean flow has attracted considerable interest, since it offers high throughput and efficiency. Dean flow, the secondary vortices transverse to the fluid flow direction in a curved pipe, was first observed by the British Scientist W. R. Dean (Dean 1928). The Dean Number (De) is the characteristic parameter used to describe the secondary flow in curved channels (Eq. 7.1):

$$De = Re\sqrt{\frac{D}{2R_c}} \tag{7.1}$$

where Re is the Reynolds number, D is the width of the channel, and R is the radius of the channel curvature. Dean Number indicates the ratio of the inertial forces to the viscous forces and can characterize the strength of secondary flow or Dean vortices.

In Dean flow devices, equilibrium position-based separation was explored for membrane-free filtration, and enriching particulate matter in suspension depend on their sizes (Figure 7.5). Suspended particles are driven through the microchannel with spiral or asymmetrical structures. Initially, randomly distributed in the fluid, particles above a critical size migrate to equilibrium positions. Strategically placed outlet channels or bifurcations separate fluid with concentrated particles of specific sizes. The spiral dean flow was first introduced for microparticle separation in 2007

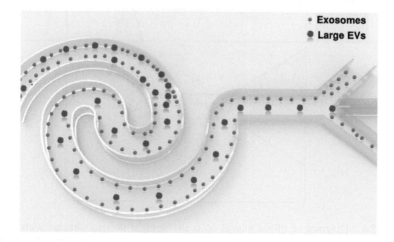

FIGURE 7.5 Schematic of the size-based elasto-inertial exosome sorting technique with input section, midstream section and trifurcated output section. (Reprinted with permission from Zhou et al. 2019. *Analytical Chemistry.* Copyright (2019) American Chemical Society.)

(Seo, Lean, and Kole 2007), demonstrating a concentration of 10.5-μm polystyrene microspheres with greater than 99% efficiency. Bhagat et al. (2008) further
demonstrated a complete separation of 7.3-μm beads from 1.9-μm beads. The spiral dean flow was applied to separate 8-μm neuroblastoma and 15-μm glioma cells
with 80% efficiency and >90% relative viability with high throughput of ~1 million
cells/min (Kuntaegowdanahalli et al. 2009). Using viscoelastic fluids, a Dean flow
device for elasto-inertial focusing and sorting of submicron particles and exosomes
(30–200 nm) was established, separating exosomes with purity >92% and recovery
rate >81% (Zhou et al. 2019).

PFF is an alternative technique to Dean flow for the continuous size separation
of particles that utilizes inertial forces with a simple laminar flow setting (Yamada
et al. 2004). As shown in Figure 7.6, PFF uses a pinched channel-segment to align
particles to one sidewall followed by a spreading flow profile to exert forces with

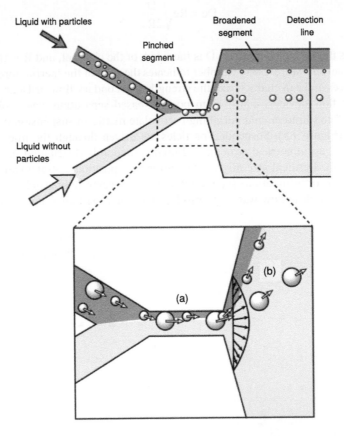

FIGURE 7.6 Diagram of PFF. Channel (a) is the straight flowing channel with different
size particles. According to the size, smaller particles are flowing into the dark segment in
the channel (b), and larger particles are flowing into the other path. (Reprinted from Yamada
et al. 2004. *Analytical Chemistry*, 76(18):5465–5471. Copyright (2004) American Chemical
Society.) PFF, pinched flow fractionation.

varied direction toward the main channel; therefore, the particles are separated perpendicularly to the flow direction according to their sizes. PFF can separate 15-μm beads from 30-μm beads with 99% recovery rate. To enhance the particle separation performance, acoustically driven microbubbles have been integrated with PFF to separate differently sized microparticles with diameters of 10 and 2 μm.

To realize size-dependent separation of particles, lateral particle migration has been achieved simply using the co-flow configuration of viscoelastic and Newtonian fluids (Ha et al. 2016, Yuan et al. 2016). As illustrated in Figure 7.7, a mixture of microparticles with different sizes in a viscoelastic fluid is loaded into the device with co-flow of a Newtonian fluid, that is, deionized water. The elastic lift forces present in the non-Newtonian fluidic pass the larger particles over to the Newtonian fluid while leaving the smaller particles in the viscoelastic fluid. This lateral migration method was applied in continuous purification of leukemic Jurkat cells (Yuan et al. 2017), which offers dilution-free and highly efficient particle separation over a wide range of flow rates.

7.2.2 ACTIVE SORTING TECHNIQUES

The active sorting technologies can effectively select and particles based on measurements using use a series of external fields, including DEP, SAW, magnetic force, and optical manipulation. By combining these active sorting techniques with on chip real-time imaging, optical detection, and electrochemical measurement, particle sorting systems with high specificity and accuracy have been established, which opens up a wide range of applications in biological and biomedical applications. Among various active sorting techniques, DEP technology is widely used

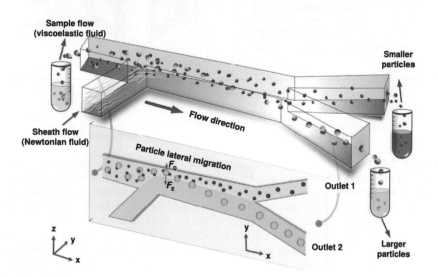

FIGURE 7.7 Diagram of isolation of different size particles using a Newtonian and co-flow fluid. (Reprinted with permission from Yuan et al. 2017. *Analytical Chemistry*, 89(17):9574–9582. Copyright (2017) American Chemical Society.)

because of its label-free advantage, high selectivity, and high efficiency. DEP is a phenomenon in which a nonuniform electric field exerted a force on a particle to induce a directional migration. DEP can be used to manipulate, transport, separate, and sort various types of cells, microorganisms, and particles. A DEP motion of particles can be induced toward the microelectrodes (positive DEP) or away from the microelectrodes (negative DEP). DEP-activated cell sorter device has been established to perform size-based fractionation and separation of platelets from whole blood (Pommer et al. 2008). Park et al. (2011) described a continuous DEP separation of bacterial cells in complex physiological samples, achieving 94.3% and 87.2% for *E. coli* in cerebrospinal fluid and whole blood, respectively (Figure 7.8a). The device can process biological samples with a flow rate of up to 800 μL/h.

In addition to direct manipulation of cells in continuous suspension, DEP can also be used to actively manipulate monodisperse droplets with picoliter to nanoliter

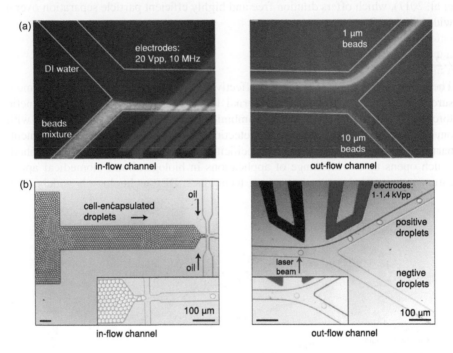

FIGURE 7.8 (a) Continuous cell separation using dielectrophoresis. Separation channel inflow (upper inlet—DI water, lower inlet—TAE buffer suspending a binary-bead mixture) and separation channel outflow shows that 1-μm beads are selectively separated from the sample stream into DI water stream while 10-μm beads remain in the sample stream. The electrodes were actuated with an AC signal of 20 V_{pp} at 10 MHz. (Adapted with permission from Park et al. 2011. *Lab on a Chip*, 11:2893–2900. Copyright (2011) Royal Society of Chemistry.) (b) Droplet loading and fluorescence-activated sorting by dielectrophoresis. The black part is negative electrode and red is positive electrode. Droplets were deflected into upper channel when the electrodes were charged with an AC electric field. When the AC electric field is off, droplets enter the lower channel, which has a lower hydraulic resistance (inset). Scale bars: 100 μm. (Adapted from Baret et al. 2009. *Lab on a Chip*, 9:1850–1858. Copyright (2009) Royal Society of Chemistry.)

volumes in an immiscible phase. Droplet-based microfluidics provide a range of convenient operations for the compartmentalization of single cells, as well as mixing, splitting, sorting, and detecting, with high throughput (Guo et al. 2012, Joensson and Andersson Svahn 2012). Among these methods, fluorescence-activated droplet sorting is a powerful approach that uses laser-induced fluorescence detection and DEP to detect and sort single-cell encapsulated droplets at high throughput (Figure 7.8b). Baret et al. (2009) first applied fluorescence-activated droplet sorting to sort *E. coli* cells expressing enzyme β-galactosidase at rates of ~300 droplets/s. The false-positive error rate of sorting was less than 1 in 10,000 droplets. There are also many other applications in the enzymes-producing microorganism sorting field, such as α-amylase (Beneyton et al. 2016), aldolase (Obexer et al. 2017), phosphotriesters (Colin et al. 2015), cellulose (Ostafe et al. 2014), lipase (Qiao et al. 2017), and sulfatase (van Loo et al. 2018).

SAW is an acoustic wave traveling along the surface of a material exhibiting elasticity, commonly excited on a piezoelectric crystal using an interdigitated electrode pattern, or an interdigital transducer (IDT). SAW has been applied in microparticle manipulation, as well as the sorting of droplets in microfluidic channel (Wixforth et al. 2004, Franke et al. 2009). The SAW devices consist of two series of IDTs in opposite directions or only one IDT in one side. SAW forces can act on particles flowing in the microfluidic chips, resulting in a change of distribution. As Shi et al. (2008) reported, beads were induced into the channels under the pressure can be focused in the center of the channel by SAW (Figure 7.9a). Franke et al. (2009) designed a microfluidic device with an IDT to pose SAW for switching the direction

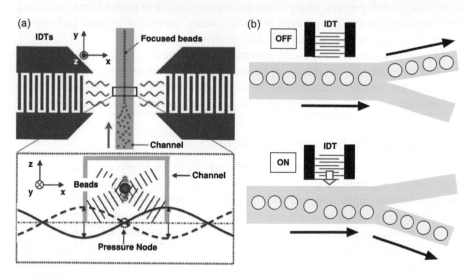

FIGURE 7.9 Schematic of the standing SAW focusing device and hybrid PDMS–SAW chip. (a) Two identical IDTs. Microchannels act as acoustic resonators. (Reprinted form Shi et al. 2008. *Lab on a Chip*, 8:221–223. Copyright (2008) Royal Society of Chemistry.) (b) Single IDT. Drive the droplets to the lower channel as the SAW power is switched. (Adapted from Franke et al. 2009. *Lab on a Chip*, 9:2625–2627. Copyright (2009) Royal Society of Chemistry.) IDT, interdigital transducer; PDMS, polydimethylsiloxane; SAW, surface acoustic wave.

of droplets in branched PDMS channels (Figure 7.9b). Based on the same mechanism, an acoustic fluorescence-activated droplet sorting system has been established (Schmid et al. 2014), achieving sorting rates of $3,000\,s^{-1}$ at acoustic power of 15 dBm.

In summary, microfluidic cell sorting is a range of technologies that separates specific cells from mixtures. With powerful and tunable external fields applied, active sorting techniques can provide ultra-high separation efficiency, accuracy, and throughput. Encapsulation of single cells or biomolecules in picoliter to nanoliter droplets can be coupled with various active droplet-sorting systems with high throughput and low cost, which can be used to address critical problems for biomedical research in the future. In light of these techniques' promising prospects, microfluidic cell sorting may play an extremely important role in biological and biomedical research. However, there are still many aspects that warrant improvement, such as sorting throughput, real-time measurement of single cells, sorting costs, and user-friendly operation.

7.3 SINGLE-CELL GENOMIC ANALYSIS

Single-cell genomics is a powerful tool to enlarge our understanding of genetics, and it elucidates the metabolic, genotypic, and evolutionary diversity and heterogeneity at the single-cell resolution. Single-cell analysis can be used to identify heterogeneous responses or to compare cell states between complex samples with unknown population structure (Gawad et al. 2016). This approach is highly useful in analyzing rare cell types such as circulating tumor cells and uncultured microbes. The application of single-cell genomic analysis has grown remarkably in recent years, indicating that single-cell genomics is robust in biology fields, especially in basic and clinical research, which are closely related to human disease diagnostics.

Microfluidics-based single-cell genomics has become a robust tool in the last decade, especially for droplet microfluidics. Figure 7.10 shows the commonly used workflow for droplet microfluidics in single-cell genomic analysis, which often includes single-cell droplet encapsulation, lysis, amplification, and sequencing.

7.3.1 SINGLE-CELL ISOLATION TECHNOLOGY

One of the major obstacles in single-cell analysis is the single-cell isolation. The most widely used methods for isolating single cells are fluorescent activated cell sorting, micromanipulation, and microfluidics. Microfluidics has been established as an enabling technology in single-cell studies due to its low contamination compared to conventional millimeter amplification. The microfluidic droplet is an appealing branch of microfluidics and has been widely used in single-cell assays. The principle of the microfluidic droplet is to tailor the volume of reactors to single-cell size, ranging from picoliter to nanoliter volume, rely on the water/oil surface tension. Microfluidic droplets are substitutions of wells and tubes but in a much smaller volume. Single-cell assays can be performed in parallel with thousands of microdroplets, which can reduce the reagent cost to the minimum. Another notable advantage is that the microdroplets can circumvent limitations of conventional microfluidics such as low throughput and high contamination (Zilionis et al. 2017). The recent

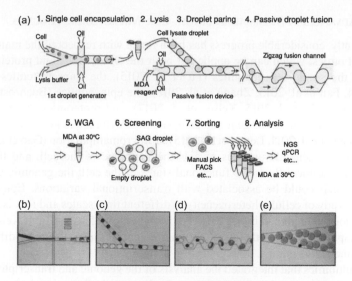

FIGURE 7.10 Typical schematic workflow of the single-cell genomic amplification using microfluidic droplets. (a) The common workflow of the droplet MDA and subsequent analysis. (b) The first droplet generator, (c) droplet paring channel, (d) zigzag channel, and (e) downstream channel in the passive fusion device. (Reprinted from Hosokawa et al. 2017. *Scientific Reports*, 7(1):5199 licensed under CC BY 4.0.) MDA, multiple displacement amplification.

single-cell multiple displacement amplification (MDA) amplification in microfluidic droplets shows high uniformity and coverage (Lan et al. 2017). The current microfluidic droplet technique mainly relies on PDMS chips (Guo et al. 2012, Joensson and Andersson Svahn 2012, Lan et al. 2017) and has met challenges, such as long-time fabrication, obligation for skilled personnel and confined to only one specific application. Finally, unlike metagenomics, which assemble genomes from a complex system, resulting in mismatches and errors, the sequences obtained from one cell are genetically linked, thereby facilitating assembly.

7.3.2 Single-Cell Genome Amplification Technology

Most single cells contain only several picograms (mammalian eukaryotic cells) or femtograms (microbes) of DNA. Whole-genome amplification (WGA) should be performed to generate micrograms of the genomic template before sequencing.

Currently, there are four types of WGA techniques: the degenerate oligonucleotide primed PCR, which preferentially amplifies specific sites in the genome, which results in low physical coverage of the genome but better uniformity of amplification (Zhang et al. 1992). The other techniques are the MDA (Dean et al. 2001), the multiple annealing and looping-based amplification cycles (Zong et al. 2012) and the newly developed Linear Amplification via Transposon Insertion (Chen et al. 2017) methods. The Linear Amplification via Transposon Insertion method shows the advantages of less amplification bias and errors. Among these techniques, MDA is now the most widely used method for single-cell genome amplification.

7.3.3 ADVANCES IN MAMMALIAN SINGLE-CELL MICROFLUIDIC ANALYSIS

Until recently, considerable progress has been made with respect to the mammalian single-cell omics, which can be applied at both the DNA, RNA and protein levels, including the single-cell genomics (Fu et al. 2015), the transcriptomics (Pollen et al. 2014, Fan et al. 2015, Zhang et al. 2018), the epigenomics (Buenrostro et al. 2015, Cusanovich et al. 2015, Kelsey et al. 2017), and proteomics (Mazutis et al. 2013, Azizi et al. 2018). For genomic analysis, flow cytometers are widely used for sorting (Navin et al. 2011, Leung et al. 2015) or micromanipulation (Gao et al. 2016). The DNA-level genomic analysis reveals the genotype of the cell, and the RNA level transcriptomic reflects the functional states of the cell; the genomic variation in a single cell could be associated with transcriptional variations. Epigenomics allows the study of cellular heterogeneity at different time scales and for discovering intrinsic molecular connectivities between the genome and its function (Kelsey et al. 2017), it captures DNA methylation, chromatin accessibility, histone modifications, chromosome conformation, and replication dynamics.

The multiomics that integrated the analysis of the genome and transcriptome and, probably the epigenome of a single cell can lead to a multidimensional measurement of heterogeneity, stratification, and phenotype regulation associated with the population of cells. By combining different layers of genomic output, the cell identity and function can be recorded simultaneously. Overall, these techniques are rapidly becoming a powerful tool in studies of cellular plasticity and diversity, as seen in stem cells and cancer (Schwartzman and Tanay 2015).

Despite the multiomics that can display cell heterogeneity at different layers, methods that aim to improve the amplification quality have also been developed. For example, high uniformity and fidelity of WGA are needed to accurately determine genomic variations, such as copy number variations and single nucleotide variations (SNVs). Conventional WGA methods are limited by unstable amplification genome yield and the false-positive and -negative errors for SNV identification. An emulsion WGA (eWGA) was developed to overcome these problems (Fu et al. 2015). This easy-to-operate approach enables simultaneous detection of copy number variations and SNVs in an individual human cell, exhibiting significantly improved amplification evenness and accuracy.

Furthermore, the direct library preparation technique can build a sequencing library from a single cancer cell genome. The direct library preparation platform shows robust, scalable, and high-fidelity microfluidic methods that use nanoliter-volume transposition reactions for single-cell whole-genome library preparation without pre-amplification (Zahn et al. 2017).

7.3.4 MICROBIAL SINGLE-CELL GENOMIC ANALYSIS

Unlike mammalian cells, the single-cell genomics in microbes encounters many more challenges. The most important one is that the genomic DNA from a single microorganism is at least 1,000 times less abundant than in mammalian cells, which have only several femtograms of genomic DNA making the isolation and amplification process more complex and are more prone to contamination. There are two primary methods to isolate microbial cells: fluorescent activated cell sorting (Rinke et al. 2014, Stepanauskas et al. 2017, Ahrendt et al. 2018) and microfluidics (Marcy et al. 2007b, Xu et al. 2016, Hosokawa et al. 2017, Lan et al. 2017). The microfluidic chips provide an ideal tool to improve the amplification uniformity and reduce the contamination in single-cell genomic analysis due to the nanoliter- and picoliter-scale volumes; this tool is especially helpful when retrieving uncultured microorganisms.

The first microfluidics device for single-cell separation and amplification was developed in 2007 (Marcy et al. 2007b). The valve-dependent device can separate eight cells at a time in small chambers. The device was used to separate TM7, an uncultivated candidate phylum. Later, the device was optimized to separate and amplify microbial genomes in nanoliter volumes, which further reduced the reagent cost with lower contamination and higher genomic uniformity (Marcy et al. 2007a). The device architecture is displayed in Figure 7.11. One disadvantage of the chamber-based microfluidic devices is their low throughput.

The high-throughput microfluidics have been developed to facilitate the isolation and amplification of single microbial cells in thousands or tens of thousands of compartments at one time. For example, the in-gel virtual microfluidics that can generate single-cell MDA products with excellent coverage uniformity and notably reduced chimaeras compared to liquid MDA reactions (Xu et al. 2016). The ultra-high–throughput microfluidic droplet barcoding technique named single-cell genomic sequencing uses picoliter to nanoliter droplet to compartmentalize, amplify, fragmentate, barcode the whole genomes of single cells followed by pooling of barcoded DNA and high-throughput sequencing (Lan et al. 2017). The ability to routinely sequence large populations of single cells will enable the deconvolution of genetic heterogeneity in diverse cell populations.

7.4 PERSPECTIVES

Microfluidic technology not only allows fast and precise measurement of chemotaxis responses of microbial and mammalian cells, but also plays a critical role in other cell-based bioanalyses, such as bacterial biofilm development (Jin et al. 2018), single-cell cultivation (Jiang et al. 2016), cancer metastasis (Boussommier-Calleja et al. 2016), and stem cell biology (Rothbauer et al. 2018), etc. The rapid development of microfluidic technology and microfluidic devices facilitates us better understanding of complex biological processes and revolutionizing high-throughput screening and drug discovery.

Considerable progress has been made in microfluidics-based single-cell sorting and analysis. However, there is still considerable progress to be made in the field. The following are some suggestions for further research direction in the field. First, to develop user friendly single-cell sorting and genomic analysis platforms to retrieve increasingly abundant and diverse microbial species and their metabolic and enzymatic resources. Second, to develop microfluidic platforms that can accomplish multiomics analysis of mammalian cells in integrated systems. Third, current single-cell analyses are mainly based on in vitro amplification, which inevitable induce amplification noises and biases. Nonamplification-based methods and more sensitive and uniform library preparation technologies are needed for more quantitative single-cell multiomics measurement.

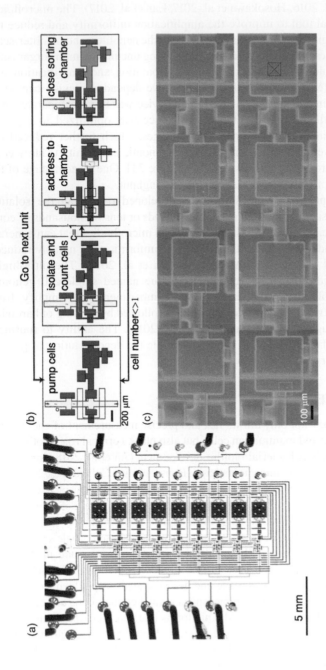

FIGURE 7.11 Microfluidic chip based on the valve single-cell sorting and amplification. (a) The chip architecture, visualized by filling the channels and chambers with blue coloring and the control lines of the valves with red coloring. Red represent for closed valves, transparent represent open valves. Green represents cells. (b) Schematic graph of the automated sorting procedure. Red represent results for each unit a color combination of a phase contrast image (gray) and a fluorescence image (green). Green squares have been placed around the cell for visualization, whereas a red crossed squares shows the absence of cell. (Reprinted from Marcy et al. 2007. *PLoS Genetics*, 3(9):e155 licensed under the Creative Commons Attribution License.)

REFERENCES

Abbott, Alison. 2003. Biology's new dimension. *Nature*, 424:870–872.

Adams, Jonathan D., Unyoung Kim, and H. Tom Soh. 2008. Multitarget magnetic activated cell sorter. *Proceedings of the National Academy of Sciences of the United States of America*, 105 (47):18165.

Adler, Julius. 1966. Chemotaxis in bacteria. *Science*, 153 (3737):708–716.

Adler, Julius, and Margaret M. Dahl. 1967. A method for measuring the motility of bacteria and for comparing random and non-random motility. *Journal of General Microbiology*, 46 (2):161–173.

Agresti, Jeremy J., Eugene Antipov, Adam R. Abate, et al. 2010. Ultrahigh-throughput screening in drop-based microfluidics for directed evolution. *Proceedings of the National Academy of Sciences of the United States of America*, 107 (9):4004–4009.

Ahmed, Tanvir, Thomas S. Shimizu, and Roman Stocker. 2010. Microfluidics for bacterial chemotaxis. *Integrative Biology*, 2 (11–12):604–629.

Ahrendt, Steven R., C. Alisha Quandt, Doina Ciobanu, et al. 2018. Leveraging single-cell genomics to expand the fungal tree of life. *Nature Microbiology*, 3:1417–1428.

Aizel, Koceila, Andrew G. Clark, Anthony Simon, et al. 2017. A tuneable microfluidic system for long duration chemotaxis experiments in a 3D collagen matrix. *Lab on a Chip*, 17 (22):3851–3861.

Applegate, Robert Jr, Jeff Squier, Tor Vestad, John Oakey, and David Marr. 2004. Optical trapping, manipulation, and sorting of cells and colloids in microfluidic systems with diode laser bars. *Optics Express*, 12 (19):4390–4398.

Atencia, Javier, Jayne Morrow, and Laurie E. Locascio. 2009. The microfluidic palette: A diffusive gradient generator with spatio-temporal control. *Lab on a Chip*, 9 (18):2707–2714.

Azizi, Elham, Ambrose J. Carr, George Plitas, et al. 2018. Single-cell map of diverse immune phenotypes in the breast tumor microenvironment. *Cell*, 174 (5):1293–1308.e36.

Baret, Jean Christophe, Oliver J. Miller, Valerie Taly, et al. 2009. Fluorescence-activated droplet sorting (FADS): Efficient microfluidic cell sorting based on enzymatic activity. *Lab on a Chip*, 9 (13):1850–1858.

Beneyton, Thomas, I. Putu Mahendra Wijaya, Prexilia Postros, et al. 2016. High-throughput screening of filamentous fungi using nanoliter-range droplet-based microfluidics. *Scientific Reports*, 6 (1):27223.

Bhagat, Ali Asgar, Sathyakumar S. Kuntaegowdanahalli, and Ian Papautsky. 2008. Continuous particle separation in spiral microchannels using Dean flows and differential migration. *Lab on a Chip*, 8 (11):1906–1914.

Boussommier-Calleja, Alexandra, Ran Li, Michelle B. Chen, Siew Cheng Wong, and Roger D. Kamm. 2016. Microfluidics: A new tool for modeling cancer–immune interactions. *Trends in Cancer*, 2 (1):6–19.

Buenrostro, Jason D., Beijing Wu, Ulrike M. Litzenburger, et al. 2015. Single-cell chromatin accessibility reveals principles of regulatory variation. *Nature*, 523 (7561):486.

Chen, Chongyi, Dong Xing, Longzhi Tan, et al. 2017. Single-cell whole-genome analyses by Linear Amplification via Transposon Insertion (LIANTI). *Science*, 356 (6334):189–194.

Chen, Dongwei, Shuang-Jiang Liu, and Wenbin Du. 2019. Chemotactic screening of imidazolinone-degrading bacteria by microfluidic SlipChip. *Journal of Hazardous Materials*, 366:512–519.

Cheng, Shing-Yi, Steven Heilman, Max Wasserman, et al. 2007. A hydrogel-based microfluidic device for the studies of directed cell migration. *Lab on a Chip*, 7 (6):763–769.

Colin, Pierre Y., Balint Kintses, Fabrice Gielen, et al. 2015. Ultrahigh-throughput discovery of promiscuous enzymes by picodroplet functional metagenomics. *Nature Communications*, 6:10008.

Cusanovich, Darren A., Riza Daza, Andrew Adey, et al. 2015. Multiplex single-cell profiling of chromatin accessibility by combinatorial cellular indexing. *Science*, 348 (6237):910–914.

de Oliveira, Sofia, Emily E. Rosowski, and Anna Huttenlocher. 2016. Neutrophil migration in infection and wound repair: Going forward in reverse. *Nature Reviews Immunology*, 16:378–391.

Dean, Frank B., John R. Nelson, Theresa L. Giesler, and Roger S. Lasken. 2001. Rapid amplification of plasmid and phage DNA using Phi29 DNA polymerase and multiply-primed rolling circle amplification. *Genome Research*, 11 (6):1095–1099.

Dean, Wiliam Reginald. 1928. Fluid motion in a curved channel. *Proceedings of the Royal Society of London. Series A, Containing Papers of a Mathematical and Physical Character*, 121 (787):402–420.

Dertinger, Stephan K. W., Xingyu Jiang, Zhiying Li, Venkatesh N. Murthy, and George M. Whitesides. 2002. Gradients of substrate-bound laminin orient axonal specification of neurons. *Proceedings of the National Academy of Sciences of the United States of America*, 99 (20):12542–12547.

Diao, Jinpian, Lincoln Young, Sue Kim, et al. 2006. A three-channel microfluidic device for generating static linear gradients and its application to the quantitative analysis of bacterial chemotaxis. *Lab on a Chip*, 6 (3):381–388.

Dong, Libing, Dong-Wei Chen, Shuang-Jiang Liu, and Wenbin Du. 2016. Automated chemotactic sorting and single-cell cultivation of microbes using droplet microfluidics. *Scientific Reports*, 6:24192.

Dormann, Dirk, and Cornelis J. Weijer. 2003. Chemotactic cell movement during development. *Current Opinion in Genetics & Development*, 13 (4):358–364.

Eisenbach, Michael. 2004. *Chemotaxis*. London: World Scientific Publishing Company.

Englert, Derek L., Michael D. Manson, and Arul Jayaraman. 2009. Flow-based microfluidic device for quantifying bacterial chemotaxis in stable, competing gradients. *Applied and Environmental Microbiology*, 75 (13):4557–4564.

Fan, Xiaoying, Xiannian Zhang, Xinglong Wu, et al. 2015. Single-cell RNA-seq transcriptome analysis of linear and circular RNAs in mouse preimplantation embryos. *Genome Biology*, 16 (1):148.

Feng, Haichao, Nan Zhang, Wenbin Du, et al. 2018. Identification of chemotaxis compounds in root exudates and their sensing chemoreceptors in plant-growth-promoting rhizobacteria *Bacillus amyloliquefaciens* SQR9. *Molecular Plant-Microbe Interactions*, 31 (10):995–1005.

Feng, Haichao, Nan Zhang, Ruixin Fu, et al. 2019. Recognition of dominant attractants by key chemoreceptors mediates recruitment of plant growth-promoting rhizobacteria. *Environmental Microbiology*, 21 (1):402–415.

Franke, Thomas, Adam R. Abate, David A. Weitz, and Achim Wixforth. 2009. Surface acoustic wave (SAW) directed droplet flow in microfluidics for PDMS devices. *Lab on a Chip*, 9 (18):2625.

Franz, Clemens M., Gareth E. Jones, and Anne J. Ridley. 2002. Cell migration in development and disease. *Developmental Cell*, 2 (2):153–158.

Fu, Yusi, Chunmei Li, Sijia Lu, et al. 2015. Uniform and accurate single-cell sequencing based on emulsion whole-genome amplification. *Proceedings of the National Academy of Sciences of the United States of America*, 112 (38):11923–11928.

Gao, Ruli, Alexander Davis, Thomas O. McDonald, et al. 2016. Punctuated copy number evolution and clonal stasis in triple-negative breast cancer. *Nature Genetics*, 48 (10):1119.

Gawad, Charles, Winston Koh, and Stephen R. Quake. 2016. Single-cell genome sequencing: Current state of the science. *Nature Reviews Genetics*, 17 (3):175.

Gossett, Daniel R., Westbrook M. Weaver, Albert J. Mach, et al. 2010. Label-free cell separation and sorting in microfluidic systems. *Analytical and Bioanalytical Chemistry*, 397 (8):3249–3267.

Guo, Mira T., Assaf Rotem, John A. Heyman, and David A. Weitz. 2012. Droplet microfluid-ics for high-throughput biological assays. *Lab on a Chip*, 12 (12):2146–2155.

Ha, Byung, Jinsoo Park, Ghulam Destgeer, Jin Ho Jung, and Hyung Jin Sung. 2016. Transfer of microparticles across laminar streams from non-Newtonian to Newtonian fluid. *Analytical Chemistry*, 88 (8):4205–4210.

Hosokawa, Masahito, Yohei Nishikawa, Masato Kogawa, and Haruko Takeyama. 2017. Massively parallel whole genome amplification for single-cell sequencing using droplet microfluidics. *Scientific Reports*, 7 (1):5199.

Huang, Lotien Richard. 2004. Continuous particle separation through deterministic lateral displacement. *Science*, 304 (5673):987–990.

Jeon, Noo Li, Stephan K. W. Dertinger, Daniel T. Chiu, et al. 2000. Generation of solution and surface gradients using microfluidic systems. *Langmuir*, 16 (22):8311–8316.

Ji, Hongmiao M., Victor Samper, Yu Chen, et al. 2008. Silicon-based microfilters for whole blood cell separation. *Biomedical Microdevices*, 10 (2):251–257.

Jiang, Cheng-Ying, Libing Dong, Jian-Kang Zhao, et al. 2016. High-throughput single-cell cultivation on microfluidic streak plates. *Applied and Environmental Microbiology*, 82 (7):2210–2218.

Jin, Zengjun J., Mengyue Y. Nie, Ran Hu, et al. 2018. Dynamic sessile-droplet habitats for controllable cultivation of bacterial biofilm. *Small*, 14 (22):e1800658.

Joensson, Haakan N., and Helene Andersson Svahn. 2012. Droplet microfluidics—A tool for single-cell analysis. *Angewandte Chemie International Edition*, 51 (49):12176–12192.

Keenan, Thomas M., Charles W. Frevert, Aileen Wu, Venus Wong, and Albert Folch. 2010. A new method for studying gradient-induced neutrophil desensitization based on an open microfluidic chamber. *Lab on a Chip*, 10 (1):116–122.

Keenan, Thomas M., Chia-Hsien Hsu, and Albert Folch. 2006. Microfluidic "jets" for gen-erating steady-state gradients of soluble molecules on open surfaces. *Applied Physics Letters*, 89 (11):114103.

Kelsey, Gavin, Oliver Stegle, and Wolf Reik. 2017. Single-cell epigenomics: Recording the past and predicting the future. *Science*, 358 (6359):69–75.

Kim, Beum Jun, and Mingming Wu. 2012. Microfluidics for mammalian cell chemotaxis. *Annals of Biomedical Engineering*, 40 (6):1316–1327.

Kim, Minseok, Su Hyun Kim, Sung Kuk Lee, and Taesung Kim. 2011. Microfluidic device for analyzing preferential chemotaxis and chemoreceptor sensitivity of bacterial cells toward carbon sources. *Analyst*, 136 (16):3238–3243.

Kuntaegowdanahalli, Sathyakumar S., Ali Asgar Bhagat, Girish Kumar, and Ian Papautsky. 2009. Inertial microfluidics for continuous particle separation in spiral microchannels. *Lab on a Chip*, 9 (20):2973–2980.

Lambert, Bennett S., Jean-Baptiste Raina, Vicente I. Fernandez, et al. 2017. A microfluidics-based in situ chemotaxis assay to study the behaviour of aquatic microbial communi-ties. *Nature Microbiology*, 2 (10):1344–1349.

Lan, Freeman, Benjamin Demaree, Noorsher Ahmed, and Adam R. Abate. 2017. Single-cell genome sequencing at ultra-high-throughput with microfluidic droplet barcoding. *Nature Biotechnology*, 35:640.

Leung, Marco L., Yong Wang, Jill Waters, and Nicholas E. Navin. 2015. SNES: Single nucleus exome sequencing. *Genome Biology*, 16 (1):55.

Li Jeon, Noo, Harihara Baskaran, Stephan K. W. Dertinger, et al. 2002. Neutrophil che-motaxis in linear and complex gradients of interleukin-8 formed in a microfabricated device. *Nature Biotechnology*, 20:826–830.

Lin, Francis, Connie Minh-Canh Nguyen, Shur-Jen Wang, et al. 2004. Effective neutro-phil chemotaxis is strongly influenced by mean IL-8 concentration. *Biochemical and Biophysical Research Communications*, 319 (2):576–581.

Mao, Hanbin, Paul S. Cremer, and Michael D. Manson. 2003. A sensitive, versatile microfluidic assay for bacterial chemotaxis. *Proceedings of the National Academy of Sciences of the United States of America*, 100 (9):5449–5454.

Marcy, Yann, Thomas Ishoey, Roger S. Lasken, et al. 2007a. Nanoliter reactors improve multiple displacement amplification of genomes from single cells. *PLoS Genetics*, 3 (9):e155.

Marcy, Yann, Cleber Ouverney, Elisabeth M. Bik, et al. 2007b. Dissecting biological "dark matter" with single-cell genetic analysis of rare and uncultivated TM7 microbes from the human mouth. *Proceedings of the National Academy of Sciences of the United States of America*, 104 (29):11889–11894.

Markx, Gerard H., Mark S. Talary, and Ronald Pethig. 1994. Separation of viable and non-viable yeast using dielectrophoresis. *Journal of Biotechnology*, 32 (1):29.

Mazutis, Linas, John Gilbert, W. Lloyd Ung, et al. 2013. Single-cell analysis and sorting using droplet-based microfluidics. *Nature Protocols*, 8:870.

Mosadegh, Bobak, Carlos Huang, Jeong Won Park, et al. 2007. Generation of stable complex gradients across two-dimensional surfaces and three-dimensional gels. *Langmuir*, 23 (22):10910–10912.

Nagy, Krisztina, Orsolya Sipos, Sándor Valkai, et al. 2015. Microfluidic study of the chemotactic response of *Escherichia coli* to amino acids, signaling molecules and secondary metabolites. *Biomicrofluidics*, 9 (4):044105.

Navin, Nicholas, Jude Kendall, Jennifer Troge, et al. 2011. Tumour evolution inferred by single-cell sequencing. *Nature*, 472 (7341):90.

Obexer, Richard, Alexei Godina, Xavier Garrabou, et al. 2017. Emergence of a catalytic tetrad during evolution of a highly active artificial aldolase. *Nature Chemistry*, 9 (1):50–56.

Ostafe, Raluca, Radivoje Prodanovic, W. Lloyd Ung, David A. Weitz, and Rainer Fischer. 2014. A high-throughput cellulase screening system based on droplet microfluidics. *Biomicrofluidics*, 8 (4):041102.

Park, Jae-Sung, and Hyo-Il Jung. 2009. Multiorifice flow fractionation: Continuous size-based separation of microspheres using a series of contraction/expansion microchannels. *Analytical Chemistry*, 81 (20):8280–8288.

Park, Seungkyung, Yi Zhang, Tza Huei Wang, and Samuel Yang. 2011. Continuous dielectrophoretic bacterial separation and concentration from physiological media of high conductivity. *Lab on a Chip*, 11 (17):2893–2900.

Pollen, Alex A., Tomasz J. Nowakowski, Joe Shuga, et al. 2014. Low-coverage single-cell mRNA sequencing reveals cellular heterogeneity and activated signaling pathways in developing cerebral cortex. *Nature Biotechnology*, 32:1053.

Pommer, Matthew S., Yanting Zhang, Nawarathna Keerthi, et al. 2008. Dielectrophoretic separation of platelets from diluted whole blood in microfluidic channels. *Electrophoresis*, 29 (6):1213–1218.

Qiao, Yuxin, Xiaoyan Zhao, Jun Zhu, et al. 2017. Fluorescence-activated droplet sorting of lipolytic microorganisms using a compact optical system. *Lab on a Chip*, 18 (1):190–196.

Rinke, Christian, Janey Lee, Nandita Nath, et al. 2014. Obtaining genomes from uncultivated environmental microorganisms using FACS–based single-cell genomics. *Nature Protocols*, 9 (5):1038.

Rothbauer, Mario, Helene Zirath, and Peter Ertl. 2018. Recent advances in microfluidic technologies for cell-to-cell interaction studies. *Lab on a Chip*, 18 (2):249–270.

Saadi, Wajeeh, Seog Woo Rhee, Francis Lin, et al. 2007. Generation of stable concentration gradients in 2D and 3D environments using a microfluidic ladder chamber. *Biomedical Microdevices*, 9 (5):627–635.

Saadi, Wajeeh, Shur-Jen Wang, Francis Lin, and Noo Li Jeon. 2006. A parallel-gradient microfluidic chamber for quantitative analysis of breast cancer cell chemotaxis. *Biomedical Microdevices*, 8 (2):109–118.

Schmid, Lothar, David A. Weitz, and Thomas Franke. 2014. Sorting drops and cells with acoustics: Acoustic microfluidic fluorescence-activated cell sorter. *Lab on a Chip*, 14 (19):3710–3718.

Schwartzman, Omer, and Amos Tanay. 2015. Single-cell epigenomics: Techniques and emerging applications. *Nature Reviews Genetics*, 16:716.

Seo, Jeonggi, Meng H. Lean, and Ashutosh Kole. 2007. Membrane-free microfiltration by asymmetric inertial migration. *Applied Physics Letters*, 91 (3):033901.

Shen, Chaohua H., Peng Xu, Zhou Huang, et al. 2014. Bacterial chemotaxis on SlipChip. *Lab on a Chip*, 14 (16):3074–3080.

Shi, Jinjie, Xiaole Mao, Daniel Ahmed, Ashley Colletti, and Tony Jun Huang. 2008. Focusing microparticles in a microfluidic channel with standing surface acoustic waves (SSAW). *Lab on a Chip*, 8 (2):221–223.

Si, Guangwei, Wei Yang, Shuangyu Bi, Chunxiong Luo, and Qi Ouyang. 2012. A parallel diffusion-based microfluidic device for bacterial chemotaxis analysis. *Lab on a Chip*, 12 (7):1389–1394.

Siyan, Wang, Yue Feng, Zhang Lichuan, et al. 2009. Application of microfluidic gradient chip in the analysis of lung cancer chemotherapy resistance. *Journal of Pharmaceutical and Biomedical Analysis*, 49 (3):806–810.

Squires, Todd M., and Stephen R. Quake. 2005. Microfluidics: Fluid physics at the nanoliter scale. *Reviews of Modern Physics*, 77 (3):977–1026.

Stepanauskas, Ramunas, Elizabeth A. Fergusson, Joseph Brown, et al. 2017. Improved genome recovery and integrated cell-size analyses of individual uncultured microbial cells and viral particles. *Nature Communications*, 8 (1):84.

Tao, Ye, Yukun K. Ren, Hui Yan, and Hongyuan Y. Jiang. 2016. Continuous separation of multiple size microparticles using alternating current dielectrophoresis in microfluidic device with acupuncture needle electrodes. *Chinese Journal of Mechanical Engineering*, 29 (2):325–331.

van Loo, Bert, Magdalena Heberlein, Philip Mair, et al. 2018. High-throughput, lysis-free screening for sulfatase activity using *Escherichia coli* autodisplay in microdroplets. *bioRxiv*:479162. doi:10.1101/479162.

Wang, Shur-Jen, Wajeeh Saadi, Francis Lin, Connie Minh-Canh Nguyen, and Noo Li Jeon. 2004. Differential effects of EGF gradient profiles on MDA-MB-231 breast cancer cell chemotaxis. *Experimental Cell Research*, 300 (1):180–189.

Wixforth, Achim, Christoph Strobl, Ch Gauer, et al. 2004. Acoustic manipulation of small droplets. *Analytical and Bioanalytical Chemistry*, 379 (7):982–991.

Wolfram, Christopher J., Gary W. Rubloff, and Xiaolong Luo. 2016. Perspectives in flow-based microfluidic gradient generators for characterizing bacterial chemotaxis. *Biomicrofluidics*, 10 (6):061301.

Wu, Jiandong D., Xun Wu, and Francis Lin. 2013. Recent developments in microfluidics-based chemotaxis studies. *Lab on a Chip*, 13 (13):2484–2499.

Xu, Liyi, Ilana L. Brito, Eric J. Alm, and Paul C. Blainey. 2016. Virtual microfluidics for digital quantification and single-cell sequencing. *Nature Methods*, 13 (9):759.

Yamada, Masumi, Megumi Nakashima, and Minoru Seki. 2004. Pinched flow fractionation: Continuous size separation of particles utilizing a laminar flow profile in a pinched microchannel. *Analytical Chemistry*, 76 (18):5465–5471.

Yuan, Dan, Say Hwa Tan, Ronald Sluyter, et al. 2017. On-chip microparticle and cell washing using coflow of viscoelastic fluid and Newtonian fluid. *Analytical Chemistry*, 89 (17):9574–9582.

Yuan, Dan, Jun Zhang, Sheng Yan, et al. 2016. Investigation of particle lateral migration in sample-sheath flow of viscoelastic fluid and Newtonian fluid. *Electrophoresis*, 37 (15–16):2147–2155.

Zahn, Hans, Adi Steif, Emma Laks, et al. 2017. Scalable whole-genome single-cell library preparation without preamplification. *Nature Methods*, 14:167.

Zhang, Lin, Xue Hong Cui, Karin Schmitt, et al. 1992. Whole genome amplification from a single cell: Implications for genetic analysis. *Proceedings of the National Academy of Sciences of the United States of America*, 89 (13):5847–5851.

Zhang, Xiannian, Tianqi Li, Feng Liu, et al. 2018. Comparative analysis of droplet-based ultra-high-throughput single-cell RNA-seq systems. *Molecular Cell*, 73(1):130–142.e5.

Zhou, Yinning, Zhichao Ma, Mahnoush Tayebi, and Ye Ai. 2019. Submicron particle focusing and exosome sorting by wavy microchannel structures within viscoelastic fluids. *Analytical Chemistry*, 91 (7):4577–4584.

Zilionis, Rapolas, Juozas Nainys, Adrian Veres, et al. 2017. Single-cell barcoding and sequencing using droplet microfluidics. *Nature Protocols*, 12 (1):44.

Zong, Chenghang, Sijia Lu, Alec R. Chapman, and X. Sunney Xie. 2012. Genome-wide detection of single-nucleotide and copy-number variations of a single human cell. *Science*, 338 (6114):1622–1626.

8 3D Printing and Bioprinting Technologies

Se-Hwan Lee, Hyoryung Nam, Bosu Jeong, and Jinah Jang
Pohang University of Science and Technology

CONTENTS

8.1 INTRODUCTION

Additive manufacturing (AM), often called rapid prototyping (RP) or three-dimensional (3D) printing, can be operated by computer-aided design/manufacturing (CAD/CAM) processes. In early 1990, AM technology was first introduced by Prof. Sachs of Massachusetts Institute of Technology (MIT) (Bose, Vahabzadeh, and Bandyopadhyay 2013, Sachs, Cima, and Cornie 1990, Sachs et al. 1993). Various types of printing heads, such as inkjet and extrusion heads, were developed for the purpose of increasing fabrication versatility. Three-dimensional printing has important advantages for tissue engineering purposes, such as its integration of nano/micro-scale functions and reproducible internal architecture (Duan et al. 2010). In particular, this technology enables the manufacture of customized architectures based on 3D medical imaging data, such as magnetic resonance imaging (MRI) and computed tomography (CT), which is an important feature to achieve the emerging therapeutic concepts for patient-specific implants (Melchels et al. 2012, Sun et al. 2004, Ballyns and Bonassar 2009).

The 3D printing technique provides accurate control over the internal and external architecture of scaffolds, which is essential for effective nutrient diffusion, cell proliferation, ingrowth, and cellular attachment (Yang et al. 2001). Moreover, porosity and pore morphology can be designed to promote efficient mass transfer throughout the structure (Miller et al. 2012). This advanced strategy could also apply various materials with encapsulated cells and/or bioactive molecules. Thus, 3D printing has become an ideal fabrication technique candidate for creating tissue-like constructs (Chua, Liu, and Chou 2011).

The 3D bioprinting technique is a highly flexible and versatile method to localize cells with bioink in a predetermined position (Park, Jung, and Min 2016, 2017). This strategy has high potential to realize a single-step fabrication process through a novel design and advanced bioinks (Murphy and Atala 2014). It offers benefits for delivering biological components, such as living cells or growth factors, to the injured tissue sites. In particular, decellularized extracellular matrix (dECM)-based bioinks are derived from various native tissue sources. Moreover, they can provide a cell-friendly microenvironment that can recapitulate the specific components and compositions of the ECM similar to target tissues (Kim, Kim, and Jung 2016, Jang et al. 2018). Therefore, dECM-based bioinks have been an attractive material for tissue regeneration.

8.2 INKJET-BASED BIOFABRICATION

Inkjet printing is a noncontact method of forming a uniform film by dispersing a liquid-like material through a nozzle onto a piece of paper or substrate surface (Calvert 2001). The first successful device using inkjet printing was invented by Elmqvist in the 1950s (Heinzl and Hertz 1985). Since then, it has made remarkable progress with the development and popularization of the computer (Tabata et al. 2013, Yoshikawa et al. 2009). Inkjet technology is classified into continuous and drop-on-demand (DOD) methods according to the ink ejection method (Derby 2010). A continuous inkjet system was developed by Richard Sweet in the 1960s. In this system, the ink ejected from the nozzle forms a liquid stream. The produced droplets are approximately 100 μm in size, and it is mainly used for labeling and marking requiring high-speed results (Derby 2010, De Gans, Duineveld, and Schubert 2004).

To generate smaller drops and increase placement accuracy, DOD was developed by Zoltan in the 1970s. The diameter of droplets generated by DOD is smaller than that of continuous inkjet droplets, approximately 20–50 μm. It is used primarily for printing text documents and graphics (Derby 2010, De Gans, Duineveld, and Schubert 2004, Sweet 1965). Recently, DOD inkjet printing has been successfully adapted to develop bioengineering products, such as biosensors (Li, Rossignol, and Macdonald 2015), medical devices (Tse and Smith 2018), tissue engineering products (Cui et al. 2014, Hewes, Wong, and Searson 2017), and drug screening and delivery devices (Choi et al. 2017, Marizza et al. 2014), because DOD has precise controllability and fewer contamination properties.

DOD is divided largely into five approaches: valve-jet, electrostatic, acoustic, thermal, and piezoelectric technologies. Among them, the thermal and piezoelectric types are widely utilized for bioprinting. These techniques use an acoustic pressure pulse that can be produced by thermal or piezoelectric methods to eject ink droplets through a nozzle (Derby 2010). In the thermal type, the ink is heated and boiled to generate bubbles. The bubbles generate pressure and discharge ink droplets from the nozzle to the bottom. The apparatus for the thermal-type dispenser is simpler than that for the piezoelectric method (Beeson 1998). This advantage increases the accessibility for users and the potential of miniaturization (Derby 2010, Cui et al. 2012). The piezoelectric type uses a piezoelectric effect that can convert electricity into physical motion. A piezoelectric element is placed on the nozzle directly, and an electric signal is applied to the piezoelectric element to push out the ink droplets. It is possible to control the size of the ink droplets more precisely with this type than with the thermal type (Derby 2010, Kim et al. 2010).

8.2.1 Bioink for Inkjet Printers

In order to apply this technique to the field of bioengineering, the ink material, called bioink, should have appropriate characteristics. A bioink is a printable material used in bioprinting, where cells, proteins, nucleic acids, biomolecules, or complex combinations of these substances are deposited in a spatially controlled pattern to fabricate bio-applications, such as tissues and organs (Hospodiuk et al. 2017). Depending on the purpose, there are cases where the ink for inkjet printing does not contain cells, which is distinguished by biomaterial ink (Ji and Guvendiren 2017).

One of the most important characteristics of a bioink is its printability. To construct desired structures, the bioink should have proper physical properties that can be used for the inkjet printing method (Hölzl et al. 2016). For inkjet printing, the required viscosity varies depending on the printing head type (de Gans and Schubert 2003, Saunders, Gough, and Derby 2008, Allain, Stratis-Cullum, and Vo-Dinh 2004). Thermal type can operate effectively when the viscosity of the ink is in the range of under 4 mPa s. (Gao et al. 2015, Gudapati, Dey, and Ozbolat 2016). In the piezoelectric type, bioinks are adaptable for dispensing in the viscosity range of 0.5–20 mPa s (Allain, Stratis-Cullum, and Vo Dinh 2004). In addition, the cells should be protected from the physical stresses that can occur during the printing process. In order to be used with cells, bioink should be biocompatible and provide a suitable environment for the attachment, proliferation, and differentiation of the printed or seeded cells as well. If the printing process is long and complex, the cells in the cartridge must be supplied nutrients and oxygen to ensure cell survival (Truby and Lewis 2016).

Hydrogels are utilized as bioink because they usually satisfy the above conditions. These materials were originally divided into natural and synthetic materials. The natural hydrogels include alginate, collagen, and fibrin, while the synthetic materials include poly(ethylene glycol) (PEG), poly(acryl amide) (PAM), and poly(vinyl alcohol) (PVA) (Chirani et al. 2015).

8.2.2 Bioprinting Using Inkjet Printers

The main goal of inkjet bioprinting is to generate droplets quickly and deliver the bioink to the target location safely with very high resolution. Inkjet printing is highly suitable for generating 2D patterns because it is a method of making a thin film by dispersing droplets. Original inkjet printers did not aim to fabricate 3D structures, but because of their accuracy and accessibility, they have been studied to build 3D constructs by controlling patterns at the single-cell level. To validate the potential of inkjet printers, 2D gradient patterns according to the cell concentration, materials, and growth factors have been tried that can alter droplet densities or sizes (Derby 2010, De Gans, Duineveld, and Schubert 2004). However, there are additional problems that need to be addressed depending on the printing head module.

The most critical problem is cell damage. The thermal type can induce cell damage by altering physiological activities and deforming molecules because of boiling at high temperatures over 300°C. To overcome this problem, 3D bio-inkjet printing usually melts the ink at room temperature (~46°C) by heating the nozzles. By doing so, researchers showed that the modified printing process did not influence the viability of printed Chinese hamster ovary cells (Cui et al. 2010). Xu et al. developed the inkjet printing process for fabricating heterogeneous cellular constructs by using various cell types. This study generated precise and reproducible patterns with various cell types, including human amniotic fluid-derived stem cells, smooth muscle cells, and bovine aortic endothelial cells (Figure 8.1a; Xu et al. 2013).

In the piezoelectric type, there is less concern over cellular damage during the printing process, as no heat source is used. However, the critical problem is the physical motion that can negatively affect cell behavior. To reduce the stress exerted at the nozzle, the driving pulse amplitude and shape can be controlled. Saunders et al. found that the physical motion could be regulated by changing the driving voltage and frequency of drop ejection and that this did not affect cell viability (Saunders et al. 2008).

The piezoelectric type can also be used to produce more sophisticated 3D constructs. Xu et al. (2012) fabricated fibroblast-laden 3D complex-shaped constructs such as zigzag tube-shaped constructs using a platform-assisted 3D inkjet printing system. The construction of the 3D cellular tube was also found to be important to increasing the scale of the 3D printing and the complexity of the architecture (Figure 8.1b; Xu et al. 2012). Christensen et al. fabricated a precise 3D organ structure with vascular trees. Vascular structures with horizontal and vertical bifurcations were generated using alginate hydrogels with mouse fibroblasts. This study showed the feasibility of the inkjet 3D organ printing process. (Figure 8.1c; Christensen et al. 2015).

8.3   LASER-BASED BIOFABRICATION

Laser-based biofabrication (LBB) technology has the advantage of creating a tissue structure with high accuracy and resolution compared to other biofabrication technologies (Figure 8.2a; Ji and Guvendiren 2017). Since the LBB method is a nozzle-free system, it completely avoids the problems of cell damage and death due to shear stress that can occur in other methods. High-resolution 3D biofabrication through LBB has made cell–cell interaction and cell patterning more controllable. It also alleviates other major problems, such as collision between the printhead and the bio-structure or an unexpected increase in the clearance between the orifice and the receiving substrate in a noncontact manner. The LBB technique can be largely

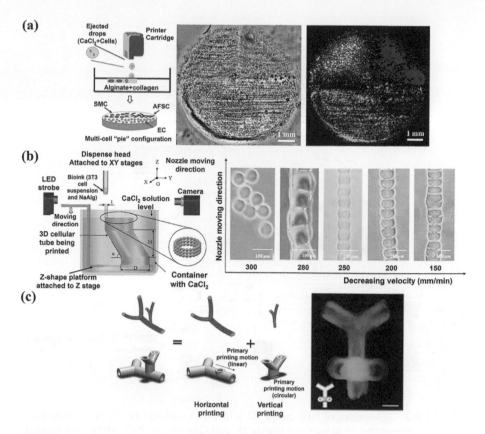

FIGURE 8.1 (a) (Left) Schematic drawing of proposed bioprinting method to fabricate multi-cell heterogeneous tissue constructs; (Right) light and fluorescence microscopic top views of complete 3D multi-cell "pie" construct before implantation. The cells that appear green are bECs labeled with PKH 26 dyes, the cells that appear blue are hAFSCs tagged with CMHC dyes, and the cells that appear red are dSMCs labeled with PKH 67 dyes. Different cells were located onto their predetermined locations after printing (Xu et al. 2013). (b) (Left) Schematic of proposed platform-assisted 3D inkjet bioprinting system; (Right) gel lines printed using 120 μm dispenser head (Xu et al. 2012). (c) (Left) Basic tubular structure of vascular network and horizontal and vertical bifurcations; (Right) inkjet-printed cellular structure with both horizontal and vertical bifurcations and insets showing structure as designed (scale bar, 3 mm) (Christensen et al. 2015). (Reproduced with permission from Xu et al. 2013, 2012, Christensen et al. 2015.)

divided into stereolithography (SLA) and laser-induced forward transfer (LIFT) depending on the operating mechanism. SLA technology, in which ultraviolet (UV) light is irradiated on photo-curable material to selectively make the structure, was initiated by Charles W. Hull in 1985. Recently, a digital mirror device (DMD) was used to improve fabrication speed exponentially (Zhu et al. 2017). In addition, the emergence of two-photon polymerization (TPP) made it possible to construct structures with high resolution (Worthington et al. 2017). LIFT uses a laser as an energy

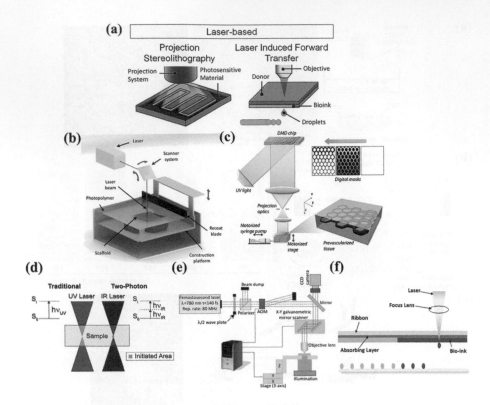

FIGURE 8.2 Schematics of laser-based 3D printing technology: (a) 3D bioprinting techniques for bioprinting of tissues and organs (Ji and Guvendiren 2017, Miller and Burdick 2016), (b) SLA (Moroni et al. 2017), (c) schematic of the bioprinting platform (Zhu et al. 2017), (d) schematic showing traditional photopolymerization versus TPP (Worthington et al. 2017, Zhu et al. 2017), (e) Two-photon polymerization (TPP) (Moroni et al. 2017); (f) laser-associated bioprinting consists of three parts: a pulsed laser source, a ribbon, and a receiving substrate. The lasers irradiate the ribbon, causing the liquid biological materials to evaporate and reach the receiving substrate in droplet form (Li et al. 2016). (Reproduced with permission from Miller and Burdick 2016, Moroni et al. 2017, Zhu et al. 2017, Worthington et al. 2017, Li et al. 2016.)

source to deposit biomaterials on substrates (Knowlton et al. 2015). This technique usually consists of three parts: a pulsed laser source, a ribbon coated with a liquid biological material attached to a metal film, and a receiving substrate.

8.3.1 STEREOLITHOGRAPHY (SLA)

A SLA system irradiates laser light (usually UV light) onto a photo-curable liquid resin to create 3D parts (Figure 8.2b; Moroni et al. 2017). The UV laser cures the liquid resin to a very thin layer (approximately several hundred micrometers depending on exposure energy (Bennett 2017), and the liquid resin solidifies due to the heat radiated from the laser beam, which is called a polymerization reaction.

The liquid is replenished and flattened systematically. The support structure is also fabricated in the same way, but generally, it is made of a microscopic microstructure. Finished parts require cleaning and curing after UV light or heat exposure. These SLA methods can be divided into direct laser writing and mask-based methods. The direct laser writing method, the first SLA method, was developed in 1986 by Charles Hull, co-founder of 3D Systems. This technique works by drawing each layer of the object using a UV laser and by using two motors driven by galvanometer mirrors to quickly aim the laser. All layers are selectively solidified according to the CAD design and attached to the underlying layer. After laser irradiation for all layers, the uncompleted portion is removed using a solvent that dissolves the monomer. This direct laser writing method uses high-energy lasers to pattern directly onto the resin surface, while mask-based SLA techniques can filter high-energy light sources through pattern-based physical or digital masks. This creates a 3D section within a single exposure without continuously tracking the laser beam. Therefore, this process is considered to have higher throughput than the direct laser writing process because the complete resin layer is polymerized in a single layer of resin to shorten the build time. The physical masks used in this type of mask-based projection photolithography are fabricated through a method developed in the field of semiconductor manufacturing. Recently, design flexibility has been improved by using a DMD that can display a wide range of patterns (Figure 8.2c). The DMD produces a pixelated image of millions of microscopic mirrors that can be rotated accurately and independently in an "on" or "off" state. Patterning and projecting light through this digital mask allows the photopolymerization of resins in precisely defined patterns. Direct laser writing and mask-based SLA methods rely heavily on computer-controlled building steps to move the 2D polymerized sections by a precisely defined amount, keeping the successive layers close together. Therefore, understanding the curing depth relationship of the resins used and the correct calibration of the light energy of the device is very important for manufacturing 3D parts through SLA.

A promising 3D micromachining method that has received considerable attention in recent years is based on TPP using an ultrafast laser. TPP is a technology capable of producing microstructured materials. TPP is essentially a 3D writing technique that does not require layer-by-layer access to create complex objects and that can create submicron-feature size microstructures due to the nonlinear optical properties of light absorption and polymerization chemistry. In general, near-infrared (NIR) emission is used to excite photosensitive materials that cause a phase change from liquid to solid through a polymerization process during light absorption. Since the resin is transparent in the spectrum of the NIR region, high numerical aperture lenses and femtosecond pulsed lasers are used to increase the probability of occurrence of multiphoton absorption events. In other words, more than one photon is absorbed simultaneously into the special molecules of the resin to produce active species that start to polymerize. Under these conditions, multiphoton absorption occurs only in the region with the highest light intensity (Figure 8.2d,e). This limits the polymerization within the volume of the focused laser beam. The 3D microstructures are created by accurately superimposing voxels by scanning the laser beam or sample around a predetermined shape.

8.3.2 Laser-Induced Forward Transfer (LIFT)

LIFT is a noncontact direct-write process that deposits biological materials onto a substrate using a pulsed laser (Figure 8.2f). In LIFT based on cell transfer, cells utilize a mechanism to transfer living cells from one location to another in a media or viscoelastic hydrogel. Key components of most LIFT systems are a pulsed laser source, bioink-coated ribbon, and receiving substrate. Nanosecond lasers with UV or near-UV wavelengths are used as sources. Ribbons are transparent to laser radiation wavelengths and are coated with glass-fiber or biotite target plates, heat-sensitive bio-links consisting of cells attached to biological polymers or uniformly encapsulated within a thin layer of hydrogel. Depending on the optical characteristics of the bioink and the laser wavelength, the system may include a laser-absorbing interlayer between the target plate and the bio-link to permit biochemical cell migration. The pulses of the laser cause rapid volatilization at the bioink interface of the ribbon and propel the high-velocity jet of the bioink containing the cells on the receiving substrate that is coated with a biopolymer or cell culture medium to maintain cell attachment and sustained growth after cell transfer from the ribbon.

8.4 EXTRUSION-BASED FABRICATION

8.4.1 Fused Deposition Modeling-Based Technology

In late 1980, the extrusion-based 3D printing technique was introduced by Crump, and it was successfully commercialized in 1990 (Crump 1992). In this technique, a prepared material is loaded in the metal or plastic syringe of a head and extruded by mechanical pressure, such as a pneumatic, and piston- or screw-driven system. The construct is fabricated along the motion of a head (Murphy and Atala 2014). The printing resolution is defined by various printing parameters, such as nozzle size, dispensing speed, and feed rate. A suitable temperature with respect to each material, including thermoplastic polymers or thermal-sensitive hydrogels, should be constantly maintained by the temperature controller. The directional motion and the material extrusion can be independently controlled by the computational motion-control command. A 3D scaffold with architecture, porosity, and pore size controlled by CAD can be fabricated through the following processes.

To date, several synthetic polymers with biocompatible, biodegradable, and thermoplastic characteristics, such as poly(lactic acid) (PLA) (Serra, Planell, and Navarro 2013), poly(glycolic acid) (PGA) (Huang and Niklason 2011), poly(lactic-*co*-glycolic acid) (PLGA) (Chia and Wu 2014), poly(caprolactone) (PCL) (Bracci, Maccaroni, and Cascinu 2013, Lee, Cho, et al. 2017), and poly-L-lactic acid (PLLA) (Spadaccio et al. 2011), have been widely utilized for tissue regeneration. These synthetic polymers are non-infectious, adjustable, and highly printable. In addition, these materials are suitable for fused deposition modeling (FDM)-based fabrication technology. In this chapter, pneumatic, piston-, and screw-driven heads are introduced as representative FDM technologies.

A typical FDM technology has an extruder in which a filament-type material is melted with high heat (Figure 8.3a; Zhang et al. 2017). The prepared filament

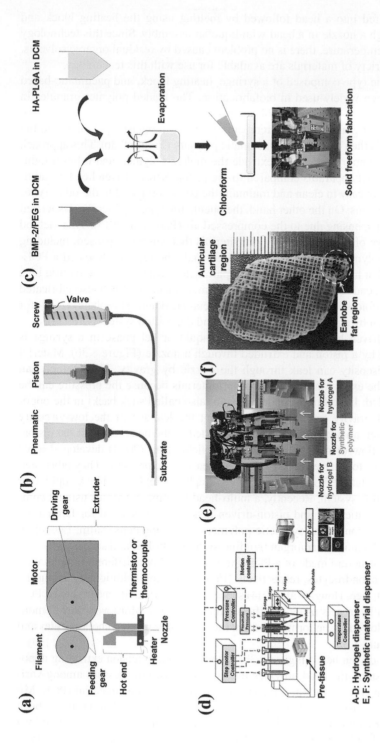

FIGURE 8.3 (a) Schematic picture of 3D printer extruder (Zhang et al. 2017), (b) microextrusion printers in the form of pneumatic, piston-driven, or screw-driven robotic dispensing systems (Knowlton et al. 2015). (c) schematic illustration of the preparation of feeding solution for SFF of tissue engineering scaffolds using an MHDS (Park et al. 2011), (d) MtoBS system: A schematic diagram (Shim, Lee, et al. 2012), (e) its dispensing parts (Shim, Lee, et al. 2012); (f) fabricated ear-shaped structure with dual hydrogel type (red color: auricular cartilage region, blue color: lobe fat region) (Lee, Hong, et al. 2012, Park et al. 2011, Shim, Lee, et al. 2012, Lee, Hong, et al. 2014, Knowlton et al. 2015, Park et al. 2011, Shim, Lee, et al. 2012, Lee, Hong, et al. 2014). (Reproduced with permission from Zhang et al. 2017, Knowlton et al. 2015, Park et al. 2011, Shim, Lee, et al. 2012, Lee, Hong, et al. 2014.)

is continuously fed into a head followed by melting using the heating block and extruding through a nozzle in a head with liquefier assembly. Since this technology uses only high temperature, there is no problem caused by residual organic solvents. In addition, a variety of materials are available for use with this technology.

The pneumatic type composed of a syringe, heating block, and pneumatic-based dispenser has been widely used in biofabrication. The loaded polymer granules in the syringe are melted by a heating block that can maintain the temperature at the specific melting (or glass transition) temperature of thermoplastic polymer, and the molten material is then extruded by pneumatic pressure (Figure 8.3b). This approach usually requires compressed gas to extrude the molten polymer, and it has a light-weight and compact apparatus compared to the piston-/screw-driven head. In addition, it is relatively easy to clean and maintain the parts compared to the other types of dispensing systems. On the other hand, the pneumatic type is generally associated with a delay in dispensing due to the compressed air (Pati et al. 2015). This method has the advantage of having a variety of materials that can be dispensed, including both synthetic polymer and hydrogel. Lee, Lee, and Cho (2014) fabricated a PCL-based scaffold that had various geometrical parameters with the lattice pattern by a pneumatic-type head for tissue regeneration. Shim, Huh, et al. (2012) also fabricated a composite scaffold with PCL:PLGA:β-TCP (beta-tricalcium phosphate) (2:2:1) for bone regeneration by a pneumatic-type multi-head deposition system (MHDS).

In a piston-driven head, a material in the liquid or gel phase in a syringe is directly pressed by a piston and extruded through a nozzle (Figure 8.3b). Material having a low viscosity can leak through the nozzle by gravity. This method can be suitable for the extrusion of low-viscosity materials because the pressure can be directly controlled. In addition, decompression (also called suck back) in the opposite direction of gravity can be used to prevent the leakage of the low-viscosity material. However, an additional actuator is needed in the head to adjust the piston. To regulate the flow of molten material, Woodfield et al. (2004) developed a 3D printer having a piston-driven head to directly measure the pressure. They fabricated 3D PEG-terephthalate (PEGT)-poly(butylene terephthalate) (PBT) block copolymer scaffolds using this system. Recently, a multi-head printing apparatus using a combination of pneumatic-type and piston-driven heads was developed for fabricating a hybrid structure containing synthetic polymers and living cells. Shim, Lee, et al. (2012) developed a multi-head/organ building system (MtoBS) and successfully fabricated a single structure made of a PCL framework and two different bioinks with osteoblasts and chondrocytes, respectively (Figures 8.3d,e). Additionally, using the same equipment, Lee, Hong, et al. (2014) successfully fabricated an ear-shaped PCL framework with hydrogels using sacrificial support (PEG). Moreover, they demonstrated that it allowed tissue formation from the separately printed chondrocytes and adipocytes (Figure 8.3f).

In the screw-driven head (Figure 8.3b), often called a precision extruding deposition (PED) system, the head provides the highest extrusion pressure among various dispensing methods, and thus, it can extrude highly viscous material (PCL, Mn 45,000) with a precision nozzle under 100 μm in diameter. In addition, it has been used for the fabrication of a scaffold with higher resolution than that fabricated by a pneumatic-type head. Among extrusion-based printing technologies, the screw type provides the most reliable decompression function. It consists of a barrel, screw, cartridge body, heating block, and extruding motor (Lee et al. 2017). The material melted by the heating block is transferred from the cartridge body by pneumatic pressure, and the transferred material is mechanically extruded by rotating the screw by the motor. Wang et al. (2004) introduced the PED system and fabricated a PCL scaffold with a 250-μm strand size. Shor et al. (2007) fabricated a PCL-HA (25 wt%) scaffold with enhanced mechanical properties and osteoconductivity compared with a pure PCL scaffold. Recently, to fabricate a structurally superior scaffold, Lee et al. (2017) and Lee et al. (2019) successfully developed PCL scaffolds with a 3D kagome pattern by using a PED head with a 50-μm nozzle. The fabricated PCL kagome scaffolds, which had a similar porosity and pore size to conventional grid scaffolds, exhibited better compressive stiffness and enhanced bending properties. Additionally, the customized kagome scaffolds were observed in a rabbit calvaria defect model for 16 weeks and showed excellent fitting ability and better osteoconductivity than conventional grid scaffolds. However, the screw part has a complicated structure, so this system is usually very heavy compared to the pneumatic and piston-driven head systems. In addition, the screw-type system requires more cleaning and maintenance compared to the pneumatic and piston-driven head systems.

8.4.2 Fabrication Method Using Solvent-Based Polymeric Material

As mentioned above, a thermal process has been widely used for the fabrication of polymeric scaffolds using 3D printers. However, this process cannot incorporate growth factors including bone morphogenetic proteins (BMPs) in the scaffolds because of the thermal protein denaturation (Park et al. 2011). To encapsulate the growth factors within polymeric scaffolds, a solvent-based deposition method could be an alternative.

Extrusion-based technologies can be utilized for the solution-based deposition method. This method contains additional bioactive ceramics, such as hydroxyapatite (HAp) and tricalcium phosphate (TCP), because it uses materials close to the liquid phase initially. In addition, since the used materials have low viscosity at room temperature, high pressure and heat are not required.

For the method using solvent-based biomaterials, the synthetic polymers mentioned in Section 8.4.1 are mainly used. They are easily dissolved by organic solvents, such as chloroform ($CHCl_3$), tetrahydrofuran (THF, C_4H_8O), and dichloromethane (DCM, CH_2Cl_2). The molecular chains in the polymer are separated by the organic solvent, and the polymer is changed into a liquid state. However, the widely used organic solvents are toxic and must be evaporated sufficiently after fabrication. For this reason, additional environments, such as a heated air blower/air knife or hood system, are necessary to safely vaporize organic solvents (Park et al. 2011). In addition, the surface coating of bioactive factors could be required to minimize the degeneration of the bioactive factors contained in the organic solvent. Park et al. (2011) successfully developed a 3D porous scaffold of PLGA chemically grafted hyaluronic acid (HA) encapsulated with BMP-2/PEG using solvent-based 3D printing. The HA-PLGA, PEG, and BMP-2 mixture was dissolved in chloroform and fabricated using an MHDS (Figure 8.3c). The differentiation of the scaffold with PEG/BMP-2 was better than that of a scaffold without PEG/BMP-2. Jakus et al. (2016) developed polymer–ceramic composite material using synthetic polymers (PCL or PLGA), Hap, and a tri-solvent mixture (dichloromethane, 2-butoxyethanol, and dibutyl phthalate). They fabricated a 3D scaffold with hyper-elasticity by solvent-based 3D printing.

8.4.3 BIOINKS FOR DIRECT CELL PRINTING

In the 1960s, Wichterle and Lim introduced the medical application of hydrogels (Wichterle and Lim 1960). However, given increasing demand in the field of tissue engineering, hydrogels homogenously mixed with desired cells or bioactive factors, classically termed bioinks, are now being researched extensively. To apply bioink to 3D cell printing, it should satisfy several requirements: (1) rheological behavior to decrease the hazardous shear stress-induced effects on cell viability, (2) sufficient printability, (3) modulus to retain the predefined structural morphology, (4) mild and cell-friendly gelation mechanism causing sol–gel transition, and (5) promotion of cellular differentiation (Fedorovich et al. 2007, 2011, Chang et al. 2011). Widely available bioinks used for 3D bioprinting can be broadly categorized into protein-based hydrogels (e.g., elastin/collagen (Boland et al. 2004), gelatin (Wang et al. 2006, Elsayed et al. 2016), and fibrin/fibronectin (Chung et al. 2015)) and polysaccharide-based hydrogels (e.g., HA (Tavana et al. 2016, Lee et al. 2018), alginate (Benning et al. 2018, Lee et al. 2013), and chitosan (Kong et al. 2012)). Additionally, growth factors, such as BMP (Park et al. 2011) and vascular endothelial growth factor (VEGF) (Kim, Lee, et al. 2016), and drugs can also be incorporated for therapeutic purposes. Furthermore, tissue-specific bioinks (i.e., dECM-based advanced bioinks inherently containing various proteins, proteoglycans and glycoproteins, and essential growth factors) were investigated as promising biomaterials (Nelson and Bissell 2006, Ozbolat and Hospodiuk 2016). In general, the dECM-based bioink for 3D bioprinting is in the liquid phase under a low temperature (10°C), and it transforms into the gel phase at a temperature (37°C) similar to that of the human body (Zhu, Qu et al. 2017). It is important to maintain the low temperature until the bioink is extruded. To maintain the liquid phase of the bioink at the low temperature until the extrusion, a cooling system could be required around the syringe or metal barrel that contains the bioink. In addition, a heating system could be required in the work area (working bed) to maintain the predefined 3D structural morphology of the construct after printing.

8.4.4 3D BIOPRINTING USING BIOINKS

Three-dimensional tissue modeling using the extrusion-based 3D printing technique along with various biomaterials and required bioink is capable of successfully fabricating the desired structure while retaining cellular activities. For instance, Gaetani et al. established the 3D printing of HA/gelatin (HA/gel)- (Gaetani et al. 2015) and alginate-based (Gaetani et al. 2012) biomaterials including cardiomyocyte (CM) progenitor cells in order to create cardiac constructs *in vitro*, having a precise pore size and microstructure, thus permitting better cell viability and engraftment over time. In another study, Mannoor et al. fabricated a bionic ear by 3D bioprinting chondrocytes in alginate along with silver nanoparticles (Mannoor et al. 2013, Ozbolat and Hospodiuk 2016). The engineered model was found to exhibit improved auditory sensing for radio frequency reception. Massa et al. (2017) applied a sacrificial bioprinting approach to fabricate a 3D vascularized liver tissue model by developing hollow micro-channels in the liver tissue model containing HepG2/C3A cells encapsulated in gelatin methacryloyl hydrogel.

Recently, Faramarzi et al. (2018) developed a bioink based on alginate with platelet-rich plasma (PRP) as a patient-specific source of autologous growth factors and fabricated 3D constructs using an extrusion-based 3D printer. The developed alginate-based bioink was reported to be efficient for the controlled release of PRP-associated growth factors that may enhance vascularization and stem cell migration. Similarly, Lee et al. (2018) developed a bioink based on HA on which angiogenic peptides and osteogenic peptides were immobilized for complex tissue. Further, the developed acrylated HA bioink with encapsulated L929 cells was printed by a piston-driven head that demonstrated excellent angiogenic and osteogenic effects *in vitro*.

In the applications using dECM-based bioinks, Pati, Jang et al. (2014) and Jang et al. (2018) introduced printable tissue-specific dECM-based bioink (fat, cartilage, and heart tissue) that can mimic the native tissue microenvironment. Further, they fabricated constructs using dECM-based bioink via the extrusion-based technique and reported that it can induce the multi-lineage targeted differentiation of stem cells (Figure 8.4a). Recently, Jang et al. (2017) developed a 3D cell-laden prevascularized cardiac patch using heart tissue-derived dECM (hdECM) in combination with human cardiac progenitor cells and mesenchymal stem cells (MSCs) and implanted it in a rat myocardial infarction model. The analyzed *in vivo* data demonstrated reduced cardiac hypertrophy and fibrosis, enhanced cardiac function, and cellular infiltration into the area of infarction (Figure 8.4b–d). Similarly, to treat ischemic regions, Gao et al. (2017) fabricated a blood vessel using dECM-based hybrid bioink (blend of vascular tissue-derived dECM (vdECM) and alginate by extrusion-based 3D bioprinting with a coaxial-type nozzle for the delivery of endothelial progenitor cells (EPCs) and microcapsules of the proangiogenic drug atorvastatin (Figure 8.5a). In another study, Lee, Han et al. (2017) reported fabricating a liver tissue model using liver tissue-derived dECM bioink that demonstrated improved functions of human bone marrow-derived MSCs and encapsulated human hepatocellular carcinoma (HepG2) cell lines (Figure 8.5b–d).

To date, the development of the convergence technologies of biology, medicine, and engineering strategies has achieved significant results for tissue engineering. In addition to 3D bioprinting technology, the development of dECM-based bioinks similar to the native tissues in composition has increased significantly, enabling the fabrication of constructs mimicking native complex tissue. However, most of the achievements have been developed only at the laboratory scale. Ultimately, in order to consider the utilization of bioprinting, it is necessary to consider 3D multi-bioprinting using a mimetic model based on real tissue data and tissue-specific bioinks.

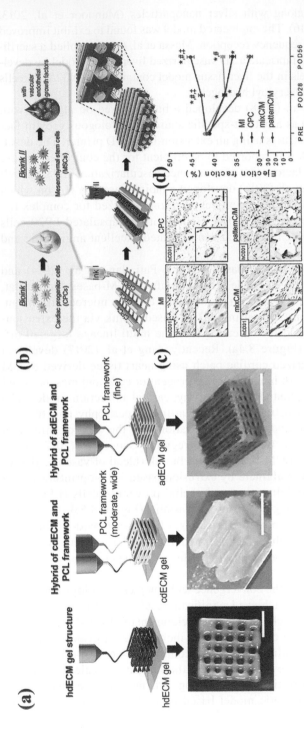

FIGURE 8.4 (a) Printing process of particular tissue constructs with dECM bioink. Heart tissue construct was printed with only heart dECM (hdECM). Cartilage and adipose tissues were printed with cartilage dECM (cdECM) and adipose dECM (adECM), respectively, and in combination with PCL framework (scale bar, 5mm) (Pati et al. 2014), (b) illustrative representation of 3D cell printing of prevascularized stem cell patch, (c) immunohistochemistry results against human-specific CD31 specificity at infarct regions. (d) Effects of prevascularized stem cell patch on the therapeutic efficacy post-MI. EF values at baseline and after four and eight weeks. Error bars represent standard errors of the mean (SEM) (*p < 0.05 compared with MI; #p < 0.05 compared with CPC; ‡p < 0.05 compared with mix C/M). MI, myocardial infarction; EF, ejection fraction; CPC, cardiac progenitor cells; C/M, both CPC and mesenchymal stem cells (MSCs); POD: postoperation day (Jang et al. 2017). (Reproduced with permission from Pati, Jang, et al. 2014, Jang et al. 2017.).

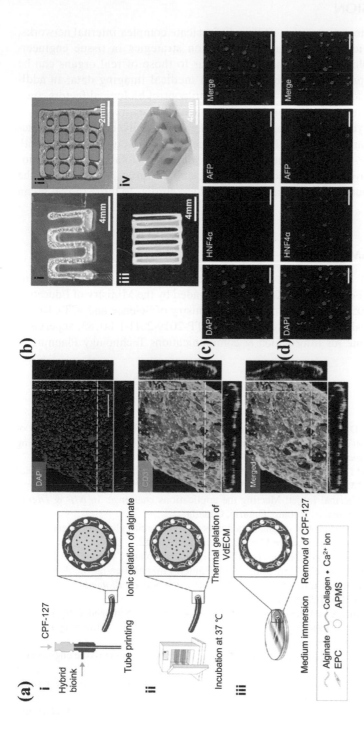

FIGURE 8.5 (a) A schematic depiction of the bio-blood vessel (BBV) fabrication process. (i) The ionic gelation of alginate in 3V2A realized BBV printing, (ii) the thermal crosslinking of collagen fibers was induced by incubation at 37°C, (iii) medium immersion dissolved and removed CPF127 to obtain BBV with a hollow tubular shape, CD31/DAPI staining indicated that the encapsulated EPCs formed a layer of fully differentiated endothelium on the BBV after culturing for 7 days (Gao et al. 2017). (b) results of (i and ii) 2D printing patterns and (iii and iv) 3D hybrid structures with the polymer. Immunofluorescence staining of representative genes on day 7: stem cells in Lee, Han, et al. (2017), (c) the collagen group and (d) the liver dECM group (scale bar of 200 μm) (Lee, Han, et al. 2017). (Reproduced with permission from Gao et al. 2017, Lee, Han, et al. 2017.)

8.5   CONCLUSION

Since 3D printing techniques can effectively fabricate complex internal networks, they have been widely utilized as biofabrication strategies in tissue engineering. Porous scaffolds with morphologies similar to those of real organs can be fabricated by 3D printing techniques based on medical imaging data. In addition, 3D printing techniques have the potential to control biological factors, such as cell deposition, migration, orientation, alignment, and aggregation. Moreover, newly developed hydrogels for bioprinting are capable of delivering bioactive factors or cells. In particular, the dECM-based bioink is attracting attention with respect to its tissue-specific potential. In recent years, the biofabrication and bioink approaches have been carried out broadly, and in future, they will be the main ways of fabricating engineered tissues that are similar to those of real organ systems.

ACKNOWLEDGMENT

This research was supported by Basic Science Research Program through the National Research Foundation of Korea (NRF) funded by the Ministry of Education (2015R1A6A3A04059015) and by the MSIT (Ministry of Science and ICT), Korea, under the ICT Consilience Creative program (IITP-2019-2011-1-00783) supervised by the IITP (Institute for Information & communications Technology Planning & Evaluation).

REFERENCES

Allain, Leonardo R, Dimitra N Stratis-Cullum, and Tuan Vo-Dinh. 2004. Investigation of microfabrication of biological sample arrays using piezoelectric and bubble-jet printing technologies. *Analytica Chimica Acta* 518 (1–2):77–85.

Ballyns, Jeffrey J, and Lawrence J Bonassar. 2009. Image-guided tissue engineering. *Journal of Cellular and Molecular Medicine* 13 (8a):1428–1436.

Beeson, Rob. 1998. Thermal inkjet: Meeting the applications challenge. In *NIP & Digital Fabrication Conference,* Toronto, Ontario, Canada.

Bennett, Joe. 2017. Measuring UV curing parameters of commercial photopolymers used in additive manufacturing. *Additive Manufacturing* 18:203–212.

Benning, Leo, Ludwig Gutzweiler, Kevin Tröndle, Julian Riba, Roland Zengerle, Peter Koltay, Stefan Zimmermann, G Björn Stark, and Günter Finkenzeller. 2018. Assessment of hydrogels for bioprinting of endothelial cells. *Journal of Biomedical Materials Research Part A* 106 (4):935–947.

Boland, Eugene D, Jamil A Matthews, Kristin J Pawlowski, David G Simpson, Gary E Wnek, and Gary L Bowlin. 2004. Electrospinning collagen and elastin: Preliminary vascular tissue engineering. *Frontiers in Bioscience* 9(2):1422–1432.

Bose, Susmita, Sahar Vahabzadeh, and Amit Bandyopadhyay. 2013. Bone tissue engineering using 3D printing. *Materials Today* 16 (12):496–504.

Bracci, Raffaella, Elena Maccaroni, and Stefano Cascinu. 2013. Bioresorbable airway splint created with a three-dimensional printer.

Calvert, Paul. 2001. Inkjet printing for materials and devices. *Chemistry of Materials* 13 (10):3299–3305.

Chang, Carlos C, Eugene D Boland, Stuart K Williams, and James B Hoying. 2011. Direct-write bioprinting three-dimensional biohybrid systems for future regenerative therapies. *Journal of Biomedical Materials Research Part B: Applied Biomaterials* 98 (1):160–170.

Chia, Helena N, and Benjamin M Wu. 2014. High-resolution direct 3D printed PLGA scaffolds: Print and shrink. *Biofabrication* 7 (1):015002.

Chirani, Naziha, Lukas Gritsch, Federico Leonardo Motta, and Silvia Fare. 2015. History and applications of hydrogels. *Journal of Biomedical Sciences* 4 (2):13.

Choi, Moonhyun, Jangsun Hwang, Jonghoon Choi, and Jinkee Hong. 2017. Multicomponent high-throughput drug screening via inkjet printing to verify the effect of immunosuppressive drugs on immune T lymphocytes. *Scientific Reports* 7 (1):6318.

Christensen, Kyle, Changxue Xu, Wenxuan Chai, Zhengyi Zhang, Jianzhong Fu, and Yong Huang. 2015. Freeform inkjet printing of cellular structures with bifurcations. *Biotechnology and Bioengineering* 112 (5):1047–1055.

Chua, Chee Kai, Mei Jun Jolene Liu, and Siaw Meng Chou. 2011. Additive manufacturing-assisted scaffold-based tissue engineering. In Paulo J Bártolo *Innovative Developments in Virtual and Physical Prototyping, Proceedings of the 5th International Conference on Advanced Research in Virtual and Rapid Prototyping, Leiria, Portugal,* 13–21. CRC Press, London.

Chung, Eunna, Julie A Rytlewski, Arjun G Merchant, Kabir S Dhada, Evan W Lewis, and Laura J Suggs. 2015. Fibrin-based 3D matrices induce angiogenic behavior of adipose-derived stem cells. *Acta Biomaterialia* 17:78–88.

Crump, S Scott. 1992. Apparatus and method for creating three-dimensional objects. Google Patents.

Cui, Xiaofeng, Thomas Boland, Darryl D D'Lima, and Martin K Lotz. 2012. Thermal inkjet printing in tissue engineering and regenerative medicine. *Recent Patents on Drug Delivery and Formulation* 6 (2):149–155.

Cui, Xiaofeng, Delphine Dean, Zaverio M Ruggeri, and Thomas Boland. 2010. Cell damage evaluation of thermal inkjet printed Chinese hamster ovary cells. *Biotechnology and Bioengineering* 106 (6):963–969.

Cui, Xiaofeng, Guifang Gao, Tomo Yonezawa, and Guohao Dai. 2014. Human cartilage tissue fabrication using three-dimensional inkjet printing technology. *Journal of Visualized Experiments* 10 (88): doi:10.3791/51294.

de Gans, Berend-Jan, Paul C Duineveld, and Ulrich S Schubert. 2004. Inkjet printing of polymers: State of the art and future developments. *Advanced Materials* 16 (3):203–213.

de Gans, Berend-Jan, and Ulrich S Schubert. 2003. Inkjet printing of polymer microarrays and libraries: Instrumentation, requirements, and perspectives. *Journal of Macromolecular Rapid Communications* 24 (11):659–666.

Derby, Brian 2010. Inkjet printing of functional and structural materials: Fluid property requirements, feature stability, and resolution. *Annual Review of Materials Research* 40:395–414.

Duan, Bin, Min Wang, Wen You Zhou, Wai Lam Cheung, Zhao Yang Li, and William W Lu. 2010. Three-dimensional nanocomposite scaffolds fabricated via selective laser sintering for bone tissue engineering. *Acta Biomaterialia* 6 (12):4495–4505.

Elsayed, Yahya, Constantina Lekakou, Fatima Labeed, Paul Tomlins. 2016. Fabrication and characterisation of biomimetic, electrospun gelatin fibre scaffolds for tunica media-equivalent, tissue engineered vascular grafts. *Journal of Materials Science and Engineering: C* 61:473–483.

Faramarzi, Negar, Iman K Yazdi, Mahboubeh Nabavinia, Andrea Gemma, Adele Fanelli, Andrea Caizzone, Leon M Ptaszek, Indranil Sinha, Ali Khademhosseini, and Jeremy N Ruskin. 2018. Patient-specific bioinks for 3D bioprinting of tissue engineering scaffolds. *Journal of Advanced Healthcare Materials* 7 (11):1701347.

Fedorovich, Natalja E, Jacqueline Alblas, Joost R de Wijn, Wim E Hennink, Ab J Verbout, and Wouter J Dhert. 2007. Hydrogels as extracellular matrices for skeletal tissue engineering: State-of-the-art and novel application in organ printing. *Tissue Engineering* 13 (8):1905–1925.

Fedorovich, Natalja E, Jacqueline Alblas, Wim E Hennink, F Cumhur Öner, and Wouter J Dhert. 2011. Organ printing: The future of bone regeneration? *Trends in Biotechnology* 29 (12):601–606.

Gaetani, Roberto, Peter A Doevendans, Coirna H Metz, Jacqueline Alblas, Elisa Messina, Alessandro Giacomello, and Joost P Sluijter. 2012. Cardiac tissue engineering using tissue printing technology and human cardiac progenitor cells. *Biomaterials* 33 (6):1782–1790. doi:10.1016/j.biomaterials.2011.11.003.

Gaetani, Ronerto, Dries A Feyen, Vera Verhage, Rolf Slaats, Elisa Messina, Karen L Christman, Alessandro Giacomello, Pieter A Doevendans, and Joost P Sluijter. 2015. Epicardial application of cardiac progenitor cells in a 3D-printed gelatin/hyaluronic acid patch preserves cardiac function after myocardial infarction. *Biomaterials* 61:339–348. doi:10.1016/j.biomaterials.2015.05.005.

Gao, Ge, Jun Hee Lee, Jinah Jang, Dong Han Lee, Jeong-Sik Kong, Byoung Soo Kim, Yeong-Jin Choi, Woong Bi Jang, Young Joon Hong, Sang-Mo Kwon, and Dong-Woo Cho. 2017. Tissue engineered bio-blood-vessels constructed using a tissue-specific bioink and 3D coaxial cell printing technique: A novel therapy for ischemic disease. *Advanced Functional Materials* 27 (33):1700798. doi:10.1002/adfm.201700798.

Gao, Guifang, Tomo Yonezawa, Karen Hubbell, Guohao Dai, and Xiaofeng Cui. 2015. Inkjet-bioprinted acrylated peptides and PEG hydrogel with human mesenchymal stem cells promote robust bone and cartilage formation with minimal printhead clogging. *Biotechnology Journal* 10 (10):1568–1577.

Gudapati, Hemanth, Madhuri Dey, and Ibrahim Ozbolat. 2016. A comprehensive review on droplet-based bioprinting: Past, present and future. *Biomaterials* 102:20–42.

Heinzl, Johann, and Carl Hellmuth Hertz. 1985. Ink-jet printing. Advances in Electronics and Electron Physics 65:91–171.

Hewes, Sarah, Andrew D Wong, and Peter C Searson. 2017. Bioprinting microvessels using an inkjet printer. *Bioprinting* 7:14–18.

Hölzl, Katja, Shengmao Lin, Liesbeth Tytgat, Sandra Van Vlierberghe, Linxia Gu, and Aleksandr Ovsianikov. 2016. Bioink properties before, during and after 3D bioprinting. *Biofabrication* 8 (3):032002.

Hospodiuk, Monika, Madhuri Dey, Donna Sosnoski, and Ibrahim T Ozbolat. 2017. The bioink: A comprehensive review on bioprintable materials. *Biotechnology Advances* 35 (2):217–239.

Huang, Angela H, and Laura E Niklason. 2011. Engineering biological-based vascular grafts using a pulsatile bioreactor. *Journal of Visualized Experiments* 14 (52). doi:10.3791/2646.

Jakus, Adam E, Alexandra L Rutz, Sumanas W Jordan, Abhishek Kannan, Sean M Mitchell, Chawon Yun, Katie D Koube, Sung C Yoo, Herbert E Whiteley, and Claus-Peter Richter. 2016. Hyperelastic bone: A highly versatile, growth factor–free, osteoregenerative, scalable, and surgically friendly biomaterial. *Science Translational Medicine* 8 (358):358ra127.

Jang, Jinah, Ju Young Park, Ge Gao, and Dong-Woo Cho. 2018. Biomaterials-based 3D cell printing for next-generation therapeutics and diagnostics. *Biomaterials* 156:88–106.

Jang, Jinah, Hun-Jun Park, Seok-Won Kim, Heejin Kim, Ju Young Park, Soo Jin Na, Hyeon Ji Kim, Moon Nyeo Park, Seung Hyun Choi, and Sun Hwa Park. 2017. 3D printed complex tissue construct using stem cell-laden decellularized extracellular matrix bioinks for cardiac repair. *Biomaterials* 112:264–274.

Ji, Shen, and Murat Guvendiren. 2017. Recent advances in bioink design for 3D bioprinting of tissues and organs. *Frontiers in Bioengineering and Biotechnology* 5:23.

Kim, Jae Dong, Ji Suk Choi, Beob Soo Kim, Young Chan Choi, and Yong Woo Cho. 2010. Piezoelectric inkjet printing of polymers: Stem cell patterning on polymer substrates. *Polymer* 51 (10):2147–2154.

Kim, Ji Eun, Soo Hyun Kim, and Youngmee Jung. 2016. Current status of three-dimensional printing inks for soft tissue regeneration. *Tissue Engineering and Regenerative Medicine* 13 (6):636–646.

Kim, Sook Kyoung, Jaeyeon Lee, Myeongjin Song, Mirim Kim, Soon Jung Hwang, Hwanseok Jang, and Yongdoo Park. 2016. Combination of three angiogenic growth factors has synergistic effects on sprouting of endothelial cell/mesenchymal stem cell-based spheroids in a 3D matrix. *Journal of Biomedical Materials Research Part B: Applied Biomaterials* 104 (8):1535–1543.

Knowlton, Stephanie, Sevgi Onal, Chu Hsiang Yu, Jean J Zhao, and Savas Tasoglu. 2015. Bioprinting for cancer research. *Journal of Trends in Biotechnology* 33 (9):504–513.

Kong, Xiaoying, Baoqin Han, Haixia Wang, Hui Li, Wenhua Xu, and Wanshun Liu. 2012. Mechanical properties of biodegradable small-diameter chitosan artificial vascular prosthesis. *Journal of Biomedical Materials Research Part A* 100 (8):1938–1945.

Lee, Se-Hwan, Yong Sang Cho, Myoung Wha Hong, Bu-Kyu Lee, Yongdoo Park, Sang-Hyug Park, Young Yul Kim, and Young-Sam Cho. 2017. Mechanical properties and cell-culture characteristics of a polycaprolactone kagome-structure scaffold fabricated by a precision extruding deposition system. *Biomedical Materials* 12 (5):055003.

Lee, Hyungseok, Wonil Han, Hyeonji Kim, Dong-Heon Ha, Jinah Jang, Byoung Soo Kim, and Dong-Woo Cho. 2017. Development of liver decellularized extracellular matrix bioink for three-dimensional cell printing-based liver tissue engineering. *Biomacromolecules* 18 (4):1229–1237.

Lee, Jung-Seob, Jung Min Hong, Jin Woo Jung, Jin-Hyung Shim, Jeong-Hoon Oh, and Dong-Woo Cho. 2014. 3D printing of composite tissue with complex shape applied to ear regeneration. *Biofabrication* 6 (2):024103.

Lee, Se-Hwan, A Ra Jo, Ghi Pyoung Choi, Chang Hee Woo, Seung Jae Lee, Beom-Su Kim, Hyung-Keun You, Young-Sam Cho. 2013. Fabrication of 3D alginate scaffold with interconnected pores using wire-network molding technique. *Tissue Engineering and Regenerative Medicine* 10 (2):53–59.

Lee, Se-Hwan, Jun Hee Lee, and Young-Sam Cho. 2014. Analysis of degradation rate for dimensionless surface area of well-interconnected PCL scaffold via in-vitro accelerated degradation experiment. *Tissue Engineering and Regenerative Medicine* 11 (6):446–452.

Lee, Se-Hwan, Kang-Gon Lee, Jong-Hyun Hwang, Yong Sang Cho, Kang-Sik Lee, Hun-Jin Jeong, Sang-Hyug Park, Yongdoo Park, Young-Sam Cho, and Bu-Kyu Lee. 2019. Evaluation of mechanical strength and bone regeneration ability of 3D printed kagome-structure scaffold using rabbit calvarial defect model. *Materials Science and Engineering: C* 98:949–959.

Lee, Jaeyeon, Se-Hwan Lee, Byung Soo Kim, Young-Sam Cho, and Yongdoo Park. 2018. Development and evaluation of hyaluronic acid-based hybrid bio-ink for tissue regeneration. *J Tissue Engineering and Regenerative Medicine* 15 (6) 761–769.

Lee, Jaeyeon, Se-Hwan Lee, Bu-Kyu Lee, Sang-Hyug Park, Young-Sam Cho, and Yongdoo Park. 2018. Fabrication of microchannels and evaluation of guided vascularization in biomimetic hydrogels. *Tissue Engineering and Regenerative Medicine* 15 (4):403–413.

Li, Jipeng, Mingjiao Chen, Xianqun Fan, and Huifang Zhou. 2016. Recent advances in bioprinting techniques: Approaches, applications and future prospects. *Journal of Translational Medicine* 14 (1):271.

Li, Jia, Fabrice Rossignol, and Joanne Macdonald. 2015. Inkjet printing for biosensor fabrication: Combining chemistry and technology for advanced manufacturing. *Lab on a Chip* 15 (12):2538–2558.

Mannoor, Manu S, Ziwen Jiang, Teena James, Yong Lin Kong, Karen A Malatesta, Winston O Soboyejo, Naveen Verma, David H Gracias, and Michael C McAlpine. 2013. 3D printed bionic ears. *Nano Letters* 13 (6):2634–2639.

Marizza, Paolo, Stephan S Keller, Anette Müllertz, and Anja Boisen. 2014. Polymer-filled microcontainers for oral delivery loaded using supercritical impregnation. *Journal of Controlled Release* 173:1–9.

Massa, Solange, Mahmoud Ahmed Sakr, Jungmok Seo, Praveen Bandaru, Andrea Arneri, Simone Bersini, Elaheh Zare-Eelanjegh, Elmira Jalilian, Byung-Hyun Cha, and Silvia Antona. 2017. Bioprinted 3D vascularized tissue model for drug toxicity analysis. *Biomicrofluidics* 11 (4):044109.

Melchels, Ferry PW, Marco AN Domingos, Travis J Klein, Jos Malda, Paulo J Bartolo, and Dietmar W Hutmacher. 2012. Additive manufacturing of tissues and organs. *Progress in Polymer Science* 37 (8):1079–1104.

Miller, Jordan S, and Jason A Burdick. 2016. Special issue on 3D printing of biomaterials. ACS Publications.

Miller, Jordan S, Kelly R Stevens, Michael T Yang, Brendon M Baker, Duc-Huy T Nguyen, Daniel M Cohen, Esteban Toro, Alice A Chen, Peter A Galie, and Xiang Yu. 2012. Rapid casting of patterned vascular networks for perfusable engineered three-dimensional tissues. *Nature Materials* 11 (9):768.

Moroni, Lorenzo, Thomas Boland, Jason A Burdick, Carmelo De Maria, Brian Derby, Gabor Forgacs, Jürgen Groll, Qing Li, Jos Malda, and Vladimir A Mironov. 2017. Biofabrication: A guide to technology and terminology. *Trends in Biotechnology* 36(4):384–402.

Murphy, Sean V, and Anthony Atala. 2014. 3D bioprinting of tissues and organs. *Nature Biotechnology* 32 (8):773.

Nelson, Celeste M, and Mina J Bissell. 2006. Of extracellular matrix, scaffolds, and signaling: Tissue architecture regulates development, homeostasis, and cancer. *Annual Review of Cell and Developmental Biology* 22:287–309. doi:10.1146/annurev.cellbio.22.010305.104315.

Ozbolat, Ibrahim T, and Monika Hospodiuk. 2016. Current advances and future perspectives in extrusion-based bioprinting. *Biomaterials* 76:321–343.

Park, Jeong Hun, Jinah Jang, Jung-Seob Lee, and Dong-Woo Cho. 2017. Three-dimensional printing of tissue/organ analogues containing living cells. *Annals of Biomedical Engineering* 45 (1):180–194.

Park, Sang-Hyug, Chi Sung Jung, Byoung-Hyun Min. 2016. Advances in three-dimensional bioprinting for hard tissue engineering. *Tissue Engineering and Regenerative Medicine* 13 (6):622–635.

Park, Jung Kyu, Jin-Hyung Shim, Kyung Shin Kang, Junseok Yeom, Ho Sang Jung, Jong Young Kim, Keum Hong Lee, Tae-Ho Kim, Shin-Yoon Kim, and Dong-Woo Cho. 2011. Solid free-form fabrication of tissue-engineering scaffolds with a poly (lactic-co-glycolic acid) grafted hyaluronic acid conjugate encapsulating an intact bone morphogenetic protein–2/poly (ethylene glycol) complex. *Advanced Functional Materials* 21 (15):2906–2912.

Pati, Falguni, Jinah Jang, Dong-Heon Ha, Sung Won Kim, Jong-Won Rhie, Jin-Hyung Shim, Deok-Ho Kim, and Dong-Woo Cho. 2014. Printing three-dimensional tissue analogues with decellularized extracellular matrix bioink. *Nature Communications* 5:3935.

Pati, Falguni, Jinah Jang, Jin Woo Lee, and Dong-Woo Cho. 2015. Extrusion bioprinting. In Anthony Atala, and James J Yoo *Essentials of 3D Biofabrication and Translation*, 123–152. Elsevier, San Diego, CA.

Sachs, Emanuel, Michael Cima, and James Cornie. 1990. Three-dimensional printing: Rapid tooling and prototypes directly from a CAD model. *CIRP Annals-Manufacturing Technology* 39 (1):201–204.

Sachs, Emanuel M, John S Haggerty, Michael J Cima, and Paul A Williams. 1993. Three-dimensional printing techniques. Google Patents.

Saunders, Rachel E, Julie E Gough, and Brian Derby. 2008. Delivery of human fibroblast cells by piezoelectric drop-on-demand inkjet printing. *Biomaterials* 29 (2):193–203.

Serra, Tiziano, Josep A Planell, and Melba Navarro. 2013. High-resolution PLA-based composite scaffolds via 3-D printing technology. *Acta Biomaterialia* 9 (3):5521–5530.

Shim, Jin-Hyung, Jung-Bo Huh, Ju Young Park, Young-Chan Jeon, Seong Soo Kang, Jong Young Kim, Jong-Won Rhie, and Dong-Woo Cho. 2012. Fabrication of blended poly-caprolactone/poly (lactic-co-glycolic acid)/β-tricalcium phosphate thin membrane using solid freeform fabrication technology for guided bone regeneration. *Tissue Engineering Part A* 19 (3–4):317–328.

Shim, Jin-Hyung, Jung-Seob Lee, Jong Young Kim, Dong-Woo Cho. 2012. Bioprinting of a mechanically enhanced three-dimensional dual cell-laden construct for osteochondral tissue engineering using a multi-head tissue/organ building system. *Journal of Micromechanics and Microengineering* 22 (8):085014.

Shor, Lauren, Selçuk Güçeri, Xuejun Wen, Milind Gandhi, and Wei Sun. 2007. Fabrication of three-dimensional polycaprolactone/hydroxyapatite tissue scaffolds and osteoblast-scaffold interactions in vitro. *Biomaterials* 28 (35):5291–5297.

Spadaccio, Cristiano, Albertp Rainer, Marcella Trombetta, Matteo Centola, Mario Lusini, Massimo Chello, Elvio Covino, Federico De Marco, Raffaella Coccia, Yoshiya Toyoda, and Jorge A Genovese. 2011. A G-CSF functionalized scaffold for stem cells seeding: A differentiating device for cardiac purposes. *Journal of Cellular and Molecular Medicine* 15 (5):1096–108.

Sun, Wei, Andrew Darling, Binil Starly, Jae Nam. 2004. Computer-aided tissue engineering: Overview, scope and challenges. *Biotechnology and Applied Biochemistry* 39 (1):29–47.

Sweet, Richard G 1965. High frequency recording with electrostatically deflected ink jets. *Review of Scientific Instruments* 36 (2):131–136.

Tabata, Kunio, Atsushi Oshima, Osamu Shinkawa, Toshiyuki Suzuki, and Tomoki Hatano. 2013. Head drive device of inkjet printer and ink jet printer. Google Patents.

Tavana, Somayeh, Mahnaz Azarnia, Mojtaba Rezazadeh Valojerdi, and Abdolhossein Shahverdi. 2016. Hyaluronic acid-based hydrogel scaffold without angiogenic growth factors enhances ovarian tissue function after autotransplantation in rats. *Biomedical Materials* 11 (5):055006.

Truby, Ryan L, and Jennifer A Lewis. 2016. Printing soft matter in three dimensions. *Nature* 540 (7633):371.

Tse, Christopher Chi Wai, and Patrick J Smith. 2018. Inkjet Printing for Biomedical Applications. In Peter Ertl, and Mario Rothbauer *Cell-Based Microarrays*, 107–117. Springer, New York.

Wang, Fang, Lauren Shor, Andrew Darling, Saif Khalil, Wei Sun, S Güçeri, and Alan Lau. 2004. Precision extruding deposition and characterization of cellular poly-ε-caprolactone tissue scaffolds. *Rapid Prototyping Journal* 10 (1):42–49.

Wang, Xiaohong, Yongnian Yan, Yuqiong Pan, Zhuo Xiong, Haixia Liu, Jie Cheng, Feng Liu, Feng Lin, Rendong Wu, and Renji Zhang. 2006. Generation of three-dimensional hepatocyte/gelatin structures with rapid prototyping system. *Tissue Engineering* 12 (1):83–90.

Wichterle, Otto, and Drahoslav Lim. 1960. Hydrophilic gels for biological use. *Nature* 185 (4706):117.

Woodfield, Tim BF, Jos Malda, J De Wijn, Fabienne Peters, Jens Riesle, and Clemens A van Blitterswijk. 2004. Design of porous scaffolds for cartilage tissue engineering using a three-dimensional fiber-deposition technique. *Biomaterials* 25 (18):4149–4161.

Worthington, Kristan S, Luke A Wiley, Emily E Kaalberg, Malia M Collins, Robert F Mullins, Edwin M Stone, and Budd A Tucker. 2017. Two-photon polymerization for production of human iPSC-derived retinal cell grafts. *Acta Biomaterialia* 55:385–395.

Xu, Changxue, Wenxuan Chai, Yong Huang, Roger R Markwald. 2012. Scaffold-free inkjet printing of three-dimensional zigzag cellular tubes. *Biotechnology and Bioengineering* 109 (12):3152–3160.

Xu, Tao, Weixin Zhao, Jian-Ming Zhu, Mohammad Z Albanna, James J Yoo, and Anthony Atala. 2013. Complex heterogeneous tissue constructs containing multiple cell types prepared by inkjet printing technology. *Biomaterials* 34 (1):130–139.

Yang, Shoufeng, Kah-Fai Leong, Zhaohui Du, and Chee-Kai Chua. 2001. The design of scaffolds for use in tissue engineering. Part I. Traditional factors. *Tissue Engineering* 7 (6):679–689.

Yoshikawa, Hirokazu, Hiroshi Tajika, Hitoshi Nishikori, Daisaku Ide, Takeshi Yazawa, Atsuhiko Masuyama, Akiko Maru, and Hideaki Takamiya. 2009. Inkjet recording apparatus and maintenance method thereof. Google Patents.

Zhang, Jiaxiang, Xin Feng, Hemlata Patil, Roshan V Tiwari, and Michael A Repka. 2017. Coupling 3D printing with hot-melt extrusion to produce controlled-release tablets. *International Journal of Pharmaceutics* 519 (1–2):186–197.

Zhu, Wei, Xin Qu, Jie Zhu, Xuanyi Ma, Sherrina Patel, Justin Liu, Pengrui Wang, Cheuk Sun Edwin Lai, Maling Gou, and Yang Xu. 2017. Direct 3D bioprinting of prevascularized tissue constructs with complex microarchitecture. *Biomaterials* 124:106–115.

Part III

Pathomimetic Disease Modeling

9 Microengineered Models of Human Gastrointestinal Diseases

Woojung Shin, Landon A. Hackley,
and Hyun Jung Kim
The University of Texas at Austin

CONTENTS

9.1 INTRODUCTION

The human intestine primarily contributes to digestion and nutrient absorption (Goodman 2010). It also serves critical physiological functions such as regulation of local immunity (Mowat and Agace 2014), metabolism of orally administered drugs and chemicals (Yang et al. 2007) in coordination with intestinal bacteria, and orchestration of mental health factors via the enteric nervous system (Dalile et al. 2019, Cryan and O'Mahony 2011). The multicellular interactions controlled by contributing cells of the intestinal microenvironment have been known to contribute significantly to the maintenance of intestinal homeostasis, paradoxically suggesting that the impairment of homeostasis in the gastrointestinal (GI) tract is often directly related to GI diseases such as inflammatory bowel disease (IBD) (Xavier and Podolsky 2007), colorectal cancer (CRC) (Terzić et al. 2010), and pathogenic infection (Deitch et al. 1990).

To investigate underlying etiologies or new therapeutic strategies for GI diseases, both *in vitro* and *in vivo* experimental models have been widely utilized. Traditional *in vitro* models have leveraged static two-dimensional (2D) or 3D cell culture methods in a Petri dish, on a porous membrane insert (e.g., Transwell), or inside a hydrogel (e.g., organoid or spheroid cultures). Although these *in vitro* cell culture models are relatively simple and robust, a lack of biological complexity and mechanical dynamics prevents proper emulation of the *in vivo*-relevant pathophysiology components of GI diseases. Animal models have been extensively used in both basic research laboratories and pharmaceutical industries as a "gold standard". However, the structural and physiological discrepancies between animals and humans seriously hinder the reliability and reproducible validity of newly developed drug candidates. It is noted that drug metabolism, pharmacokinetics (PK) and pharmacodynamics (PD), and clearance profiles in animals are considerably different from those of humans (Nguyen et al. 2015). Animal vivisection and the resource-intensive nature of animal testing have been ethical concerns in the scientific community. More importantly, animal models do not correctly reflect the genotypic and phenotypic backgrounds of individual patients because of species differences (Seok et al. 2013). Thus, the development of a new experimental model that recapitulates the structural, mechanical, chemical, biological, and pathological features of humans remains an urgent, unmet need.

In recent years, efforts have been made to better emulate the intestinal physiology and pathology of various GI diseases using a biomimetic microengineering approach (Bein and Shin et al. 2018). Microfluidics-based modeling of GI diseases has allowed researchers to demonstrate a controllable microenvironment, fluid shear mechanics and hydrodynamics, directional accessibility (e.g., apical vs. basolateral), and spatiotemporal visualization of cellular and molecular interactions that occur in GI diseases. These enabling technologies and implementable breakthroughs allow researchers to better mimic and display the pathophysiology and complex multicellular crosstalks of pathogenesis. In this chapter, we overview the *in vitro* methodologies to recapitulate GI diseases using biomimetic engineering approaches and current pathomimetic GI disease models.

9.2 *IN VITRO* METHODOLOGIES FOR RECAPITULATING THE PATHOPHYSIOLOGY OF GI DISEASES

To recapitulate the pathophysiological microenvironment of GI diseases, essential components of illness should be considered in a defined 3D structure. The intestinal epithelium is the most critical component to build an intestinal mucosal microenvironment, where an intact epithelial barrier forms two compartments: a luminal and an abluminal (i.e., lamina propria, capillaries, or mesenchyme) side. Because the GI tract is continuously exposed to the external environment, a physical barrier is necessary to protect the body from foreign invaders (Peterson and Artis 2014). Since barrier dysfunction prevalently occurs in most GI diseases (Turner 2009), demonstrating the integrity of the epithelial barrier function in a model is the first and foremost prerequisite for mimicking a GI disease. Both innate and adaptive immune cells are necessary components for inducing immune responses that are involved in GI diseases (Mowat and Agace 2014). An imbalanced population of the commensal gut microbes (i.e., dysbiosis) plays a crucial role in the initiation and pathogenesis of GI diseases (Round and Mazmanian 2009). Dysbiosis is remarkably implicated in IBD (Tamboli et al. 2004) and colorectal cancer (Sobhani et al. 2011). Furthermore, specific bacteria are known to be involved in the pathogenesis of certain GI diseases, such as *Helicobacter pylori* in gastric cancer (Uemura et al. 2001) and *Fusobacterium nucleatum* in CRC (Mima et al. 2016). Thus, human microbiome must be taken into account to accurately model GI diseases.

Here, we categorize *in vitro* intestine models into static and microfluidic culture formats. Static culture methods include conventional Petri dish, well plate, and Transwell cultures. Emerging static culture methods also include 3D organoid culture and biomaterials-based hydrogel scaffolds to emulate intestinal epithelial structures. Microfluidic culture formats include cell culture approaches that utilize microchannels, in which culture medium flows under laminar flow, to introduce fluid shear stress and mechanical deformation to recapitulate the physical microenvironment of the human intestine.

9.2.1 STATIC CELL CULTURE MODELS

Static cell culture models are generally straightforward and are the most popular of *in vitro* model systems to study GI diseases. A culture medium supports cell growth in a static culture condition, where epithelial cells often grow in a 2D planar monolayer.

The Transwell model is an advanced cell culture system compared to the conventional Petri dish culture as it compartmentalizes apical (AP) and basolateral (BL) surfaces of the epithelium (Figure 9.1a). Immune and microbial cells can be

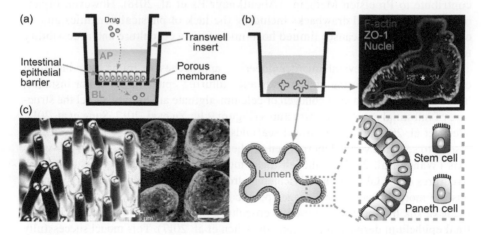

FIGURE 9.1 Static culture models of human intestine. (a) A schematic that describes the structural compartmentalization of the Transwell cultures. A porous insert that contains a monolayer of human intestinal epithelium forms an intact, tight junction, by which a drug compound can be introduced into the apical (AP) side of the Transwell to estimate the apparent permeability by measuring the transported amount in the basolateral (BL) side. (b) Organoid culture is performed in a well plate, where organoids embedded inside the hydrogel (top left) form a 3D morphology with physiological cytodifferentiation (top right) and structural villus/crypt domains (bottom left). Intestinal stem and Paneth cells are found in the crypt domain (bottom right). An asterisk in the top right image indicates the enclosed lumen in a single organoid. (c) An array of hydrogel-based scaffold that mimics the structure of intestinal villi (left) lined by a monolayer of intestinal epithelium (right). Blue: nuclei; Green: actin. Scale bars: 100 μm.

co-cultured by directionally seeding each cell type independently to the AP or BL side (Bisping et al. 2001). Because of this physical compartmentalization, epithelial barrier function can be measured along with apparent permeability (P_{app}) (Gao et al. 2000). Intestinal barrier function can usually be assessed by measuring transepithelial electrical resistance (TEER), where two Ag/AgCl electrodes are inserted separately into the apical and basolateral sides to quantify the resistance across the epithelial monolayer. Apparent permeability of a drug compound or a food substrate can be measured by quantifying the concentration of a molecule that is added to one chamber and passed to the other side over time, following the equation P_{app} (cm/s) = $(\Delta Q/\Delta t)/(AC_0)$, where Q is the amount transported (mol), t is time (s), A is surface area (cm^2), and C_0 is initial concentration (mol/cm^3) (Gao et al. 2000). The Transwell model can easily be applied to multiple different well plates (6-, 12-, 24-, and 96-wells), where transport assays of a drug compound can be robustly performed.

Organoid culture technology has revolutionized static *in vitro* cell cultures by prolonging the culture period of primary cells and maintaining the stem cell population and differentiated epithelial cell types (Figure 9.1b; Sato et al. 2011). Because of the ability to culture primary cells long-term, patient samples can be robustly used to create disease-associated organoids, which enables the pursuit of patient-specific modeling of disease. Since the patient-derived organoid can reflect the genetic susceptibility of individual patients, the organoid system can contribute to Precision Medicine (Aboulkheyr Es et al. 2018). However, organoid culture has noted drawbacks including the lack of physical dynamics and an enclosed lumen that causes limited host–microbiome co-culture and accessibility to the apical side.

There are also several microengineering approaches to mimicking intestinal epithelial structures using biomaterials and culturing epithelial cells. For instance, hydrogel scaffolds made of collagen or calcium-alginate are used to model the structure of intestinal crypt–villus structures (Figure 9.1c; Yu et al. 2012, Sung et al. 2011, Wang et al. 2017). These hydrogel scaffolds can be placed on modified Transwell inserts to generate vertical morphogen gradients through the engineered crypt–villus axis (Wang et al. 2017). Although these models may better reflect tissue-specific structures of the GI tract, the fabricated hydrogel scaffold presents neither spontaneous 3D morphogenesis nor lineage-specific cell differentiation. A tubular scaffold made of silk hydrogel can be used to co-culture intestinal myofibroblasts and intestinal epithelium derived from organoids (Chen et al. 2017). This model successfully mimicked the tubular shape of the intestine and underlying connective tissue where myofibroblasts exist, and this tubular shape also enabled the generation of an oxygen gradient across the hydrogel scaffold.

Aforementioned static *in vitro* models can easily be scalable using conventional well plates for the high-throughput screening (HTS). However, paradoxically, the rudimentary nature of these models substantially limits demonstration of complicated intercellular crosstalks of the GI microenvironment. The duration of co-culture with gut bacteria is restricted to less than 24 h because of bacterial overgrowth and subsequent cell death due to the buildup of microbial wastes and depletion of nutrients (Park et al. 2017).

9.2.2 MICROFLUIDIC CELL CULTURE MODELS

The most representative GI models with microfluidic setups possess two microchannels layered on top of each other, separated by a porous membrane. Epithelial cells are cultured on one of the channels, microbiome organisms are co-cultured on top of the epithelium, and other components, such as immune cells and endothelial cells, can be co-cultured on the other channel to add complexity to the microenvironment, depending on experimental purposes. All of the different cell types participate in intercellular crosstalk, forming *in vivo*-relevant microenvironment. There are various designs of the two-channel microfluidic chips. One example, known as NutriChip, was developed to assess the anti-inflammatory effects of dairy food components in the presence of immune cells (Ramadan et al. 2013). This model uses Caco-2 intestinal epithelial cells derived from colon carcinoma as the epithelial component and a lymphoma cell line as the immune cell component, both of which are cultured in different channels. A similar approach was applied to assess molecular transport in a two-channel microfluidic device (Kimura et al. 2008). However, most of these models had a monolayer of epithelial cells rather than emulating the differentiated 3D structure of intestinal epithelium. They also did not include microbiome cells in their culture systems, limiting their utility to study microbiome-associated intestinal diseases.

A recently emerged human organ-on-a-chip technology opened up a new avenue to model human tissues and organs (Bhatia and Ingber 2014). A representative model is the microfluidic gut-on-a-chip where continuous fluid flow and mechanical forces are applied to closely mimic the physical dynamics *in vivo* (Figure 9.2a; Kim et al. 2016, 2012, Kim and Ingber 2013). There are two apposed microfluidic channels separated by a stretchable, extracellular matrix (ECM)-coated porous membrane (Figure 9.2a,b). A silicone polymer, polydimethylsiloxane (PDMS) that is gas permeable, crystal clear, and elastic has been widely used to build the body and the membrane of a gut-on-a-chip (Figure 9.2b). The medium flowing through the microchannels in the device not only supplies fresh nutrients and oxygen that are essential for cell growth but also exerts fluid-induced shear stresses to cultured cells. These shear stresses are crucial for inducing the cytodifferentiation necessary to mimic the physiological functions of the human gut epithelium. Peristaltic mechanical movements are especially necessary to accurately recapitulate the gut bowel movement. To induce these biomechanics, vacuum chambers on opposite sides of the central cell channel are linked to a computer-controlled vacuum generator that applies cyclic, rhythmical motions (Figure 9.2c). Using a soft lithographic method, layer-by-layer microfabrication was applied to build a gut-on-a-chip (Figure 9.2d; Kim et al. 2012, Huh et al. 2013).

The gut-on-a-chip was developed by growing a Caco-2 cell line (Kim et al. 2012), but primary intestinal epithelial cells derived from human intestinal organoids can also be utilized (Shin et al. 2019, Kasendra et al. 2018). Both Caco-2 and primary intestinal epithelial cells cultured in a gut-on-a-chip undergo spontaneous villous morphogenesis and develop villus–crypt 3D characteristics (Figure 9.3). This feature is of great importance for modeling the intestine or GI diseases because 3D epithelial structures affect the efficiency of nutrient absorption as well as provide a spatial

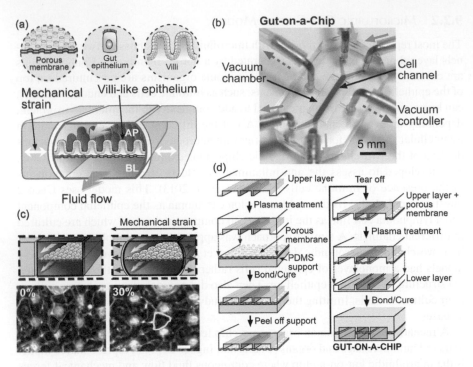

FIGURE 9.2 A human gut-on-a-chip. (a) A schematic that describes the structural, biological, mechanical, and physiological components of a gut-on-a-chip. (b) A photograph of a gut-on-a-chip. Colored arrows indicate the direction of flow inside the upper (blue) and the lower (yellow) microchannels. Gray dashed arrows show the direction of vacuum suction that creates peristalsis-like motions. (c) A set of schematics and phase contrast images show the cell strain response to the vacuum suction. A magenta and a cyan contour show the cellular elongation at 0% and 30%, respectively. Scale bar: 20 μm. (d) A flow schematic that describes the microfabrication procedure of a gut-on-a-chip.

niche for bacterial colonization, compartmentalized stem cell niches, and spatially organized proliferating and differentiating cells. A Caco-2 line (Kim et al. 2012), human small intestine organoid line (Kasendra et al. 2018), human large intestine organoid line (Shin et al. 2019), and an organoid line derived from induced pluripotent stem cells (Workman et al. 2018) have all been tested to emulate epithelial 3D morphogenesis in the gut-on-a-chip. A recent study discovered that the Wnt signaling and flow-dependent physical cues could initiate the intestinal morphogenesis in the gut-on-a-chip (Shin et al. 2019).

Interestingly, the gut-on-a-chip microphysiological cultures allow Caco-2 cells to be differentiated into four epithelial cell lineages, including absorptive, enteroendocrine, mucus-producing goblet, and Paneth cells (Kim and Ingber 2013). The microengineered gut-on-a-chip also demonstrates physiological functions such as mucus production, drug metabolizing P450 activity, and glucose reuptake. Either capillary or lymphatic endothelial cells can also be co-cultured on the other side of the porous membrane to constitute blood or lymphatic vessels (Kim et al. 2016).

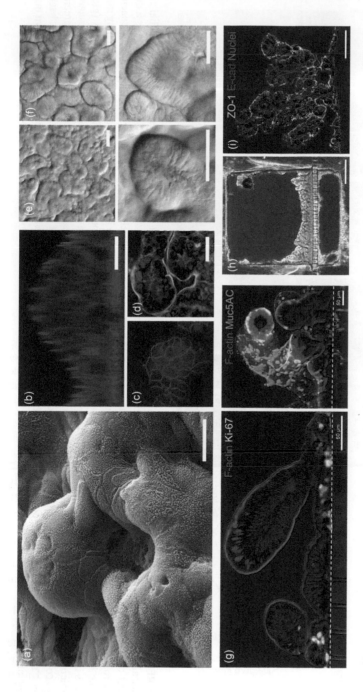

FIGURE 9.3 Intestinal 3D morphogenesis in the gut-on-a-chip. (a) A scanning electron micrograph (SEM) of the microengineered villi. A side (b) and a top-down (c) view of the villi stained with ZO-1 tight junction protein. (d) A horizontal cross-cut view of the villi that display the brush border (green) and nuclei (blue). Differential interference contrast (DIC) micrographs of the microengineered villi grown by the Caco-2 cells (e) and primary colonic organoid-derived epithelium (f) at low-power (on top) and high-power magnifications (below). (g) Small intestinal organoid-derived villi grown in the gut-on-a-chip. A vertical cross-cut view of the sectioned gut-on-a-chip (h) and an immunofluorescence micrograph (i) of the iPS-organoid-derived villi cultured in a gut-on-a-chip. Scale bars: 25 μm (a), 50 μm (b–g) and (i), and 250 μm (h).

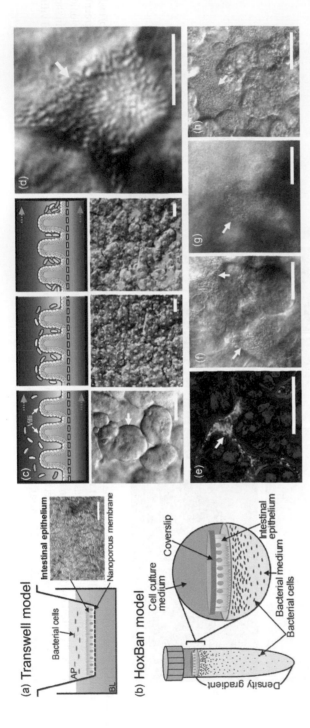

FIGURE 9.4 Models for demonstrating host–microbe co-cultures *in vitro*. (a) A Transwell model lined by a monolayer of intestinal epithelial cells. Bacterial cells are inoculated in the AP side in a static condition. (b) A "Human Oxygen-Bacteria Anaerobic (HoxBan)" system that induces a direct contact of the monolayer of intestinal epithelium grown on the porous membrane with bacterial cells embedded in the solid agar under an anaerobic or microaerobic condition. (c) The schematics (top) of the sequence of host–microbiome co-culture in the gut-on-a-chip and the corresponding micrographic images (bottom) that overlay DIC and fluorescent images. Green, green fluorescence protein (GFP)-labeled *Escherichia coli* cells (green). Arrows in the schematic indicate the applied fluid flow in the gut-on-a-chip. (d) A DIC snapshot of a living microcolony of VSL#3 probiotic bacteria in the microengineered villi grown in the gut-on-a-chip. (e) An overlaid immunofluorescence (IF) image shows the co-culture of GFP *E. coli* (green) on the Caco-2 villi grown in the gut-on-a-chip. Magenta: brush border; blue: nuclei. (f) A DIC image of the microcolonies of the VSL#3 probiotic bacteria on the Caco-2 villi grown in the gut-on-a-chip. (g) A biofilm (*L. rhamnosus* GG) formed inside the villi grown gut-on-a-chip. (h) An overgrown microcolony of the EIEC. White arrows indicate the formed microcolony. Scale bars: 20 μm (d), 50 μm (a), and 100 μm (c, e and h).

Most importantly, the human gut microbiome has been stably co-cultured for more than a week to emulate host–microbiome ecosystems *in vitro* (Figure 9.4; Kim et al. 2012, 2016).

The host–microbe co-culture *in vitro* has been considerably limited because the optimal culture conditions for bacterial and intestinal epithelial cells are quite different (e.g., nutrient composition, oxygen level). Uncontrolled microbial overgrowth is another technical challenge, which restricts the co-culture to less than 24 h when using a conventional static cell culture method (Park et al. 2017). In contrast, the gut-on-a-chip successfully demonstrated the co-culture with various microbial cells such as commensal (Shin et al. 2019, Shin and Kim 2018), infectious (Kim et al. 2016, Tovaglieri et al. 2019), and probiotic bacteria (Kim et al. 2012, Shin and Kim 2018) for a much longer period of time.

The human intestinal lumen is hypoxic (<5% O_2) or even anoxic, where the oxygen molecules diffuse from the intestinal capillary vasculature in the lamina propria toward the lumen (Zeitouni et al. 2016). Accordingly, transepithelial oxygen gradients exist in the intestine, and most of the gut microbiome is anaerobic (Walsh et al. 2014). There have been several approaches to recapitulate the oxygen gradient in microfluidic devices (Figure 9.5). The human–microbial crosstalk (HuMiX) microdevice contains three microchambers that are independently controlled (Figure 9.5a), in which each compartment is segregated by semipermeable polycarbonate membranes (Shah et al. 2016). Caco-2 cells were chosen as an epithelial cell component, and *Lactobacillus rhamnosus* GG (LGG) and *Bacteroides caccae* were grown in the device. HuMiX uses the anoxic medium in the top chamber and oxic medium in the epithelial cell chamber (Figure 9.5b). The HuMiX model sophisticatedly recreated host–microbiome interactions incorporated with an oxygen sensing module and a TEER measuring module in one setup. However, a limited co-culture period of approximately 24 h and a lack of 3D epithelial structures and mechanical movements limit this model and leave room for future model improvement.

The gut-on-a-chip provides two independent approaches to reflect the anoxic gradient in the culture setup. One approach is to put the entire experimental culture setup in an anaerobic chamber and flow oxic medium to the capillary channel (Figure 9.5c; Jalili-Firoozinezhad et al. 2019). Microbiome derived from human or human-microbiome-transplanted mice are then co-cultured in the device. A more diverse bacterial population was shown in an anaerobic setup compared to the aerobic setup. Using this approach, over 200 unique operational taxonomic units (OTUs) could be co-cultured *in vitro* and analyzed by 16S rRNA sequencing method. Another approach is relatively simple, where anoxic culture medium is perfused through the lumen microchannel, and the oxic medium is flowed through the capillary channel and named anoxic-oxic gradient (AOI)-on-a-chip (Figure 9.5d; Shin et al. 2019). The formulation of the AOI condition was verified using both computer simulation and actual experiments (Figure 9.5e). Under this condition, two commensal gut bacteria, *Bifidobacterium adolescentis* and *Eubacterium hallii*, were co-cultured in an AOI for a week, maintaining intact epithelial barrier function and cell viability. (Figure 9.5f).

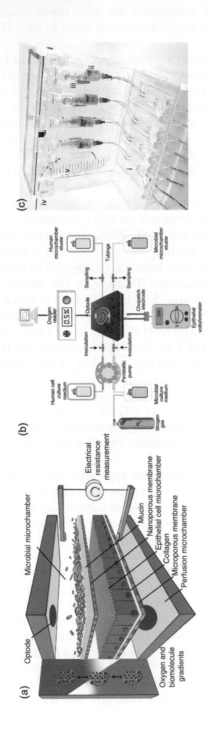

FIGURE 9.5 *In vitro* co-cultures of anaerobic bacteria with host intestinal epithelium. (a) A human–microbial crosstalk (HuMiX) system that supports the co-culture of anaerobic human gut bacteria with Caco-2 epithelial cells. (b) A schematic of the anaerobic conditioning module applied in the HuMiX system. (c) An anaerobic–aerobic co-culture system to support multi-species human fecal microbiome with either Caco-2 or organoid-derived intestinal epithelial cells.

(*Continued*)

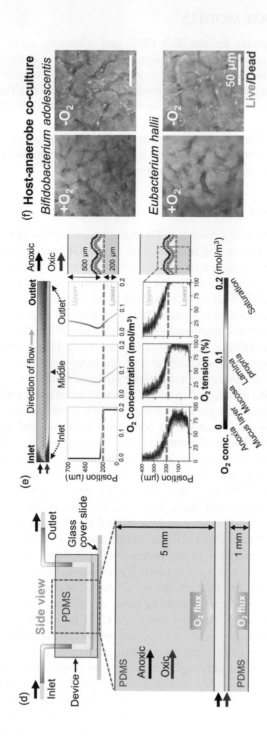

FIGURE 9.5 (CONTINUED) (d) A schematic of the gut-on-a-chip microdevice conditioned to form an anoxic–oxic interface (AOI) by flowing anoxic and oxic culture mediums through the microchannels. (e) Computational simulation of the AOI created in the gut-on-a-chip (a heat map, top), a theoretical profile of O_2 concentration at different locations in the microchannel (middle upper), an actual experimental profile of O_2 tension (middle lower), and the level of O_2 in the gut (bottom). (f) Overlaid micrographs of the Caco-2 villi with anaerobic commensal human gut bacteria in the presence or the absence of oxygen in the gut-on-a-chip.

9.3 INTESTINAL DISEASE MODELS

Representative intestinal diseases include IBD, CRC, and infections, all of which involve epithelial components, immune components, and microbiome components. Thus, proficient intestinal disease models need to encompass all or some of these elements. One approach to developing a disease model is to first build a healthy-state setup of an epithelial barrier with necessary components and induce the disease by combinatorially introducing a disease trigger. The other approach is to use the samples originated from a patient to reconstruct a diseased microenvironment. Both methods can be useful for modeling the different levels of GI pathophysiology and therefore provide strong versatility in research methodologies.

9.3.1 INFLAMMATORY BOWEL DISEASE (IBD)

IBD is a collective term for Crohn's disease (CD) and ulcerative colitis (UC), both of which are characterized by chronic inflammation in the GI tract (Xavier and Podolsky 2007). Although the exact cause of IBD has not been fully understood, genetic susceptibility (Khor, Gardet, and Xavier 2011), intestinal epithelial barrier dysfunction (McGuckin et al. 2009), and dysregulated immune functions (Xu et al. 2014) have been suspected as possible triggers of IBD pathogenesis. Furthermore, it has been reported that the gut microbiome and its population also affect the progression of chronic inflammation (Kostic, Xavier, and Gevers 2014). Thus, biomimetic models of IBD need to incorporate the intestinal mucosal layer, immune elements, and gut microbiome to more accurately reflect pathophysiological features.

As previously discussed, the gut-on-a-chip microsystem or equivalent microfluidic model can be used to recapitulate the pathophysiology of IBD. By introducing luminal (e.g., gut bacteria, pathogenic bacteria, or bacterial endotoxin) and immune elements (e.g., peripheral blood-derived immune cells), the healthy gut epithelium formed with the Caco-2 villi can be challenged in a spatiotemporal manner (Kim et al. 2016). Other types of intestinal inflammatory cells, including endothelial cells, gut microbiome, and immune cells, were introduced and co-cultured in the system to induce intestinal inflammatory responses. For instance, enteroinvasive *E. coli* (EIEC); lipopolysaccharide (LPS) endotoxin; or a defined set of proinflammatory cytokines such as tumor necrosis factor (TNF)-α, interleukin (IL)-1β, IL-6, or IL-8 can be added to trigger inflammation in the presence of peripheral blood mononuclear cells (PBMCs) in the vascular microchannel. This study successfully demonstrated that the immune system and microbiome are essential for eliciting intestinal inflammation. However, one limitation is that the cause of inflammatory milieu is artificial (e.g., the introduction of an excessive amount of LPS), suggesting that a better model that reflects an IBD patient's etiology and pathophysiology is still required.

The colitis model induced with dextran sodium sulfate (DSS) has been widely used in IBD animal models because the pathological outcomes are similar to human IBD (Perse and Cerar 2012). The gut-on-a-chip device can be harnessed to develop a human gut inflammation-on-a-chip model by challenging the intact healthy epithelial barrier with DSS (Figure 9.6a,b; Shin and Kim 2018). This study successfully demonstrated that a microphysiological disease model could be utilized to identify the disease

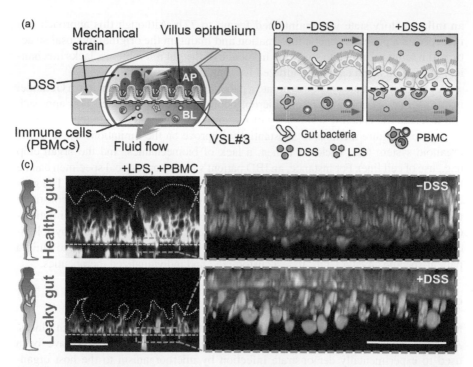

FIGURE 9.6 A pathomimetic gut inflammation-on-a-chip model. (a) A schematic that describes the complex intercellular interactions manipulated by the spatiotemporal introduction of luminal (DSS, LPS, or probiotic gut bacteria) and capillary (immune cells) elements in the gut-on-a-chip. (b) A schematic that displays the pathophysiology of intestinal barrier dysfunction perturbed by the addition of DSS. (c) The immunofluorescence micrographs that highlight the villus epithelial structure (grey) and the recruited PBMCs (green) in the absence (Healthy gut) or the presence (Leaky gut) of DSS. Scale bars: 50 μm.

trigger, where independent decoupling and combinatorial recoupling of inflammatory elements revealed that the healthy intestinal barrier is necessary and sufficient to maintain "homeostatic tolerance" in the gut. Under the epithelial barrier disruption state due to the DSS treatment, which is reminiscent of the "leaky gut syndrome" in IBD patients, the compromised epithelial barrier is remarkably vulnerable to normal physiological conditions (e.g., physiological level of LPS in the lumen), leading to the loss of homeostatic tolerance and the initiation of gut inflammation (Figure 9.6c). This model was validated by applying health-promoting probiotic strains, in which the dysfunctional barrier caused the transmigration of probiotic bacteria into the vascular compartment, suggesting sepsis-like pathogenicity. This study suggested that even "good bugs" may cause a bad outcome when the epithelial barrier is damaged or injured, providing a new insight to develop a bacteria-based therapy for IBD.

Another approach to modeling IBD is to integrate *ex vivo* tissue of an IBD patient with a microfluidic setup (Dawson et al. 2016). This method used a strategy similar to the Ussing chamber (Clarke 2009), where a dissected piece of tissue was intercalated between PDMS compartments containing microfluidic channels, and it showed an inflammatory state was maintained for up to 72 h. Although this approach can incorporate a patient's genetic information and maintain the original mucosal structure, the setup is limited to study the influence of the gut microbiome, lacks mechanical movements, and cannot be utilized for longitudinal studies.

An innovative strategy to incorporate a patient's genetic information into a GI model is to integrate 3D organoids into a microphysiological intestine model. This approach can leverage the enabling technology of the gut-on-a-chip and engage the advantage of organoid culture methods to substantially improve on the limitations above in the organoid system (e.g., enclosed lumen, a lack of biomechanics) and the organ chip (e.g., use of cell line). For instance, an IBD patient's biopsy or surgical specimen can be harvested and used to generate intestinal organoid lines (Suzuki et al. 2018). The generated organoids can be cultured in biomimetic models of the intestine to reconstitute the IBD patient-specific intestinal epithelium. Autologous matches with other patient-derived cells (e.g., PBMCs, microbiome) can be co-cultured in the microphysiological intestine model. This patient-oriented approach can potentially be used to build a patient-specific "Avatar Chip" to test customized and individualized therapeutics.

9.3.2 MICROBIAL INFECTION

Microbial infection with bacteria or virus is one of the major causes of human illness. As Koch's postulates describe (Evans 1976), multiple animal models have been used to experimentally demonstrate infection by microorganism to the host organism, specifically to host tissues or cells, and retroactive infection to other uninfected animals. Despite its prevalence, it has been challenging to model infectious diseases *in vitro* because infectious microorganisms are prone to rapidly overgrow and destroy the experimental host surrogate. Thus, *in vitro* infection models have typically been used in a relatively short period under the defined multiplicity of infection (MOI) that dictates the ratio of infectious agents to target cells (Schierack et al. 2006). These infection models are used to demonstrate the pathophysiology of microbial infection, and the host responses before the host cells/tissues are destroyed by the infectious agent. In the manifestation of infection, recruitment of neutrophils (Kusek et al. 2014) or phagocytic interaction of macrophages (Noel et al. 2017) has been observed.

To mimic intestinal infection *in vitro*, Transwell cultures have been predominantly used to show a short-term simulation of infectious bacteria pathogenesis (typically, <12 h). As previously discussed, a Transwell culture provides a robust, reproducible, and easy-to-access model to run multiple sets of experiments (Bisping et al. 2001). However, due to the limited culture period of bacterial cells to avoid microbial overgrowth, the microbial infection has been restrictively carried out in *in vitro* cultures. When a cell monolayer of the intestinal epithelium is formed with a reasonable TEER value, an infectious microbial species is added into the apical side of the epithelial monolayer grown on the porous Transwell insert, which corresponds to the luminal side of the gut. Reactions to the induced infection are then observed. Optionally, immune cells (e.g., peripheral blood mononuclear cells (Haller et al. 2000) or immortalized immune cell lines (Noel et al. 2017)) can be introduced to elicit infection-mediated innate immune responses for observation (Kampfer et al. 2017). This increases the robustness of the model by more closely simulating the *in vivo* environment.

Organoid culture methods have also been used to demonstrate microbial infection and pathogenesis. The major pitfall of the 3D organoid system in host–microbiome co-culture is the enclosed lumen inside the organoid body (Park et al. 2017) by which the structural hindrance seriously compromises the robust use of the organoid culture system in infection studies. To overcome this challenge, a microinjection method has been considered to introduce the viable population of microbial cells, such as *Salmonella* sp. (Yin and Zhou 2018, Forbester et al. 2015), *H. pylori* (McCracken et al. 2014), *Clostridium difficile* (Engevik et al. 2015), or viruses (Wilson et al. 2017), to the organoid lumens. However, this approach is still considered limited because of the static nature of the organoids that lack mechanical deformations and physiological fluid flow. This limitation incredibly jeopardizes the ability to longitudinally co-culture infectious microbes with organoids to determine disease pathogenesis. Alternatively, the 3D organoid epithelium can be transformed on the surface of an ECM gel (Wang et al. 2017) or a Transwell insert (In et al. 2016) to form a polarized epithelial tissue interface. This approach resolves the limitation of compartmentalization of luminal epithelial surface in 3D organoids. The polarized planar epithelial monolayer reformed from 3D organoids presents accessibility to apical and basolateral sides of the epithelial tissue. This makes the introduction of infectious microbial cells to the apical side (i.e., lumen-like compartment) very feasible and straightforward. This approach allows researchers to induce infection and host–pathogen interactions (e.g., infection of colonic epithelial monolayer with enterohemorrhagic *E. coli* (EHEC) (In et al. 2016), for a short-term period (<18 h); however, the static nature remains a critical limitation for performing a long-term host–pathogen co-culture (Mills and Estes 2016).

In recent years, the microfluidic human gut-on-a-chip model has been used to demonstrate microbial infection via pathogenic bacteria. In terms of the bacterial infection, EIEC strain was tested to induce rapid infection and elicit subsequent immune responses on the villi-like intestinal epithelial cells in the gut-on-a-chip (Figure 9.4h; Kim et al. 2016). Because of the continuous flow of fresh culture medium and the dynamic, mechanical peristaltic-like movements, a relatively long-term co-culture of living microbial species and an epithelial cell layer was possible (Kim et al. 2012). Similarly, the introduction of infectious microbes to the luminal side that mimics food poisoning or foodborne illness demonstrated disruption of the apical brush border, decreased epithelial barrier function, increased immune cell recruitment, and elevated production of proinflammatory cytokines into the basolateral side (Kim et al. 2016). Compared to the static culture method, infection models in the gut-on-a-chip were promising in their demonstration of the pathogenicity and pathophysiological outcomes of infected host cells, partly because static models (e.g., infections in Transwell cultures) often quickly lose their host cells, which results in termination of the ongoing study. A similar approach was adopted in the human primary Colon Chip model lined by a colonoid-derived 3D colonic epithelial cell layer that underwent physiological differentiation. This study's purpose was to identify the modulating microbial metabolites in the EHEC pathogenesis (Tovaglieri et al. 2019). This approach can be also replicated to demonstrate the infection of coxsackie B1 virus (CVB1) as a prototype enterovirus strain in the gut-on-a-chip grown with Caco-2 villi (Villenave et al. 2017), suggesting that a possible study regarding the contribution of the crosstalk between human gut virome, gut microbiome, and host cells is possible.

9.4 FUTURE OUTLOOK

Microengineered GI models are highly versatile and can be used for the advancement of both pharmaceutical and clinical fields. Through the use of highly transformative and implementable gut-on-a-chip models, many novel drug therapies can be tested quickly at relatively low cost for the benefit of pharmaceutical practices. In the following sub-sections, we discuss the potential applications of microengineered GI models in the pharmaceutical and clinical fields.

9.4.1 PHARMACEUTICAL APPLICATIONS

Pathomimetic GI disease models can be applied in the pharmaceutical industry. The major driving force that led to the development of human organ chip models was a strong demand for "humanized models" of GI diseases to validate the efficacy and toxicity of a new drug compound that targets per oral administration. Since experimental animal models have poorly reflected drug absorption, efflux, metabolism, conjugation, and transporter-mediated excretion in the human body (Hatton et al. 2015), development of a human cell-based *in vitro* disease model has been a critically unmet need. Microfluidic model systems are the first solution of their kind to meet this need. Microfluidic model systems like gut-on-a-chip more closely reflect the pathophysiology of human GI diseases and are more cost-, labor-, and time-effective than animal models. They are much more effective for validating the efficacy and safety of a new drug. The cost and number of new drug candidates in the pharmaceutical pipeline are enormously high (DiMasi, Grabowski, and Hansen 2016). Thus, the need for a more efficient GI model is great. Microengineered GI models incorporate fast loading times, high accessibility, the use of patient-derived cells and body fluids, multiplexed analysis, and detailed readouts including biochemical, genomic, transcriptomic, proteomic, metabolomic, and physiomic outcomes to accomplish more in a shorter time for a lower cost. One important application is discerning drug metabolism events that occur by either the cytochrome P450 enzyme in the intestinal epithelium or by the microbial enzymes in the gut microbiome using the GI model (Kim and Ingber 2013). Another example is the testing of irinotecan, an anti-cancer chemotherapeutic drug that undergoes a serious enzymatic transformation in the gut and exerts chronic GI toxicity, which elucidates the importance of a humanized GI model that can emulate a complex host–microbiome interplay during drug metabolism *in vivo* (Wallace et al. 2010). Due to the burgeoning awareness of the impacts of the gut microbiome on drug metabolism and host–drug interactions (Spanogiannopoulos et al. 2016), it is necessary for prospective GI models to provide the potential for longitudinal co-culture with host cells.

9.4.2 CLINICAL APPLICATIONS

Because microengineered gut-on-a-chip models support stable co-culture with the gut microbiome, efficacy and safety of microbiome-based therapeutic options can be tested. For instance, fecal microbiota transplantation (FMT) is a promising therapeutic method for IBD and potentially other GI diseases (Borody and Khoruts 2012).

However, its efficacy has not been fully validated because there has been no appropriate experimental surrogate. Prebiotics, probiotics, and synbiotics, which refer to the combined administration of pre- and probiotics, are also anticipated to be effective to control IBD (Quigley 2019) but have been minimally effective in clinical trials, although cell culture and animal testing demonstrated their efficacy (Claes et al. 2011). This result is likely due to the gap in knowledge that exists in host microbiome–drug interactions. Until microengineered gut-on-a-chip models were invented, these important interactions have not been testable – the microbiome of animal models varies too greatly in comparison to the human microbiome and therefore did not accurately reflect drug interactions that would occur when moving on to clinical trials.

In conclusion, patient-specific modeling using microfluidic *in vitro* platforms will open a new avenue for developing personalized therapeutic options and allow researchers to effectively predict drug efficacy and safety before reaching clinical trials. Utilization of patient-derived cells, including epithelial, fibroblast, immune, and microbiome cells, can more accurately reflect a patient's microenvironment and thus lead to more representative experimental results.

REFERENCES

Aboulkheyr Es, Hamidreza, Leila Montazeri, Amir R. Aref, Massoud Vosough, and Hossein Baharvand. 2018. Personalized cancer medicine: An organoid approach. *Trends in Biotechnology* 36 (4):358–371. doi:10.1016/j.tibtech.2017.12.005.

Bein, Amir, Woojung Shin, Sasan Jalili-Firoozinezhad, Min Hee Park, Alexandra Sontheimer-Phelps, Alessio Tovaglieri, Angeliki Chalkiadaki, Hyun Jung Kim, and Donald E. Ingber. 2018. Microfluidic organ-on-a-chip models of human intestine. *Cellular and Molecular Gastroenterology and Hepatology* 5 (4):659–668. doi:10.1016/j.jcmgh.2017.12.010.

Bhatia, Sangeeta N., and Donald E. Ingber. 2014. Microfluidic organs-on-chips. *Nature Biotechnology* 32 (8):760–772. doi:10.1038/nbt.2989.

Bisping, Guido, Norbert Lügering, Stefan Lütke-Brintrup, Hans-Gerd Pauels, Guido Schürmann, Wolfram Domschke, and Torsten Kucharzik. 2001. Patients with inflammatory bowel disease (IBD) reveal increased induction capacity of intracellular interferon-gamma (IFN-γ) in peripheral CD8+ lymphocytes co-cultured with intestinal epithelial cells. *Clinical & Experimental Immunology* 123 (1):15–22. doi: 10.1046/j.1365-2249.2001.01443.x.

Borody, Thomas J., and Alexander Khoruts. 2012. Fecal microbiota transplantation and emerging applications. *Nature Reviews Gastroenterology & Hepatology* 9 (2):88–96. doi:10.1038/nrgastro.2011.244.

Chen, Ying, Wenda Zhou, Terrence Roh, Mary K. Estes, and David L. Kaplan. 2017. In vitro enteroid-derived three-dimensional tissue model of human small intestinal epithelium with innate immune responses. *PLoS ONE* 12 (11):e0187880. doi: 10.1371/journal.pone.0187880.

Claes, Ingmar J. J., Sigrid C. J. De Keersmaecker, Jos Vanderleyden, and Sarah Lebeer. 2011. Lessons from probiotic-host interaction studies in murine models of experimental colitis. *Molecular Nutrition & Food Research* 55 (10):1441–1453. doi:10.1002/mnfr.201100139.

Clarke, Lane L. 2009. A guide to using chamber studies of mouse intestine. *American Journal of Physiology-Gastrointestinal and Liver Physiology* 296 (6):G1151–G1166. doi:10.1152/ajpgi.90649.2008.

Cryan, John F., and Siobhain M. O'Mahony. 2011. The microbiome-gut-brain axis: From bowel to behavior. *Neurogastroenterology & Motility* 23 (3):187–192. doi: 10.1111/j.1365-2982.2010.01664.x.

Dalile, Boushra, Lukas Van Oudenhove, Bram Vervliet, and Kristin Verbeke. 2019. The role of short-chain fatty acids in microbiota–gut–brain communication. *Nature Reviews Gastroenterology & Hepatology* 16(8):461–478. doi: 10.1038/s41575-019-0157-3.

Dawson, Amy, Charlotte Dyer, John Macfie, Joanna Davies, Laszlo Karsai, John Greenman, and Morten Jacobsen. 2016. A microfluidic chip based model for the study of full thickness human intestinal tissue using dual flow. *Biomicrofluidics* 10 (6):064101. doi:10.1063/1.4964813.

Deitch, Edwin A., William M. Bridges, Jing Wen Ma, Li Ma, Rodney D. Berg, and Robert D. Specian. 1990. Obstructed intestine as a reservoir for systemic infection. *The American Journal of Surgery* 159 (4):394–401. doi:10.1016/S0002-9610(05)81280-2.

DiMasi, Joseph A., Henry G. Grabowski, and Ronald W. Hansen. 2016. Innovation in the pharmaceutical industry: New estimates of R&D costs. *Journal of Health Economics* 47:20–33. doi:10.1016/j.jhealeco.2016.01.012.

Engevik, Melinda A., Mary Beth Yacyshyn, Kristen A. Engevik, Jiang Wang, Benjamin Darien, Daniel J. Hassett, Brice R. Yacyshyn, and Roger T. Worrell. 2015. Human *Clostridium difficile* infection: Altered mucus production and composition. *American Journal of Physiology-Gastrointestinal and Liver Physiology* 308 (6):G510–G524. doi:10.1152/ajpgi.00091.2014.

Evans, Alfred S. 1976. Causation and disease: The Henle-Koch postulates revisited. *Yale Journal of Biology and Medicine* 49 (2):175–195.

Forbester, Jessica L., David Goulding, Ludovic Vallier, Nicholas Hannan, Christine Hale, Derek Pickard, Subhankar Mukhopadhyay, and Gordon Dougan. 2015. Interaction of *Salmonella enterica* Serovar Typhimurium with intestinal organoids derived from human induced pluripotent stem cells. *Infection and Immunity* 83 (7):2926–2934. doi:10.1128/IAI.00161-15.

Gao, Jinnian, Erin D. Hugger, Melissa S. Beck-Westermeyer, and Ronald T. Borchardt. 2000. Estimating intestinal mucosal permeation of compounds using Caco-2 cell monolayers. *Current Protocols in Pharmacology* 8 (1):7.2. 1–7.2. 23. doi: 10.1002/0471141755. ph0702s08.

Goodman, Barbara E. 2010. Insights into digestion and absorption of major nutrients in humans. *Advances in Physiology Education* 34 (2):44–53. doi: 10.1152/advan.00094.2009.

Haller, Dirk, Carolin Bode, Walter P. Hammes, Andrea M. Pfeifer, Eduardo J. Schiffrin, and Stephanie Blum. 2000. Non-pathogenic bacteria elicit a differential cytokine response by intestinal epithelial cell/leucocyte co-cultures. *Gut* 47 (1):79–87. doi:10.1136/ gut.47.1.79.

Hatton, Grace B., Vipul Yadav, Abdul W. Basit, and Hamid A. Merchant. 2015. Animal farm: Considerations in animal gastrointestinal physiology and relevance to drug delivery in humans. *Journal of Pharmaceutical Sciences* 104 (9):2747–2776. doi:10.1002/ jps.24365.

Huh, Dongeun, Hyun Jung Kim, Jacob P. Fraser, Daniel E. Shea, Mohammed Khan, Anthony Bahinski, Geraldine A. Hamilton, and Donald E. Ingber. 2013. Microfabrication of human organs-on-chips. *Nature Protocols* 8 (11):2135–2157. doi:10.1038/ nprot.2013.137.

In, Julie, Jennifer Foulke-Abe, Nicholas C. Zachos, Anne Marie Hansen, James B. Kaper, Harris D. Bernstein, Marc Halushka, Sarah Blutt, Mary K. Estes, Mark Donowitz, and Olga Kovbasnjuk. 2016. Enterohemorrhagic *Escherichia coli* reduces mucus and intermicrovillar bridges in human stem cell-derived colonoids. *Cellular and Molecular Gastroenterology and Hepatology* 2 (1):48–62. doi:10.1016/j. jcmgh.2015.10.001.

Jalili-Firoozinezhad, Sasan, Francesca S. Gazzaniga, Elizabeth L. Calamari, Diogo M. Camacho, Cicely W. Fadel, Amir Bein, Ben Swenor, Bret Nestor, Michael J. Cronce, Alessio Tovaglieri, Oren Levy, Katherine E. Gregory, David T. Breault, Joaquim M. S. Cabral, Dennis L. Kasper, Richard Novak, and Donald E. Ingber. 2019. A complex human gut microbiome cultured in an anaerobic intestine-on-a-chip. *Nature Biomedical Engineering.* doi:10.1038/s41551-019-0397-0.

Kampfer, Angela A. M., Patricia Urban, Sabrina Gioria, Nilesh Kanase, Vicki Stone, and Agnieszka Kinsner-Ovaskainen. 2017. Development of an in vitro co-culture model to mimic the human intestine in healthy and diseased state. *Toxicology in Vitro* 45 (Pt 1):31–43. doi:10.1016/j.tiv.2017.08.011.

Kasendra, Magdalena, Alessio Tovaglieri, Alexandra Sontheimer-Phelps, Sasan Jalili-Firoozinezhad, Amir Bein, Angeliki Chalkiadaki, William Scholl, Cheng Zhang, Hannah Rickner, Camilla A. Richmond, Hu Li, David T. Breault, and Donald E. Ingber. 2018. Development of a primary human Small Intestine-on-a-Chip using biopsy-derived organoids. *Scientific Reports* 8 (1):2871. doi:10.1038/s41598-018-21201-7.

Khor, Bernard, Agnes Gardet, and Ramnik J. Xavier. 2011. Genetics and pathogenesis of inflammatory bowel disease. *Nature* 474 (7351):307–317. doi:10.1038/nature10209.

Kim, Hyun Jung, Dongeun Huh, Geraldine Hamilton, and Donald E Ingber. 2012. Human gut-on-a-chip inhabited by microbial flora that experiences intestinal peristalsis-like motions and flow. *Lab on a Chip* 12 (12):2165–2174. doi: 10.1039/c2lc40074j.

Kim, Hyun Jung and Donald E. Ingber. 2013. Gut-on-a-Chip microenvironment induces human intestinal cells to undergo villus differentiation. *Integrative Biology* 5 (9):1130–1140. doi:10.1039/C3IB40126J.

Kim, Hyun Jung, Hu Li, James J. Collins, and Donald E. Ingber. 2016. Contributions of microbiome and mechanical deformation to intestinal bacterial overgrowth and inflammation in a human gut-on-a-chip. *Proceedings of the National Academy of Sciences of the United States of America* 113 (1):E7–E15. doi:10.1073/pnas.1522193112.

Kimura, Hiroshi, Takatoki Yamamoto, Hitomi Sakai, Yasuyuki Sakai, and Teruo Fujii. 2008. An integrated microfluidic system for long-term perfusion culture and on-line monitoring of intestinal tissue models. *Lab on a Chip* 8 (5):741–746. doi:10.1039/b717091b.

Kostic, Aleksandar D., Ramnik J. Xavier, and Dirk Gevers. 2014. The microbiome in inflammatory bowel disease: Current status and the future ahead. *Gastroenterology* 146 (6):1489–1499. doi:10.1053/j.gastro.2014.02.009.

Kusek, Mark E., Michael A. Pazos, Waheed Pirzai, and Bryan P. Hurley. 2014. In vitro coculture assay to assess pathogen induced neutrophil trans-epithelial migration. *Journal of Visualized Experiments* 6(83):e50823. doi:10.3791/50823.

McCracken, Kyle W., Emily M. Cata, Calyn M. Crawford, Katie L. Sinagoga, Michael Schumacher, Briana E. Rockich, Yu Hwai Tsai, Christopher N. Mayhew, Jason R. Spence, Yana Zavros, and James M. Wells. 2014. Modelling human development and disease in pluripotent stem-cell-derived gastric organoids. *Nature* 516 (7531):400–404. doi:10.1038/nature13863.

McGuckin, Michael A., Rajaraman Eri, Lisa A. Simms, Timothy H. Florin, and Graham Radford-Smith. 2009. Intestinal barrier dysfunction in inflammatory bowel diseases. *Inflammatory Bowel Diseases* 15 (1):100–113. doi:10.1002/ibd.20539.

Mills, Melody, and Mary K. Estes. 2016. Physiologically relevant human tissue models for infectious diseases. *Drug Discovery Today* 21 (9):1540–1552. doi:10.1016/j. drudis.2016.06.020.

Mima, Kosuke, Reiko Nishihara, Zhi Rong Qian, Yin Cao, Yasutaka Sukawa, Jonathan A Nowak, Juhong Yang, Ruoxu Dou, Yohei Masugi, and Mingyang Song. 2016. Fusobacterium nucleatum in colorectal carcinoma tissue and patient prognosis. *Gut* 65 (12):1973–1980. doi: 10.1136/gutjnl-2015-310101.

Mowat, Allan M., and William W. Agace. 2014. Regional specialization within the intestinal immune system. *Nature Reviews Immunology* 14 (10):667. doi: 10.1038/nri3738.

Nguyen, Thi Loan Anh, Sara Vieira-Silva, Adrian Liston, and Jeroen Raes. 2015. How informative is the mouse for human gut microbiota research? *Disease Models & Mechanisms* 8 (1):1. doi: 10.1242/dmm.017400.

Noel, Gaelle, Nicholas W. Baetz, Janet F. Staab, Mark Donowitz, Olga Kovbasnjuk, Marcela F. Pasetti, and Nicholas C. Zachos. 2017. A primary human macrophage-enteroid co-culture model to investigate mucosal gut physiology and host-pathogen interactions. *Scientific Reports* 7:45270. doi:10.1038/srep45270.

Park, Gun-Seok, Min Hee Park, Woojung Shin, Connie Zhao, Sameer Sheikh, So Jung Oh, and Hyun Jung Kim. 2017. Emulating host-microbiome ecosystem of human gastrointestinal tract in vitro. *Stem Cell Reviews and Reports* 13 (3):321–334. doi:10.1007/s12015-017-9739-z.

Perse, Martina, and Anton Cerar. 2012. Dextran sodium sulphate colitis mouse model: Traps and tricks. *Journal of Biomedicine and Biotechnology* 2012:718617. doi:10.1155/2012/718617.

Peterson, Lance W., and David Artis. 2014. Intestinal epithelial cells: Regulators of barrier function and immune homeostasis. *Nature Reviews Immunology* 14 (3):141. doi: 10.1038/nri3608.

Quigley, Eamonn M. M. 2019. Prebiotics and probiotics in digestive health. *Clinical Gastroenterology and Hepatology* 17 (2):333–344. doi:10.1016/j.cgh.2018.09.028.

Ramadan, Qasem, Hamideh Jafarpoorchekab, Chaobo Huang, Paolao Silacci, Sandro Carrara, Gozen Koklu, Julien Ghaye, Jeremy Ramsden, Christine Ruffert, Guy Vergeres, and Martin A. Gijs. 2013. NutriChip: Nutrition analysis meets microfluidics. *Lab on a Chip* 13 (2):196–203. doi:10.1039/c2lc40845g.

Round, June L., and Sarkis K. Mazmanian. 2009. The gut microbiota shapes intestinal immune responses during health and disease. *Nature Reviews immunology* 9 (5):313. doi:10.1038/nri2515.

Sato, Toshiro, Daniel E. Stange, Marc Ferrante, Robert G. J. Vries, Johan H. Van Es, Stieneke Van Den Brink, Winan J. Van Houdt, Apollo Pronk, Joost Van Gorp, and Peter D. Siersema. 2011. Long-term expansion of epithelial organoids from human colon, adenoma, adenocarcinoma, and Barrett's epithelium. *Gastroenterology* 141 (5):1762–1772. doi:10.1053/j.gastro.2011.07.050.

Schierack, Peter, Marcel Nordhoff, Miroslav Pollmann, Karl D. Weyrauch, Salah Amasheh, Ulrike Lodemann, Joerg Jores, Babila Tachu, Sylvia Kleta, Anthony Blikslager, Karsten Tedin, and Lothar H. Wieler. 2006. Characterization of a porcine intestinal epithelial cell line for in vitro studies of microbial pathogenesis in swine. *Histochemistry and Cell Biology* 125 (3):293–305. doi:10.1007/s00418-005-0067-z.

Seok, Junhee, H. Shaw Warren, Alex G. Cuenca, Michael N. Mindrinos, Henry V. Baker, Weihong H. Xu, Daniel R. Richards, Grace P. McDonald-Smith, Hong Gao, Laura Hennessy, Celeste C. Finnerty, Cecilia M. Lopez, Shari Honari, Ernest E. Moore, Joseph P. Minei, Joseph Cuschieri, Paul E. Bankey, Jeffrey L. Johnson, Jason Sperry, Avery B. Nathens, Timothy R. Billiar, Michael A. West, Marc G. Jeschke, Matthew B. Klein, Richard L. Gamelli, Nicole S. Gibran, Bernard H. Brownstein, Carol Miller-Graziano, Steve E. Calvano, Philip H. Mason, J. Perren Cobb, Laurence G. Rahme, Stephen F. Lowry, Ronald V. Maier, Lyle L. Moldawer, David N. Herndon, Ronald W. Davis, Wenzhong Z. Xiao, Ronald G. Tompkins, and Inflammation Host Response Injury. 2013. Genomic responses in mouse models poorly mimic human inflammatory diseases. *Proceedings of the National Academy of Sciences of the United States of America* 110 (9):3507–3512. doi:10.1073/pnas.1222878110.

Shah, Pranjul, Joëlle V. Fritz, Enrico Glaab, Mahesh S. Desai, Kacy Greenhalgh, Audrey Frachet, Magdalena Niegowska, Matthew Estes, Christian Jäger, Carole Seguin-Devaux, Frederic Zenhausern, and Paul Wilmes. 2016. A microfluidics-based in vitro model of the gastrointestinal human–microbe interface. *Nature Communications* 7:11535. doi:10.1038/ncomms11535.

Shin, Woojung, Christopher D. Hinojosa, Donald E. Ingber, and Hyun Jung Kim. 2019. Human intestinal morphogenesis controlled by transepithelial morphogen gradient and flow-dependent physical cues in a microengineered gut-on-a-chip. *iScience* 15:391–406. doi:10.1016/j.isci.2019.04.037.

Shin, Woojung and Hyun Jung Kim. 2018. Intestinal barrier dysfunction orchestrates the onset of inflammatory host–microbiome cross-talk in a human gut inflammation-on-a-chip. *Proceedings of the National Academy of Sciences of the United States of America* 115 (45):E10539. doi:10.1073/pnas.1810819115.

Shin, Woojung, Alexander Wu, Miles W. Massidda, Charles Foster, Newin Thomas, Dong-Woo Lee, Hong Koh, Youngwon Ju, Joohoon Kim, and Hyun Jung Kim. 2019. A robust longitudinal co-culture of obligate anaerobic gut microbiome with human intestinal epithelium in an anoxic-oxic interface-on-a-chip. *Frontiers in Bioengineering and Biotechnology* 7:13. doi:10.3389/fbioe.2019.00013.

Sobhani, Iradj, Julien Tap, Françoise Roudot-Thoraval, Jean P Roperch, Sophie Letulle, Philippe Langella, Gerard Corthier, Jeanne Tran Van Nhieu, and Jean P Furet. 2011. Microbial dysbiosis in colorectal cancer (CRC) patients. *PLoS ONE* 6 (1):e16393. doi:10.1371/journal.pone.0016393.

Spanogiannopoulos, Peter, Elizabeth N. Bess, Rachel N. Carmody, and Peter J. Turnbaugh. 2016. The microbial pharmacists within us: A metagenomic view of xenobiotic metabolism. *Nature Reviews Microbiology* 14 (5):273–287. doi:10.1038/nrmicro.2016.17.

Sung, Jong Hwan, Jiajie Yu, Dan Luo, Michael L Shuler, and John C March. 2011. Microscale 3-D hydrogel scaffold for biomimetic gastrointestinal (GI) tract model. *Lab on a Chip* 11 (3):389–392. doi: 10.1039/c0lc00273a.

Suzuki, Kohei, Tatsuro Murano, Hiromichi Shimizu, Go Ito, Toru Nakata, Satoru Fujii, Fumiaki Ishibashi, Ami Kawamoto, Sho Anzai, Reiko Kuno, Konomi Kuwabara, Junichi Takahashi, Minami Hama, Sayaka Nagata, Yui Hiraguri, Kento Takenaka, Shiro R. Yui, Kiichiro Tsuchiya, Tetsuya Nakamura, Kazuo Ohtsuka, Mamoru Watanabe, and Ryuichi Okamoto. 2018. Single cell analysis of Crohn's disease patient-derived small intestinal organoids reveals disease activity-dependent modification of stem cell properties. *Journal of Gastroenterology* 53 (9):1035–1047. doi:10.1007/s00535-018-1437-3.

Tamboli, Cyrus P., Christel Neut, Pierre Desreumaux, and Jean F. Colombel. 2004. Dysbiosis in inflammatory bowel disease. *Gut* 53 (1):1–4. doi:10.1136/gut.53.1.1.

Terzić, Janoš, Sergei Grivennikov, Eliad Karin, and Michael Karin. 2010. Inflammation and colon cancer. *Gastroenterology* 138 (6):2101.e5–2114.e5. doi:10.1053/j.gastro.2010.01.058.

Tovaglieri, Alessio, Alexandra Sontheimer-Phelps, Annelies Geirnaert, Rachelle Prantil-Baun, Diogo M. Camacho, David B. Chou, Sasan Jalili-Firoozinezhad, Tomas de Wouters, Mark Kasendra, Michael Super, Mark J. Cartwright, Camilla A. Richmond, David T. Breault, Christophe Lacroix, and Donald E. Ingber. 2019. Species-specific enhancement of enterohemorrhagic *E-coli* pathogenesis mediated by microbiome metabolites. *Microbiome* 7(1):43. doi:10.1186/s40168-019-0650-5.

Turner, Jerrold R. 2009. Intestinal mucosal barrier function in health and disease. *Nature Reviews Immunology* 9 (11):799. doi: 10.1038/nri2653.

Uemura, Naomi, Shiro Okamoto, Soichiro Yamamoto, Nobutoshi Matsumura, Shuji Yamaguchi, Michio Yamakido, Kiyomi Taniyama, Naomi Sasaki, and Ronald J. Schlemper. 2001. Helicobacter pylori infection and the development of gastric cancer. *New England Journal of Medicine* 345 (11):784–789. doi: 10.1056/NEJMoa001999.

Villenave, Remi, Samantha Q. Wales, Tiama Hamkins-Indik, Efstathia Papafragkou, James C. Weaver, Thomas C. Ferrante, Anthony Bahinski, Christopher A. Elkins, Michael Kulka, and Donald E. Ingber. 2017. Human gut-on-a-chip supports polarized infection of coxsackie B1 virus in vitro. *PLoS ONE* 12 (2):e0169412. doi:10.1371/journal.pone.0169412.

Wallace, Bert D., Hongwei W. Wang, Kimberly T. Lane, John E. Scott, Jillian Orans, Ja Seol Koo, Madhukumar Venkatesh, Christian Jobin, Li An Yeh, Sridhar Mani, and Matthew R. Redinbo. 2010. Alleviating cancer drug toxicity by inhibiting a bacterial enzyme. *Science* 330 (6005):831–835. doi:10.1126/science.1191175.

Walsh, Calum J., Caitriona M. Guinane, Paul W. O'Toole, and Paul D. Cotter. 2014. Beneficial modulation of the gut microbiota. *FEBS Letters* 588 (22):4120–4130. doi:10.1016/j.febslet.2014.03.035.

Wang, Yuli, Dulan B. Gunasekara, Mark I. Reed, Matthew DiSalvo, Scott J. Bultman, Christopher E. Sims, Scott T. Magness, and Nancy L. Allbritton. 2017. A microengineered collagen scaffold for generating a polarized crypt-villus architecture of human small intestinal epithelium. *Biomaterials* 128:44–55. doi:10.1016/j.biomaterials.2017.03.005.

Wilson, Sarah S., Beth A. Bromme, Mayumi K. Holly, Mayim E. Wiens, Anshu P. Gounder, Youngmee Sul, and Jason G. Smith. 2017. Alpha-defensin-dependent enhancement of enteric viral infection. *PLoS Pathogens* 13 (6):e1006446. doi:10.1371/journal.ppat.1006446.

Workman, Michael J., John P. Gleeson, Elissa J. Troisi, Hannah Q. Estrada, S. Jordan Kerns, Christopher D. Hinojosa, Geraldine A. Hamilton, Stephen R. Targan, Clive N. Svendsen, and Robert J. Barrett. 2018. Enhanced utilization of induced pluripotent stem cell-derived human intestinal organoids using microengineered chips. *Cellular and Molecular Gastroenterology and Hepatology* 5 (4):669.e2–677.e2. doi:10.1016/j.jcmgh.2017.12.008.

Xavier, Ramnik J., and Daniel K. Podolsky. 2007. Unravelling the pathogenesis of inflammatory bowel disease. *Nature* 448 (7152):427–434. doi:10.1038/nature06005.

Xu, Xiao R., Chang Qin Liu, Bai Sui Feng, and Zhanju J. Liu. 2014. Dysregulation of mucosal immune response in pathogenesis of inflammatory bowel disease. *World Journal of Gastroenterology* 20 (12):3255–3264. doi:10.3748/wjg.v20.i12.3255.

Yang, Jiansong, Masoud Jamei, Karen R Yeo, Geoffrey T Tucker, and Amin Rostami-Hodjegan. 2007. Prediction of intestinal first-pass drug metabolism. *Current Drug Metabolism* 8 (7):676–684. doi: 10.2174/138920007782109733.

Yin, Yuebang, and Daoguo Zhou. 2018. Organoid and enteroid modeling of salmonella infection. *Frontiers in Cellular and Infection Microbiology* 8:102. doi:10.3389/fcimb.2018.00102.

Yu, Jiajie, Songming Peng, Dan Luo, and John C. March. 2012. In vitro 3D human small intestinal villous model for drug permeability determination. *Biotechnology and Bioengineering* 109 (9):2173–2178. doi:10.1002/bit.24518.

Zeitouni, Nathalie E., Sucheera Chotikatum, Maren von Kockritz-Blickwede, and Hassan Y. Naim. 2016. The impact of hypoxia on intestinal epithelial cell functions: Consequences for invasion by bacterial pathogens. *Molecular and Cellular Pediatrics* 3 (1):14. doi:10.1186/s40348-016-0041-y.

10 Respiratory Pathophysiology

Microphysiological Models of Human Lung

Brian F. Niemeyer, Alexander J. Kaiser,
and Kambez H. Benam
University of Colorado

CONTENTS

10.1 BURDEN OF RESPIRATORY CONDITIONS

Respiratory diseases represent an immense threat to populations worldwide, incurring millions of deaths annually and imposing a significant socioeconomic burden. In 2016, the World Health Organization reported that chronic obstructive pulmonary disease (COPD), lower respiratory tract infections (LRTIs), lung cancers, and tuberculosis constitute four of the top ten global causes of death, while other common respiratory illnesses such as asthma affect millions of people, detailed in Figure 10.1 (World Health Organization 2018, Forum of International Respiratory Societies 2017, GBD 2015 LRI Collaborators 2017). Recent reports suggest that COPD alone caused 3.17 million fatalities in 2015, which was 5% of the annual global deaths (Forum of International Respiratory Societies 2017). Meanwhile, there were over 10 million new tuberculosis diagnoses, which had a successful treatment rate of only 83%

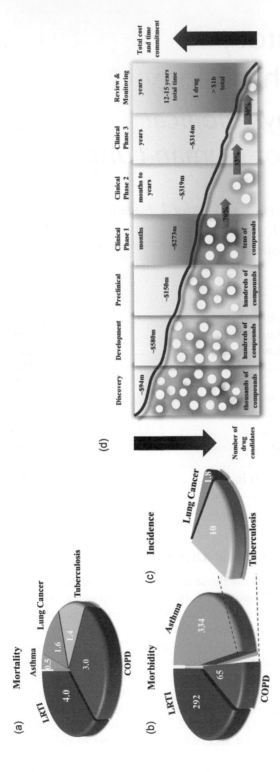

FIGURE 10.1 Global burden of leading respiratory diseases and the therapeutic development process. (a) A diagram indicating the global rates of mortality associated with five of the most common respiratory conditions: chronic obstructive pulmonary disease (COPD), lower respiratory tract infections (LRTI), asthma, tuberculosis, and lung cancer. Numbers listed indicate the fatalities associated with each disease in the millions. (b) Rates of morbidity for COPD, LRTI, asthma, tuberculosis, and lung cancer. Numbers indicate the number of people in millions. (c) Incidence rates for lung cancer and tuberculosis in 2012 and 2015, respectively. Numbers indicate the number of new cases in millions. Data was compiled from references (Forum of International Respiratory Societies 2017, Collaborators 2017, World Health Organization 2018). (d) Overview of the drug discovery process and associated cost. The first phases, known as the discovery and development phases, identify potential candidate therapeutics. These phases typically require high-throughput screening to rapidly assay thousands of compounds. Following this, preclinical tests are performed to determine initial drug efficacy and safety. Three phases of clinical trials are then performed to validate safety and efficacy in human subjects. Finally, they are monitored for post-marketing effects. Combined, the costs associated with the drug discovery process can be over one billion US dollars for one therapeutic compound and take over 15 years to complete. Input data shown here were obtained from references (DiMasi, Grabowski, and Hansen 2016, Paul et al. 2010). (Reprinted with permission from SAGE Publications (SLAS Discovery) (Niemeyer et al. 2018, US Food and Drug Administration 2015).)

(Forum of International Respiratory Societies 2017, World Health Organization 2017). With respect to latent tuberculosis infection, it is estimated that 2 billion people (roughly 33% of the global population) are infected with *Mycobacterium tuberculosis*, the etiological agent of tuberculosis (Kling et al. 2014). Lung cancers are the most common cause of death among all cancers, with an estimated 234,000 cases predicted to be diagnosed in the United States and 1.8 million new cases globally (Forum of International Respiratory Societies 2017, Vyfhuis et al. 2018). From this, 85% of lung cancers diagnosed in the United States will be non-small-cell lung cancer, which has a 5 year survival rate of 44% across all stages; however, out of the non-small-cell lung cancers, 58% of patients will be diagnosed with stage IV which has a survival rate of only 5% (Vyfhuis et al. 2018).

The leading cause of upper respiratory tract infections, such as strep throat, are from Group A Streptococcus (GAS); in the United States alone, strep throat emerges in an estimated 1–2.6 million individuals annually (Centers for Disease Control 2013). GAS is also the leading cause of necrotizing fasciitis which has a fatality of 25%–35% (Centers for Disease Control 2013). With respect to children and infants, respiratory syncytial virus (RSV) is the leading cause of lower respiratory tract infections and is estimated to result in 10,000 deaths annually, 18% of all respiratory illnesses in children under 5 years old, 20% of all hospitalizations, 18% of emergency room visits, and 15% of all pediatric visits in the United States (Kling et al. 2014). RSV in 2005 was estimated to globally infect 33.8 million children under the age of 5 and causes 2.8–4.3 million hospitalizations with 66,000–199,000 fatalities (Nair et al. 2010).

Despite the high prevalence and fatalities resulting from respiratory disease, there is an extensive unmet need in the development of respiratory medicines, as the cost and time to develop lifesaving drugs is increasing and the success rate of the drug development process is decreasing, as shown in Figure 11.1 (Mestre-Ferrandiz, Sussex, and Towse 2012). In 2001, the success rate of respiratory drugs passing clinical trials was 12%; in 2010, it dropped to 5% and has been reported as low as 3%, which is among the lowest out of the many therapeutic categories investigated (Mestre-Ferrandiz, Sussex, and Towse 2012). Additionally, the cost of respiratory drug production has been estimated to reach $1.46 billion–$1.8 billion and is expected to increase (Mestre-Ferrandiz, Sussex, and Towse 2012, Paul et al. 2010). With respect to total capitalized cost, there has been an increase from $179–$413 million from the 1970s to early 1990s, a 2.3-fold increase. Following this, from the early 1990s to mid-2000s, there was a 2.53-fold increase, with another 2.45-fold from mid-2000s to mid-2010s (DiMasi, Grabowski, and Hansen 2016). For each approved new drug, capitalized clinical cost was estimated at $1.45 billion and out-of-pocket clinical costs were estimated at $965 million, which over the last 10 years is a 2.4-fold and 2.6-fold increase, respectively (DiMasi, Grabowski, and Hansen 2016). Additionally, in this same 10 year period, the time between clinical testing and submission of a new drug application or biological license application with the FDA has increased 12%, and the time between clinical testing and marketing approval increased by 7% (DiMasi, Grabowski, and Hansen 2016).

The increase in cost and time to generate new therapeutic agents for respiratory pathologies can in part be attributed to high drug attrition rates during the drug development process. It has been estimated that drug candidates have only a 10% likelihood of receiving approval with a lack of safety and lack of efficiency being the leading causes of drug attrition (Arrowsmith and Miller 2013, Mullard 2016). The safety and efficacy failures of many of these drugs are a result of preclinical disease models that insufficiently recreate human respiratory disease, including animal models and two-dimensional (2D) cell culture models. Many animal models are limited by significant differences in lung tissue microstructures, airway branching, cellular distribution, and immune cell composition and function between animals and humans (Henderson et al. 2013, Wright, Cosio, and Churg 2008). On the other hand, classic *in vitro* systems using 2D cell culturing techniques fail to faithfully recreate the microenvironment of the lung and lack the means to incorporate integral intricacies, such as complex three-dimensional (3D) structures or mechanical stress from breathing or vascular shear, and often inaccurately predict drug activity in preclinical tests (Bhatia and Ingber 2014). Thus, the development of more sophisticated *in vitro* systems using human-derived cells may further our current understanding of respiratory disease pathogenesis as well as promote therapeutic developments through improved preclinical screening.

In this chapter, we will provide a brief overview of widely used model systems, including animal models and conventional *in vitro* systems which formed the basis of our current understanding of respiratory pathologies. We will then discuss in more depth advanced microengineered biomimetic systems recently developed to model specific pulmonary conditions with a focus on microfluidic and organ-on-chip technologies.

10.2   ANIMAL MODELS OF HUMAN LUNG DISEASES

The use of animal models has been instrumental in driving our current understanding of the biology of respiratory diseases. With respect to pulmonary research, many different animal models have been employed to replicate various physiological and pathological responses, some of which include mice, rats, ferrets, guinea pigs, and non-human primates (Sacco, Durbin, and Durbin 2015, Thangavel and Bouvier 2014, Williams and Orme 2016, Zhan et al. 2017). Generally speaking, rodents such as mice are the most commonly used animal models to study respiratory pathologies; however, translating data from mice to human subjects is impeded due to significant biological differences, as well as altered disease severity and pathogen susceptibility in mice. For example, in influenza studies, susceptibility of mice to human influenza virus strains often varies based off of both mouse and virus strains; rather adapted virus strains have been generated to allow study of infection in mice with these viruses (Bouvier and Lowen 2010). Moreover, disease severity for other common respiratory pathologies such as COPD in mice can manifest as a more mild pathology, poorly reflecting the severity observed in humans (Wright, Cosio, and Churg 2008).

Alternative animal models such as ferret and guinea pig often exhibit respiratory pathophysiological responses more representative of humans and in certain instances more faithfully replicate the human diseases. For instance, many of the hallmarks of *M. tuberculosis* infection cannot be recreated within the mouse, although guinea pigs infected with tuberculosis exhibit many responses seen in the human forms of the disease such as lung necrosis, lymphadenopathy, and disease dissemination (Williams and Orme 2016). Ferrets closely mimic human influenza infections, as they are readily susceptible to infection without modifications or adaptations; symptoms found in ferrets include fever, nasal discharge, lethargy, weakness, anorexia, and sneezing (Thangavel and Bouvier 2014). Guinea pigs can also be readily infected with non-adapted human influenza strains; however, guinea pigs often present less severe symptoms which makes studying the efficacy of vaccines difficult (Thangavel and Bouvier 2014). Unfortunately, many of these non-rodent animal models lack readily available immunological reagents, have incomplete genome sequencing, lack transgenic models, and can cause logistical housing issues making widespread use in research unrealistic (Sacco, Durbin, and Durbin 2015, Thangavel and Bouvier 2014).

10.3 *IN VITRO* MODELS OF HUMAN LUNG PATHOPHYSIOLOGY

To better recapitulate healthy and pathologic functions found *in vivo* using human samples, *in vitro* systems are constantly being developed and improved upon. Perhaps one of the larger problems with pulmonary 2D cell culture is recreating the air–liquid interface (ALI) at the epithelium to promote ciliated differentiation. Adaptations to *in vitro* culture devices have been implemented to study asthma and various other pathologies by creating a two-compartment well that is separated by a porous membrane, known as a transwell or transwell insert. The membrane of the transwell can be coated in extracellular matrix proteins, allowing for adherence of a multitude of primary human cells including airway epithelial cells, endothelial cells, and fibroblasts. Further, transwell inserts support differentiation of primary basal airway epithelial cells into ciliated or mucus producing cells by recreating the air–liquid interface. ALI is achieved by removing media from the apical side of the insert, leaving media only in the basal side of the insert. After inclusion of specific factors, such as retinoic acid, and induction of ALI, basal airway epithelial cells will undergo differentiation, creating milieu of cells, and environmental conditions (ALI) which better replicate the human airway than conventional 2D cultures (Pezzulo et al. 2011).

Transwell systems are commonly used to incorporate increased complexity through co-culture conditions by seeding airway epithelial cells on apical side of the transwell while additional cell types such as fibroblasts or endothelial cells are cultured in the basal compartment. In addition to this, a triple cell culture model was created by culturing airway epithelial cells on the apical side of the transwell followed by addition of macrophages and dendritic cells (Lehmann et al. 2011, Rothen-Rutishauser, Kiama, and Gehr 2005). Comparative analysis of primary human airway epithelial cells derived from healthy and asthmatic donors using transwell systems has led to incredible insights regarding the dysregulation of pathways involved in extracellular matrix production, airway remodeling, immune cell activation, and epithelial wound repair in people with asthma (Reeves et al. 2014, Jiao et al. 2017).

While transwell inserts have added many attributes to monolayer cell culture that better replicates what is found *in vivo*, there are many 3D characteristics that are still deficient such as 3D cell–cell interactions and vascularized perfusion of nutrients and oxygen. Spheroids are multicellular spheroid cultures which allow for complex cell–cell interactions through co-culture and are often seeded onto 3D scaffolds

(Fang and Eglen 2017, Amann et al. 2014). Lung cancer has been extensively studied using spheroid models, for example, tumor-stroma cell interactions have been investigated through multiple non-small-cell lung cancer cell lines cultured with lung fibroblast cell lines (Chimenti et al. 2017). Spheroids have also been used to help better understand how different microenvironments within the lungs affect the phenotypical expression, in order to better recreate the ideal cell culture microenvironments (Chimenti et al. 2017). Another similar but distinct *in vitro* culture system is organoids, which are cultures of organ-specific cell types, derived from stem cell or organ progenitors which form a 3D tissue that shows microanatomy through self-organizing mechanisms (Fang and Eglen 2017). Organoids have great potential to improve the understanding of organ-level differentiation and development, which is only further enhanced through new technologies such as CRISPR-Cas9 which allows for modifications of genes related to self-renewal, differentiation, tissue morphogenesis, and disease phenotypes in pulmonary medicine (Nikolic and Rawlins 2017). Additionally, many lung organoids have been designed through cell differentiation of primary cells, cell lines, and human pluripotent stem cells to recreate many functional components of the lungs including proximal airway epithelium, distal airway epithelium, and bronchioalveolar epithelium (Nadkarni, Abed, and Draper 2016).

While spheroids and organoids have improved the understanding of complex cell–cell interaction and cellular differentiation found in the lungs, both still lack essential anatomical structures such as vascularization and heterogenous hydrogel gradients, which has been addressed through 3D bioprinting. 3D bioprinting utilizes placement of various hydrogels, both cellular and acellular, in 3D arrangements in conjunction with various polymerization strategies to allow rigidity in the biomimetic block (El-Sherbiny and Yacoub 2013). Like conventional 3D printers, a digital model of the structure is converted into a file which programs the printer to deposit the appropriate material onto the structure being printed in a layer-by-layer fashion (Murphy and Atala 2014). Some of the most common 3D bioprinting deposition methods are inkjet, microextrusion, and laser-assisted bioprinting, each having various trade-offs (Gao et al. 2018, Jose et al. 2016, Murphy and Atala 2014, Wust, Muller, and Hofmann 2011). A large benefit of bioprinting is the ability to print a wide range of hydrogels, such as collagen, gelatin, fibrin, chitosan, agarose, alginate, decellularized tissue, hyaluronic acid, polyethylene glycol (PEG), poly hydroxyethyl methacrylate (pHEMA), and Pluronics (Kolesky et al. 2016, 2014, Murphy and Atala 2014, Wust, Muller, and Hofmann 2011). With respect to Pluronics, channels can be printed that will later be evacuated to allow for hollow lumens to be seeded with cells, forming vascularization, and as a result allowing for bioprinters to generate thick 3D cell culture tissue analogues (Kolesky et al. 2016, 2014).

Although bioprinting allows for new capabilities in engineering *in vitro* cell culture systems, there are important considerations related to the hydrogels being used such as printability, biocompatibility, cell adhesion and proliferation, degradation kinetics, structural and mechanical properties, and the biomimicry; additionally the system must be carefully controlled as to not create environments in cellular hydrogels that can kill cells before, during, and after printing, such as high pressures resulting in damaging shear forces (Murphy and Atala 2014). Unfortunately with respect to pulmonary science, there has been a limited number of bioprinting-related research, with one of the more well-known studies being the printing of an air–blood barrier consisting of single layers of endothelial cells, basement membrane, and epithelial cells (Horvath et al. 2015). It was shown that bioprinting the tissue construct resulted in thinner and more homogenous cell layers, in addition to reduced translocation of blue dextran when compared to traditional pipetting methods (Horvath et al. 2015).

The mentioned technologies all have trade-offs associated with recreating an anatomically and physiologically representative model of the human lung, which have been detailed in Table 10.1; however, often there is a disconnect from including both cellular and organ-level responses, which has in part been addressed by

TABLE 10.1
Comparison of *In Vitro* Model Systems

Model System	Advantages	Disadvantages	References
Transwell inserts	Generates discreet compartments Compatible with co-culture	Static culture system Provides limited 3D cell–cell interaction	Pezzulo et al. (2011), Lehmann et al. (2011), Rothen-Rutishauser, Kiama, and Gehr (2005)
Precision-cut lung slices	Maintains 3D microanatomy of lung Functional responses of organs are maintained	Historically, cultivation has been achieved for a limited duration Difficulties preserving arteriole and smooth muscle cell morphology	Sanderson (2011), Morin et al. (2013), Neuhaus et al. (2017)
Spheroids	High degree of consistency in results Formats are suitable for therapeutic testing Allows co-culture and complex cell–cell interactions	Limited architecture complexity Cell ratios in co-culture can be difficult to control Challenges with regulating spheroid size	Fang and Eglen (2017)
Organoids	Derived from relevant tissues Replicates natural architecture Mimics some organ function and complexity	Does not contain key organ components such as vasculature Often only early stages of organ development are achieved	Fang and Eglen (2017)
3D bioprinting	Can print multiple cell types for co-culture Creates extremely well-defined architecture in a layer-by-layer fashion Prints a wide array of hydrogels creating heterogenous gradients	Challenges with cells and materials in regards to mechanicals forces, degradation, and viability during cell extrusion	Fang and Eglen (2017), El-Sherbiny and Yacoub (2013), Gao et al. (2018), Wust, Muller, and Hofmann (2011), Kolesky et al. (2016)

ex vivo systems such as precision-cut lung slices (PCLSs) (Sanderson 2011). PCLSs have demonstrated effective biomimicry in the fields of mucociliary transport, vascular and airway reactivity, xenobiotic biotransformation, polyamine transport, chemical toxicology, and complex aerosol exposure (Morin et al. 2013). While PCLSs are often only cultivated for short periods of time, recently PCLS have been maintained for 15 days while maintaining functionality such as bronchoconstriction and the release of pro-inflammatory cytokines when exposed to lipopolysaccharides (LPS) of *Escherichia coli*, despite functionality decreasing with time (Neuhaus et al. 2017). This can allow for analysis of slowly metabolized therapeutics or chronic exposure to various stimuli (Neuhaus et al. 2017). While having shorter culturing times, another study also evaluated immunomodulatory effects by evaluating the change in cytokine production and cell surface marker expression through exposure to LPS, macrophage-activating lipopeptide-2, interferon gamma, and dexamethasone (Henjakovic et al. 2008). The study demonstrated characteristic responses to the pro-inflammatory stimuli, in addition to enhanced expression of major histocompatibility complex class II alloantigen within the lung tissue when cultured with LPS and a reduction with dexamethasone (Henjakovic et al. 2008). However PCLS, like other mentioned technologies, cannot completely recreate all attributes seen *in vivo*; for example, there are major challenges with preserving arteriole morphology as arterioles often have strong irreversible contractions during slicing; a similar situation is observed for airway smooth muscle cells, but the cells often still stay attached to the parenchyma (Sanderson 2011). Additionally there are challenges related to drug delivery due to the cut at the epithelial barrier, dethatched neural networks, and the inability to perform single-cell biochemical, protein, or gene analysis (Sanderson 2011).

Together, these above-mentioned animal and cell culture model systems have been instrumental in developing our current understanding of the basic biology of respiratory system as well as the pathology of many respiratory illnesses. Despite this, clear improvements could be made to better mimic the nature of respiratory conditions and the microenvironment of the lung. The static nature of the above *in vitro* systems significantly hinders the biomimetic nature of many of these model systems. Advances in microfluidic technologies and microengineering techniques has allowed researchers to overcome these limitations and better recreate human respiratory disease outside of the living body. For the remainder of this chapter, we will discuss some of the most common respiratory illnesses and the innovative, microfluidic, biomimetic systems used to advance our understanding of disease and promote therapeutic discovery.

10.4 ASTHMA

Asthma is a highly prevalent respiratory pathology affecting more than 300 million people globally, with rates of incidence on the rise (Olin and Wechsler 2014, Blume and Davies 2013). Despite the increasing prevalence, pathogenesis of the disease is poorly understood with limited advancement of therapeutics due in part to its multifactorial nature with genetic, epigenetic, and environmental factors each contributing to development of the disease (Holgate et al. 2015). Hallmarks of asthma include airway remodeling, bronchoconstriction, mucus production, and inflammation mediated through several factors (e.g., IL-4, IL-5, IL-13, and IL-17) and cell types such as airway epithelial cells, dendritic cells, T cells, B cells, eosinophils, mast cells, neutrophils, and smooth muscle cells (Olin and Wechsler 2014). Given the limited understanding of gene–gene and gene–environment interactions involved in the development of asthma, model systems have thus far only been able to replicate some, rather than all, aspects of disease. A wide array of animal models have been employed including mice, rats, guinea pigs, dogs, cats, pigs, horses, and primates, and while these systems have been instrumental in our current understanding of asthma, their use is limited by the fact that the majority of animals do not spontaneously develop asthma (Blume and Davies 2013). Rather, these animals require sensitization using a combination of many different antigens and adjuvants, resulting in a striking amount of variability observed between model systems depending on the different methods of sensitization and antigens employed (Blume and Davies 2013). This, coupled with numerous biological differences in airway structure and immune response between humans and animals, necessitates great caution when translating data from animals to humans.

To circumvent some of the challenges found in animal models, engineered organ-on-chip systems have been extensively adapted to recreate the lung microenvironment for use in modeling numerous respiratory pathologies, including asthma, *in vitro* (Konar et al. 2016). Previously, Benam et al. recreated the human small airway microenvironment in the Small Airway-on-a-Chip model, as shown in Figure 10.2 (Benam, Villenave, et al. 2016). This chip, fabricated through soft lithography technology using polydimethylsiloxane (PDMS), was comprised of two meticulously aligned microchannels separated by a thin, porous polyester membrane coated in type I collagen. After both fabrication and collagen coating were completed, the sides of the membrane in the upper channels were seeded with primary human airway epithelial cells (AECs), which were cultured in an ALI and differentiated into pseudostratified, mucociliary epithelial cells. After airway epithelial cell differentiation, primary human microvascular endothelial cells were seeded to the opposing side of the membrane in the lower channels where media was constantly perfused, mimicking vascular shear forces. Meanwhile, the upper AEC containing channel was exposed to air, recreating the lumen of the small airway complete with functional epithelium capable of ciliary beating and mucus secretion, as highlighted in Figure 10.2 (Benam, Villenave, et al. 2016).

The Small Airway-on-a-Chip provides an elegant model system for studying asthma using patient-derived primary cells. These chips were first used to recreate phenotypes associated with asthma through treatment with IL-13, a cytokine proven to be necessary for induction of asthma in animal models and which directly effects airway epithelium, airway inflammation, goblet cell hyperplasia, and mucus secretion (Wills-Karp 2004). After treating the chips with IL-13 for eight days, the authors observed an increase in the secretion of inflammatory cytokines G-CSF and GM-CSF into the vascular channel, an increase in number of goblet cells, and a decrease in the cilia beating frequency, which has previously been reported from asthmatic individuals (Thomas et al. 2010). To further the use of this model system, Benam et al. investigated whether the Small Airway-on-a-Chip could be utilized for drug discovery and therapeutic intervention studies with asthma. As IL-13 signals

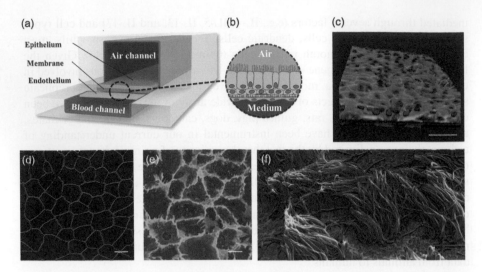

FIGURE 10.2 Human lung Small Airway-on-a-Chip. (a) Depiction of a cross-section of the Small Airway-on-a-Chip with upper airway lumen containing airway epithelial cells and a lower blood channel lined with endothelial cells. (b) Use of this device enables culture of differentiated, mucociliated, pseudostratified human bronchiolar epithelium in close apposition with human lung microvascular endothelial cells on opposing sides of a porous polyester membrane. (c) A 3D reconstruction of the Small Airway-on-a-Chip using immunofluorescent staining for F-Actin of epithelial cells (green) and endothelial cells (red) as well as nuclei (DAPI, blue). Scale bar is 30 µm. (d and e) Immunostaining of epithelial cells via ZO-1 (red) and endothelial cells through CD31 (green) showing tight junction formation in chips. Scale bars are each 20 µm. (f) Pseudocolored scanning electron micrograph of cilia (magenta) as well as microvilli (cyan) on well-differentiated bronchiolar epithelial cells. (Reprinted with permission from Elsevier, Inc. (Cell Press) and Macmillan Publishers Limited (Springer Nature) (Benam, Novak, et al. 2016, Benam, Villenave, et al. 2016).)

through the JAK-STAT pathway, efficacy of the drug tofacitinib, a JAK inhibitor, was evaluated in ameliorating IL-13-induced asthmatic symptoms (Pernis and Rothman 2002, Benam, Villenave, et al. 2016). Healthy AECs cultured in the chip were first exposed to IL-13 to induce asthmatic changes and then subsequently were exposed to tofacitinib. It was found that tofacitinib restored normal cilia beating frequency, diminished secretion of G-CSF and GM-CSF, and reduced goblet cell hyperplasia (Benam, Villenave, et al. 2016). These results were consistent with other reports that inhibition of JAK signaling may represent a new therapeutic avenue for the treatment of allergic airway diseases (Kudlacz et al. 2008).

This work clearly demonstrates the substantial strengths of the Small Airway-on-a-Chip system for modeling pathogenesis and therapeutic intervention of asthma *in vitro*; however, it is not the only microfluidic system used for asthma research and biomarker discovery. Sackmann et al. previously created a microfluidic neutrophil chemotaxis platform, which was then adapted into a diagnostic chip to assay neutrophil function in asthmatic patients (Sackmann et al. 2014, 2012). The diagnostic chip contained two primary components composed of a base with a functionalized

migration channel coated in p-selectin and a multifunction lid containing chemoattractants and preventing evaporation from the microfluidic system. Whole blood
was pumped through the channel where, by nature of the p-selectin coating, neutrophils were captured and retained, while remaining blood components were removed
through laminar flow during subsequent washes. After neutrophil capture, the lid
was placed on the base bringing the chemoattractants in contact with the channel
resulting in neutrophil chemotaxis. Three measurements were used to characterize
neutrophil function: the absolute migration speed (independent of direction), chemotaxis velocity (speed of migration toward increasing concentration of chemoattractant), and the chemotactic index (the displacement of the cell divided by its total
path length) (Sackmann et al. 2014). By applying blood samples from healthy and
asthmatic patients to this system, comparative analysis of healthy and asthmatic neutrophils showed that neutrophils from asthmatic patients migrate significantly slower
than those from non-asthmatic patients. Importantly, this work both highlights neutrophil chemotaxis velocity as a potential biomarker for asthma in a clinical setting,
and this validates the potential of this system as an extremely useful tool for clinicians given that accurate diagnosis of asthma can be challenging (Bakirtas 2017).

10.5 LUNG CANCER

Lung cancer is an extremely complex and often fatal disease, which develops in a
heterogeneous milieu of cells, signaling molecules, and matrix proteins (Sung and
Beebe 2014). This environment, known as the tumor microenvironment, plays a crucial role in multiple stages of disease progression including local drug resistance,
immune evasion, sustained growth, as well as metastasis (Chen et al. 2015). While
many individual aspects of cancer and the tumor microenvironment can be recreated
using 2D culture systems, researchers are finding these models are too simplistic and
cannot effectively recreate key interactions in disease progression, ultimately limiting their use for advanced studies.

To more faithfully recreate the various hallmarks of cancer progression, several bioengineered devices have been employed. One such device, created by
Anguiano et al., allowed for the characterization of lung cancer cell migration in
3D (Anguiano et al. 2017). This device was fabricated in a similar manner to other
organ-on-chip devices, where cylinders embedded with central chambers and inlets
were fabricated out of PDMS. The chambers of these devices were then filled with
varying combinations of hydrogels, seeded with lung adenocarcinoma cancer cells,
and subsequently used to analyze migration of the lung cancer cells. The authors
found that different lung cancer migration phenotypes could be achieved by varying
the composition of the hydrogel. For example, in matrices comprised solely of collagen (similar to connective tissue), the cancer cells exhibited a mesenchymal type
of migration whereas matrices that more closely resemble a disorganized basement
membrane (consisting of a collagen:matrigel mixture) yielded a lobopodial migration phenotype (Anguiano et al. 2017).

Apart from gaining a greater understanding of cancer biology, microengineered devices have also been utilized as a drug screening platform for lung cancer therapeutics. A multichambered microfluidic device was created to model

epithelial–mesenchymal transition (EMT) by seeding lung cancer spheroid cells in a 3D hydrogel in one chamber of the device while endothelial cells were cultured in an adjacent channel (Aref et al. 2013, Esch, Bahinski, and Huh 2015). The cancer cells were induced to migrate when cultured in the device while a variety of drugs targeting EMT pathways were flowed through the vascular channel of the device. The effective dose of the drug was determined by the concentration of drug needed to inhibit cancer cell dispersion. It was found that compared to conventional 2D and monoculture systems, use of the 3D microfluidic device resulted in significantly different drug responses in that the effective dose for the drugs tested were much lower (Aref et al. 2013).

10.6 CHRONIC OBSTRUCTIVE PULMONARY DISEASE (COPD)

COPD is a progressive lung disorder characterized by chronic bronchitis, emphysema, and poorly reversible obstruction of the small airways (Barnes et al. 2015). COPD represents a major global public health burden as it is the third leading cause of death worldwide and one of the highest ranked pathologies in terms of disease burden (Gershon et al. 2011). Despite its high prevalence and disease burden, mechanisms of COPD pathogenesis are poorly understood, and there is a dearth of model systems for studying COPD with *in vitro* studies primarily relying on transwell insert systems (Benam et al. 2015).

To improve upon available model systems, Benam et al. adapted the Small Airway-on-a-Chip technology described above to model COPD exacerbation by lining the small airway chips with human airway epithelial cell (hAECs) from either healthy or COPD donors (Benam, Villenave, et al. 2016). After seeding the chips, the epithelial cells were exposed to either poly(I:C) or LPS to mimic viral or bacterial infection respectively, as pathogenic infections often exacerbate COPD. Stimulation with poly(I:C) and LPS resulted in upregulation of M-CSF and IL-8, respectively, in the COPD chips but not the healthy chips. These findings were consistent with previous studies suggesting that bronchial epithelial cells from COPD patients are hyper-responsive to LPS (Deslee et al. 2007). Moreover, this work identified M-CSF as a potentially new biomarker for COPD exacerbation.

In addition to infectious agents, cigarette smoke is known to be a major contributor to exacerbation of COPD in patients (Laniado-Laborin 2009). To recreate physiologically relevant cigarette smoke exposure in COPD, the Small Airway-on-a-Chip platform was connected to a smoking machine capable of "breathing" whole cigarette smoke onto hAECs from COPD patients functionally creating a Breathing-Smoking Airway-on-a-Chip, as observed in Figure 10.3 (Benam, Novak, et al. 2016, Benam and Ingber 2016). Briefly, a microrespirator was constructed to cyclically breathe microliter volumes of air in and out of the upper airway channel of the Small Airway-on-a-Chip. A programmable smoking machine was utilized to recreate smoking behaviors including puff duration, volume, and puffs per cigarette. By "breathing" whole cigarette smoke over the hAECs in the airway chip, the authors were able to successfully recapitulate many hallmarks of smoking including increased oxidative stress and smoke-induced ciliary dysfunctions (Benam,

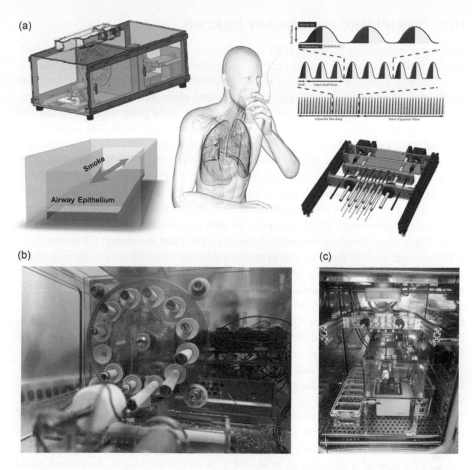

FIGURE 10.3 Smoking-Breathing Human Small Airway-on-a-Chip. (a) Recreation of smoking in the Small Airway-on-a-Chip. Cigarettes are placed in a cigarette smoking machine (upper left) capable producing whole cigarette smoke. A programmable microrespirator (bottom right) can reproduce human breathing and smoking profiles including normal respiration cycles, cigarette puff durations, and time between cigarette puffs (upper right). By connecting the microrespirator to the Small Airway-on-a-Chip, hAECs can be exposed to whole cigarette smoke (bottom right). (b) Image of the cigarette smoking machine "smoking" multiple cigarettes. (c) Image of the entire smoking machine and microrespirator inserted into an incubator. (Reprinted with permission from Elsevier, Inc. (Cell Press) (Benam, Novak, et al. 2016).)

Novak, et al. 2016). Further, when these chips were lined with either healthy or COPD epithelium and exposed to cigarette smoke, it was found that the COPD epithelium alone had a large increase in the secretion of IL-8, a finding that is consistent with clinical reports of elevated IL-8 levels in COPD patients who smoke compared to healthy individuals (Keatings et al. 1996, Benam, Novak, et al. 2016).

10.7   IDIOPATHIC PULMONARY FIBROSIS

Idiopathic pulmonary fibrosis (IPF) is a chronic, progressive lung disease which results from deregulated wound healing and inflammation leading to lung fibrosis and scarring. Fibrosis of the lungs in turn prevents normal, healthy lung function leading to impaired breathing, gas exchange, and ultimately death due to respiratory failure. The prognosis for patients with IPF is quite poor with a median life expectancy of 3 years after diagnosis (Sundarakrishnan et al. 2018). The incidence of IPF has been estimated to be 130,000 in the United States, 300,000 in Europe, and 640,000 in East Asia although the prevalence is on the rise (Sundarakrishnan et al. 2018, Martinez et al. 2017). Although the exact cause IPF has yet to be determined, several factors associated with increased risk of disease have been identified including exposure to cigarette smoke, viral infection, altered bacterial loads and composition, microaspiration of gastric contents, as well as genetic susceptibility (Martinez et al. 2017). Over time, several animal models have been developed for the study of IPF; however, each of these animal systems is derived by artificially triggering fibrosis using various compounds, which may not accurately reflect IPF found in humans as the etiological agent is unknown (Sundarakrishnan et al. 2018, Chua, Gauldie, and Laurent 2005). Further, no current animal model is able to recapitulate all of the aspects of IPF in humans (Chua, Gauldie, and Laurent 2005). The discrepancies between IPF in humans and the current animal models simply highlights the need for complementary *in vitro* systems capable of replicating the disease in humans.

To recreate IPF *in vitro*, Felder et al. (2014) created a multichannel microfluidic platform that mimics wound healing from microinjuries in the alveolus. Initially wounds were created in monolayers of epithelial cells like A549 cells using 0.5% trypsin-EDTA. The trypsin-EDTA solution was administered to the cells using a previously described hydrodynamic flow focusing technique so that the wounding agent was maintained in the center of the channel while media was perfused on the surrounding sides of the central stream (Felder et al. 2012). With this technique, the authors could control the microwound width, location, and length of time to generate the wound. To improve the physiological relevance of the model system, the trypsin-EDTA wounding agent was replaced with acidified pepsin by incorporating pepsin-HCl, these are gastric components that the respiratory system is normally exposed to via acid reflux (Farrell et al. 2006, Ward et al. 2005, Bathoorn et al. 2011). This effectively allowed the study of the effects of microaspirations of acid reflux in alveolar epithelium and the potential contribution of this to microwound formation associated with IPF (Felder et al. 2014, Lee et al. 2010). Felder et al. found they were able to recapitulate alveolar wounding in a reproducible manner and showed that pepsin-HCl synergistically damages alveolar epithelial cells in a manner that directly supports a causal relation between acid reflux observed with gastroesophageal reflux disease and IPF.

10.8   PULMONARY EDEMA

Pulmonary edema is a life-threatening disease comprised of abnormal accumulation of intravascular fluid in the alveolar air spaces and interstitial tissues of the lung mediated by increased lung vascular and epithelial cell permeability and elevated vascular pressure (Matthay 2014). The increased vascular pressure and permeability observed in pulmonary edema can be a result of numerous cardiovascular and respiratory diseases as well as dose-limiting drug toxicities as seen in cancer patients receiving systemic IL-2 treatment (Huh, Leslie, et al. 2012). With the increasing use of organ-on-chip technology to recreate human lung diseases *in vitro*, a lung Alveolus-on-a-Chip model was adapted to study drug-induced pulmonary edema. Briefly, the lung Alveolus-on-a-Chip was fabricated through soft lithography technology and chemical etching and comprises two closely aligned microchannels separated by a thin porous membrane fabricated from PDMS. The PDMS membrane was coated with the ECM proteins fibronectin and collagen; then the two opposing channels were seeded with human alveolar epithelial cells and microvascular endothelial cells. This device not only recreated the local cellular environment of the alveolus but could mimic cyclic mechanical strain experienced through breathing motions by use of a vacuum force on two adjacent microchannels (Huh et al. 2010). To reconstruct drug-induced pulmonary edema, clinically relevant doses of IL-2 were perfused through the vascular channel resulting in fluid leaking from the vascular channel into the air-filled epithelial channel (Huh, Leslie, et al. 2012). A crucial finding from this study was that the mechanical forces associated with breathing were critical to vascular leaking while circulating immune cells seemed to have minimal influence on development of edema. Further, this platform identified two new potential therapeutics, angiopoietin-1 and transient receptor potential vanilloid 4 ion channel inhibitor GSK2193874, in amelioration of pulmonary edema.

10.9   PULMONARY THROMBOSIS

Pulmonary thrombosis is a major cause of mortality in patients, which is instigated through activation of platelets and inflammatory stimulation of pulmonary endothelial cells. This can occur in the context of numerous respiratory pathologies including COPD and pulmonary fibrosis (Akpinar et al. 2014, Sprunger et al. 2012). Over the years, research endeavors have provided greater understanding of the pathogenesis of pulmonary thrombosis; however, shortcomings of model systems have stymied advancement of therapeutic discoveries. To improve *in vitro* systems, Jain and collaborators adapted the lung Alveolus-on-a-Chip for the study of pulmonary thrombosis pathogenesis and therapeutic intervention (Jain et al. 2018). They replaced the lung alveolar epithelial cell line A549 with primary human lung alveolar epithelial cells, coated on all four side of the vascular microchannel with human vascular endothelial cells and perfused whole human blood in the vascular channel rather than culture media. To recreate endotoxin-induced acute lung injury and pulmonary thrombosis, LPS was added either to the endothelial channel with no upper epithelial cells or to the epithelial lumen with intact chips. This led to the finding that LPS indirectly stimulates thrombosis by activating alveolar epithelium rather than acting directly on endothelial cells. When LPS was added to the epithelial cells, the endothelial cell–cell junctions were disrupted, endothelial cell activation occurred, and large aggregates of fibrin containing platelets formed. These same changes were not found when LPS was added to endothelial cells alone in the lower vascular channel (Jain et al. 2018).

To further expand the use of this system, Jain et al. (2018) used the pulmonary thrombosis chip model to test the anti-thrombotic and anti-inflammatory activities of the potential drug candidate pamodulin-2 (PM2). PM2 is a potent inhibitor of protease-activated receptor-1 (PAR-1) which has been shown to mediate tissue inflammation and thrombosis (Abdel-Latif and Smyth 2012). When either blood or the endothelial cells lining the vascular channel were pretreated with PM2, LPS-induced vascular permeability and platelet binding were reduced. Similar results were achieved in a mouse model where LPS-induced lung injury was achieved through intratracheal LPS delivery and intravenous delivery of PM2 (Jain et al. 2018, Conti et al. 2010). Taken together, this work shows a great potential for microengineered organ systems in both the study and treatment of pulmonary thrombosis.

10.10 RESPIRATORY INFECTIONS

Respiratory tract infections are the leading cause of death in developing countries with infants and children being particularly susceptible to infection with pathogens (Ferkol and Schraufnagel 2014). Moreover, fungal, bacterial, and viral infections are widely recognized as factors that exacerbate other respiratory diseases including asthma, COPD, and cystic fibrosis (Britto et al. 2017, Knutsen et al. 2012, Parkins and Floto 2015). Animals models have been widely used to model pathogen infection in the lungs; however, these are limited in their translation to humans due to significant biological differences including immune cell composition, airway structure, and susceptibility to infection by human pathogens (Niemeyer et al. 2018). With advances in microfluidics and microengineering, *in vitro* culture systems can now incorporate the high levels of complexity required to study respiratory infection including pathogen–host interaction, activation of innate signaling pathways, and recruitment of immune cells (Niemeyer et al. 2018, Benam et al. 2015).

Aspergillus fumigatus is a potentially lethal airborne fungal pathogen which can infect both the upper and lower respiratory tract (deShazo, Chapin, and Swain 1997). Neutrophils are known to be required for effective control of *A. fumigatus* infection; however, their precise contribution to clearance of infection is unclear. To further elucidate the roles of neutrophils in controlling fungal infection, Jones et al. (2016) recreated neutrophil migration and interaction with *A. fumigatus* on a microscale through two complementary platforms. The first platform consisting of an array of wells containing chemoattractant, neutrophils, and fungus was used to measure chemoattractant priming of neutrophils during *A. fumigatus* infection. The second platform assayed migration of neutrophils toward *A. fumigatus* by establishing a chemokine gradient within a microchamber. Using these two systems, it was discovered that neutrophils have an enhanced ability to inhibit *A. fumigatus* growth in the presence of chemokine N-formyl-methionyl-leucyl-phenylalanine and that growth inhibition was further elevated when neutrophils were exposed to chemokine gradients. Further, it was found that neutrophils from immunosuppressed individuals exhibited an impaired capacity to inhibit *A. fumigatus* growth (Jones et al. 2016).

Barkal, Procknow, et al. (2017) developed a human organotypic lung model to reconstruct the human bronchiole *in vitro* and utilized this system to investigate inflammatory cytokine production during *A. fumigatus* pathogenesis. The organotypic model of the human bronchiole was comprised of three compartments lined with cells and a 3D matrix of collagen and pulmonary fibroblasts. The central compartment was lined with primary human bronchial epithelial cells and was filled with air, recreating the ALI naturally found in bronchioles. The remaining compartments were lined with lung microvascular endothelial cells. Wild-type *A. fumigatus* or a less virulent strain lacking the immunomodulating protein LaeA was introduced to the central airway chamber to establish infection. The less virulent mutant strain showed greater inflammatory cytokine response and recruitment of polymorphonuclear leukocytes to the site of infection (Barkal, Procknow, et al. 2017). The lung microenvironment is comprised of several different types of microbes, such as *Pseudomonas aeruginosa*, which often influence each other and host cells through production of volatile compounds (Barkal, Berthier, et al. 2017). To further advance the organotypic model system, the researchers developed a co-culture microbial insert that sets into the airway bronchiole chamber preventing direct contact between the host cells and pathogens but still allowing for contact of volatile compounds. Using this system, the airway bronchioles were exposed to volatile compounds from either monomicrobial cultures of *A. fumigatus*, *P. aeruginosa*, or multimicrobial cultures containing both *A. fumigatus* and *P. aeruginosa*. Interestingly, the inflammatory response of the bronchiole to the volatile compounds produced by *A. fumigatus* was altered by the presence of *P. aeruginosa* in that higher levels of GM-CSF, IL-8, IL-6, and IFN-gamma were secreted in multimicrobial conditions compared to monomicrobial with *A. fumigatus* (Barkal, Procknow, et al. 2017).

Microengineered models are not limited to use in fungal and bacterial pathogen studies; these systems have also been employed for analysis of respiratory viral infections. The Small Airway-on-a-Chip system described by Benam, Villenave, et al. (2016) and Benam et al. (2017) was adapted to study inflammatory responses of primary human airway epithelia during virus infection. Viral infections were mimicked by applying poly(I:C) to the airway lumen of the chips resulting in stimulation of epithelial cells. This resulted in the induction of inflammatory cytokines including RANTES, IL-6, and IP-10, which were further elevated in the presence of underlying endothelial cells. Moreover, adhesion molecules E-selectin and VCAM-1 were upregulated on the endothelial cells, a critical requirement for effective adhesion and rolling of recruited neutrophils during inflammation (Benam, Villenave, et al. 2016).

10.11 TRANSFUSION-ASSOCIATED VASCULAR INJURY

Red blood cell (RBC) transfusion is a widely practiced and generally accepted form of care in critically ill patients; however, recent reports suggest that lung damage from the transfusion process may result in an increased risk of lung pathologies such as acute respiratory distress syndrome (ARDS) (Zilberberg et al. 2007). Extensive research efforts have shown that endothelial cells of the microvasculature in the distal lungs is a critical component of transfusion-associated lung damage whereby activation of endothelial cells by RBCs leads to elevated reactive oxygen species, inflammatory signals, and cell death (Ware 2014). Prior to the advent of microfluidic and organ-on-chip technologies, further research efforts were limited due to the technical challenges of modeling human lung vasculature *in vitro* as static culture conditions fail to recreate critical aspects of the endothelial cell environment such as shear stress (Huh, Torisawa, et al. 2012, Tarbell 2010).

To recreate the human lung microvasculature on a platform that incorporates dynamic vascular flow, Seo et al. (2017) fabricated a microfluidic chip containing a single channel which they then lined with human endothelial cells. RBC transfusion was mimicked by perfusing the vascular channel with RBCs for four hours at physiologically relevant rates. As read out of vascular injury, the authors measured extracellular levels of the DNA-binding protein HMGB1, which were previously associated with necroptosis of cells, immune cell activation, and systemic inflammation (Pisetsky 2014). Although the endothelial cells remained adherent, the authors found that extracellular HMGB1 levels were significantly increased during transfusion of RBCs when compared with control samples perfused with media alone or the RBCs alone prior to transfusion (Seo et al. 2017). The authors then went on to measure the effects of shear stress on their system by repeating their experiments with considerably reduced shear stress. Interestingly, reduced shear stress lead to an exacerbation of HMGB1 release, suggesting that reduced shear force is associated with increased transfusion-associated vascular damage (Seo et al. 2017). These results were consistent with previous clinical findings that patients with increased susceptibility to transfusion-associated injuries exhibit reduced blood flow velocity (Donati et al. 2014).

10.12 CONCLUDING REMARKS

Technological advances in microfluidic and microfabrication techniques have paved the way for the development of highly complex biomimetic model systems, which are integral to the advancement of understanding the pathogenesis and treatment of respiratory diseases. These platforms have recreated tissue- and organ-level physiology using multicellular components and well-defined architecture in ways which cannot be accomplished by more primitive culture systems (Bhatia and Ingber 2014). Further, the applications of these systems are wide ranging with the ability to analyze cell signaling pathways, dissect cell–tissue and tissue–tissue interactions, and test therapeutic intervention (Bhatia and Ingber 2014, Esch, Bahinski, and Huh 2015, Niemeyer et al. 2018). Aside from their use in a research setting, some of these systems have also shown clear value in a clinical setting for disease diagnosis (Sackmann et al. 2014, 2012).

Despite the many strengths these models have, there are instances where improvements could be made to either better recreate the native environment of the human lung or to better recapitulate respiratory disease conditions. In several of the model systems outlined, cancerous or immortalize cell lines were used in lieu of primary cells (Felder et al. 2014, Huh, Leslie, et al. 2012). Cell lines derived from tumors or generated through immortalization often have significantly altered signaling pathways and phenotypes when compared to cells naturally found in a healthy respiratory system, which may provide additional difficulties when trying to interpret data. Although some organ-on-chip systems use primary donor cells (e.g., Small Airway-on-a-Chip, microfluidic thrombosis model, and organotypic lung model), they still lack certain cell populations normally found in the lung milieu such as tissue-resident immune cells, smooth muscle cells, and lung mesenchymal progenitor cells (Barkal, Procknow, et al. 2017, Benam, Novak, et al. 2016, Benam, Villenave, et al. 2016, Jain et al. 2018). These cell types have been found to play critical roles in the pathogenesis of numerous respiratory diseases such as asthma, COPD, and respiratory infection (Zuyderduyn et al. 2008, Shaykhiev and Crystal 2013, Iwasaki, Foxman, and Molony 2017). So far one major hurdle for incorporating these missing cells is the widely different growth conditions and culture media required for maintenance of these diverse cell types; however, with advances in our understanding of cell growth requirements as well as advances in cell culture technologies, this problem should be resolved imminently (Benam, Villenave, et al. 2016). Further, these new model systems cannot be mass produced and often involve lengthy fabrication processes, resulting in low to medium-throughput and limiting some of their applications. Many of these systems have shown extreme promise in the drug discovery process with biomarker identification, drug efficacy testing, and drug safety testing capabilities; however, rapid screening of several drug candidates necessitates high-throughput platforms (Benam, Villenave, et al. 2016, Aref et al. 2013, Niemeyer et al. 2018, Esch, Bahinski, and Huh 2015, Jain et al. 2018, Huh, Leslie, et al. 2012, Bhatia and Ingber 2014). As many of the microfabrication technologies become more streamlined and advanced, these systems could reach a more high-throughput capacity expanding their use in drug discovery.

While biomimetic microengineered systems have been utilized to recreate the numerous pathologies listed here, there still exist several respiratory pathologies that have either inadequate *in vitro* culture models or no cell culture model at all. There is a desperate need for improved models of cystic fibrosis, a disease that manifests in several organs including the lung where it results in elevated mucus viscosity, impaired airway clearance, airway inflammation, and recurring respiratory infections (Elborn 2016). It should be possible to adapt current systems to recreate cystic fibrosis using donor-derived cells from patients suffering from the disease in a similar manner to what was reported for patients with COPD (Benam, Novak, et al. 2016). Since the etiology of cystic fibrosis has been well established as known mutations in a gene encoding for a transmembrane ion channel, it may also be possible to adapt current microengineered systems using novel gene editing techniques. Other respiratory pathologies which currently have **no** human *in vitro* system include diffuse alveolar damage, emphysema, sarcoidosis, chronic interstitial lung disease, pulmonary hypertension (Nichols et al. 2014). Creating biomimetic microengineered model systems for these diseases would fill a void that currently exists in these fields and would help to advance our understanding and treatment of these pathologies.

REFERENCES

Abdel-Latif, A., and S. S. Smyth. 2012. Preventing platelet thrombosis with a PAR1 pepducin. *Circulation* 126 (1):13–5.

Akpinar, E. E., D. Hosgun, S. Akpinar, G. K. Atac, B. Doganay, and M. Gulhan. 2014. Incidence of pulmonary embolism during COPD exacerbation. *J Bras Pneumol* 40 (1):38–45.

Amann, A., M. Zwierzina, G. Gamerith, M. Bitsche, J. M. Huber, G. F. Vogel, M. Blumer, S. Koeck, E. J. Pechriggl, J. M. Kelm, W. Hilbe, and H. Zwierzina. 2014. Development of an innovative 3D cell culture system to study tumour-stroma interactions in non-small cell lung cancer cells. *PLoS One* 9 (3):e92511.

Anguiano, M., C. Castilla, M. Maska, C. Ederra, R. Pelaez, X. Morales, G. Munoz-Arrieta, M. Mujika, M. Kozubek, A. Munoz-Barrutia, A. Rouzaut, S. Arana, J. M. Garcia-Aznar, and C. Ortiz-de-Solorzano. 2017. Characterization of three-dimensional cancer cell migration in mixed collagen-Matrigel scaffolds using microfluidics and image analysis. *PLoS One* 12 (2):e0171417.

Aref, A. R., R. Y. Huang, W. Yu, K. N. Chua, W. Sun, T. Y. Tu, J. Bai, W. J. Sim, I. K. Zervantonakis, J. P. Thiery, and R. D. Kamm. 2013. Screening therapeutic EMT blocking agents in a three-dimensional microenvironment. *Integr Biol (Camb)* 5 (2):381–9.

Arrowsmith, J., and P. Miller. 2013. Trial watch: Phase II and phase III attrition rates 2011–2012. *Nat Rev Drug Discov* 12 (8):569.

Bakirtas, A. 2017. Diagnostic challenges of childhood asthma. *Curr Opin Pulm Med* 23 (1):27–33.

Barkal, L. J., E. Berthier, A. B. Theberge, N. P. Keller, and D. J. Beebe. 2017. Multikingdom microscale models. *PLoS Pathog* 13 (8):e1006424.

Barkal, L. J., C. L. Procknow, Y. R. Alvarez-Garcia, M. Niu, J. A. Jimenez-Torres, R. A. Brockman-Schneider, J. E. Gern, L. C. Denlinger, A. B. Theberge, N. P. Keller, E. Berthier, and D. J. Beebe. 2017. Microbial volatile communication in human organotypic lung models. *Nat Commun* 8 (1):1770.

Barnes, P. J., P. G. Burney, E. K. Silverman, B. R. Celli, J. Vestbo, J. A. Wedzicha, and E. F. Wouters. 2015. Chronic obstructive pulmonary disease. *Nat Rev Dis Primers* 1:15076.

Bathoorn, E., P. Daly, B. Gaiser, K. Sternad, C. Poland, W. Macnee, and E. M. Drost. 2011. Cytotoxicity and induction of inflammation by pepsin in Acid in bronchial epithelial cells. *Int J Inflam* 2011:569416.

Benam, K. H., S. Dauth, B. Hassell, A. Herland, A. Jain, K. J. Jang, K. Karalis, H. J. Kim, L. MacQueen, R. Mahmoodian, S. Musah, Y. S. Torisawa, A. D. van der Meer, R. Villenave, M. Yadid, K. K. Parker, and D. E. Ingber. 2015. Engineered in vitro disease models. *Annu Rev Pathol* 10:195–262.

Benam, K. H., and D. E. Ingber. 2016. Commendation for exposing key advantage of organ chip approach. *Cell Syst* 3 (5):411.

Benam, K. H., M. Mazur, Y. Choe, T. C. Ferrante, R. Novak, and D. E. Ingber. 2017. Human lung small airway-on-a-chip protocol. *Methods Mol Biol* 1612:345–365.

Benam, K. H., R. Novak, J. Nawroth, M. Hirano-Kobayashi, T. C. Ferrante, Y. Choe, R. Prantil-Baun, J. C. Weaver, A. Bahinski, K. K. Parker, and D. E. Ingber. 2016. Matched-comparative modeling of normal and diseased human airway responses using a microengineered breathing lung chip. *Cell Syst* 3 (5):456–466 e4.

Benam, K. H., R. Villenave, C. Lucchesi, A. Varone, C. Hubeau, H. H. Lee, S. E. Alves, M. Salmon, T. C. Ferrante, J. C. Weaver, A. Bahinski, G. A. Hamilton, and D. E. Ingber. 2016. Small airway-on-a-chip enables analysis of human lung inflammation and drug responses in vitro. *Nat Methods* 13 (2):151–7.

Bhatia, S. N., and D. E. Ingber. 2014. Microfluidic organs-on-chips. *Nat Biotechnol* 32 (8):760–72.

Blume, C., and D. E. Davies. 2013. In vitro and ex vivo models of human asthma. *Eur J Pharm Biopharm* 84 (2):394–400.

Bouvier, N. M., and A. C. Lowen. 2010. Animal models for influenza virus pathogenesis and transmission. *Viruses* 2 (8):1530–63.

Britto, C. J., V. Brady, S. Lee, and C. S. Dela Cruz. 2017. Respiratory viral infections in chronic lung diseases. *Clin Chest Med* 38 (1):87–96.

Centers for Disease Control. 2013. Antibiotic restistance threats in the United States, 2013. www.cdc.gov/drugresistance/pdf/ar-threats-2013-508.pdf.

Chen, F., X. Zhuang, L. Lin, P. Yu, Y. Wang, Y. Shi, G. Hu, and Y. Sun. 2015. New horizons in tumor microenvironment biology: Challenges and opportunities. *BMC Med* 13:45.

Chimenti, I., F. Pagano, F. Angelini, C. Siciliano, G. Mangino, V. Picchio, E. De Falco, M. Peruzzi, R. Carnevale, M. Ibrahim, G. Biondi-Zoccai, E. Messina, and G. Frati. 2017. Human lung spheroids as in vitro niches of lung progenitor cells with distinctive paracrine and plasticity properties. *Stem Cells Transl Med* 6 (3):767–777.

Chua, F., J. Gauldie, and G. J. Laurent. 2005. Pulmonary fibrosis: Searching for model answers. *Am J Respir Cell Mol Biol* 33 (1):9–13.

Collaborators, GBD LRI. 2017. Estimates of the global, regional, and national morbidity, mortality, and aetiologies of lower respiratory tract infections in 195 countries: A systematic analysis for the Global Burden of Disease Study 2015. *Lancet Infect Dis* 17 (11):1133–1161.

Conti, G., S. Tambalo, G. Villetti, S. Catinella, C. Carnini, F. Bassani, N. Sonato, A. Sbarbati, and P. Marzola. 2010. Evaluation of lung inflammation induced by intratracheal administration of LPS in mice: Comparison between MRI and histology. *MAGMA* 23 (2):93–101.

deShazo, R. D., K. Chapin, and R. E. Swain. 1997. Fungal sinusitis. *N Engl J Med* 337 (4):254–9.

Deslee, G., S. Dury, J. M. Perotin, D. Al Alam, F. Vitry, R. Boxio, S. C. Gangloff, M. Guenounou, F. Lebargy, and A. Belaaouaj. 2007. Bronchial epithelial spheroids: An alternative culture model to investigate epithelium inflammation-mediated COPD. *Respir Res* 8:86.

DiMasi, J. A., H. G. Grabowski, and R. W. Hansen. 2016. Innovation in the pharmaceutical industry: New estimates of R&D costs. *J Health Econ* 47:20–33.

Donati, A., E. Damiani, M. Luchetti, R. Domizi, C. Scorcella, A. Carsetti, V. Gabbanelli, P. Carletti, R. Bencivenga, H. Vink, E. Adrario, M. Piagnerelli, A. Gabrielli, P. Pelaia, and C. Ince. 2014. Microcirculatory effects of the transfusion of leukodepleted or non-leukodepleted red blood cells in patients with sepsis: A pilot study. *Crit Care* 18 (1):R33.

El-Sherbiny, I. M., and M. H. Yacoub. 2013. Hydrogel scaffolds for tissue engineering: Progress and challenges. *Glob Cardiol Sci Pract* 2013 (3):316–42.

Elborn, J. S. 2016. Cystic fibrosis. *Lancet* 388 (10059):2519–2531.

Esch, E. W., A. Bahinski, and D. Huh. 2015. Organs-on-chips at the frontiers of drug discovery. *Nat Rev Drug Discov* 14 (4):248–60.

Fang, Y., and R. M. Eglen. 2017. Three-dimensional cell cultures in drug discovery and development. *SLAS Discov* 22 (5):456–472.

Farrell, S., C. McMaster, D. Gibson, M. D. Shields, and W. A. McCallion. 2006. Pepsin in bronchoalveolar lavage fluid: A specific and sensitive method of diagnosing gastro-oesophageal reflux-related pulmonary aspiration. *J Pediatr Surg* 41 (2):289–93.

Felder, M., P. Sallin, L. Barbe, B. Haenni, A. Gazdhar, T. Geiser, and O. Guenat. 2012. Microfluidic wound-healing assay to assess the regenerative effect of HGF on wounded alveolar epithelium. *Lab Chip* 12 (3):640–6.

Felder, M., A. O. Stucki, J. D. Stucki, T. Geiser, and O. T. Guenat. 2014. The potential of microfluidic lung epithelial wounding: Towards in vivo-like alveolar microinjuries. *Integr Biol (Camb)* 6 (12):1132–40.

Ferkol, T., and D. Schraufnagel. 2014. The global burden of respiratory disease. *Ann Am Thorac Soc* 11 (3):404–6.

Forum of International Respiratory Societies. 2017. *The Global Impact of Respiratory Disease- Second Edition*: Sheffield, European Respiratory Society.

Gao, G. F., Y. Huang, A. F. Schilling, K. Hubbell, and X. F. Cui. 2018. Organ bioprinting: Are we there yet? *Advanced Healthcare Materials* 7 (1).

Gershon, A. S., L. Warner, P. Cascagnette, J. C. Victor, and T. To. 2011. Lifetime risk of developing chronic obstructive pulmonary disease: A longitudinal population study. *The Lancet* 378 (9795):991–6.

Henderson, V. C., J. Kimmelman, D. Fergusson, J. M. Grimshaw, and D. G. Hackam. 2013. Threats to validity in the design and conduct of preclinical efficacy studies: A systematic review of guidelines for in vivo animal experiments. *PLoS Med* 10 (7):e1001489.

Henjakovic, M., K. Sewald, S. Switalla, D. Kaiser, M. Muller, T. Z. Veres, C. Martin, S. Uhlig, N. Krug, and A. Braun. 2008. Ex vivo testing of immune responses in precision-cut lung slices. *Toxicol Appl Pharmacol* 231 (1):68–76.

Holgate, S. T., S. Wenzel, D. S. Postma, S. T. Weiss, H. Renz, and P. D. Sly. 2015. Asthma. *Nat Rev Dis Primers* 1:15025.

Horvath, L., Y. Umehara, C. Jud, F. Blank, A. Petri-Fink, and B. Rothen-Rutishauser. 2015. Engineering an in vitro air-blood barrier by 3D bioprinting. *Sci Rep* 5:7974.

Huh, D., D. C. Leslie, B. D. Matthews, J. P. Fraser, S. Jurek, G. A. Hamilton, K. S. Thorneloe, M. A. McAlexander, and D. E. Ingber. 2012. A human disease model of drug toxicity-induced pulmonary edema in a lung-on-a-chip microdevice. *Sci Transl Med* 4 (159):159ra147.

Huh, D., B. D. Matthews, A. Mammoto, M. Montoya-Zavala, H. Y. Hsin, and D. E. Ingber. 2010. Reconstituting organ-level lung functions on a chip. *Science* 328 (5986):1662–8.

Huh, D., Y. S. Torisawa, G. A. Hamilton, H. J. Kim, and D. E. Ingber. 2012. Microengineered physiological biomimicry: Organs-on-chips. *Lab Chip* 12 (12):2156–64.

Iwasaki, A., E. F. Foxman, and R. D. Molony. 2017. Early local immune defences in the respiratory tract. *Nat Rev Immunol* 17 (1):7–20.

Jain, A., R. Barrile, A. D. van der Meer, A. Mammoto, T. Mammoto, K. De Ceunynck, O. Aisiku, M. A. Otieno, C. S. Louden, G. A. Hamilton, R. Flaumenhaft, and D. E. Ingber. 2018. Primary human lung alveolus-on-a-chip model of intravascular thrombosis for assessment of therapeutics. *Clin Pharmacol Ther* 103 (2):332–40.

Jiao, D., C. K. Wong, M. S. Tsang, I. M. Chu, D. Liu, J. Zhu, M. Chu, and C. W. Lam. 2017. Activation of eosinophils interacting with bronchial epithelial cells by antimicrobial peptide LL-37: Implications in allergic asthma. *Sci Rep* 7 (1):1848.

Jones, C. N., L. Dimisko, K. Forrest, K. Judice, M. C. Poznansky, J. F. Markmann, J. M. Vyas, and D. Irimia. 2016. Human neutrophils are primed by chemoattractant gradients for blocking the growth of *Aspergillus fumigatus*. *J Infect Dis* 213 (3):465–75.

Jose, R. R., M. J. Rodriguez, T. A. Dixon, Fiorenzo Omenetto, and D. L. Kaplan. 2016. Evolution of bioinks and additive manufacturing technologies for 3D bioprinting. *ACS Biomater. Sci. Eng.* 2 (10):1662–1678.

Keatings, V. M., P. D. Collins, D. M. Scott, and P. J. Barnes. 1996. Differences in interleukin-8 and tumor necrosis factor-alpha in induced sputum from patients with chronic obstructive pulmonary disease or asthma. *Am J Respir Crit Care Med* 153 (2):530–4.

Kling, H. M., G. J. Nau, T. M. Ross, T. G. Evans, K. Chakraborty, K. M. Empey, and J. L. Flynn. 2014. Challenges and future in vaccines, drug development, and immunomodulatory therapy. *Ann Am Thorac Soc* 11 Suppl 4:S201–10.

Knutsen, A. P., R. K. Bush, J. G. Demain, D. W. Denning, A. Dixit, A. Fairs, P. A. Greenberger, B. Kariuki, H. Kita, V. P. Kurup, R. B. Moss, R. M. Niven, C. H. Pashley, R. G. Slavin, H. M. Vijay, and A. J. Wardlaw. 2012. Fungi and allergic lower respiratory tract diseases. *J Allergy Clin Immunol* 129 (2):280–91; quiz 292–3.

Kolesky, D. B., K. A. Homan, M. A. Skylar-Scott, and J. A. Lewis. 2016. Three-dimensional bioprinting of thick vascularized tissues. *Proc Natl Acad Sci U S A* 113 (12):3179–84.

Kolesky, D. B., R. L. Truby, A. S. Gladman, T. A. Busbee, K. A. Homan, and J. A. Lewis. 2014. 3D bioprinting of vascularized, heterogeneous cell-laden tissue constructs. *Adv Mater* 26 (19):3124–30.

Konar, D., M. Devarasetty, D. V. Yildiz, A. Atala, and S. V. Murphy. 2016. Lung-on-a-chip technologies for disease modeling and drug development. *Biomed Eng Comput Biol* 7 (Suppl 1):17–27.

Kudlacz, E., M. Conklyn, C. Andresen, C. Whitney-Pickett, and P. Changelian. 2008. The JAK-3 inhibitor CP-690550 is a potent anti-inflammatory agent in a murine model of pulmonary eosinophilia. *Eur J Pharmacol* 582 (1–3):154–61.

Laniado-Laborin, R. 2009. Smoking and chronic obstructive pulmonary disease (COPD). Parallel epidemics of the 21 century. *Int J Environ Res Public Health* 6 (1):209–24.

Lee, J. S., H. R. Collard, G. Raghu, M. P. Sweet, S. R. Hays, G. M. Campos, J. A. Golden, and T. E. King, Jr. 2010. Does chronic microaspiration cause idiopathic pulmonary fibrosis? *Am J Med* 123 (4):304–11.

Lehmann, A. D., N. Daum, M. Bur, C. M. Lehr, P. Gehr, and B. M. Rothen-Rutishauser. 2011. An in vitro triple cell co-culture model with primary cells mimicking the human alveolar epithelial barrier. *Eur J Pharm Biopharm* 77 (3):398–406.

Martinez, F. J., H. R. Collard, A. Pardo, G. Raghu, L. Richeldi, M. Selman, J. J. Swigris, H. Taniguchi, and A. U. Wells. 2017. Idiopathic pulmonary fibrosis. *Nat Rev Dis Primers* 3:17074.

Matthay, M. A. 2014. Resolution of pulmonary edema. Thirty years of progress. *Am J Respir Crit Care Med* 189 (11):1301–8.

Mestre-Ferrandiz, J., J. Sussex, and A. Towse. 2012. *The R&D Cost of a New Medicine.* United Kingdom: Office of Health Economics.

Morin, J. P., J. M. Baste, A. Gay, C. Crochemore, C. Corbiere, and C. Monteil. 2013. Precision cut lung slices as an efficient tool for in vitro lung physio-pharmacotoxicology studies. *Xenobiotica* 43 (1):63–72.

Mullard, A. 2016. Parsing clinical success rates. *Nat Rev Drug Discov* 15 (7):447.

Murphy, S. V., and A. Atala. 2014. 3D bioprinting of tissues and organs. *Nat Biotechnol* 32 (8):773–85.

Nadkarni, R. R., S. Abed, and J. S. Draper. 2016. Organoids as a model system for studying human lung development and disease. *Biochem Biophys Res Commun* 473 (3):675–82.

Nair, H., D. J. Nokes, B. D. Gessner, M. Dherani, S. A. Madhi, R. J. Singleton, K. L. O'Brien, A. Roca, P. F. Wright, N. Bruce, A. Chandran, E. Theodoratou, A. Sutanto, E. R. Sedyaningsih, M. Ngama, P. K. Munywoki, C. Kartasasmita, E. A. Simoes, I. Rudan, M. W. Weber, and H. Campbell. 2010. Global burden of acute lower respiratory infections due to respiratory syncytial virus in young children: a systematic review and meta-analysis. *Lancet* 375 (9725):1545–55.

Neuhaus, V., D. Schaudien, T. Golovina, U. A. Temann, C. Thompson, T. Lippmann, C. Bersch, O. Pfennig, D. Jonigk, P. Braubach, H. G. Fieguth, G. Warnecke, V. Yusibov, K. Sewald, and A. Braun. 2017. Assessment of long-term cultivated human precision-cut lung slices as an ex vivo system for evaluation of chronic cytotoxicity and functionality. *J Occup Med Toxicol* 12:13. doi:10.1186/s12995-017-0158-5.

Nichols, J. E., J. A. Niles, S. P. Vega, L. B. Argueta, A. Eastaway, and J. Cortiella. 2014. Modeling the lung: Design and development of tissue engineered macro- and microphysiologic lung models for research use. *Experimental Biology and Medicine* 239 (9):1135–69.

Niemeyer, B. F., P. Zhao, R. M. Tuder, and K. H. Benam. 2018. Advanced Microengineered Lung Models for Translational Drug Discovery. *SLAS Discov* 23 (8):777–789. doi: 10.1177/2472555218760217.

Nikolic, M. Z., and E. L. Rawlins. 2017. Lung organoids and their use to study cell–cell interaction. *Curr Pathobiol Rep* 5 (2):223–31.

Olin, J. T., and M. E. Wechsler. 2014. Asthma: Pathogenesis and novel drugs for treatment. *BMJ* 349:g5517. doi:10.1136/bmj.g5517.

Parkins, M. D., and R. A. Floto. 2015. Emerging bacterial pathogens and changing concepts of bacterial pathogenesis in cystic fibrosis. *J Cyst Fibros* 14 (3):293–304.

Paul, S. M., D. S. Mytelka, C. T. Dunwiddie, C. C. Persinger, B. H. Munos, S. R. Lindborg, and A. L. Schacht. 2010. How to improve R&D productivity: The pharmaceutical industry's grand challenge. *Nat Rev Drug Discov* 9 (3):203–14.

Pernis, A. B., and P. B. Rothman. 2002. JAK-STAT signaling in asthma. *J Clin Invest* 109 (10):1279–83.

Pezzulo, A. A., T. D. Starner, T. E. Scheetz, G. L. Traver, A. E. Tilley, B. G. Harvey, R. G. Crystal, P. B. McCray, Jr., and J. Zabner. 2011. The air-liquid interface and use of primary cell cultures are important to recapitulate the transcriptional profile of in vivo airway epithelia. *Am J Physiol Lung Cell Mol Physiol* 300 (1):L25–31.

Pisetsky, D. S. 2014. The expression of HMGB1 on microparticles released during cell activation and cell death in vitro and in vivo. *Mol Med* 20:158–63.

Reeves, S. R., T. Kolstad, T. Y. Lien, M. Elliott, S. F. Ziegler, T. N. Wight, and J. S. Debley. 2014. Asthmatic airway epithelial cells differentially regulate fibroblast expression of extracellular matrix components. *J Allergy Clin Immunol* 134 (3):663–670 e1.

Rothen-Rutishauser, B. M., S. G. Kiama, and P. Gehr. 2005. A three-dimensional cellular model of the human respiratory tract to study the interaction with particles. *Am J Respir Cell Mol Biol* 32 (4):281–9.

Sacco, R. E., R. K. Durbin, and J. E. Durbin. 2015. Animal models of respiratory syncytial virus infection and disease. *Curr Opin Virol* 13:117–22.

Sackmann, E. K., E. Berthier, E. A. Schwantes, P. S. Fichtinger, M. D. Evans, L. L. Dziadzio, A. Huttenlocher, S. K. Mathur, and D. J. Beebe. 2014. Characterizing asthma from a drop of blood using neutrophil chemotaxis. *Proc Natl Acad Sci U S A* 111 (16):5813–8.

Sackmann, E. K., E. Berthier, E. W. Young, M. A. Shelef, S. A. Wernimont, A. Huttenlocher, and D. J. Beebe. 2012. Microfluidic kit-on-a-lid: A versatile platform for neutrophil chemotaxis assays. *Blood* 120 (14):e45–53.

Sanderson, M. J. 2011. Exploring lung physiology in health and disease with lung slices. *Pulm Pharmacol Ther* 24 (5):452–65.

Seo, J., D. Conegliano, M. Farrell, M. Cho, X. Ding, T. Seykora, D. Qing, N. S. Mangalmurti, and D. Huh. 2017. A microengineered model of RBC transfusion-induced pulmonary vascular injury. *Sci Rep* 7 (1):3413.

Shaykhiev, R., and R. G. Crystal. 2013. Innate immunity and chronic obstructive pulmonary disease: A mini-review. *Gerontology* 59 (6):481–9.

Sprunger, D. B., A. L. Olson, T. J. Huie, E. R. Fernandez-Perez, A. Fischer, J. J. Solomon, K. K. Brown, and J. J. Swigris. 2012. Pulmonary fibrosis is associated with an elevated risk of thromboembolic disease. *Eur Respir J* 39 (1):125–32.

Sundarakrishnan, A., Y. Chen, L. D. Black, B. B. Aldridge, and D. L. Kaplan. 2018. Engineered cell and tissue models of pulmonary fibrosis. *Adv Drug Deliv Rev* 129:78–94.

Sung, K. E., and D. J. Beebe. 2014. Microfluidic 3D models of cancer. *Adv Drug Deliv Rev* 79–80:68–78.

Tarbell, J. M. 2010. Shear stress and the endothelial transport barrier. *Cardiovasc Res* 87 (2):320–30.

Thangavel, R. R., and N. M. Bouvier. 2014. Animal models for influenza virus pathogenesis, transmission, and immunology. *J Immunol Methods* 410:60–79.

Thomas, B., A. Rutman, R. A. Hirst, P. Haldar, A. J. Wardlaw, J. Bankart, C. E. Brightling, and C. O'Callaghan. 2010. Ciliary dysfunction and ultrastructural abnormalities are features of severe asthma. *J Allergy Clin Immunol* 126 (4):722–729 e2.

US Food and Drug Administration. 2015. The drug development process. www.fda.gov/ forpatients/approvals/drugs/.

Vyfhuis, M. A. L., N. Onyeuku, T. Diwanji, S. Mossahebi, N. P. Amin, S. N. Badiyan, P. Mohindra, and C. B. Simone, II. 2018. Advances in proton therapy in lung cancer. *Ther Adv Respir Dis* 12:1753466618783878.

Ward, C., I. A. Forrest, I. A. Brownlee, G. E. Johnson, D. M. Murphy, J. P. Pearson, J. H. Dark, and P. A. Corris. 2005. Pepsin like activity in bronchoalveolar lavage fluid is suggestive of gastric aspiration in lung allografts. *Thorax* 60 (10):872–4.

Ware, L. B. 2014. Transfusion-induced lung endothelial injury: A DAMP death? *Am J Respir Crit Care Med* 190 (12):1331–2.

Williams, A., and I. M. Orme. 2016. Animal models of tuberculosis: An overview. *Microbiol Spectr* 4 (4).

Wills-Karp, M. 2004. Interleukin-13 in asthma pathogenesis. *Immunol Rev* 202:175–90.

World Health Organization. 2017. Global Tuberculosis Report 2017. www.who.int/tb/publications/factsheet_global.pdf.

World Health Organization. 2018. The top 10 causes of death. Last Modified 24 May 2018. www.who.int/news-room/fact-sheets/detail/the-top-10-causes-of-death.

Wright, J. L., M. Cosio, and A. Churg. 2008. Animal models of chronic obstructive pulmonary disease. *Am J Physiol Lung Cell Mol Physiol* 295 (1):L1–15.

Wust, S., R. Muller, and S. Hofmann. 2011. Controlled positioning of cells in biomaterials-approaches towards 3D tissue printing. *J Funct Biomater* 2 (3):119–54.

Zhan, L., J. Tang, M. Sun, and C. Qin. 2017. Animal models for tuberculosis in translational and precision medicine. *Front Microbiol* 8:717.

Zilberberg, M. D., C. Carter, P. Lefebvre, M. Raut, F. Vekeman, M. S. Duh, and A. F. Shorr. 2007. Red blood cell transfusions and the risk of acute respiratory distress syndrome among the critically ill: A cohort study. *Crit Care* 11 (3):R63.

Zuyderduyn, S., M. B. Sukkar, A. Fust, S. Dhaliwal, and J. K. Burgess. 2008. Treating asthma means treating airway smooth muscle cells. *Eur Respir J* 32 (2):265–74.

11 *In Vitro* Alzheimer's Disease Modeling Using Stem Cells

Hyun-Ji Park, Song Ih Ahn, Jeong-Kee Yoon, Hyunjung Lee, and YongTae Kim
Georgia Institute of Technology

CONTENTS

11.1 INTRODUCTION

Alzheimer's disease (AD) is the most common neurodegenerative disease that causes dementia in elder people. As the aging trend of humans becomes a global phenomenon, the number of AD patients is also increasing worldwide. The increasing risk of AD has led many researchers to study the AD pathophysiology for understanding the mechanism and finding treatments for the AD. In this occasion, there is a great need for tools to study AD pathophysiology *in vitro* to understand the molecular and physiological mechanisms of AD progress and to emphasize the challenges of testing new drugs, whether delaying or curing AD progression.

Because of the easy and precise observation, researchers have relied on model systems to investigate cellular and molecular mechanisms of AD instead of directly observing the patients using a variety of imaging techniques such as PEG or MRI.

Animal models have taught a lot about the basic principles of disease progression, but there are challenges remained in using the animal models such as translational gap between animals and human physiology and the ethical issues. In this context, stem cell technology, which enables patient-specific AD mimicry of pathological condition, can be used as an excellent research tool to reproduce all known theories of AD pathology, from the amyloid cascade hypothesis to neuroinflammation. In particular, as the development of human-induced pluripotent stem cell (PSC) technology allows us to investigate the pathogenesis of patient-specific diseases, studies are currently underway to embody the AD model using stem cell *in vitro*.

This chapter discusses the novel strategies for an efficient modeling of AD *in vitro*. Effective and rapid clinical translation is highly dependent on the ability to adequately model the disease for comprehensive assessment of potential therapies and in-depth understanding of underlying mechanisms. In addition, this chapter provides an overview of potential uses of stem cell technology and other bioengineered strategies (i.e. organoids and microfluidic platform) for the more *in vivo*-like AD model.

11.2 ALZHEIMER'S DISEASE (AD)

AD was first introduced in 1906 by a German neuropathologist and psychiatrist Alois Alzheimer, by presenting his observation of neurofibrillary tangles (NFTs) in the brain of Auguste D. postmortem, who had suffered from memory loss, delusions, hallucinations, and focal symptoms (Goedert and Spillantini 2006, Chow et al. 2008). Since then, AD has been considered as the most common form of dementia. The prevalence of AD increases with age as the structure and function of the brain progressively deteriorates, causing cognitive impairment and susceptibility that compromises the intellectual and social skills of AD patients (Maurer et al. 2006). AD is known debatably with several hypotheses to be caused by a combination of genetic and environmental factors that affect the brain over time (Blennow, de Leon, and Zetterberg 2006). Although the primary cause of AD has yet to be fully elucidated, its effect on the brain is evident as selective degeneration of cholinergic neurons in the forebrain (Whitehouse et al. 1982). The gradual death of neurons in AD patients leads to several neuropathophysiological symptoms including depression, insomnia, aphasia, apraxia, agnosia, and other cognitive impairment such as impaired decision-making (Blennow, de Leon, and Zetterberg 2006). Scientists have not clearly understood the pathogenesis of this fatal neurodegenerative disease but have uncovered several shreds of evidence including undesired protein accumulation and chronic neuroinflammation. One prime evidence is a small brain protein fragment called β-amyloid (Aβ), a sticky compound that is normally generated in brain at lower concentrations between the range of 200 and 1,000 pM that is continuously produced and cleared by neurons and glia (Cirrito et al. 2003). Its normal function is not fully elucidated yet; however, it is believed to have important physiological functions including memory formation (Garcia-Osta and Alberini 2009, Senechal, Larmet, and Dev 2006), cholesterol transport regulation (Yao and Papadopoulos 2002, Igbavboa et al. 2009), and antimicrobial function (Kumar et al. 2016). However, when Aβ accumulates in the brain unless under controlled clearance, this protein interrupts communication between brain cells, as well as restricts the supply of nutrients and oxygen to them, eventually leading to fatal neurodegeneration. Many researchers believe that defects in the processes of production or degradation of Aβ are a main cause of AD.

11.2.1 Amyloid Cascade Hypothesis

Since Alois Alzheimer introduced the AD, researchers have tried to find out the etiological cause of dementia. However, over the eight decades after Alois Alzheimer's findings, there is little progress made in understanding the accurate pathology of the AD in the brain or other organs. In the mid-1960s, researchers used electron microscopy and X-ray fiber diffraction studies to show that Aβ consists of β-pleated sheets (Kidd 1964, Terry 1963, Eanes and Glenner 1968) fueling a controversy over the role of Aβ in the neuropathology of the AD. Later, Glenner (1983) found that more than 90% of AD patients have severe deposition of Aβ in their brain and since then "amyloid cascade hypothesis" has become a leading explanation for the AD pathogenesis until the early 2000s (Masters et al. 1985).

Aβ is a small polypeptide that consists of about 40 amino acid residues, a fragment of a large amyloid precursor protein (APP; Figure 11.1; Kametani and Hasegawa 2018). The APP is a transmembrane protein associated with neuronal outgrowth, axonal transport, and neuronal development (Kang et al. 1987). In the normal physiological status, APP is degraded into Aβ fragments by β-secretase 1 (BACE 1) and γ-secretase (a complex containing presenilin 1), a process known as regulated intercellular proteolysis. Then Aβ fragments are secreted outside the cell and degraded rapidly (Kametani and Hasegawa 2018). In detail, the extracellular domain of APP is cleaved by α-secretase (TACE/ADAM) and BACE 1 (Vassar et al. 1999, Lammich et al. 1999), which produce N-terminal fragments of soluble APP, named sAPPα and sAPPβ, respectively. Following the cleavage, the membrane-bound C-terminal residues, C83 and C99, are further cleaved at three different sites (γ, ε, and ζ) by γ-secretase complex, including Aph-1, Pen 2, presenilin-1, and nicastrin, with the incision at the γ-site, finally releasing Aβ peptides from the membrane (Francis et al. 2002, De Strooper et al. 1998, Takasugi et al. 2003, Yu et al. 2000). The γ-secretase complex is responsible for the production of released Aβ peptides with the different numbers of amino acid residues. Specifically, the mutations of presenilin 1/2 affect the formation and processing of Aβ residues (Xu et al. 2016). In normal condition, Aβ 40 is the main product that is released outside the cell and consequently digested. However, with aging or under pathological conditions, the abnormal incision of 42 amino acid residues increases the production of Aβ 42 instead of Aβ 40. The genetic mutations of APP, which are discovered in the early-onset familial AD, are clustered near β- or γ-secretase cleavage sites and associated with an increase in the production of Aβ 42 (Kang et al. 1987). Other familial AD (fAD) mutations have been found in presenilin 1/2, a component of γ-secretase complex (Svedruzic, Popovic, and Sendula-Jengic 2015). The mutation of presenilin 1/2 leads to an increase of APP C-terminal fragments, mostly Aβ 42. Since Aβ 42 is more hydrophobic than Aβ 40, it has the prominent ability to oligomerize and accumulate to form the toxic amyloid fibrils.

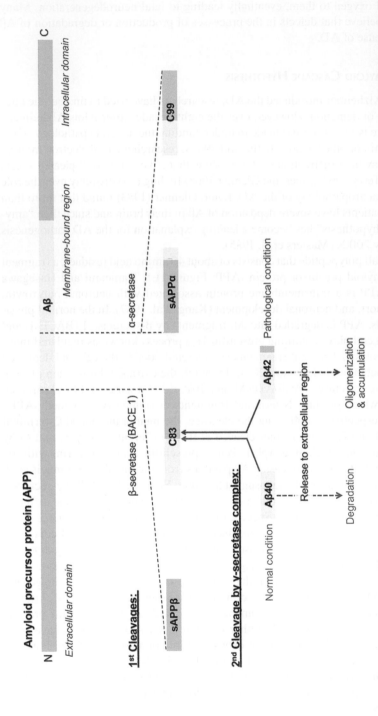

FIGURE 11.1 Ab synthesis from APP by sequential protease activities. APP can be processed through two different pathways: (1) Cleavage of APP by a-secretase that produces peptide that do not form Ab and (2) cleavage of APP by b-secretase 1 and r-secretase that generates amyloid, which forms oligomers and accumulates in pathological condition.

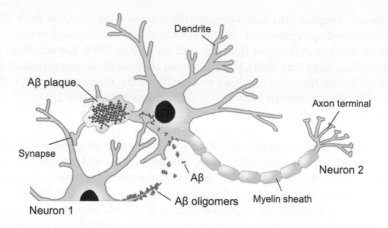

FIGURE 11.2 The Ab cascade hypothesis. Ab oligomers and plaques impair both synaptic structure and function of neurons.

Several pieces of evidence suggest that Aβ accumulation affects the volume loss of hippocampus, a brain region responsible for memory. One well-known hypothesis is that soluble Aβs deteriorate the synaptic function of the AD neurons (Figure 11.2; LaFerla, Green, and Oddo 2007). The soluble Aβs accumulated in several cellular compartments, including Golgi apparatus, endoplasmic reticulum (ER), and endosomal-lysosomal systems, were shown to induce long-term disruption of synaptic plasticity in neurons, consequently leading to cell death (Cleary et al. 2005, Billings et al. 2005, Townsend et al. 2006). However, the accurate mechanisms by which such synaptic plasticity changes remain unclear. Another view on the effect of Aβ accretion is that Aβ peptides are responsible for mitochondrial damage in synaptic neurons (Mungarro-Menchaca et al. 2002, Glabe and Kayed 2006, Manczak et al. 2006, Reddy and Beal 2008). During APP cleavage, Aβ monomers were built up to form fine oligomeric Aβ fibrils in the synaptic terminal while the Aβ fibrils were penetrating the membrane of cellular compartments. As a result, the cytochrome oxidase activity in neurons decreased whereas the production of free radical and carbonyl proteins increased. Further, Aβ oligomers observed in synaptosomal mitochondrial fractions reduced energy metabolism (Gillardon et al. 2007).

11.2.2 Tau Axis Hypothesis

According to Dr. Alzheimer's observation, the substantial deposition of NFTs is one pathological hallmark of AD. Currently, NFTs correspond to a self-assembled aggregation of the hyperphosphorylated tau protein, which is called as the "Tau hypothesis". Tau is one of the microtubule-associated proteins (MAPs) that originally regulate the stability of tubulin assemblies in neural cells. There are six isoforms of tau in the human brain, and their expressions are regulated by mRNA alternative splicing and post-translational modifications such as phosphorylation and acetylation (Cohen et al. 2011, Martin, Latypova, and Terro 2011). Among the six isoforms, 3-repeat (3R) and 4-repeat (4R) taus (4R tau contains exon 10, the microtubule-binding region while 3R tau does not), are accumulated in hyperphosphorylated states in AD brains (Goedert and Spillantini 2017, Serrano-Pozo et al. 2011, Iqbal, Liu, and Gong 2016). Since those tau proteins form unique twisted fibrils with paired helical filaments or related straight filaments, the accumulated tau fibrils in neuronal cells are referred to as NFTs (Zempel and Mandelkow 2014).

Although the tau hypothesis has begun to get attention later than the amyloid cascade hypothesis, recent results show that the tau hypothesis has a strong correlation with cognitive impairment. With aging or under neurodegeneration, tau spreads into the transentorhinal region, limbic region, and neocortical region spatiotemporally (Braak and Braak 1991), whose expression pattern is related to neurodegenerative orders in the AD patients (Bejanin et al. 2017, Okamura and Yanai 2017). In particular, the cognitive decline of the AD patients is closely associated with the early tauopathy in the temporal lobe, especially in the entorhinal cortex (Buckley et al. 2017, Scholl et al. 2016, Schwarz et al. 2016). Furthermore, tau lesions have been reported to occur earlier than Aβ accumulation (Braak and Del Tredici 2014, Johnson et al. 2016), findings that complicate the existing hypotheses of the AD pathogenesis. The mechanism with which tau aggregates transfer from one cell to another remains to be investigated, but the tau pathology in the brain may have a closer correlation with the clinical stages of AD, and even further with the clinical features of dementia.

Scientists explain in two ways how tau accumulation induces neurodegeneration described in Figure 11.3. One is that tauopathy causes extensive damage to the cytoskeletal structures and transport system of neurons (Kametani and Hasegawa 2018). In AD, abnormal production of Aβ causes mislocalization of tau, which is connected

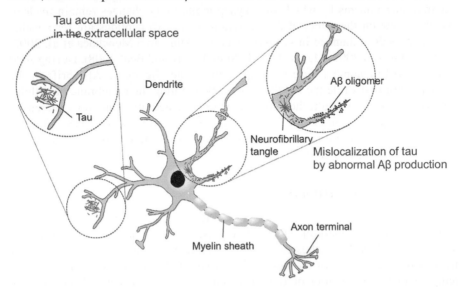

FIGURE 11.3 Illustration of the tau hypothesis. The two pathways associated with tau accumulation in neurodegenerative diseases are: (1) Abnormal production of Ab around neurons and (2) accumulation of tau proteins in the extracellular space.

to increased phosphorylation and aggregation of tau and impaired microtubule interactions (Zempel and Mandelkow 2014). The other is that the stability of tau in the extracellular space including cerebrospinal fluid (CSF) and brain interstitial fluid (ISF) increases by point mutation or hyperphosphorylation (Yamada et al. 2015). The interaction between stabilized tau and fibrous actin leads to cytoskeletal and synaptic impairment (Fulga et al. 2007, Cabrales Fontela et al. 2017, Zhou et al. 2017) and defects in mitochondrial integrity (DuBoff, Gotz, and Feany 2012). Recent experiments showed that APP rather than Aβ promotes tau accumulation and propagation in the brain regardless of fAD mutations (Takahashi et al. 2015).

11.2.3 NEUROINFLAMMATION

When he first described the AD, Dr. Alzheimer observed that the presence of glial cells was highly associated with brain inflammation involved in Aβ plaques and NFTs. Neuroinflammation has since been regarded as a secondary response in AD pathophysiology, which is passively activated by increasing senile plaques and NFTs mostly in late-to-end stages of AD. However, since the late 1980s, growing evidence has indicated that the neuroinflammation is an essential cause of AD pathogenesis in both early and late onset of the disease (Heneka et al. 2015, Wyss-Coray and Rogers 2012). The causal role of neuroinflammation in AD is supported by the increased levels of inflammatory mediators in the brain and CSF of AD patients (Akiyama et al. 2000). Moreover, recent studies also showed that neuroinflammation initiated by Aβ deposition in the early stage of AD precedes NFT formation (Kitazawa et al. 2005, Lee and Landreth 2010) and that early immunosuppression may reduce the tau pathology (Yoshiyama et al. 2007). The combined results suggest that innate immunity in the brain driven by neuroinflammation may represent a link between Aβ deposition and tau pathology in the AD.

In the AD brain, the continuous increase in Aβ deposition and aggregation induces a chronic reaction of the brain immune system which may lead to severe damage to the brain. These inflammatory responses in the brain are dominated by microglia, primary resident immune cells in the brain (Figure 11.4; Hanisch and Kettenmann 2007). Microglia change their morphological and functional phenotypes to reactive states after detecting Aβ through their pattern recognition receptors including Toll-like receptor 2 (TLR2), TLR4, and TLR6 and their co-receptors such as CD36, CD14, and CD47 (Weggen et al. 2001). These reactive microglia release an excessive amount of pro-inflammatory cytokines, interleukin-1β (IL-1β) family including IL-1β and IL-18, and TNF-α. These pro-inflammatory cytokines consequently induce neurotoxic astrocytes (Liddelow et al. 2017) and cause impaired neuronal function and structural changes, ultimately leading to neuronal death (Heppner, Ransohoff, and Becher 2015, Heneka, Kummer, and Latz 2014, Block, Zecca, and Hong 2007). In addition, microglia, phagocytic immune defenders in the brain, are responsible for removing cellular debris and aggregated proteins. Thus, their impaired phagocytic activity directly contributes to AD development by disrupting the clearance of Aβ or elimination of stressed-but-functional neurons (Brown and Vilalta 2015). In addition, abnormal synaptic pruning in the postnatal brain caused by impaired microglial activity leads to disease progression with synaptic loss (Paolicelli et al. 2011, Schafer et al. 2012).

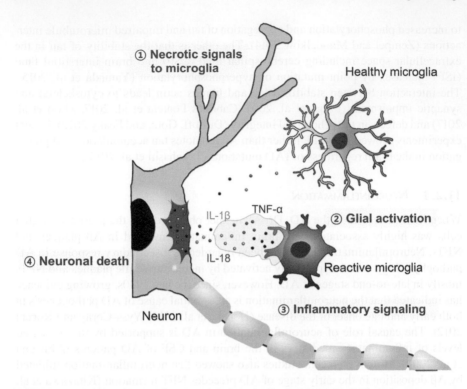

FIGURE 11.4 Microglia-mediated neuroinflammation in AD brain. Proinflammatory cytokines secreted by reactive microglia induce inflammatory signaling in neurons, ultimately leading to neuronal deaths.

11.2.4 GENETIC ETIOLOGY

Unlike the fAD, the sporadic, late-onset AD (sAD), which mostly occurs to the patient over age 60, is closely related to the apolipoprotein E, ε4 (apoE4) allele (Huang and Mucke 2012, Corder et al. 1993). Apolipoprotein E (apoE) is one of the major apolipoproteins that are produced by astrocytes and contributes to the brain lipid homeostasis (Ahn et al. 2018). During development or after neuronal damage occurs in the CNS, apoE is transmitted to neurons via apoE receptors (Holtzman, Herz, and Bu 2012). Among the apoEs, apoE4 is known to cause severe damage to neurons by disrupting Aβ clearance pathways (Robert et al. 2017, Zlokovic 2013) and by modulating Aβ-induced effects on inflammatory signaling (Chan et al. 2015, Tai et al. 2015). A medical statistics research revealed that the age of late-onset AD decreases and the rate of AD onset increases along with an increase in the number of apoE4 genes (Maestre et al. 1995). In addition, a recent finding using human apoE knock-in and knockout in tau transgenic mouse models revealed that apoE4 affects neurodegeneration independent of Aβ in the context of tauopathy (Shi et al. 2017).

In addition, recent findings indicate that one of the receptor proteins expressed on the membrane of microglia, triggering receptor expressed on myeloid cells 2 (TREM2), plays a critical role in AD pathology related to apoE (Figure 11.5).

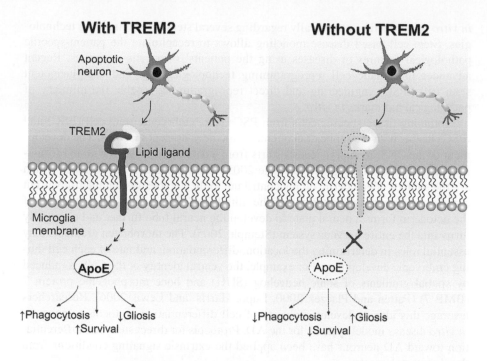

FIGURE 11.5 TREM2-apoE signaling pathway for microglial activity in AD brain. Microglia express TREM2 that binds to lipoproteins expressed in apoptotic cells for phagocytosis. TREM2 deficiency reduces microglial response to the pathological condition.

In response to brain damage, microglia recognize lipoproteins expressed in apoptotic cells via TREM2 for phagocytosis (Krasemann et al. 2017). On the contrary, recent experiments showed that TREM2 deficiency with the AD tauopathy not only decreases gliosis and neuroinflammation but also protects the brain from atrophy, indicating that TREM2 promotes a microglial role in response to tau pathology (Bemiller et al. 2017, Leyns et al. 2017). These results are consistent with the findings of markedly reduced inflammation and neurodegeneration in mice without apoE as described above. Thus, the TREM2-apoE pathway is crucial to promote microglial response to brain injury (Krasemann et al. 2017).

11.3 *IN VITRO* AD MODELING USING HUMAN STEM CELLS

Cell culture systems are an essential tool for basic research and extensive clinical studies. However, conventional two-dimensional (2D) cell cultures cannot mimic the accurate physiological conditions of living organisms, a translational gap that undermines the reliability and importance of data obtained from *in vitro* modeling studies. Three-dimensional (3D) cell culture systems using extracellular matrix-mimetic hydrogels provide more physiological environments. Recent approaches using human organs-on-chips offer more pathophysiologically relevant and predictive *in vitro* human disease models. This section discusses the current status of *in vitro* AD modeling, specifically regarding several stem cell engineering technologies. Stem cell-based disease modeling allows to recapitulate the patient-specific pathological features of diseases using the patients' own cells or tissues. Recent advances in somatic cell reprogramming technologies [e.g., induced-pluripotent stem cell (iPSC) engineering and direct reprogramming] enable the mimicry of pathological hallmarks *in vitro*.

Obtaining tissue-specific cells from PSCs [e.g., embryonic stem cell (ESC) and iPSC] begins with the replication of embryonic development in a dish. The development of the CNS in embryogenesis starts from a trilaminar embryo with the formation of notochord and somites (Stemple 2005). Both notochord and somites, which are the sources of ectoderm, do not contribute to the nervous system themselves but are associated with the patterning of the initial CNS formation. The central part of the ectoderm forms a neural plate to develop the neural tube further and eventually turns into the entire nervous system (Stemple 2005). The morphogen gradients play essential roles in determining the location, differentiation, and fate of each cell during embryonic development. For example, the ventral identity of the CNS is induced by spatial gradients of sonic hedgehog (SHH) and bone morphogenic protein 7 (BMP 7) (Patten and Placzek 2000, Lupo, Harris, and Lewis 2006). Researchers leverage this idea to develop a variety of cell differentiation protocols to generate *in vitro* disease model systems for the AD. Protocols for direct stem cell differentiation toward AD neurons have been applied the extrinsic signaling conditions from developmental events.

11.3.1 GENERATION OF HUMAN STEM CELL-DERIVED NEURONS FOR AD MODELING

The success of stem cell-based disease modeling depends on the efficient differentiation of stem cells into the desired cell type with high purity. Since the neocortex and hippocampus are the primary tissues affected by the AD, stem cells should be differentiated into the generation of pyramidal-shaped glutamatergic neurons for their use in AD modeling (Mann 1996). Conventionally, *in vitro* stem cell neurogenesis consists of three steps: neural induction, lineage specification, and terminal differentiation. Dual SMAD inhibition is the most widely accepted methods to differentiate human PSCs into early neuroectoderm (Chambers et al. 2009, Morizane et al. 2011, Arber, Lovejoy, and Wray 2017). Since SMADs are utilized as transducers for the BMP signaling and transforming growth factor β (TGF-β) signaling pathways, inhibition of SMAD promotes neural induction (Chambers et al. 2009).

In consequence, octamer-binding transcription factor 4 (Oct 4, a hallmark of pluripotency) disappears whereas paired box protein 6 (Pax 6, a marker of an early neuroepithelium) begins to be expressed to produce either neural rosettes or neural stem cells (NSCs) (Gaspard et al. 2008). Neural rosette is a multicellular structure which is a common intermediate observed during organogenesis as well as *in vitro* differentiation. By changing the extrinsic signaling conditions, neural rosettes and NSCs are specified to cortical-specific neural progenitor cells (NPCs). The combination of the SHH signal inhibition and the Wnt signal acceleration leads dorsalization of early neuroepithelium with high efficiency (Li et al. 2009, van de Leemput et al. 2014). This signaling regulation could be recapitulated in stem cell culture by using cytokines and antagonist treatment. Finally, these lineage-specified neural progenitors subsequently undergo long-term differentiation into mature neurons, which is known as corticogenesis. The cerebral cortex in human consists of six layers with distinctive features and marker expression to each other. During corticogenesis, the innermost layer is formed first, and the outer layers are formed in a time-dependent manner by cell migration. *In vitro* neurogenesis recapitulates these processes by differentiating stem cells, and the neurons with the same characteristics in the uppermost layer are produced later, requiring a differentiation period of more than 3 months to produce electrically functional neurons (Shi, Kirwan, Smith, Robinson, et al. 2012). Recently, several strategies to accelerate neural maturation like the corticogenesis process have been developed (Zhang et al. 2013, Qi et al. 2017). Surprisingly, treatment of antagonist cocktails targeting Wnt, ERK, Notch, and FGF signaling pathways generates cortical neurons within 13 days (Qi et al. 2017). Overexpression of Neurogenin 2 (Ngn 2) in human stem cells also accelerates neural maturation (Zhang et al. 2013).

11.3.2 3D BRAIN ORGANOID: ADVANCED *IN VITRO* MODEL SYSTEM FOR AD

Although neurons were cultured two-dimensionally (2D) in many disease models due to the ease of culture and observation, these 2D culture models cannot accurately mimic the neuronal function in human brain because of the mechano-physiological differences between 2D and three-dimensional (3D) microenvironments (Bosi et al. 2015). In addition, 3D models also allow for a better co-culture space of multiple neuronal cells, providing a more physiologically relevant environment. One way to create a 3D disease model is to encapsulate neurons into scaffolding materials. While rigid scaffolds such as hydrogels incorporated with carbon nanotubes (Shin et al. 2017) support and guide the neurite outgrowth along the scaffold surface (Loh and Choong 2013), soft scaffolds mimicking a brain extracellular matrix facilitate the processing of neurites into the soft gel (Uemura et al. 2010). In soft materials such as Matrigel or tunable porous polydimethylsiloxane (PDMS) scaffolds, neurons express longer neurites and pass through the scaffolds. In addition, primary neurons or the neurons differentiated from PSCs cultured in 3D soft matrix show round cell bodies with the neurite processes in all directions, closely mimicking the complex neural architecture *in vivo* (Loh and Choong 2013, Slonska and Cymerys 2017).

The use of various scaffolding systems mentioned above shows the importance of 3D tissue-like environment to cells; however, these are still far from the real brain tissue due to the neuron monoculture. Most *in vitro* AD models have focused on either neuron or glia, not on their interaction, although the interaction between neurons and glia is critical in the AD progression including neuroinflammation. This is due to the plasticity and instability of immature neurons derived from stem cells. When stem cell-derived neurons are exposed to serum, they stop differentiation process and are transformed into glial cells, making it difficult to conduct specific pathological studies for AD. Recent advances in the development of organoids using PSCs have shown more physiologically relevant AD models (Edmondson, Adcock, and Yang 2016). Neural differentiation in 3D follows the same three-step process as 2D cultured neurons: neural induction, lineage specification, and terminal differentiation, with enhancement of neural fate-committed cells by using SMAD inhibitors (Morizane et al. 2011, Watanabe et al. 2005, Smith et al. 2008). For the organoid development, PSCs must first form the small 3D structure called the embryoid bodies (EBs). The SMAD inhibition in EBs helps cells to be apart from the pluripotent state and enhance the selection of neural states (Watanabe et al. 2005, Smith et al. 2008). In addition, either EB or organoid culture enables intrinsic spatiotemporal signaling between cells, and these intercellular communications cause migration, polarization of the neuronal cortex, and the production of a series of neuroglial subtypes to form self-organized heterogenous nerve tissues. These self-organizing phenotypes give more *in vivo* relevance to the 3D structures using scaffolding materials such as Matrigel.

Development of human organoids is the most advanced technology in the stem cell research field. In brain organoid models, the radial glia and neurons construct the most similar structure to the developing human cortex (Qian et al. 2016). Glia develop with early neurons and play a crucial role in modeling the onset and progression of AD (Rodriguez-Arellano et al. 2016). Recent brain organoid studies focus on the cortex region. During spontaneous differentiation, endogenous morphogen patterning during organoid maturation generates hierarchical structures of cortical layers with functional neurons (Eiraku et al. 2008). Additional extrinsic signaling further patterns the other brain regions of interest (Smith et al. 2008, Muguruma et al. 2015, Jo et al. 2016). The brain organoids have been successfully maintained for up to 1 year (Lancaster and Knoblich 2014). During the culture period, brain organoids get bigger for the first 6 months; however, when the size of organoid exceeds 6–10 mm, cells in the center region undergo necrosis due to the limited nutrient supply and increased cell stress (Quadrato, Brown, and Arlotta 2016, Lancaster and Knoblich 2014). Suppressing the Notch signaling enhances terminal differentiation of organoids by inhibiting proliferation of the organoid (Nakano et al. 2012, Tieng et al. 2014). The use of spinner flasks also increases the survival rate of organoids with promoted nutrient and oxygen diffusion (Pasca et al. 2015).

However, the high heterogeneity of brain organoids, as well as the batch variability, remains a challenge. Recent bioreactor approaches have gained attention due to relatively decreased variability (Qian et al. 2016, Hartley and Brennand 2017). In addition, terminal differentiation of brain organoid needs over several months. Despite these limitations, brain organoid technology holds great promise to be a new model in AD research. It is the only existing model that recapitulates the cortical folding and crosstalk between neurons and glia on a dish (Edmondson, Adcock, and Yang 2016, Li et al. 2017).

11.3.3 Studying Normal Functions of AD-Related Proteins Using Stem Cell-Derived Neuron

Stem cell-derived neurons enable the study of both normal and pathological function of AD-associated proteins in human brain. For example, the role of APP and Aβ in healthy brain is yet to be fully understood. APP is programmed to express in neurons at the time of cell proliferation and differentiation; however, there is strong evidence that sAPP and Aβ are involved in the proliferation and differentiation of NSCs and

NPCs (Demars et al. 2011, Baratchi et al. 2012). Inhibition of α-secretase activity in NPCs reduced proliferation, which is recovered by the addition of sAPPα in association with ERK and MAPK signaling pathways (Demars et al. 2011). In addition, both sAPPα and sAPPβ treatment to brain subgranular zone-derived NPCs not only increases cell proliferation but also promotes glial differentiation over neuronal differentiation (Baratchi et al. 2012).

11.3.4 STUDYING AD MECHANISMS USING STEM CELL AD MODELS

Stem cell-derived neurons offer a new insight to AD research. Stem cells can supply human neurons limitlessly that represent the physiological characteristics in AD. The fact that these cells express similar phenotypes to human neurons in the brain offers great advantages to AD research. Many studies showed that neurons generated from AD patient-derived cells produce AD-associated proteins such as Aβ and NFTs, which could consequently recapitulate disease state *in vitro* (Israel et al. 2012, Shi, Kirwan, Smith, MacLean, et al. 2012, Kondo et al. 2013, Mahairaki et al. 2014, Hossini et al. 2015). In detail, neurons derived from both sAD and fAD patient-specifc iPSCs produce high levels of Aβ42 oligomers and lower levels of Aβ40 (Israel et al. 2012, Shi, Kirwan, Smith, MacLean, et al. 2012, Kondo et al. 2013, Mahairaki et al. 2014, Hossini et al. 2015). And the abnormal ratio of Aβ42/Aβ40 leads to its accumulation in the ER, endosomes, and synapses. The Aβ accumulation induces cellular stresses such as ER stress and oxidative stress due to reactive oxygen species (ROS) generation (Kondo et al. 2013) that involves in the impaired postsynaptic AMPA signaling (Nieweg et al. 2015). In addition, Aβs accumulate specifically to the glutamatergic neurons rather than GABAergic neurons (Vazin et al. 2014), which reinforce the relevance between *in vitro* stem cell models and *in vivo* AD pathologies.

Physiological relevance of stem cell AD models provides new insights into studying AD molecular mechanisms. Indeed, patient-derived neurons are widely used to investigate AD pathological events. Several researchers showed that Aβs, mostly Aβ42, are accumulated in the ER and early endosomes in AD patient-specific iPSC-derived neurons, leading to high levels of ER stress and oxidative stress and impairing the function of neurons and astrocytes (Kondo et al. 2013). In sAD patient-derived neurons, the same cellular stress and cell death pathways were induced along with the activation of Wnt signaling pathway. Stem cell AD models could demonstrate the effects of risk-associated mutations. In addition to the mutations in PSEN and APP, apoE3/E4 mutation is one of the risk genotypes in both fAD and sAD patients. In PSC-derived neurons, apoE stimulates APP and subsequent Aβ production via ERK/MAPK signaling pathway (Huang et al. 2017). Among transductions of various apoE alleles to the PSC-derived neurons, apoE4 overexpression stimulates more APP subsequent Aβ production than other apoE alleles. These findings in stem cell AD models are fit to the mouse AD model, in which apoE disrupts mitochondrial respiration in neurons (Chen et al. 2011).

Recently, researchers attempt to reproduce AD conditions *in vitro* using microfluidic technology (Park et al. 2018). The microfluidic system allows the monitoring of the clusters of multiple brain cells at cellular and molecular levels. For example, human iPSC-derived neurons and astrocytes genetically engineered to reproduce

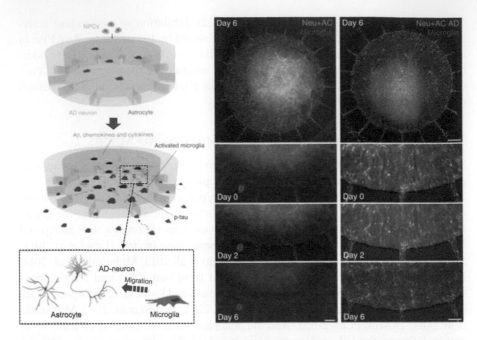

FIGURE 11.6 A 3D microfluidic human AD culture model with AD neurons, astrocytes differentiated from human NPCs, and human immortalized microglia. (Reproduced with permission from the RightsLink: Springer Nature, Nature Neuroscience [A 3D human triculture system modeling neurodegeneration and neuroinflammation in Alzheimer's disease, Park et al.), Copyright 2018.]

the AD pathology were implanted to the microfluidic platform (Figure 11.6). The compartmentalized structure facilitated the observation of microglial recruitment under inflammatory conditions in AD. To mimic the AD environment, human iPSC-derived NPCs were transduced by lentiviral particles containing the fAD mutation. The fAD-mutated human iPSCs cultured in the central chamber of the microfluidic platform secreted soluble Aβ42 and pro-inflammatory cytokines and induced microglial migration toward the central chamber. Moreover, the recruited microglia induced neuronal death through TLR4 and interferon-γ signaling cascades. These results show the potential of the combination of stem cell engineering and microfluidic technology as a tool for studying the molecular mechanism of the AD progression and pathogenesis *in vitro*.

11.4 CONCLUSIONS

As the life span increases with the advancement of medical science and technology, AD turns into the most disastrous version among neurodegenerative disease with could reach epidemic levels of prevalence in the near future. However, due to the brain's natural features with very complex and hierarchical system, there is no appropriate research model to properly recapitulate the AD pathology. In this point of view, stem cell-based AD model systems with various bioengineering approaches can offer alternatives to study the precise mechanism of early onset and progression of AD. *In vitro* AD models generated from patient-specific stem cells clearly provide many advantages over conventional models that could not properly mimic the physiology of human brain. By studying neurogenesis to understand precise and clear mechanisms of AD and using the stem cell-based *in vitro* model system for drug development, it would be possible to make the drug development process more efficient, which consequently prevents the translational failure of costly clinical trials.

REFERENCES

Ahn, S. I., H. J. Park, J. Yom, T. Kim, and Y. Kim. 2018. High-density lipoprotein mimetic nanotherapeutics for cardiovascular and neurodegenerative diseases. *Nano Research* 11 (10):5130–5143.

Akiyama, H., S. Barger, S. Barnum, B. Bradt, J. Bauer, G. M. Cole, N. R. Cooper, P. Eikelenboom, M. Emmerling, B. L. Fiebich, C. E. Finch, S. Frautschy, W. S. Griffin, H. Hampel, M. Hull, G. Landreth, L. Lue, R. Mrak, I. R. Mackenzie, P. L. McGeer, M. K. O'Banion, J. Pachter, G. Pasinetti, Plata-Salaman, J. Rogers, R. Rydel, Y. Shen, W. Streit, R. Strohmeyer, I. Tooyoma, F. L. Van Muiswinkel, R. Veerhuis, D. Walker, S. Webster, B. Wegrzyniak, G. Wenk, and T. Wyss-Coray. 2000. Inflammation and Alzheimer's disease. *Neurobiol Aging* 21 (3):383–421.

Arber, C., C. Lovejoy, and S. Wray. 2017. Stem cell models of Alzheimer's disease: Progress and challenges. *Alzheimers Res Ther* 9 (1):42.

Baratchi, S., J. Evans, W. P. Tate, W. C. Abraham, and B. Connor. 2012. Secreted amyloid precursor proteins promote proliferation and glial differentiation of adult hippocampal neural progenitor cells. *Hippocampus* 22 (7):1517–1527.

Bejanin, A., D. R. Schonhaut, R. La Joie, J. H. Kramer, S. L. Baker, N. Sosa, N. Ayakta, A. Cantwell, M. Janabi, M. Lauriola, J. P. O'Neil, M. L. Gorno-Tempini, Z. A. Miller, H. J. Rosen, B. L. Miller, W. J. Jagust, and G. D. Rabinovici. 2017. Tau pathology and neurodegeneration contribute to cognitive impairment in Alzheimer's disease. *Brain* 140 (12):3286–3300.

Bemiller, S. M., T. J. McCray, K. Allan, S. V. Formica, G. Xu, G. Wilson, O. N. Kokiko-Cochran, S. D. Crish, C. A. Lasagna-Reeves, R. M. Ransohoff, G. E. Landreth, and B. T. Lamb. 2017. TREM2 deficiency exacerbates tau pathology through dysregulated kinase signaling in a mouse model of tauopathy. *Mol Neurodegener* 12 (1):74.

Billings, L. M., S. Oddo, K. N. Green, J. L. McGaugh, and F. M. LaFerla. 2005. Intraneuronal Abeta causes the onset of early Alzheimer's disease-related cognitive deficits in transgenic mice. *Neuron* 45 (5):675–688.

Blennow, K., M. J. de Leon, and H. Zetterberg. 2006. Alzheimer's disease. *Lancet* 368 (9533):387–403.

Block, M. L., L. Zecca, and J. S. Hong. 2007. Microglia-mediated neurotoxicity: Uncovering the molecular mechanisms. *Nat Rev Neurosci* 8 (1):57–69.

Bosi, S., R. Rauti, J. Laishram, A. Turco, D. Lonardoni, T. Nieus, M. Prato, D. Scaini, and L. Ballerini. 2015. From 2D to 3D: Novel nanostructured scaffolds to investigate signalling in reconstructed neuronal networks. *Sci Rep* 5:9562.

Braak, H., and E. Braak. 1991. Neuropathological staging of Alzheimer-related changes. *Acta Neuropathol* 82 (4):239–259.

Braak, H., and K. Del Tredici. 2014. Are cases with tau pathology occurring in the absence of Abeta deposits part of the AD-related pathological process? *Acta Neuropathol* 128 (6):767–772.

Brown, G. C., and A. Vilalta. 2015. How microglia kill neurons. *Brain Res* 1628 (Pt B):288–297.

Buckley, R. F., B. Hanseeuw, A. P. Schultz, P. Vannini, S. L. Aghjayan, M. J. Properzi, J. D. Jackson, E. C. Mormino, D. M. Rentz, R. A. Sperling, K. A. Johnson, and R. E. Amariglio. 2017. Region-specific association of subjective cognitive decline with tauopathy independent of global beta-amyloid burden. *JAMA Neurol* 74 (12):1455–1463.

Cabrales Fontela, Y., H. Kadavath, J. Biernat, D. Riedel, E. Mandelkow, and M. Zweckstetter. 2017. Multivalent cross-linking of actin filaments and microtubules through the microtubule-associated protein Tau. *Nat Commun* 8 (1):1981.

Chambers, S. M., C. A. Fasano, E. P. Papapetrou, M. Tomishima, M. Sadelain, and L. Studer. 2009. Highly efficient neural conversion of human ES and iPS cells by dual inhibition of SMAD signaling. *Nat Biotechnol* 27 (3):275–280.

Chan, E. S., C. Chen, G. M. Cole, and B. S. Wong. 2015. Differential interaction of Apolipoprotein-E isoforms with insulin receptors modulates brain insulin signaling in mutant human amyloid precursor protein transgenic mice. *Sci Rep* 5:13842.

Chen, H. K., Z. S. Ji, S. E. Dodson, R. D. Miranda, C. I. Rosenblum, I. J. Reynolds, S. B. Freedman, K. H. Weisgraber, Y. Huang, and R. W. Mahley. 2011. Apolipoprotein E4 domain interaction mediates detrimental effects on mitochondria and is a potential therapeutic target for Alzheimer disease. *J Biol Chem* 286 (7):5215–5221.

Chow, T. W., C. Binder, S. Smyth, S. Cohen, and A. Robillard. 2008. 100 years after Alzheimer: Contemporary neurology practice assessment of referrals for dementia. *Am J Alzheimers Dis Other Demen* 23 (6):516–527.

Cirrito, J. R., P. C. May, M. A. O'Dell, J. W. Taylor, M. Parsadanian, J. W. Cramer, J. E. Audia, J. S. Nissen, K. R. Bales, S. M. Paul, R. B. DeMattos, and D. M. Holtzman. 2003. In vivo assessment of brain interstitial fluid with microdialysis reveals plaque-associated changes in amyloid-beta metabolism and half-life. *J Neurosci* 23 (26):8844–8853.

Cleary, J. P., D. M. Walsh, J. J. Hofmeister, G. M. Shankar, M. A. Kuskowski, D. J. Selkoe, and K. H. Ashe. 2005. Natural oligomers of the amyloid-beta protein specifically disrupt cognitive function. *Nat Neurosci* 8 (1):79–84.

Cohen, T. J., J. L. Guo, D. E. Hurtado, L. K. Kwong, I. P. Mills, J. Q. Trojanowski, and V. M. Lee. 2011. The acetylation of tau inhibits its function and promotes pathological tau aggregation. *Nat Commun* 2:252.

Corder, E. H., A. M. Saunders, W. J. Strittmatter, D. E. Schmechel, P. C. Gaskell, G. W. Small, A. D. Roses, J. L. Haines, and M. A. Pericak-Vance. 1993. Gene dose of apolipoprotein E type 4 allele and the risk of Alzheimer's disease in late onset families. *Science* 261 (5123):921–923.

De Strooper, B., P. Saftig, K. Craessaerts, H. Vanderstichele, G. Guhde, W. Annaert, K. Von Figura, and F. Van Leuven. 1998. Deficiency of presenilin-1 inhibits the normal cleavage of amyloid precursor protein. *Nature* 391 (6665):387–390.

Demars, M. P., A. Bartholomew, Z. Strakova, and O. Lazarov. 2011. Soluble amyloid precursor protein: A novel proliferation factor of adult progenitor cells of ectodermal and mesodermal origin. *Stem Cell Res Ther* 2 (4):36.

DuBoff, B., J. Gotz, and M. B. Feany. 2012. Tau promotes neurodegeneration via DRP1 mislocalization in vivo. *Neuron* 75 (4):618–632.

Eanes, E. D., and G. G. Glenner. 1968. X-ray diffraction studies on amyloid filaments. *J Histochem Cytochem* 16 (11):673–677.

Edmondson, R., A. F. Adcock, and L. Yang. 2016. Influence of matrices on 3D-cultured prostate cancer cells' drug response and expression of drug-action associated proteins. *PLoS One* 11 (6):e0158116.

Eiraku, M., K. Watanabe, M. Matsuo-Takasaki, M. Kawada, S. Yonemura, M. Matsumura, T. Wataya, A. Nishiyama, K. Muguruma, and Y. Sasai. 2008. Self-organized formation of polarized cortical tissues from ESCs and its active manipulation by extrinsic signals. *Cell Stem Cell* 3 (5):519–532.

Francis, R., G. McGrath, J. Zhang, D. A. Ruddy, M. Sym, J. Apfeld, M. Nicoll, M. Maxwell, B. Hai, M. C. Ellis, A. L. Parks, W. Xu, J. Li, M. Gurney, R. L. Myers, C. S. Himes, R. Hiebsch, C. Ruble, J. S. Nye, and D. Curtis. 2002. aph-1 and pen-2 are required for Notch pathway signaling, gamma-secretase cleavage of betaAPP, and presenilin protein accumulation. *Dev Cell* 3 (1):85–97.

Fulga, T. A., I. Elson-Schwab, V. Khurana, M. L. Steinhilb, T. L. Spires, B. T. Hyman, and M. B. Feany. 2007. Abnormal bundling and accumulation of F-actin mediates tau-induced neuronal degeneration in vivo. *Nat Cell Biol* 9 (2):139–148.

Garcia-Osta, A., and C. M. Alberini. 2009. Amyloid beta mediates memory formation. *Learn Mem* 16 (4):267–272.

Gaspard, N., T. Bouschet, R. Hourez, J. Dimidschstein, G. Naeije, J. van den Ameele, I. Espuny-Camacho, A. Herpoel, L. Passante, S. N. Schiffmann, A. Gaillard, and P. Vanderhaeghen. 2008. An intrinsic mechanism of corticogenesis from embryonic stem cells. *Nature* 455 (7211):351–357.

Gillardon, F., W. Rist, L. Kussmaul, J. Vogel, M. Berg, K. Danzer, N. Kraut, and B. Hengerer. 2007. Proteomic and functional alterations in brain mitochondria from Tg2576 mice occur before amyloid plaque deposition. *Proteomics* 7 (4):605–616.

Glabe, C. G., and R. Kayed. 2006. Common structure and toxic function of amyloid oligomers implies a common mechanism of pathogenesis. *Neurology* 66 (2 Suppl 1):S74–S78.

Glenner, G. G. 1983. Alzheimer's disease. The commonest form of amyloidosis. *Arch Pathol Lab Med* 107 (6):281–282.

Goedert, M., and M. G. Spillantini. 2006. A century of Alzheimer's disease. *Science* 314 (5800):777–781.

Goedert, M., and M. G. Spillantini. 2017. Propagation of Tau aggregates. *Mol Brain* 10 (1):18.

Hanisch, U. K., and H. Kettenmann. 2007. Microglia: Active sensor and versatile effector cells in the normal and pathologic brain. *Nat Neurosci* 10 (11):1387–1394.

Hartley, B. J., and K. J. Brennand. 2017. Neural organoids for disease phenotyping, drug screening and developmental biology studies. *Neurochem Int* 106:85–93.

Heneka, M. T., M. P. Kummer, and E. Latz. 2014. Innate immune activation in neurodegenerative disease. *Nat Rev Immunol* 14 (7):463–477. doi:10.1038/nri3705.

Heneka, M. T., M. J. Carson, J. E. Khoury, G. E. Landreth, F. Brosseron, D. L. Feinstein, A. H. Jacobs, T. Wyss-Coray, J. Vitorica, R. M. Ransohoff, K. Herrup, S. A. Frautschy, B. Finsen, G. C. Brown, A. Verkhratsky, K. Yamanaka, J. Koistinaho, E. Latz, A. Halle, G. C. Petzold, T. Town, D. Morgan, M. L. Shinohara, V. H. Perry, C. Holmes, N. G. Bazan, D. J. Brooks, S. Hunot, B. Joseph, N. Deigendesch, O. Garaschuk, E. Boddeke, C. A. Dinarello, J. C. Breitner, G. M. Cole, D. T. Golenbock, and M. P. Kummer. 2015. Neuroinflammation in Alzheimer's disease. *Lancet Neurol* 14 (4):388–405.

Heppner, F. L., R. M. Ransohoff, and B. Becher. 2015. Immune attack: The role of inflammation in Alzheimer disease. *Nat Rev Neurosci* 16 (6):358–372.

Holtzman, D. M., J. Herz, and G. Bu. 2012. Apolipoprotein E and apolipoprotein E receptors: Normal biology and roles in Alzheimer disease. *Cold Spring Harb Perspect Med* 2 (3):a006312.

Hossini, A. M., M. Megges, A. Prigione, B. Lichtner, M. R. Toliat, W. Wruck, F. Schroter, P. Nuernberg, H. Kroll, E. Makrantonaki, C. C. Zouboulis, and J. Adjaye. 2015. Induced pluripotent stem cell-derived neuronal cells from a sporadic Alzheimer's disease donor as a model for investigating AD-associated gene regulatory networks. *BMC Genomics* 16:84.

Huang, Y. A., B. Zhou, M. Wernig, and T. C. Sudhof. 2017. ApoE2, ApoE3, and ApoE4 differentially stimulate APP transcription and abeta secretion. *Cell* 168 (3):427–441 e21.

Huang, Y., and L. Mucke. 2012. Alzheimer mechanisms and therapeutic strategies. *Cell* 148 (6):1204–1222.

Igbavboa, U., G. Y. Sun, G. A. Weisman, Y. He, and W. G. Wood. 2009. Amyloid beta-protein stimulates trafficking of cholesterol and caveolin-1 from the plasma membrane to the Golgi complex in mouse primary astrocytes. *Neuroscience* 162 (2):328–338.

Iqbal, K., F. Liu, and C. X. Gong. 2016. Tau and neurodegenerative disease: The story so far. *Nat Rev Neurol* 12 (1):15–27. doi:10.1038/nrneurol.2015.225.

Israel, M. A., S. H. Yuan, C. Bardy, S. M. Reyna, Y. Mu, C. Herrera, M. P. Hefferan, S. Van Gorp, K. L. Nazor, F. S. Boscolo, C. T. Carson, L. C. Laurent, M. Marsala, F. H. Gage, A. M. Remes, E. H. Koo, and L. S. Goldstein. 2012. Probing sporadic and familial Alzheimer's disease using induced pluripotent stem cells. *Nature* 482 (7384):216–220.

Jo, J., Y. Xiao, A. X. Sun, E. Cukuroglu, H. D. Tran, J. Goke, Z. Y. Tan, T. Y. Saw, C. P. Tan, H. Lokman, Y. Lee, D. Kim, H. S. Ko, S. O. Kim, J. H. Park, N. J. Cho, T. M. Hyde, J. E. Kleinman, J. H. Shin, D. R. Weinberger, E. K. Tan, H. S. Je, and H. H. Ng. 2016. Midbrain-like organoids from human pluripotent stem cells contain functional dopaminergic and neuromelanin-producing neurons. *Cell Stem Cell* 19 (2):248–257.

Johnson, K. A., A. Schultz, R. A. Betensky, J. A. Becker, J. Sepulcre, D. Rentz, E. Mormino, J. Chhatwal, R. Amariglio, K. Papp, G. Marshall, M. Albers, S. Mauro, L. Pepin, J. Alverio, K. Judge, M. Philiossaint, T. Shoup, D. Yokell, B. Dickerson, T. Gomez-Isla, B. Hyman, N. Vasdev, and R. Sperling. 2016. Tau positron emission tomographic imaging in aging and early Alzheimer disease. *Ann Neurol* 79 (1):110–119.

Kametani, F., and M. Hasegawa. 2018. Reconsideration of amyloid hypothesis and tau hypothesis in Alzheimer's disease. *Front Neurosci* 12:25.

Kang, J., H. G. Lemaire, A. Unterbeck, J. M. Salbaum, C. L. Masters, K. H. Grzeschik, G. Multhaup, K. Beyreuther, and B. Muller-Hill. 1987. The precursor of Alzheimer's disease amyloid A4 protein resembles a cell-surface receptor. *Nature* 325 (6106):733–736.

Kidd, M. 1964. Alzheimer's disease—An electron microscopical study. *Brain* 87:307–320.

Kitazawa, M., S. Oddo, T. R. Yamasaki, K. N. Green, and F. M. LaFerla. 2005. Lipopolysaccharide-induced inflammation exacerbates tau pathology by a cyclin-dependent kinase 5-mediated pathway in a transgenic model of Alzheimer's disease. *J Neurosci* 25 (39):8843–8853.

Kondo, T., M. Asai, K. Tsukita, Y. Kutoku, Y. Ohsawa, Y. Sunada, K. Imamura, N. Egawa, N. Yahata, K. Okita, K. Takahashi, I. Asaka, T. Aoi, A. Watanabe, K. Watanabe, C. Kadoya, R. Nakano, D. Watanabe, K. Maruyama, O. Hori, S. Hibino, T. Choshi, T. Nakahata, H. Hioki, T. Kaneko, M. Naitoh, K. Yoshikawa, S. Yamawaki, S. Suzuki, R. Hata, S. Ueno, T. Seki, K. Kobayashi, T. Toda, K. Murakami, K. Irie, W. L. Klein, H. Mori, T. Asada, R. Takahashi, N. Iwata, S. Yamanaka, and H. Inoue. 2013. Modeling Alzheimer's disease with iPSCs reveals stress phenotypes associated with intracellular Abeta and differential drug responsiveness. *Cell Stem Cell* 12 (4):487–496.

Krasemann, S., C. Madore, R. Cialic, C. Baufeld, N. Calcagno, R. El Fatimy, L. Beckers, E. O'Loughlin, Y. Xu, Z. Fanek, D. J. Greco, S. T. Smith, G. Tweet, Z. Humulock, T. Zrzavy, P. Conde-Sanroman, M. Gacias, Z. Weng, H. Chen, E. Tjon, F. Mazaheri, K. Hartmann, A. Madi, J. D. Ulrich, M. Glatzel, A. Worthmann, J. Heeren, B. Budnik, C. Lemere, T. Ikezu, F. L. Heppner, V. Litvak, D. M. Holtzman, H. Lassmann, H. L. Weiner, J. Ochando, C. Haass, and O. Butovsky. 2017. The TREM2-APOE pathway drives the transcriptional phenotype of dysfunctional microglia in neurodegenerative diseases. *Immunity* 47 (3):566–581 e9.

Kumar, D. K., S. H. Choi, K. J. Washicosky, W. A. Eimer, S. Tucker, J. Ghofrani, A. Lefkowitz, G. McColl, L. E. Goldstein, R. E. Tanzi, and R. D. Moir. 2016. Amyloid-beta peptide protects against microbial infection in mouse and worm models of Alzheimer's disease. *Sci Transl Med* 8 (340):340ra72.

LaFerla, F. M., K. N. Green, and S. Oddo. 2007. Intracellular amyloid-beta in Alzheimer's disease. *Nat Rev Neurosci* 8 (7):499–509.

Lammich, S., E. Kojro, R. Postina, S. Gilbert, R. Pfeiffer, M. Jasionowski, C. Haass, and F. Fahrenholz. 1999. Constitutive and regulated alpha-secretase cleavage of Alzheimer's amyloid precursor protein by a disintegrin metalloprotease. *Proc Natl Acad Sci U S A* 96 (7):3922–3927.

Lancaster, M. A., and J. A. Knoblich. 2014. Generation of cerebral organoids from human pluripotent stem cells. *Nat Protoc* 9 (10):2329–2340. doi:10.1038/nprot.2014.158.

Lee, C. Y., and G. E. Landreth. 2010. The role of microglia in amyloid clearance from the AD brain. *J Neural Transm (Vienna)* 117 (8):949–960.

Leyns, C. E. G., J. D. Ulrich, M. B. Finn, F. R. Stewart, L. J. Koscal, J. Remolina Serrano, G. O. Robinson, E. Anderson, M. Colonna, and D. M. Holtzman. 2017. TREM2 deficiency attenuates neuroinflammation and protects against neurodegeneration in a mouse model of tauopathy. *Proc Natl Acad Sci U S A* 114 (43):11524–11529.

Li, X. J., X. Zhang, M. A. Johnson, Z. B. Wang, T. Lavaute, and S. C. Zhang. 2009. Coordination of sonic hedgehog and Wnt signaling determines ventral and dorsal telencephalic neuron types from human embryonic stem cells. *Development* 136 (23):4055–4063.

Li, Y., J. Muffat, A. Omer, I. Bosch, M. A. Lancaster, M. Sur, L. Gehrke, J. A. Knoblich, and R. Jaenisch. 2017. Induction of expansion and folding in human cerebral organoids. *Cell Stem Cell* 20 (3):385–396 e3.

Liddelow, S. A., K. A. Guttenplan, L. E. Clarke, F. C. Bennett, C. J. Bohlen, L. Schirmer, M. L. Bennett, A. E. Munch, W. S. Chung, T. C. Peterson, D. K. Wilton, A. Frouin, B. A. Napier, N. Panicker, M. Kumar, M. S. Buckwalter, D. H. Rowitch, V. L. Dawson, T. M. Dawson, B. Stevens, and B. A. Barres. 2017. Neurotoxic reactive astrocytes are induced by activated microglia. *Nature* 541 (7638):481–487.

Loh, Q. L., and C. Choong. 2013. Three-dimensional scaffolds for tissue engineering applications: Role of porosity and pore size. *Tissue Eng Part B Rev* 19 (6):485–502.

Lupo, G., W. A. Harris, and K. E. Lewis. 2006. Mechanisms of ventral patterning in the vertebrate nervous system. *Nat Rev Neurosci* 7 (2):103–114. doi:10.1038/nrn1843.

Maestre, G., R. Ottman, Y. Stern, B. Gurland, M. Chun, M. X. Tang, M. Shelanski, B. Tycko, and R. Mayeux. 1995. Apolipoprotein E and Alzheimer's disease: Ethnic variation in genotypic risks. *Ann Neurol* 37 (2):254–259.

Mahairaki, V., J. Ryu, A. Peters, Q. Chang, T. Li, T. S. Park, P. W. Burridge, C. C. Talbot, Jr., L. Asnaghi, L. J. Martin, E. T. Zambidis, and V. E. Koliatsos. 2014. Induced pluripotent stem cells from familial Alzheimer's disease patients differentiate into mature neurons with amyloidogenic properties. *Stem Cells Dev* 23 (24):2996–3010.

Manczak, M., T. S. Anekonda, E. Henson, B. S. Park, J. Quinn, and P. H. Reddy. 2006. Mitochondria are a direct site of A beta accumulation in Alzheimer's disease neurons: Implications for free radical generation and oxidative damage in disease progression. *Hum Mol Genet* 15 (9):1437–1449.

Mann, D. M. 1996. Pyramidal nerve cell loss in Alzheimer's disease. *Neurodegeneration* 5 (4):423–427.

Martin, L., X. Latypova, and F. Terro. 2011. Post-translational modifications of tau protein: Implications for Alzheimer's disease. *Neurochem Int* 58 (4):458–471.

Masters, C. L., G. Simms, N. A. Weinman, G. Multhaup, B. L. McDonald, and K. Beyreuther. 1985. Amyloid plaque core protein in Alzheimer disease and Down syndrome. *Proc Natl Acad Sci U S A* 82 (12):4245–4249.

Maurer, K., I. McKeith, J. Cummings, D. Ames, and A. Burns. 2006. Has the management of Alzheimer's disease changed over the past 100 years? *Lancet* 368 (9547):1619–1621.

Morizane, A., D. Doi, T. Kikuchi, K. Nishimura, and J. Takahashi. 2011. Small-molecule inhibitors of bone morphogenic protein and activin/nodal signals promote highly efficient neural induction from human pluripotent stem cells. *J Neurosci Res* 89 (2):117–126.

Muguruma, K., A. Nishiyama, H. Kawakami, K. Hashimoto, and Y. Sasai. 2015. Self-organization of polarized cerebellar tissue in 3D culture of human pluripotent stem cells. *Cell Rep* 10 (4):537–550.

Mungarro-Menchaca, X., P. Ferrera, J. Moran, and C. Arias. 2002. beta-Amyloid peptide induces ultrastructural changes in synaptosomes and potentiates mitochondrial dysfunction in the presence of ryanodine. *J Neurosci Res* 68 (1):89–96.

Nakano, T., S. Ando, N. Takata, M. Kawada, K. Muguruma, K. Sekiguchi, K. Saito, S. Yonemura, M. Eiraku, and Y. Sasai. 2012. Self-formation of optic cups and storable stratified neural retina from human ESCs. *Cell Stem Cell* 10 (6):771–785.

Nieweg, K., A. Andreyeva, B. van Stegen, G. Tanriover, and K. Gottmann. 2015. Alzheimer's disease-related amyloid-beta induces synaptotoxicity in human iPS cell-derived neurons. *Cell Death Dis* 6:e1709.

Okamura, N., and K. Yanai. 2017. Brain imaging: Applications of tau PET imaging. *Nat Rev Neurol* 13 (4):197–198.

Paolicelli, R. C., G. Bolasco, F. Pagani, L. Maggi, M. Scianni, P. Panzanelli, M. Giustetto, T. A. Ferreira, E. Guiducci, L. Dumas, D. Ragozzino, and C. T. Gross. 2011. Synaptic pruning by microglia is necessary for normal brain development. *Science* 333 (6048):1456–1458.

Park, J., I. Wetzel, I. Marriott, D. Dreau, C. D'Avanzo, D. Y. Kim, R. E. Tanzi, and H. Cho. 2018. A 3D human triculture system modeling neurodegeneration and neuroinflammation in Alzheimer's disease. *Nat Neurosci* 21 (7):941–951.

Pasca, A. M., S. A. Sloan, L. E. Clarke, Y. Tian, C. D. Makinson, N. Huber, C. H. Kim, J. Y. Park, N. A. O'Rourke, K. D. Nguyen, S. J. Smith, J. R. Huguenard, D. H. Geschwind, B. A. Barres, and S. P. Pasca. 2015. Functional cortical neurons and astrocytes from human pluripotent stem cells in 3D culture. *Nat Methods* 12 (7):671–678.

Patten, I., and M. Placzek. 2000. The role of Sonic hedgehog in neural tube patterning. *Cell Mol Life Sci* 57 (12):1695–1708.

Qi, Y., X. J. Zhang, N. Renier, Z. Wu, T. Atkin, Z. Sun, M. Z. Ozair, J. Tchieu, B. Zimmer, F. Fattahi, Y. Ganat, R. Azevedo, N. Zeltner, A. H. Brivanlou, M. Karayiorgou, J. Gogos, M. Tomishima, M. Tessier-Lavigne, S. H. Shi, and L. Studer. 2017. Combined small-molecule inhibition accelerates the derivation of functional cortical neurons from human pluripotent stem cells. *Nat Biotechnol* 35 (2):154–163.

Qian, X. Y., H. N. Nguyen, M. M. Song, C. Hadiono, S. C. Ogden, C. Hammack, B. Yao, G. R. Hamersky, F. Jacob, C. Zhong, K. J. Yoon, W. Jeang, L. Lin, Y. J. Li, J. Thakor, D. A. Berg, C. Zhang, E. Kang, M. Chickering, D. Nauen, C. Y. Ho, Z. X. Wen, K. M. Christian, P. Y. Shi, B. J. Maher, H. Wu, P. Jin, H. L. Tang, H. J. Song, and G. L. Ming. 2016. Brain-region-specific organoids using mini-bioreactors for modeling ZIKV exposure. *Cell* 165 (5):1238–1254.

Quadrato, G., J. Brown, and P. Arlotta. 2016. The promises and challenges of human brain organoids as models of neuropsychiatric disease. *Nat Med* 22 (11):1220–1228.

Reddy, P. H., and M. F. Beal. 2008. Amyloid beta, mitochondrial dysfunction and synaptic damage: Implications for cognitive decline in aging and Alzheimer's disease. *Trends Mol Med* 14 (2):45–53.

Robert, J., E. B. Button, B. Yuen, M. Gilmour, K. Kang, A. Bahrabadi, S. Stukas, W. Zhao, I. Kulic, and C. L. Wellington. 2017. Clearance of beta-amyloid is facilitated by apolipoprotein E and circulating high-density lipoproteins in bioengineered human vessels. *Elife* 6.

Rodriguez-Arellano, J. J., V. Parpura, R. Zorec, and A. Verkhratsky. 2016. Astrocytes in physiological aging and Alzheimer's disease. *Neuroscience* 323:170–182.

Schafer, D. P., E. K. Lehrman, A. G. Kautzman, R. Koyama, A. R. Mardinly, R. Yamasaki, R. M. Ransohoff, M. E. Greenberg, B. A. Barres, and B. Stevens. 2012. Microglia sculpt postnatal neural circuits in an activity and complement-dependent manner. *Neuron* 74 (4):691–705.

Scholl, M., S. N. Lockhart, D. R. Schonhaut, J. P. O'Neil, M. Janabi, R. Ossenkoppele, S. L. Baker, J. W. Vogel, J. Faria, H. D. Schwimmer, G. D. Rabinovici, and W. J. Jagust. 2016. PET imaging of Tau deposition in the aging human brain. *Neuron* 89 (5):971–982.

Schwarz, A. J., P. Yu, B. B. Miller, S. Shcherbinin, J. Dickson, M. Navitsky, A. D. Joshi, M. D. Devous, Sr., and M. S. Mintun. 2016. Regional profiles of the candidate tau PET ligand 18F-AV-1451 recapitulate key features of Braak histopathological stages. *Brain* 139 (Pt 5):1539–1550.

Senechal, Y., Y. Larmet, and K. K. Dev. 2006. Unraveling in vivo functions of amyloid precursor protein: Insights from knockout and knockdown studies. *Neurodegener Dis* 3 (3):134–147.

Serrano-Pozo, A., M. P. Frosch, E. Masliah, and B. T. Hyman. 2011. Neuropathological alterations in Alzheimer disease. *Cold Spring Harb Perspect Med* 1 (1):a006189.

Shi, Y., P. Kirwan, J. Smith, G. MacLean, S. H. Orkin, and F. J. Livesey. 2012. A human stem cell model of early Alzheimer's disease pathology in Down syndrome. *Sci Transl Med* 4 (124):124ra29.

Shi, Y., P. Kirwan, J. Smith, H. P. Robinson, and F. J. Livesey. 2012. Human cerebral cortex development from pluripotent stem cells to functional excitatory synapses. *Nat Neurosci* 15 (3):477–486, S1.

Shi, Y., K. Yamada, S. A. Liddelow, S. T. Smith, L. Zhao, W. Luo, R. M. Tsai, S. Spina, L. T. Grinberg, J. C. Rojas, G. Gallardo, K. Wang, J. Roh, G. Robinson, M. B. Finn, H. Jiang, P. M. Sullivan, C. Baufeld, M. W. Wood, C. Sutphen, L. McCue, C. Xiong, J. L. Del-Aguila, J. C. Morris, C. Cruchaga, Alzheimer's Disease Neuroimaging Initiative, A. M. Fagan, B. L. Miller, A. L. Boxer, W. W. Seeley, O. Butovsky, B. A. Barres, S. M. Paul, and D. M. Holtzman. 2017. ApoE4 markedly exacerbates tau-mediated neurodegeneration in a mouse model of tauopathy. *Nature* 549 (7673):523–527.

Shin, J., E. J. Choi, J. H. Cho, A. N. Cho, Y. Jin, K. Yang, C. Song, and S. W. Cho. 2017. Three-dimensional electroconductive hyaluronic acid hydrogels incorporated with carbon nanotubes and polypyrrole by catechol-mediated dispersion enhance neurogenesis of human neural stem cells. *Biomacromolecules* 18 (10):3060–3072.

Slonska, A., and J. Cymerys. 2017. Application of three-dimensional neuronal cell cultures in the studies of mechanisms of neurodegenerative diseases. *Postepy Hig Med Dosw (Online)* 71 (0):510–519.

Smith, J. R., L. Vallier, G. Lupo, M. Alexander, W. A. Harris, and R. A. Pedersen. 2008. Inhibition of Activin/Nodal signaling promotes specification of human embryonic stem cells into neuroectoderm. *Dev Biol* 313 (1):107–117.

Stemple, D. L. 2005. Structure and function of the notochord: An essential organ for chordate development. *Development* 132 (11):2503–2512.

Svedruzic, Z. M., K. Popovic, and V. Sendula-Jengic. 2015. Decrease in catalytic capacity of gamma-secretase can facilitate pathogenesis in sporadic and Familial Alzheimer's disease. *Mol Cell Neurosci* 67:55–65.

Tai, L. M., S. Ghura, K. P. Koster, V. Liakaite, M. Maienschein-Cline, P. Kanabar, N. Collins, M. Ben-Aissa, A. Z. Lei, N. Bahroos, S. J. Green, B. Hendrickson, L. J. Van Eldik, and M. J. LaDu. 2015. APOE-modulated Abeta-induced neuroinflammation in Alzheimer's disease: Current landscape, novel data, and future perspective. *J Neurochem* 133 (4):465–488.

Takahashi, M., H. Miyata, F. Kametani, T. Nonaka, H. Akiyama, S. Hisanaga, and M. Hasegawa. 2015. Extracellular association of APP and tau fibrils induces intracellular aggregate formation of tau. *Acta Neuropathol* 129 (6):895–907.

Takasugi, N., T. Tomita, I. Hayashi, M. Tsuruoka, M. Niimura, Y. Takahashi, G. Thinakaran, and T. Iwatsubo. 2003. The role of presenilin cofactors in the gamma-secretase complex. *Nature* 422 (6930):438–441.

Terry, R. D. 1963. The fine structure of neurofibrillary tangles in Alzheimer's disease. *J Neuropathol Exp Neurol* 22:629–642.

Tieng, V., L. Stoppini, S. Villy, M. Fathi, M. Dubois-Dauphin, and K. H. Krause. 2014. Engineering of midbrain organoids containing long-lived dopaminergic neurons. *Stem Cells Dev* 23 (13):1535–1347.

Townsend, M., G. M. Shankar, T. Mehta, D. M. Walsh, and D. J. Selkoe. 2006. Effects of secreted oligomers of amyloid beta-protein on hippocampal synaptic plasticity: A potent role for trimers. *J Physiol* 572 (Pt 2):477–492.

Uemura, M., M. M. Refaat, M. Shinoyama, H. Hayashi, N. Hashimoto, and J. Takahashi. 2010. Matrigel supports survival and neuronal differentiation of grafted embryonic stem cell-derived neural precursor cells. *J Neurosci Res* 88 (3):542–551.

van de Leemput, J., N. C. Boles, T. R. Kiehl, B. Corneo, P. Lederman, V. Menon, C. Lee, R. A. Martinez, B. P. Levi, C. L. Thompson, S. Yao, A. Kaykas, S. Temple, and C. A. Fasano. 2014. CORTECON: A temporal transcriptome analysis of in vitro human cerebral cortex development from human embryonic stem cells. *Neuron* 83 (1):51–68.

Vassar, R., B. D. Bennett, S. Babu-Khan, S. Kahn, E. A. Mendiaz, P. Denis, D. B. Teplow, S. Ross, P. Amarante, R. Loeloff, Y. Luo, S. Fisher, J. Fuller, S. Edenson, J. Lile, M. A. Jarosinski, A. L. Biere, E. Curran, T. Burgess, J. C. Louis, F. Collins, J. Treanor, G. Rogers, and M. Citron. 1999. Beta-secretase cleavage of Alzheimer's amyloid precursor protein by the transmembrane aspartic protease BACE. *Science* 286 (5440):735–741.

Vazin, T., K. A. Ball, H. Lu, H. Park, Y. Ataeijannati, T. Head-Gordon, M. M. Poo, and D. V. Schaffer. 2014. Efficient derivation of cortical glutamatergic neurons from human pluripotent stem cells: A model system to study neurotoxicity in Alzheimer's disease. *Neurobiol Dis* 62:62–72.

Watanabe, K., D. Kamiya, A. Nishiyama, T. Katayama, S. Nozaki, H. Kawasaki, Y. Watanabe, K. Mizuseki, and Y. Sasai. 2005. Directed differentiation of telencephalic precursors from embryonic stem cells. *Nat Neurosci* 8 (3):288–296.

Weggen, S., J. L. Eriksen, P. Das, S. A. Sagi, R. Wang, C. U. Pietrzik, K. A. Findlay, T. E. Smith, M. P. Murphy, T. Bulter, D. E. Kang, N. Marquez-Sterling, T. E. Golde, and E. H. Koo. 2001. A subset of NSAIDs lower amyloidogenic Abeta42 independently of cyclooxygenase activity. *Nature* 414 (6860):212–216.

Whitehouse, P. J., D. L. Price, R. G. Struble, A. W. Clark, J. T. Coyle, and M. R. Delon. 1982. Alzheimer's disease and senile dementia: Loss of neurons in the basal forebrain. *Science* 215 (4537):1237–1239.

Wyss-Coray, T., and J. Rogers. 2012. Inflammation in Alzheimer disease-a brief review of the basic science and clinical literature. *Cold Spring Harb Perspect Med* 2 (1):a006346.

Xu, T. H., Y. Yan, Y. Kang, Y. Jiang, K. Melcher, and H. E. Xu. 2016. Alzheimer's disease-associated mutations increase amyloid precursor protein resistance to gamma-secretase cleavage and the Abeta42/Abeta40 ratio. *Cell Discov* 2:16026.

Yamada, K., T. K. Patel, K. Hochgrafe, T. E. Mahan, H. Jiang, F. R. Stewart, E. M. Mandelkow, and D. M. Holtzman. 2015. Analysis of in vivo turnover of tau in a mouse model of tauopathy. *Mol Neurodegener* 10:55.

Yao, Z. X., and V. Papadopoulos. 2002. Function of beta-amyloid in cholesterol transport: A lead to neurotoxicity. *FASEB J* 16 (12):1677–1679.

Yoshiyama, Y., M. Higuchi, B. Zhang, S. M. Huang, N. Iwata, T. C. Saido, J. Maeda, T. Suhara, J. Q. Trojanowski, and V. M. Lee. 2007. Synapse loss and microglial activation precede tangles in a P301S tauopathy mouse model. *Neuron* 53 (3):337–351.

Yu, G., M. Nishimura, S. Arawaka, D. Levitan, L. Zhang, A. Tandon, Y. Q. Song, E. Rogaeva, F. Chen, T. Kawarai, A. Supala, L. Levesque, H. Yu, D. S. Yang, E. Holmes, P. Milman, Y. Liang, D. M. Zhang, D. H. Xu, C. Sato, E. Rogaev, M. Smith, C. Janus, Y. Zhang, R. Aebersold, L. S. Farrer, S. Sorbi, A. Bruni, P. Fraser, and P. St George-Hyslop. 2000. Nicastrin modulates presenilin-mediated notch/glp-1 signal transduction and betaAPP processing. *Nature* 407 (6800):48–54.

Zempel, H., and E. Mandelkow. 2014. Lost after translation: Missorting of Tau protein and consequences for Alzheimer disease. *Trends Neurosci* 37 (12):721–732.

Zhang, Y., C. Pak, Y. Han, H. Ahlenius, Z. Zhang, S. Chanda, S. Marro, C. Patzke, C. Acuna, J. Covy, W. Xu, N. Yang, T. Danko, L. Chen, M. Wernig, and T. C. Sudhof. 2013. Rapid single-step induction of functional neurons from human pluripotent stem cells. *Neuron* 78 (5):785–798.

Zhou, L., J. McInnes, K. Wierda, M. Holt, A. G. Herrmann, R. J. Jackson, Y. C. Wang, J. Swerts, J. Beyens, K. Miskiewicz, S. Vilain, I. Dewachter, D. Moechars, B. De Strooper, T. L. Spires-Jones, J. De Wit, and P. Verstreken. 2017. Tau association with synaptic vesicles causes presynaptic dysfunction. *Nat Commun* 8:15295.

Zlokovic, B. V. 2013. Cerebrovascular effects of apolipoprotein E: Implications for Alzheimer disease. *JAMA Neurol* 70 (4):440–444. doi:10.1001/jamaneurol.2013.2152.

Part IV

Towards Translational Application
and Precision Medicine

12 Manufacturing and Assembly of Micro- and Nanoscale Devices and Interfaces Using Silk Proteins

Zhitao Zhou and Tiger H. Tao
Chinese Academy of Sciences

CONTENTS

12.1 INTRODUCTION

12.1.1 BACKGROUND

Efficient process platforms and tools are highly desired for manufacturing and assembly of hierarchical and heterogeneous structures and systems via the two- and three-dimensional integration of components and systems across multiple length scales. Decent progresses have been made using directed self-assembly (Kim et al. 2012), nanoimprint lithography (Brenckle et al. 2013), template replication with Si wafer technologies, or high-rate roll-to-roll (R2R)-based production tools (Burghoorn et al. 2013) to yield materials and devices with many important applications including flexible hybrid electronics (Khan et al. 2016), biomimetic hierarchical (e.g., super-hydrophobic) surfaces (Yang et al. 2018), and engineered cell/tissue microenvironments (Custódio, Reis, and Mano 2014).

The fundamental challenge of fusing current IC-oriented nanofabrication technologies with biological systems is to alleviate the innate mismatch between biological (soft-wet) (Balabin, Beratan, and Skourtis 2008) and non-biological (hard-dry) (Perry and Kolb 2004) components. Numerous research efforts have invested in the precise placement of biological components and the controlled construction of functional nanostructures. However, since most biological molecules are fragile and only functional in aqueous environments, there are severe constraints on the integration of the molecules into the conventional fabrication technologies optimized for their inorganic counterparts. This includes both bottom-up and top-down approaches such as DNA origami via self-assembly and lithography-based nanopatterning techniques (Wang and Xia 2004), with most methods only yielding 2D and pseudo-3D structures with restricted geometries and functionalities (Figure 12.1). For example, none of the existing methods allow for the systematic tuning of important physicochemical cues of 3D scaffolds (e.g., topography, stiffness, and solvable factors) in a concurrent way with internal structures and ligand presentation spanning from nano- to microscale, akin to in vivo conditions. New strategies and innovations in both materials and manufacturing technologies have yet to be explored.

12.1.2 SILK PROTEINS

Silks have an abundant history as materials for biomanufacturing. Before the "micro-" and "nano-"manufacturing technology has been developed, silk proteins have been used mostly in their natural forms, i.e., fibers and tissues produced by animals, for millennia. In the last a few decades—given the advances in science and technology and the increasing desire for finding materials to interface with no harm to cells and to different tissues in the human bodies—silks have also found wide application in biomedical fields, including wound healing, drug delivery, tissue engineering, and regenerative medicine. Silk proteins possess the advantages of mechanical strength, non-toxicity, biodegradation, support of cells growth, and differentiation and biomimetism (Altman et al. 2003) that are difficult to find in synthetic materials and that enable the use of silks in many "biocompatible systems" (Williams 2014). Silk protein-based biomedical devices have been fabricated using

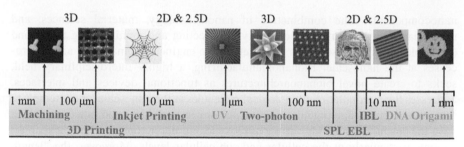

FIGURE 12.1 Depiction of main manufacturing processes with polymer-based biomaterials. Note: 2D, arbitrary patterns in x-y directions; 2.5D, greyscale patterns or repeating (or random) structures in the z direction; 3D, arbitrary geometries in both x-y and z directions. (From right to left: DNA origami: Reproduced with permission Rothemund (2006). Copyright 2006, Springer Nature. IBL: Reproduced with permission Jiang et al. (2010). Copyright 2008, Wiley-VCH. EBL: Reproduced with permission Qin et al. (2016). Copyright 2016, Nature Communications. SPL: Reproduced with permission Gan et al. (2016). Copyright 2016, The Royal Society of Chemistry. Two-photo: Reproduced with permission Spivey et al. (2013). Copyright 2013, Wiley-VCH. UV: Reproduced with permission Liu, Zhou, et al. (2017). Copyright 2017, Wiley-VCH. Inkjet printing: Reproduced with permission Tao et al. (2015). Copyright 2015, Wiley-VCH. 3D Printing: Reproduced with permission Schacht et al. (2015). Copyright 2015, Wiley-VCH. Machining: Reproduced with permission Li et al. (2016). Copyright 2016 Elsevier Ltd.)

materials both in their natural forms (e.g. silk fiber-made ligament grafts) or in reprocessed formats, including hydrogels, films, and nanofibers (Rockwood et al. 2011).

Among structural proteins, silk is well known for its outstanding mechanical properties (Dickerson et al. 2013), inspiring the design of high-performing textile materials. Silk can be produced by several types of arthropods, including silkworms and spiders. These arthropods have different spinning systems and produce silk in different ways for different purposes. While silkworms spin silk into cocoons—and in much less common circumstances, also into webs—to protect themselves from all kinds of threats and to regulate the environment to help conserving/blocking water and gases during the pupal stage, spiders—a polymath with expertise spanning architecture, materials science, engineering mechanics and even camouflage—create a silk web to capture prey to feed themselves. Spiders can produce and store silk proteins in glands as well, which allows them to spin fibers during their whole lifetime. Though nontrivial differences exist between the sequences of amino acids of silks produced by silkworms and spiders, their silk fiber formation mechanisms have much in common and they share similar structure–properties relationships as well as the transition behaviors between conformational structures (Jin and Kaplan 2003).

12.1.3 Bionanomanufacturing

Over the last few decades, manufacturing has undergone tremendous development, pushing the miniaturization of (mostly electronic) components to the nanometer length scale and enabling the controlled placement and integration of those nanocomponents. The combination of nanotechnology, material sciences, and molecular biology opens the possibility of detecting and manipulating atoms and (bio)molecules, and also offers new methods to engineer biomaterials and bioprocesses at the nanoscale. Bionanomanufacturing, a highly interdisciplinary field, seeks to create novel bionanoarchitectures as functional devices and interfaces, addresses the manipulation of biological molecules at the nanoscale, and attempts to integrate inorganic and organic components for new properties and functions, which have the potential for a wide variety of biological research topics and medical applications, particularly at the cellular and sub-cellular levels. Moreover, the "length scale gap" between 3D biostructures that can be readily fabricated by DNA origami via molecular assembly (relatively comfort zone: 1–100 nm) and multiphoton lithography using biomaterials (relatively comfort zone: >500 nm) can be fulfilled by the proposed Protein LEGO method with reduced process complexity and enhanced flexibility of functionalization. The proposed research and educational endeavors will establish a firm foundation for bridging the manufacturing of functional biomaterials with their applications and the education of students in the interdisciplinary nanoengineering field.

12.2 MANUFACTURING USING SILK PROTEINS: FROM NANOSCALE TO MICROSCALE

Numerous research efforts have been invested in the precise placement of biological components and the controlled construction of functional bionanostructures, which opens up significant opportunities with applications from biointerfaces (Liu and Wang 2014), biosensing (Kabashin et al. 2009), tissue engineering (Lawrence et al. 2009), to regenerative medicine (Rising 2014). Bionanomanufacturing, or nanomanufacturing in general, utilizes many of the fabrication techniques inherent to IC manufacturing and microelectromechanical (MEMS) fabrication including, but not limited to, lithography, etching, and deposition. Bionanomanufacturing can be largely classified into two main categories: parallel bionanomanufacturing and serial bionanomanufacturing. This includes both bottom-up and top-down approaches such as DNA origami via self-assembly (Douglas et al. 2009) and lithography-based nanopatterning techniques (e.g., e-beam lithography, EBL (Qin et al. 2016), nanoimprinting lithography, NIL (Ding et al. 2015), soft lithography (Dong, Younan, and George M 2010), and laser machining (Li et al. 2016)), where most methods yield only 2D and pseudo-3D structures with restricted geometries and functions (Figure 12.2).

12.2.1 ELECTRON BEAM LITHOGRAPHY

Due to the compatibility with various manufacturing techniques that are used in modern semiconductor industry (Tao, Kaplan, and Omenetto 2012), silk protein has shown great potential for being a functional material for a wide range of applications, which in turn promotes the engineering of silk protein on different length scales.

In the nanometer size domain, the water solubility of regenerated silk fibroin is determined by its secondary structure. The crystalline structure (i.e., β-sheet rich) is water insoluble, while the amorphous state is soluble. Interestingly, energy

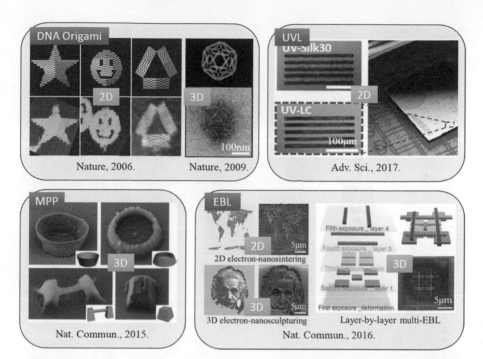

FIGURE 12.2 Some existing 2D and 3D bionanomanufacturing techniques. (Top images, from left to right: DNA origami (2D): Reproduced with permission Rothemund (2006). Copyright 2006, Springer Nature. DNA origami (3D): Reproduced with permission Douglas et al. (2009). Copyright 2009, Springer Nature. UVL: Reproduced with permission Liu, Zhou, et al. (2017). Copyright 2017, Wiley-VCH. Bottom images, from left to right: MPP: Reproduced with permission Sun et al. (2015). Copyright 2015, Springer Nature. EBL: Reproduced with permission Qin et al. (2016) Copyright 2016, Nature Communications.)

inputs such as electron irradiation can change and even adjust the secondary structures (i.e., the solubility) of silk protein (Qin et al. 2016, Kim et al. 2014), offering a pathway of using silk as a resist material for lithographic processes such as electron beam lithography (EBL). The silk protein undergoes cycles of conformational transitions with progressive degradation as the dosages of electron irradiation varies. Either the amorphous or crystalline silk can be crosslinked or de-crosslinked by the electron bombardment (Figure 12.3a). Due to the different electron penetration depth of silk protein in different conformations (i.e., crystalline or amorphous), the conformational transitions occur from top to bottom on the crystalline silk film (positive tone) while from bottom to top on the amorphous silk film (negative tone). Therefore, nanosculpturing and nanosintering can be achieved by using silk with different crystallinity (Figure 12.3b,c). Furthermore, the 3D nanostructure can be also fabricated by the layer-by-layer EBL writing process, which is realized by precisely controlling the dosage of electron irradiation (Figure 12.3d). This manufacturing paradigm is biofriendly since only water is involved in the storage and development step.

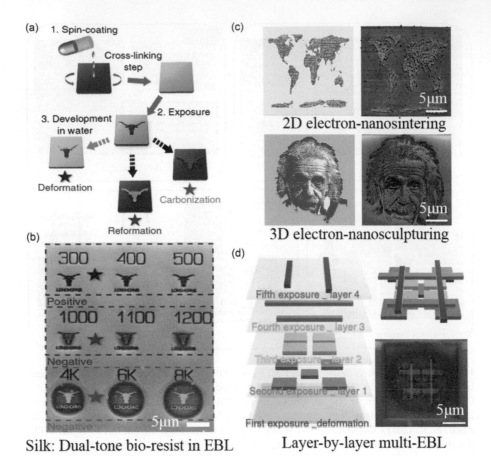

FIGURE 12.3 Nanostructuring of silk using EBL writing. (a) The silk fibroin conformation is modulated by the exposure to electron irradiation. (b) By controlling the electron irradiation dosage, both positive and negative EBL can be realized on the same silk substrate. (c) Additional regulation of the initial fibroin conformation, both nano-sintering and nanosculpturing can be achieved as well. (d) A 3D nanostructure is realized by the layer-by-layer EBL writing process. (Reproduced with permission Qin et al. (2016). Copyright 2016, Nature Communications.)

12.2.2 ION BEAM LITHOGRAPHY

Ion beam lithography (IBL) is another extensively used lithographic technique in semiconductor industry for defining nanoscale patterns. Compared with EBL, the ion beam implants its energy very efficiently in well-defined volumes with no backscattering when it is incident on the resist material, which enables exceptional pattern precision. In addition, the genetically engineered spider silk with precisely controlled protein sequences and molecular weights further improve the lithographic resolution and sharpness compared to the natural proteins.

IBL used to fabricate nanostructures on genetically engineered spider silk has recently been reported to create 2D biofunctional patterns (Figure 12.4; Jiang et al. 2018). This is realized because the manipulation of silk secondary structures can also be accomplished by ion irradiation. The IBL writing has two effects on the silk film. The ions can etch off molecules on the surface of the silk film, as also observed in other types of IBL-resist materials. This creates a positive tone patterns on the surface of the silk film. On the other hand, the ions that penetrate into the film carries energy that can crosslink the amorphous silk protein just underneath the etched areas. Once the ion irradiation dosage is high enough, the crosslinked structure anchors on the bottom of the sample and the patterns can stay on the substrate even after water development. Therefore, two complementary types of structures are formed at the same time during the fabrication process (Figure 12.4a). As a proof-of-concept demonstration, dot and line arrays have been fabricated on 30 nm thick spider silk film using IBL, with the minimum feature sizes of 13.2 and 13.7 nm, respectively (Figure 12.4b). Furthermore, spider silk doped with bioactive materials can also be fabricated into structures, and the ion beam has little influence on the bioactivity of the dopants. The fabricated mesh structures of spider silk doped with collagen or temozolomide (TMZ) can guide the proliferation of neural cells or inhibit the growth of cancer cells (Figure 12.4c). This new bio-nanofabrication paradigm provides a new alternative for engineering protein structures that could be constructed into functional nanodevices.

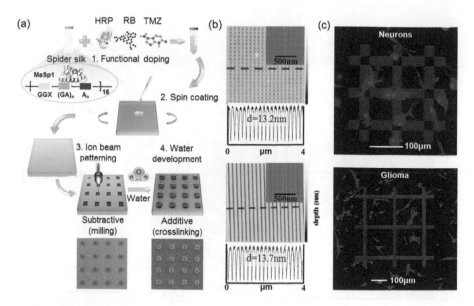

FIGURE 12.4 Bionanofabrication of recombinant spider silk using IBL. (a) Schematic illustration of the process of bionanomanufacturing of the genetically engineered spider silk using IBL. The spider silk can be further modified via doping. Both sculpturing and nanosintering can be achieved. (b) The fabricated dot and line arrays on spider silk film using IBL, whose minimum feature sizes are 13.2 and 13.7 nm, respectively. (c) The IBL-fabricated spider silk structures doped with collagen or TMZ can guide the proliferation of neural cells or inhibit the growth of cancer cells. (Reproduced with permission Jiang et al. (2018). Copyright 2018, Wiley-VCH.)

12.2.3 SELF-ASSEMBLY

Silk fibroin is kind of protein extracted from silkworm cocoon and has been successfully used in a variety of areas, such as optical sensing and implantable medical device. In fact, silk cocoon mainly contains two types of proteins: silk fibroin and sericin. Sericin also shows valuable bioactivities, such as cell adhesion, low immunogenicity, biocompatibility, and biodegradability, which makes it widely utilized in drug delivery, regenerative medicine and tissue engineering.

Recently, research demonstrates that the chemical and physical characteristics of sericin make it especially suitable for mesoporous silica nanoparticles (MSNs)-based drug delivery. Sericin is coated onto MSNs by self-assembly, serving as a functional shell to improve MSNs' capping efficiency and drug release. Taking the delivery of doxorubicin (DOX) for the efficient killing of tumor cells as an example, DOX is loaded by aldehyde-modified MSNs and then coated with sericin. The DOX-loaded sericin-coated MSN delivery system prevents the premature leakage of encapsulated DOX in extracellular environment and transports DOX to the designated location. In vitro experiments indicate sericin-coated MSNs can greatly promote the cellular internalization of the silica nanoparticles (Figure 12.5; Liu, Li, Zhang, et al. 2017). Thanks to the natural cell adhesive property, sericin has been explored in sericin-capped particles or even fabricated as self-assembled nanoparticles for cancer treatment.

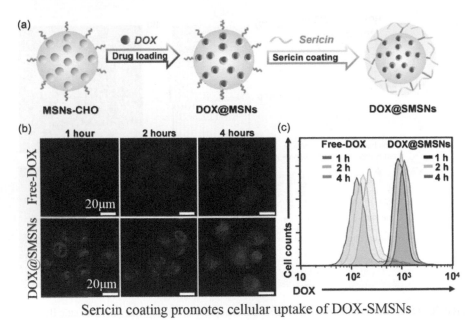

FIGURE 12.5 (a) Synthesis procedure of DOX-loaded sericin-coated MSNs (DOX@SMSNs). (b) Fluorescence images of MCF-7/ADR cells treated with free-DOX or DOX@SMSNs for 1, 2, and 4 h. (c) Flow cytometry analysis of MCF-7/ADR cells treated with free-DOX and DOX@SMSNs for 1, 2, and 4 h. (Reproduced with permission Liu, Li, Zhang, et al. (2017). Copyright 2016, Wiley-VCH.)

12.2.4 MULTIPHOTON LITHOGRAPHY

The abovementioned EBL or IBL can precisely pattern silk fibroin into desired micro/nanostructure in two and "pseudo" three dimensions (2D and pseudo-3D), which is accomplished through electron/ion-induced structural transitions of silk protein. However, electron/ion beam-based manufacturing techniques are fairly difficult for the implementation of complex 3D nano/microstructures. As a biocompatible material, silk fibroin is a promising candidate for biological scaffold, which is demand for an arbitrarily customized 3D fabrication method.

Femtosecond laser direct writing (FsLDW) is a laser-based precise 3D micro/nanofabrication method based on two-photon polymerization. It has been demonstrated as a noncontact, maskless, and fast technology for true 3D manufacturing (Lee, Moon, and West 2008). Aqueous multiphoton lithography (MPL) of various silk fibroin-based inks using FsLDW has been reported. In the experiments, silk fibroin mixing with proper methylene blue (as a photo-sensitizer) is FsLDW-crosslinked into arbitrary fine 3D micro/nanostructures with remarkable mechanical characteristics to maintain the complex structures. In addition, a silk/Ag composite microwire is also written out by FsLDW MPL and exhibits good electrical properties, which provides a simple method for the fabrication of multifunctional metal/biomaterial-based electronic micro/nanodevices used for electrical bioengineering (Figure 12.6; Sun et al. 2015). The aqueous FsLDW MPL paves the way for

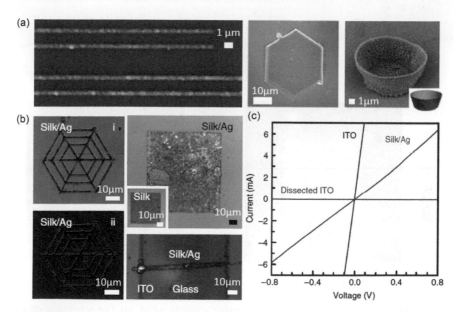

FIGURE 12.6 (a) Scanning electron microscopy (SEM) images of all-silk-based 1D, 2D, and 3D micro/nanostructures. (b) Confocal microscopic images of a silk/Ag composite "microcobweb" (left), metallographic optical microscopy image of a femtosecond laser direct writing (FsLDW)-fabricated silk/Ag composite microsquare (upper right) and microwire between two ITO electrodes (lower right). (c) Current–voltage curves of the as-fabricated silk/Ag composite microwire. (Reproduced with permission Sun et al. (2015). Copyright 2015, Springer Nature.)

a powerful development of silk-based 3D micro/nanoprocessing techniques, which will make silk ideal for a wider range of applications. It will open up many new opportunities for fields of silk-based micro/nano-level tissue engineering and biosensing in the future.

12.2.5 Scanning Probe Lithography

Scanning probe lithography (SPL) is a widely used type of lithographic technique for defining high precision patterns. The usage of SPL-based techniques on biomaterials such as dip-pen nanolithography (Salaita, Wang, and Mirkin 2007), nano-shaving (Shi, Chen, and Cremer 2008), and nano-grafting (Xu et al. 1999) has been studied extensively. This is because the SPL is usually performed at ambient environment and therefore can maximally preserve the bioactivity of the functional materials. In particular, silk micro-patterns have been created using an in situ silk solution-based atomic force microscopy (AFM) patterning technique (Figure 12.7;

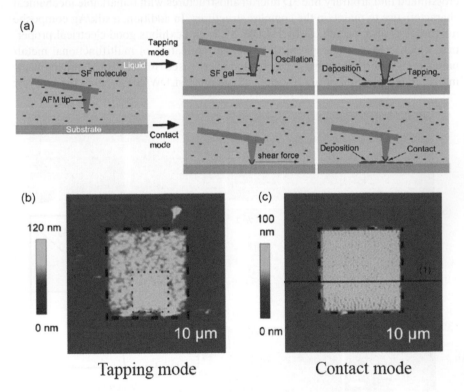

FIGURE 12.7 Tip-induced silk patterning using solution AFM. (a) The schematic of silk deposition onto the substrate by either the tapping mode or the contact mode of a solution AFM. In the tapping mode, silk micro-gel is formed due to the local oscillation while in the contact mode, the gelation is induced by the shear force between the AFM tip and the substrate. Experimental results confirm that both (b) tapping mode and (c) contact mode AFM can be used to create the microstructures. (Reproduced with permission Zhong et al. (2013). Copyright 2012, American Chemical Society.)

Zhong et al. 2013). The hydrophobic silk micro-gel can be deposited onto a hydrophilic substrate in both tapping and contact AFM mode. In the tapping mode, the local oscillation of the AFM tip induces sol–gel transition of silk protein, which is then deposited onto the substrate by the tip tapping. On the other hand, during the contact mode operation, the silk gelation is induced from the shear force between the scanning AFM tip and the substrate. In both cases, the heights of the silk micro-patterns deposited are dependent on the number of scans, which then provides an important guideline for fabricating grayscale patterns where the heights of the structure are simply controlled by the number of scans. The aqueous environment of the SPL process significantly improves the preservation of biocompatibility of the material itself as well as functional dopants that are mixed in the solution. However, as with most SPL pattering techniques, this method suffers from a very low yield since it involves a raster and repeated scanning on the patterned areas.

12.2.6 ULTRAVIOLET LITHOGRAPHY

Ultraviolet lithography (UVL) is one of the most commonly used techniques in semiconductor microfabrication to define the patterns on silicon wafer, which can rapidly fabricate high fidelity micro/nanopatterns in parallel. To make silk-based biomanufacturing be compatible with semiconductor microfabrication, silk-based UVL has been exploited. Patterning of silk microstructures using UVL has been successfully demonstrated where either silk fibroin is chemically modified to be photoreactive and then served as the photoresist.

The modification on the regenerated silk fibroin (SF) solution has better flexibility since the SF solution serves as the starting point for fabricating the SF protein into other material formats such as film, hydrogel, and foam (Rockwood et al. 2011). The amino acid side chains in SF protein can be conveniently conjugated with a variety of chemical groups in the aqueous environment. A considerable amount of efforts in this regime are devoted to the surface modification of silk fibers and silk nanostructures to alter cell adhesion and proliferation. In addition, chemical modification also facilitates the silk-based bio-nanomanufacturing. A photo-sensitized silk light chain protein (Figure 12.8) has been recently demonstrated, providing a new alternative to the high-resolution green photoresist (Liu, Zhou, et al. 2017). The silk light chain segment is extracted in the formic acid which breaks the disulfide bond between the silk light chain and silk heavy chain and dissolves the light chain segment (Wadbua et al. 2010). Then, photo-sensitization is achieved by conjugating photo-reactive agent 2-isocyanatoethyl methacrylate (IEM) onto the hydroxyl group of the silk light chain in a water-free environment. The product can then be used as a photoresist after mixing with a photoinitiator (Figure 12.8a). The chemical bonding between the IEM and the silk light chain is verified using Fourier transform infrared spectroscopy (FTIR) (Figure 12.8b). The silk light chain can overcome the intrinsic limitation of regenerated silk fibroin-based resist material where the molecular length of the fibroin protein exhibits a wide distribution due to the degumming process. The uniform molecular structure of the silk light chain improves the lithographic resolution. Additionally, the fabrication condition of silk light chain resist preserves the biocompatibility of this material and can be used for biological applications (Figure 12.8c).

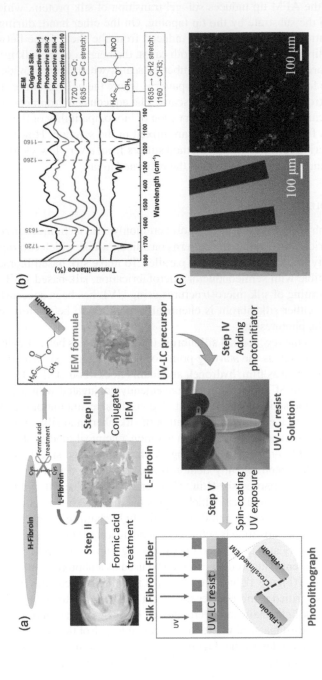

FIGURE 12.8 Photo-sensitized silk light chain used as photoresist. (a) The photo-sensitized silk light chain can be produced by light chain extraction, photo-sensitization, and photoinitiation. (b) The IEM conjugation can be verified using FTIR. (c) The lithographic resolution of the silk light chain is improved thanks to the uniform molecular length distribution. Also, the biocompatibility of the photo-sensitive silk light chain is maintained during the fabrication process. (Reproduced with permission Liu, Zhou, et al. (2017). Copyright 2017, Wiley-VCH.)

12.2.7 INKJET PRINTING

In some cases, high accuracy is not necessary while large-scale fabrication is more important; therefore, the abovementioned high-accuracy manufacturing techniques including EBL, IBL, MPL, or SPL are not suitable for these applications. Considering the proper viscosity (which can be also adjusted by diluting or condensing) of silk fibroin solution, a drop-on-demand, piezoelectric-based inkjet printing process is utilized for silk patterning, where pure or doped silk solution is used as printing ink. Silk fibroin can be doped with all kinds of dopants, from inorganic components such as gold nanoparticles, quantum dots, to organic components such as enzymes and antibiotics.

Different functional silk inks are generated by the addition of nanoparticles, enzymes, antibiotics, growth factors, and antibodies to silk solution. Silk fibroin is selected as the base material for functional ink, due to stabilization of water-soluble labile molecules (Kluge et al. 2016), aqueous solubility (Ambrose et al. 1951), proper viscosity (Moriya et al. 2008), non-toxicity (Sangkert et al. 2016), green chemistry (Aramwit et al. 2014), and polymorphic (Ling et al. 2017) features. The as-prepared silk inks can be then loaded into printer cartridges and inkjet-printed into spider-web patterns on numerous surfaces for use in sensing, therapeutics, and regenerative medicine (Figure 12.9; Tao et al. 2015). The ease of preparing silk inks in a water-based process offers plenty of opportunities by doping silk solution with other functional molecules such as carbohydrates, lipids, and nucleic acids, which will provide a compelling platform for the rapid design of sensors and assays. Thus, inkjet printing of silk fibroin will play a vital role as a green manufacturing technique to fabricate silk-based devices.

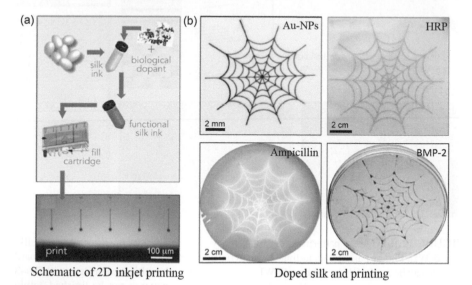

Schematic of 2D inkjet printing   Doped silk and printing

FIGURE 12.9 Deposition of silk solution to form arbitrary 2D patterns using inkjet printing. (a) Illustration of preparing silk inks by doping silk solution with functional molecules and patterned by inkjet printing. (b) Examples of patterned functional silk inks. (Reproduced with permission Tao et al. (2015). Copyright 2015, Wiley-VCH.)

12.2.8 Water Lithography

Due to the amphiphilic nature of silk protein chains, solution drops of silk can be generated in the pico-liter to nano-liter orders, enabling it to be used as inks in ink-jet printing technologies without modifying its rheological properties (Tao et al. 2015). On the other hand, the amorphous silk protein is water soluble, paired with the aqueous inks (i.e., water-based solution) of inkjet printing process, a biofriendly and versatile "print-to-pattern" scheme—termed silk-based water lithography (WL) is generated.

WL is a lithography technique that uses pure/functionalized water as the "exposure source" and water-soluble silk fibroin as the "resist" to generate desired patterns, without introducing any pollutional chemicals. By increasing the number of nozzles, the patterning rate can be improved. A series of microstructures, including dot matrix, line arrays, and fluorescent checkerboard pattern, have been fabricated using WL with a resolution of 16 μm. The resolution can be improved a lot by using smaller nozzles (Figure 12.10a; Liu et al. 2018). Furthermore, by using

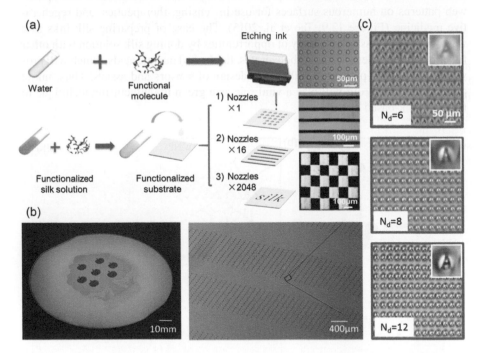

FIGURE 12.10 Direct patterning of silk film by WL. (a) Schematic illustration of silk-based WL. Planar patterns including dot matrix, line arrays, and fluorescent checkerboard pattern can be easily fabricated by silk WL. (b) The setup with multiple cartridges with 2,048 nozzles, respectively, can print in parallel, and wafer-scale patterns can be finished within 10 ms. (c) Optical images of letter "A" obtained by the fabricated silk concave microlens arrays with different curvature using WL. (Reproduced with permission Liu et al. (2018). Copyright 2018, Wiley-VCH.)

2,048 printing nozzles which can be controlled and functionalized individually, parallelized wafer-scale printing is achieved and only about 10 ms is required for patterning an entire wafer of 4 in. (Figure 12.10b). Thanks to the natural formation of microholes through WL and excellent optical performance of silk, silk-based optical devices such as microlens arrays can be easily fabricated (Figure 12.10c). In addition, various types of water-soluble functional molecules, such as fluorescent dyes, growth factors, and therapeutic molecules, can be doped into the water ink for WL, which provides an ideal technique for fabricating bioactive structures within a mild manufacturing condition.

12.2.9 3D PRINTING

Over the years, additive manufacturing has been extensively studied thanks to its capability to create freeform and complex 3D structures. The printing of "bioinks" has enabled a new platform to create biomimetic structures for tissue engineering (Lawrence et al. 2009), cell scaffolding (Wang et al. 2006), and regenerative medicine (Rising 2014). Using this method, the bioactive components (biomolecules and even cells) are doped in the scaffold materials in the solution phase, which is favorable to maintain their bioactivities. The cells can then be directly printed into desired shape along with the scaffold material instead of being seeded onto a pre-fabricated mold. Therefore, more complex structures can be achieved. In this context, a few types of materials are considered as candidates for "bioinks" with hydrogel being the most promising candidate. These materials include both synthetic ones, such as poly(acrylic acid) (PAA), poly(ethylene glycol) (PEG), and poly(ethylene oxide) (PEO), and naturally derived ones, such as agarose, alginate, chitosan, collagen, or silk (Armentano et al. 2010). In particular, genetically engineered spider silk, owing to its low toxicity, modifiability, and superior mechanical properties, is also investigated (Figure 12.11; Schacht et al. 2015). Cell-embedded and highly concentrated spider silk solution can be printed via a print-head, followed by immediate gelation due to the shear force and dehydration in the extrusion process. The fabricated silk hydrogel structures show a Young's modulus of 0.2 kPa, which is comparable to that of the human tissues. The cells also exhibit solid attachment and appropriate proliferation on recombinant spider silk with an arginine-glycine-aspartic acid (RGD) motif. The printed structures maintain structural integrity with up to 16 layers (approximately 3 mm in height) stacked together.

12.2.10 FIBER SPINNING

Silk fibers produced by silkworms and spiders arouse extensively attentions due to their exceptional mechanical properties and diverse applications (Tao, Kaplan, and Omenetto 2012, Omenetto and Kaplan 2010). Numerous efforts have been made to mimic the fiber production process of the insects using regenerated silk solutions (Koeppel and Holland 2017). However, the as-spun regenerated silk fibers (RSFs) are usually brittle, have poor mechanical properties, and require complex posttreatment process, such as dehydration and crystallization processes (Madurga et al. 2017).

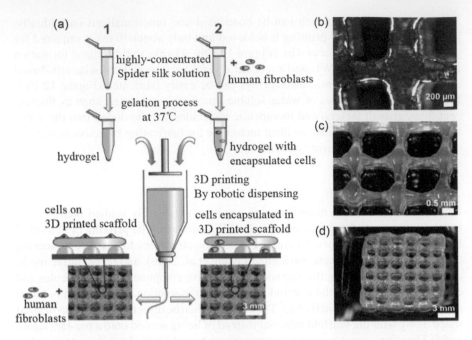

FIGURE 12.11   The 3D bio-fabrication of cell-loaded recombinant spider silk. (a) The cell-loaded and highly concentrated spider silk hydrogel can be used as the "bioink" to directly print 3D structures. Stereomicroscopy and digital images of (b) two-layer eADF4(C16) and (c and d) eight-layer eADF4(C16) scaffolds. Up to 16 layers (approximately 3 mm in height) of the cell-loaded spider silk "bioink" can be printed on top of each other without collapsing. (Reproduced with permission Schacht et al. (2015). Copyright 2015, Wiley-VCH.)

The key to obtaining RSF with high performance is the spinning process as well as the preparation of the spinning dope. In the highly concentrated (~25 wt%) natural silk spinning dope, the silk proteins are assembled into micelle-like structure with liquid-crystalline properties (Jin and Kaplan 2003). Thanks to shear stress and dehydration, the liquid crystallinity allows better alignment of the molecules. A method to prepare silk spinning dope with high concentration and liquid crystalline has been proposed (Figure 12.12; Ling et al. 2017). The degummed silk fibers are gradually and partially dissolved in hexafluoroisopropanol (HFIP), yielding a highly viscous but uniform solution that contains suspended microfibrils with diameters of 5–10 μm and length of several hundreds to thousands of micrometers. In the following extrusion process, this nematic liquid-crystal-like solution is acted as the spinning dope. Without any posttreatment, the resultant RSF can still retain the hierarchical structures of the natural silk fibers and possess mechanical properties as good as the spider silk fiber. By mixing with carbon nanotubes, the RSF can be further functionalized to act as humidity and temperature sensors. In addition, polymorphic RSF has also been created with pre-designed 3D structures, providing a powerful platform to obtain complex structures beyond mimicking natural fiber construction.

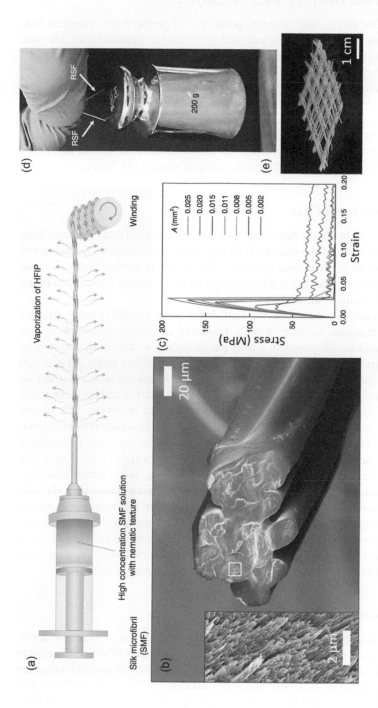

FIGURE 12.12 Bio-inspired spinning of RSF. (a) The spinning dope of RSF is obtained by a gradual and partial dissolution of silk fiber in HFIP, while it yields a uniform and viscous solution with suspended microfibrils. The nematic liquid-crystalline dope can then be extruded into fiber without the need of posttreatment. (b) The cross section of the as-spun fiber shows the preservation of hierarchical structures of the RSF. Scale bar: 20 μm (2 μm inset). (c) The mechanical strength of the fiber is correlated with the cross-sectional area. (d) A single RSF fiber with a weight of 7 mg and a length of 15 cm can hold a weight of 200 g. (e) The RSF can be easily functionalized by doping fluorescent dyes. (Reproduced with permission Ling et al. (2017). Copyright 2017, Springer Nature.)

12.3 SILK-BASED DEVICES AND INTERFACES

Due to the superior mechanical strength, good optical performance, excellent bio-compatibility, ease of functionalization, and controllable degradability, the silk becomes a promising candidate material in many areas. In particular, the mechanical strength of the silk protein ensures its capability to maintain structural integrity as a functional material (Kim et al. 2009). Good optical performance makes it suitable to fabricate optical devices (Dogru et al. 2017). Excellent biocompatibility and controllable biodegradability make it a favorable building material in transient devices (Hwang et al. 2012). Furthermore, with multi-scale manufacturing (from nanoscale to macroscale), the silk protein can be fabricated into various structures with complex functionalities beyond the capability of natural silk fibers.

12.3.1 ORGANIC SOLAR CELLS

Silk fibroin film can be easily patterned by a variety of techniques and has been applied in various sorts of devices. Silk film is particularly eligible for being the nanoimprinting mold used in organic solar cells, due to its advantageous material properties, such as little interaction with conjugated polymer, conformal imprinting on curved surface (Ding et al. 2015).

The organic solar cells based on the bulk heterojunction (BHJ) have attracted a great deal of attention because of their favorable characteristics (such as low cost, flexibility, and simple process), whose performance is mainly determined by the nanoscale surface structure of heterojunction within active layer. An efficient approach for the promotion of organic solar cells' performance is to construct an ordered bulk heterojunction (OBHJ) morphology between donor and acceptor materials, which enables both efficient exaction separation and transport. Silk fibroin film is employed as the nanoimprinting mold to transfer desired patterns onto conjugated polymer P3HT, which is served as donor materials for the fabricated organic solar cells. Comparing the measured photovoltaic responses (i.e., external quantum efficiency (EQE) curve and current density versus voltage curve under illumination) of solar cells, the OBHJ solar cell shows a better performance than the planar bulk heterojunction (PBHJ, of which the P3HT thin film is unimprinted) device (Figure 12.13; Ding et al. 2015). The nanoimprinting process with patterned silk fibroin mold at room temperature can perfectly transfer patterns to active layer of organic solar cell without damaging the charge transportation, which indicates that it is fully compatible with the fabrication process of organic solar cell.

12.3.2 TRIBOELECTRIC NANOGENERATOR

Silk fiber is an ancient material and has been long-known for the triboelectric application, for it is easy to lose electrons and exhibit significant triboelectric effects (Zhang, Brugger, and Kim 2016, Liu, Li, Che, et al. 2017). Recent research indicates that silk fibroin extracted from silk silkworm cocoon shows similar triboelectric property when it is cast into film. On the other hand, silk fibroin is biodegradable and biocompatible. The combination of these properties makes silk fibroin a feasible

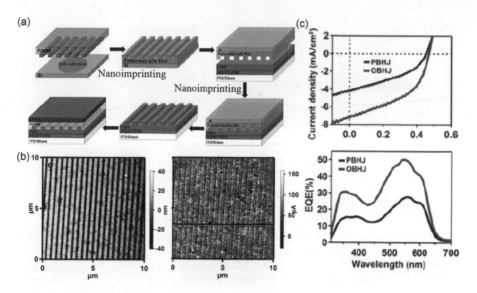

FIGURE 12.13 (a) Fabrication process of OBHJ solar cell device. (b) C-AFM height and current images of nanoimprinted P3HT film. (c) Device performance characteristics of the fabricated solar cell. (Reproduced with permission Ding et al. (2015). Copyright 2015, Springer.)

and promising candidate material for biodegradable power source, such as transient triboelectric nanogenerator (T^{2}ENG).

The silk-based T^{2}ENG is fabricated from silk and soluble magnesium (Mg), and silk has an electronegativity considerably lower than that of Mg (Figure 12.14a; Zhang et al. 2018). Applying periodic external force on the T^{2}ENG to make the silk layer and Mg layer periodically contact and separate, then the silk film loses electrons and is positively charged, while the Mg film collects electrons and is negatively charged. Therefore, biomechanical energy is generated. At original state (i.e., before the T^{2}ENG degrades), the open-circuit voltage of the T^{2}ENG could reach up to ≈60 V, and the short-circuit current (absolute value) could be as high as ≈1 μA, demonstrating the remarkable triboelectric property of silk (Figure 12.14b). Having finished in vitro characterization of the silk T^{2}ENG, it is then encapsulated by silk film and implanted in the subdermal region of Sprague-Dawley mice. After 3 weeks, the silk-based T^{2}ENG with lower crystallinity completely disappears, while the one with higher crystallinity still has some residuals in the implant region (Figure 12.14c). The implanted T^{2}ENG device gives stable outputs of ≈6 V for 6 h when it is stimulated by a repeatedly applied constant external force. To simulate the in vivo degradation process of silk T^{2}ENG, 10 mL physiological saline solution is locally injected at the implanted region, which triggers the degradation of the T^{2}ENG and results in the loss of triboelectric function quickly within 30 min (Figure 12.14d). The development of silk-based T^{2}ENG demonstrates a promising strategy for transient power source, relieving the restriction of the power supply of implantable medical electronic devices.

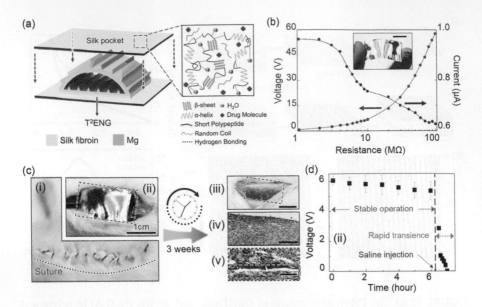

FIGURE 12.14 (a) Schematic illustration of the silk-based transient triboelectric nanogenerator (T²ENG). (b) The dependence of output voltage and current on the resistor connected to the T²ENG. (c) The subcutaneous implantation of a T²ENG in the dorsal region of a Sprague-Dawley rat. (d) The in vivo triboelectric outputs of the T²ENG at different degradation stages. (Reproduced with permission Zhang et al. (2018). Copyright 2018, Wiley-VCH.)

12.3.3 DIFFRACTIVE OPTICAL ELEMENTS

Silk film is highly transparent to the visible light and mechanically robust (Perry et al. 2010), has excellent surface smoothness, and can accurately replicate microstructures from other inorganic molds pre-patterned with desired surface features using a "cast-and-peel" soft lithography technique, all of which make it suitable for a wide range of optical devices. Furthermore, the controllable biodegradation rate of silk-based device offers a variety of sensing applications based on its transient behavior.

Recently, silk films processed into diffractive optical elements (DOEs) have been used as a multifunctional sensing platform, based on the fact that the diffraction pattern of the DOE is highly sensitive to its surrounding environments and its structural integrity. A set of bioactive DOEs microfabricated using functionalized silk fibroin films have been tailored for hydration sensing, biologically enabled data concealment, therapeutic treatment, and in vitro and in vivo drug release monitoring upon degradation. The HRP-doped silk DOE is immersed in a proteinase k solution with a concentration of 0.1 mg/mL to mimic the in vitro drug release process under enzymatic degradation. The release of the embedded HRP molecules is quantitatively characterized by a standard enzyme-linked immunosorbent assay (ELISA) process and real-time monitored by the decrement of the integrity of the diffraction patterns (Figure 12.15; Zhou et al. 2017). The silk-based optical devices are attractive for high sensitivity in sensing application and controllable biodegradability, which may open up opportunities for an emerging field: transient optics.

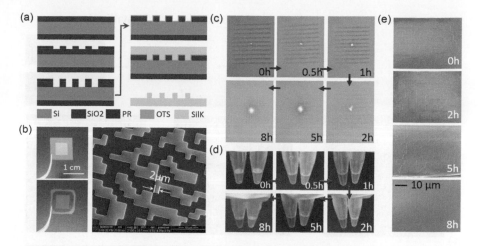

FIGURE 12.15 (a) Graphical representation of the fabrication process of silk DOEs. (b) Photograph of the silicon mold and silk DOE. (c–e) In vitro drug release monitoring upon degradation of the silk DOE. (Reproduced with permission Zhou et al. (2017). Copyright 2017, Wiley-VCH.)

12.3.4 ENVIRONMENTALLY RESPONSIVE SENSORS

In nature system, the living organisms that exhibit a unique combination of mechanical, chemical and transport properties are always consisted of materials that are organized hierarchically. These dominant structures are formed from evolved processes that govern directed self-assembly. Materials such as extracellular matrix, nacre, and bone have heterogeneous and hierarchical structures that encompass multiple length scales (Ortiz and Boyce 2008, Wegst et al. 2015, Loren, Nyden, and Hermansson 2009, Jose et al. 2012, Bellail et al. 2004).

Inspired from the natural material engineering process, a combination of top-down manufacturing and self-assembly process has been applied to silk fibroin. The proposed method utilizes a polydimethylsiloxane (PDMS) mold with predesigned geometries and topologies to orient and constraint the formation of porous fibrillary silk hydrogel networks (Figure 12.16; Tseng et al. 2017). The resulting hierarchical structures with the anisotropic microfibers orient along the direction of applied mechanical stress, which are investigated by a polarization microscopy. Thanks to the micro-arrangement of the silk protein, the bulk material has anisotropic mechanical properties and transport characteristics, assisting the accomplishment of biomimetic functions. Due to the water swelling of silk fibroin, this technique is readily adaptable to create moisture-sensitive structures that are environmentally responsive (i.e., humidity responsive). The environmental responsive constructs are tested by exposing them to 75% and 95% humidity, demonstrating the feasibility of environmentally actuating such geometries. By predicting the swelling behavior of the aligned silk fibroin nanofibrils, the transformation of the fabricated structures upon the change of humidity can be designed. In addition, the mild processing conditions of this technique make the integration of bioactive components in the silk

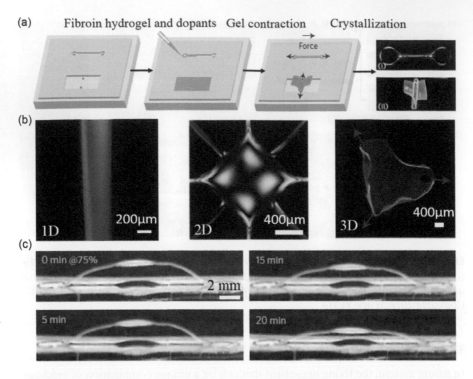

FIGURE 12.16 The directed assembly of bio-inspired silk hierarchical structures. (a) The hierarchical silk structure can be manufactured through a combination of top-down lithography and self-assembly process. The application of mechanical force directs the alignment of the silk microfibrils. (b) Transformation of silk nanofibrillar shapes in 1D, 2D, and 3D. (c) The structural evolution of the fabricated silk structure in high humidity environment. (Reproduced with permission Tseng et al. (2017). Copyright 2017, Springer Nature.)

solution easier, opening up opportunities for even wider range of functionalities. Using this technique, the fundamental relations between the nano- and microscale morphology and the mechanical and physical behavior of macrostructures can be conveniently investigated, paving the way for engineering biomaterials with more complex functionalities.

12.4 CONCLUSIONS

As one of the most ancient materials, silk has gone through a series of revolutions. Nowadays, silk fibers have achieved a perfect balance between strength and toughness in a light-weighted and flexible fashion, which is the initial driving force and ultimate challenge for many researchers who are fascinated by the strength and beauty of silk. Many scientific and commercial efforts have been invested, which however have yet to replicate silk fibers for scaled-up production. But even so, as a sustainable material, silk still holds great promise in a wide range of applications, such as scientific research, national economy, and environmental conservation, that benefit both people and the planet.

Over a long period in the past, silk was mainly used to make textiles for its lightness, softness, and heat preservation. Until about a century ago, due to the robust mechanical properties, silk began to be used as a suturing material, which was its early attempt for entry in biomedical applications. Recent research indicates that silk has outstanding biocompatibility, controllable degradation, and aqueous mild environmental conditions during processing. Therefore, more research efforts have been made on silk to exploit its new application areas. Fortunately, leading-edge applications in electronic devices, transient optics, tissue engineering, regenerative medicine, and medical implants have been implemented. In addition, a series of advanced manufacturing techniques, from nanoscale to macroscale, from 2D to 2.5D to 3D (such as EBL, IBL, MPL, SPL, UVL, WL, and 3D bioprinting), have also been developed specifically for silk.

Silk-based materials are presently quite abundant, their all aspects of performance have been improved, and the manufacturing technology has also been gradually maturing; it can be imagined that further availability for large-scale applications of silk can be expected soon. In the future, the opportunities offered by silk—an ancient material on earth—are only limited by our imaginations.

REFERENCES

Altman, Gregory H, Frank Diaz, Caroline Jakuba, Tara Calabro, Rebecca L Horan, Jingsong Chen, Helen Lu, John Richmond, and David L Kaplan. 2003. Silk-based biomaterials. *Biomaterials* no. 24 (3):401–416.

Ambrose, Edmund J, Clement H Bamford, Arthur Elliott, and William E Hanby. 1951. Water-soluble silk: An α-protein. *Nature* no. 167 (4242):264–265.

Aramwit, Pornanong, Nipaporn Bang, Juthamas Ratanavaraporn, and Sanong Ekgasit. 2014. Green synthesis of silk sericin-capped silver nanoparticles and their potent antibacterial activity. *Nanoscale Research Letters* no. 9 (1):1–7.

Armentano, Ilaria, Mariaserena Dottori, Elena Fortunati, Samantha Mattioli, Jose Kenny. 2010. Biodegradable polymer matrix nanocomposites for tissue engineering: A review. *Polymer Degradation & Stability* no. 95 (11):2126–2146.

Balabin, Ilya A, David N Beratan, and Spiros S Skourtis. 2008. Persistence of structure over fluctuations in biological electron-transfer reactions. *Physical Review Letters* no. 101 (15):158102.

Bellail, Anita C, Stephen B Hunter, Daniel J Brat, Chalet Tan, and Erwin G Van Meir. 2004. Microregional extracellular matrix heterogeneity in brain modulates glioma cell invasion. *International Journal of Biochemistry & Cell Biology* no. 36 (6):1046–1069.

Brenckle, Mark A, Hu Tao, Sunghwan Kim, Mark Paquette, David L Kaplan, and Fiorenzo G Omenetto. 2013. Protein-protein nanoimprinting of silk fibroin films. *Advanced Materials* no. 25 (17):2409.

Burghoorn, Marieke, Dorrit Roosen-Melsen, Joris de Riet, Sami Sabik, Zeger Vroon, Iryna Yakimets, and Pascal Buskens. 2013. Single layer broadband anti-reflective coatings for plastic substrates produced by full wafer and roll-to-roll step-and-flash nano-imprint lithography. *Materials* no. 6 (9):3710–3726.

Custódio, Catarina A, Rui L Reis, and Joao F Mano. 2014. Engineering biomolecular microenvironments for cell instructive biomaterials. *Advanced Healthcare Materials* no. 3 (6):797–810.

Dickerson, Matthew B, Scott P Fillery, Hilmar Koerner, Kristi M Singh, Katie Martinick, Lawrence F Drummy, Michael F Durstock, Richard A Vaia, Fiorenzo G Omenetto, and David L Kaplan. 2013. Dielectric breakdown strength of regenerated silk fibroin films as a function of protein conformation. *Biomacromolecules* no. 14 (10):3509–3514.

Ding, Guangzhu, Qianqian Jin, Qing Chen, Zhijun Hu, and Jieping Liu. 2015. The fabrication of ordered bulk heterojunction solar cell by nanoimprinting lithography method using patterned silk fibroin mold at room temperature. *Nanoscale Research Letters* no. 10 (1):491.

Dogru, Itir Bakis, Kyungtaek Min, Muhammad Umar, Houman Bahmani Jalali, Efe Begar, Deniz Conkar, Elif Nur Firat Karalar, Sunghwan Kim, and Sedat Nizamoglu. 2017. Single transverse mode protein laser. *Applied Physics Letters* no. 111 (23):231103.

Dong, Qin, Xia Younan, and Whitesides M George M. 2010. Soft lithography for micro- and nanoscale patterning. *Nature Protocols* no. 5 (3):491.

Douglas, Shawn M, Hendrik Dietz, Tim Liedl, and Bjorn Hogberg. 2009. Self-assembly of DNA into nanoscale three-dimensional shapes. *Nature* no. 26 (6):799.

Gan, Tiansheng, Bo Wu, Xuechang Zhou, and Guangzhao Zhang. 2016. Ultrahigh resolution, serial fabrication of three dimensionally-patterned protein nanostructures by liquid-mediated non-contact scanning probe lithography. *RSC Advances* no. 6 (55):50331–50335.

Hwang, Suk Won, Hu Tao, Dae-Hyeong Kim, Huanyu Cheng, Jun-Kyul Song, Elliott Rill, Mark A Brenckle, Bruce Panilaitis, Sang Min Won, Yun-Soung Kim, Young Min Song, Ki Jun Yu, Abid Ameen, Rui Li, Yewang Su, Miaomiao Yang, David L Kaplan, Mitchell R Zakin, Marvin J Slepian, Yonggang Huang, Fiorenzo G Omenetto, and John A Rogers. 2012. A physically transient form of silicon electronics. *Science* no. 337 (6102):1640.

Jiang, Jianjuan, Shaoqing Zhang, Zhigang Qian, Nan Qin, Wenwen Song, Long Sun, Zhitao Zhou, Zhifeng Shi, Liang Chen, Xinxin Li, Ying Mao, David L Kaplan, Stephanie N Gilbert Corder, Xinzhong Chen, Mengkun Liu, Fiorenzo G Omenetto, Xiaoxia Xia, and Tiger H Tao. 2018. Protein bricks: 2D and 3D bio-nanostructures with shape and function on demand. *Advanced Materials* no. 30 (20):1705919.

Jiang, Jie, Xiaomin Li, Wing Cheung Mak, and Dieter Trau. 2010. Integrated direct DNA/protein patterning and microfabrication by focused ion beam milling. *Advanced Materials* no. 20 (9):1636–1643.

Jin, Hyoung Joon, and David L Kaplan. 2003. Mechanism of silk processing in insects and spiders. *Nature* no. 424 (6952):1057–1061.

Jose, Rod R, Roberto Elia, Matthew A Firpo, David L Kaplan, and Robert A Peattie. 2012. Seamless, axially aligned, fiber tubes, meshes, microbundles and gradient biomaterial constructs. *Journal of Materials Science Materials in Medicine* no. 23 (11):2679.

Kabashin, Andrei V, Paul R Evans, Santa Pastkovsky, William Hendren, Gregory A Wurtz, Ronald Atkinson, Robert J Pollard, Viktor A Podolskiy, and Anatoly V Zayats. 2009. Plasmonic nanorod metamaterials for biosensing. *Nature Materials* no. 8 (11):867–871.

Khan, Yasser, Mohit Garg, Qiong Gui, Mark Schadt, Abhinav Gaikwad, Donggeon Han, Natasha A D Yamamoto, Paul Hart, Robert Welte, and William Wilson. 2016. Flexible hybrid electronics: Direct interfacing of soft and hard electronics for wearable health monitoring. *Advanced Functional Materials* no. 26 (47):8764–8775.

Kim, Dae-Hyeong, Yun-Soung Kim, Jason Amsden, Bruce Panilaitis, David L Kaplan, Fiorenzo G Omenetto, Mitchell R Zakin, and John A Rogers. 2009. Silicon electronics on silk as a path to bioresorbable, implantable devices. *Applied Physics Letters* no. 95 (13):133701.

Kim, Sunghwan, Benedetto Marelli, Mark A Brenckle, Alexander N Mitropoulos, Eun-Seok Gil, Konstantinos Tsioris, Hu Tao, David L Kaplan, and Fiorenzo G Omenetto. 2014. All-water-based electron-beam lithography using silk as a resist. *Nature Nanotechnology* no. 9 (4):306.

Kim, Sunghwan, Alexander N Mitropoulos, Joshua D Spitzberg, Tao Hu, David L Kaplan, and Fiorenzo G Omenetto. 2012. Silk inverse opals. *Nature Photonics* no. 6 (12):818–823.

Kluge, Jonathan A, Adrian B Li, Brooke T Kahn, Dominique S Michaud, Fiorenzo G Omenetto, and David L Kaplan. 2016. Silk-based blood stabilization for diagnostics. *Proceedings of the National Academy of Sciences of the United States of America* no. 113 (21):5892.

Koeppel, Andreas, and Chris Holland. 2017. Progress and trends in artificial silk spinning: A systematic review. *ACS Biomaterials Science & Engineering* no. 3 (3):226–237.

Lawrence, Brian D, Jeffrey K Marchant, Mariya A Pindrus, Fiorenzo G Omenetto, and David L Kaplan. 2009. Silk film biomaterials for cornea tissue engineering. *Biomaterials* no. 30 (7):1299–1308.

Lee, Soo Hong, James J Moon, and Jennifer L West. 2008. Three-dimensional micropatterning of bioactive hydrogels via two-photon laser scanning photolithography for guided 3D cell migration. *Biomaterials* no. 29 (20):2962.

Li, Chunmei, Blake Hotz, Shengjie Ling, Jin Guo, Dylan S Haas, Benedetto Marelli, Fiorenzo Omenetto, Samuel J Lin, and David L Kaplan. 2016. Regenerated silk materials for functionalized silk orthopedic devices by mimicking natural processing. *Biomaterials* no. 110:24–33.

Ling, Shengjie, Qin Zhao, Chunmei Li, Wenwen Huang, David L Kaplan, and Markus J Buehler. 2017. Polymorphic regenerated silk fibers assembled through bioinspired spinning. *Nature Communications* no. 8 (1):1387.

Liu, Chaoran, Jiaqian Li, Lufeng Che, Shaoqiang Chen, Zuankai Wang, and Xiaofeng Zhou. 2017. Toward large-scale fabrication of triboelectric nanogenerator (TENG) with silk-fibroin patches film via spray-coating process. *Nano Energy* no. 41:359–366.

Liu, Jia, Qilin Li, Jinxiang Zhang, Lei Huang, Chao Qi, Luming Xu, Xingxin Liu, Guobin Wang, Lin Wang, and Zheng Wang. 2017. Safe and effective reversal of cancer multidrug resistance using sericin-coated mesoporous silica nanoparticles for lysosome-targeting delivery in mice. *Small* no. 13 (9):1602567.

Liu, Wanpeng, Zhitao Zhou, Shaoqing Zhang, Zhifeng Shi, Justin Tabarini, Woonsoo Lee, Yeshun Zhang, S N Gilbert Corder, Xinxin Li, Fei Dong, Liang Cheng, Mengkun Liu, David L Kaplan, Fiorenzo G Omenetto, Guozheng Zhang, Ying Mao, and Tiger H Tao. 2017. Precise protein photolithography (P³): High performance biopatterning using silk fibroin light chain as the resist. *Advanced Science* no. 4 (9):1700191.

Liu, Xueli, and Shutao Wang. 2014. Three-dimensional nano-biointerface as a new platform for guiding cell fate. *Chemical Society Reviews* no. 43 (8):2385–2401.

Liu, Zhen, Zhitao Zhou, Shaoqing Zhang, Long Sun, Zhifeng Shi, Ying Mao, Keyin Liu, and Tiger H Tao. 2018. "Print-to-pattern": Silk-based water lithography. *Small* no. 14 (47):1802953.

Loren, Niklas, Magnus Nyden, and Anne-Marie Hermansson. 2009. Determination of local diffusion properties in heterogeneous biomaterials. *Advances in Colloid & Interface Science* no. 150 (1):5.

Madurga, Rodrigo, Alfonso M Ganan-Calvo, Gustavo R Plaza, Gustavo V Guinea, Manuel Elices, and Jose Perez-Rigueiro. 2017. Production of high performance bioinspired silk fibers by straining flow spinning. *Biomacromolecules* no. 18 (4):1127.

Moriya, Motoaki, Kosuke Ohgo, Yuichi Masubuchi, and Tetsuo Asakura. 2008. Flow analysis of aqueous solution of silk fibroin in the spinneret of *Bombyx mori* silkworm by combination of viscosity measurement and finite element method calculation. *Polymer* no. 49 (4):952–956.

Omenetto, Fiorenzo G, and David L Kaplan. 2010. New opportunities for an ancient material. *Science* no. 329 (5991):528.

Ortiz, Christine, and Mary C Boyce. 2008. Bioinspired structural materials. *Science* no. 319 (5866):1053–1054.

Perry, Hannah, Ashwin Gopinath, David L Kaplan, Luca Dal Negro, and Fiorenzo G Omenetto. 2010. Nano- and micropatterning of optically transparent, mechanically robust, biocompatible silk fibroin films & dagger. *Advanced Materials* no. 20 (16):3070–3072.

Perry, Randall S, and Vera M Kolb. 2004. Biological and organic constituents of desert varnish: Review and new hypotheses. *Proceedings of SPIE* no. 5163:202–217.

Qin, Nan, Shaoqing Zhang, Jianjuan Jiang, Stephanie Gilbert Corder, Zhigang Qian, Zhitao Zhou, Woonsoo Lee, Keyin Liu, Xiaohan Wang, Xinxin Li, Zhifeng Shi, Ying Mao, Hans A Bechtel, Michael C Martin, Xiaoxia Xia, Benedetto Marelli, David L Kaplan, Fiorenzo G Omenetto, Mengkun Liu, and Tiger H Tao. 2016. Nanoscale probing of electron-regulated structural transitions in silk proteins by near-field IR imaging and nano-spectroscopy. *Nature Communications* no. 7:13079.

Rising, Anna. 2014. Controlled assembly: A prerequisite for the use of recombinant spider silk in regenerative medicine. *Acta Biomaterialia* no. 10 (4):1627–1631.

Rockwood, Danielle N, Rucsanda C Preda, Tuna Yücel, Xiaoqin Wang, Michael L Lovett, and David L Kaplan. 2011. Materials fabrication from bombyx mori silk fibroin. *Nature Protocols* no. 6 (10):1612–1631.

Rothemund, Paul W K. 2006. Folding DNA to create nanoscale shapes and patterns. *Nature* no. 440 (7082):297.

Salaita, Khalid, Yuhuang Wang, and Chad A Mirkin. 2007. Applications of dip-pen nanolithography. *Nature Nanotechnology* no. 2 (3):145–155.

Sangkert, Supaporn, Jirut Meesane, Suttatip Kamonmattayakul, and Wen Lin Chai. 2016. Modified silk fibroin scaffolds with collagen/decellularized pulp for bone tissue engineering in cleft palate: Morphological structures and biofunctionalities. *Materials Science & Engineering C* no. 58 (1):1138.

Schacht, Kristin, Tomasz Jungst, Matthias Schweinlin, Andrea Ewald, Jurgen Groll, and Thomas Scheibel. 2015. Biofabrication of cell-loaded 3D spider silk constructs. *Angewandte Chemie International Edition* no. 54 (9):2816–2820.

Shi, Jinjun, Jixin Chen, and Paul S Cremer. 2008. Sub-100 nm patterning of supported bilayers by nanoshaving lithography. *Journal of the American Chemical Society* no. 130 (9):2718–2719.

Spivey, Eric C, Eric T Ritschdorff, Jodi L Connell, Christopher A Mclennon, Christine E Schmidt, and Jason B Shear. 2013. Multiphoton lithography of unconstrained three-dimensional protein microstructures. *Advanced Functional Materials* no. 23 (3):333–339.

Sun, Yun-Lu, Qi Li, Siming Sun, Jingchun Huang, Boyuan Zheng, Qidai Chen, Zhengzhong Shao, and Hongbo Sun. 2015. Aqueous multiphoton lithography with multifunctional silk-centred bio-resists. *Nature Communications* no. 6:8612.

Tao, Hu, David L Kaplan, and Fiorenzo G Omenetto. 2012. Silk materials—A road to sustainable high technology. *Advanced Materials* no. 24 (21):2824–2837.

Tao, Hu, Benedetto Marelli, Miaomiao Yang, Bo An, M Serdar Onses, John A Rogers, David L Kaplan, and Fiorenzo G Omenetto. 2015. Inkjet printing of regenerated silk fibroin: From printable forms to printable functions. *Advanced Materials* no. 27 (29):4273.

Tseng, Peter, Bradley Napier, Siwei Zhao, Alexander N Mitropoulos, Matthew B Applegate, Benedetto Marelli, David L Kaplan, and Fiorenzo G Omenetto. 2017. Directed assembly of bio-inspired hierarchical materials with controlled nanofibrillar architectures. *Nature Nanotechnology* no. 12 (5):474.

Wadbua, Paweena, Boonhiang Promdonkoy, Santi Maensiri, and Sineenat Siri. 2010. Different properties of electrospun fibrous scaffolds of separated heavy-chain and light-chain fibroins of *Bombyx mori*. *International Journal of Biological Macromolecules* no. 46 (5):493–501.

Wang, Yongzhong, Dominick J Blasioli, Hyeon-Joo Kim, Hyun Suk Kim, and David L Kaplan. 2006. Cartilage tissue engineering with silk scaffolds and human articular chondrocytes. *Biomaterials* no. 27 (25):4434–4442.

Wang, Yuliang, and Younan Xia. 2004. Bottom-up and top-down approaches to the synthesis of monodispersed spherical colloids of low melting-point metals. *Nano Letters* no. 4 (10):2047–2050.

Wegst, Ulrike G K, Hao Bai, Eduardo Saiz, Antoni P Tomsia, and Robert O Ritchie. 2015. Bioinspired structural materials. *Nature Materials* no. 14 (1):23–36.

Williams, David F. 2014. There is no such thing as a biocompatible material. *Biomaterials* no. 35 (38):10009–10014.

Xu, Song, Scott Miller, Paul E Laibinis, and Gang-Yu Liu. 1999. Fabrication of nanometer scale patterns within self-assembled monolayers by nanografting. *Langmuir* no. 15 (21):7244–7251.

Yang, Yang, Xiangjia Li, Xuan Zheng, Zeyu Chen, Qifa Zhou, and Yong Chen. 2018. 3D-printed biomimetic super-hydrophobic structure for microdroplet manipulation and oil/water separation. *Biomaterials* no. 30 (9):1704912.

Zhang, Xiao Sheng, Jürgen Brugger, and Beomjoon Kim. 2016. A silk-fibroin-based transparent triboelectric generator suitable for autonomous sensor network. *Nano Energy* no. 20:37–47.

Zhang, Yujia, Zhitao Zhou, Zhen Fan, Shaoqing Zhang, Faming Zheng, Keyin Liu, Yulong Zhang, Zhifeng Shi, Liang Chen, Xinxin Li, Ying Mao, Fei Wang, Yun-Lu Sun, and Tiger H Tao. 2018. Self-powered multifunctional transient bioelectronics. *Small* no. 14 (35):1802050.

Zhong, Jian, Mengjia Ma, Juan Zhou, Daixu Wei, Zhiqiang Yan, and Dannong He. 2013. Tip-induced micropatterning of silk fibroin protein using in situ solution atomic force microscopy. *ACS Applied Materials & Interfaces* no. 5 (3):737–746.

Zhou, Zhitao, Zhifeng Shi, Xiaoqing Cai, Shaoqing Zhang, Stephanie G Corder, Xinxin Li, Yeshun Zhang, Guozheng Zhang, Liang Chen, Mengkun Liu, David L Kaplan, Fiorenzo G Omenetto, Ying Mao, Zhendong Tao, and Tiger H Tao. 2017. The use of functionalized silk fibroin films as a platform for optical diffraction-based sensing applications. *Advanced Materials* no. 29 (15):1605471.

13 Microarray 3D Bioprinting for Creating Miniature Human Tissue Replicas for Predictive Compound Screening

Alexander D. Roth
Rochester Institute of Technology

Stephen Hong and Moo-Yeal Lee
Cleveland State University

CONTENTS

13.1 INTRODUCTION

One of the major issues confronting scientists and researchers today is the lack of adequate models for studying microscale 3D cultures. "Three-dimensional bioprinting" technology is difficult to implement as compared to 3D printing of plastics and metals as living cells constrain the variables through which printing can occur, and the development of tissues to mature and adequately reflect the *in vivo* conditions takes time. Most current trends have focused on large-scale replication of tissues and organs using 3D bioprinting. The theory behind this is that larger scale 3D prints can adequately mimic the behavior of the tissues *in vivo* and potentially be used in the future for partial or complete transplantation. However, the miniaturization of cell cultures has the effect of ease of modeling due to the fewer number of cells and potential cell–cell interactions involved. Additionally, with smaller volumes, multiple models can be generated rapidly to study any variation that can occur in tissue and disease physiology. While these cultures are not used in transplantation research, they can be used to further understand certain diseases and assist in the development of drugs to combat diseases.

Among various 3D cell culture models, including matrix-free spheroids (Cavnar et al. 2014), spheroids in hydrogels (Otsuka et al. 2013), organoids derived from pluripotent stem cells (Jo et al. 2015; Asai et al. 2017), and microfluidic tissue-on-chips (van der Helm et al. 2016; Bhatia and Ingber 2014), 3D bioprinting has gained great attention due to its rapid and robust printing of cell types in multiple layers to better mimic tissue structures in humans. Three-dimensional bioprinting technology is different from 3D printing in that living cells suspended in hydrogels are printed at room temperature and matured into tissue-like structures in biomimetic culture conditions over time. Most current trends of 3D bioprinting have focused on replicating relatively large scale of tissues by recapitulating tissue structures *in vivo*. However, this approach is still challenging due to a lack in our capability to recreate the vascular system, leading to cell death in the core of bioprinted tissues.

To address this issue and create *in vivo*-like miniature human tissues rapidly, this chapter will discuss "microarray 3D bioprinting", which is a robotic microsolenoid valve-driven cell printing technology demonstrated on several microarray chip platforms, including a micropillar chip, a microwell chip, a 384-pillar plate with a flat tip surface, and a 384-pillar plate with sidewalls and slits (384PillarPlate). This approach allows the dispensation of multiple layers of human cell types in biomimetic hydrogels and evaluation of hundreds of different cell culture conditions simultaneously for human tissue generation and predictive compound screening. This greatly impacts disease modeling, drug metabolism, human toxicology, and, ultimately, safer drug discovery. In this book chapter, we will introduce the machinery used for printing human cell types, chip platforms to facilitate cell printing, and various hydrogels used for cell encapsulation. This will be followed by methods of cell staining, imaging, and analysis on the microarray chip platforms. Finally, we will discuss applications of bioprinted miniature tissues in disease research.

13.2 CURRENT TRENDS IN *IN VITRO* TISSUE DEVELOPMENT

Several technologies have been developed for 3D bioprinting and microscale tissue engineering. Organovo, the pioneer of 3D bioprinting, has developed usage of the basic building blocks of tissues, such as extracellular matrices (ECMs) and growth factors as a scaffold for cells to be embedded (Zhang et al. 2015; Bertassoni et al. 2014). Organovo has developed this technology to be compatible with hepatic and renal functions for *ex vivo* research. While these tissues are complex and can mimic the tissue functionally, the moderate throughput (i.e., bioprinted tissues in 24-well plates) limits the number of replicas that can be tested in each sample. Additionally, cells must be printed on top of the scaffold, meaning the scaffold must be layered first. This means it takes time for the 3D printed tissue to mature before it becomes properly functional.

Another technology utilizes Transwell™ inserts to allow for specific cell layering. These well plate inserts are commercially available through several different vendors, and they facilitate cell growth into several layers (Iwai et al. 2018; Wagner et al. 2013). Cells can be seeded on opposite sides of the insert, and ECM components and cells can grow into each other, allowing cells to structure their own microenvironment. The advantages of this technology are its natural method for organizing cells into distinct layers, giving the user more control on the construction of tissues. However, this method is more labor-intensive and is only mid-throughput (compatible with 24-well plates). Scaling down for 96- and 384-well plate assays would be difficult due to the limited maneuverability of the Transwell inserts.

Another way to create 3D structures is through using ultra-low attachment (ULA) well plates to facilitate spheroid formation. Here, high-throughput plate platforms are made or coated with materials that are unfavorable to cell adhesion, which causes the cells to stick to each other and form spheroids (Selby et al. 2017). Generally, spheroids formed this way are uniform in size and shape, and ULA plates come in 96- and 384-well plate formats, making them compatible with high-throughput screening (HTS) technologies. The disadvantage of using ULA plates is the limited scope of cells that this can work with, and without growth factors or ECM components added to spheroid mixtures, this attachment may only be temporary, as cells have no natural molecules to adhere to other than to each other. It works only for adherent cells.

Another method that creates microscale tissue cultures is hanging droplet plates marketed by InSphero. The hanging droplet plates allow to produce microtissue models by gravity for drug screening, tumor applications, and other disease models. Tissues examined include pancreatic islets and hepatic cells in both human and murine cell lines, and all of these cultures feature primary cell co-cultures (Proctor et al. 2017; Kermanizadeh et al. 2014). The advantages are its high-throughput nature (InSphero 3D cultures are available for 96-well plates) and the variety of disease models that are available to choose from. The disadvantages include the susceptibility of hanging droplets containing cells to small mechanical shocks and instability of maintaining droplets for longer than a few days due to potential water evaporation and nutrient depletion issues over time. It is difficult to change growth media in hanging droplet plates.

Finally, another technology that utilizes 3D cell cultures is microfluidic plates such as OrganoPlate from MIMETAS. OrganoPlate combined 3D culture with perfusion channel technology to create high-throughput systems that can mimic tissues with capillary flow (Jang et al. 2015; Tibbe et al. 2018). This technology is compatible with other high-throughput platforms such as 96-well plates. Additionally, with added shear from capillary flow, nutrient transfer is more easily achieved, and this can also facilitate cells to behave differently as cells normally do near the capillary. The disadvantages to this technology are the cost and the complexity of the assays. Additional equipment is needed to image cells in the system rapidly.

13.3 METHODS OF MICROARRAY 3D BIOPRINTING

For human cells in hydrogels to be printed accurately in 3D on the microarray chip platforms, materials and equipment must be compatible to create functional tissues. Owing to the viscosity of hydrogels printed and viable cells encapsulated, printing parameters such as valve open time, nozzle diameter, syringe pressure, cell seeding density, hydrogel concentration, and additives must be optimized to maintain high cell viability. In addition, the materials that the parts in a microarray spotter are made of must be compatible with various biological samples, including nucleic acids and proteins, such that these materials do not deactivate, deteriorate, or decompose the biological materials. In this section, we describe the materials and methods associated with the creation of miniaturized tissue replicas, including the bioprinter itself, the chip platforms, and the hydrogels of choice.

13.3.1 MACHINERY

Microarray spotter-driven liquid dispensation is based on either piezoelectric inkjet printing or solenoid valve printing technology, where low-to-moderate viscosity liquids can be dispensed onto functionalized glass slides and pillar/well chip platforms (Fernandes et al. 2010; Lee 2016). Traditionally, piezoelectric microarray spotters have been used to dispense pico-liter volumes of nucleic acids and proteins to manufacture DNA and antibody microarray chips (Zarowna-Dabrowska et al. 2012). Recent advances have been made in the printing of nanoliter-scale volumes of cells encapsulated in hydrogels onto functionalized glass slides and pillar/well chip platforms to mimic miniaturized tissue-like structures.

Microarray 3D bioprinting technology is distinctively different from 3D bioprinting technology, which the former is based on micro-extrusion printing, in terms of printing volume and speed. Most of the microarray spotters compatible with cell printing are solenoid valve-driven, which can rapidly dispense nL to μL volumes of liquids by controlling valve open time, nozzle diameter, and syringe pressure. Thousands of spots can be dispensed within 5–10 min in a specific pattern (i.e., microarrays). Unlike protein and DNA printing, it is critical to maintain cell suspension in hydrogels for at least 5–10 min to reduce printing errors. Possible precipitation of cells in hydrogels due to low viscosity of hydrogels or high cell seeding density will lead to clogging of solenoid valves and ceramic tips or inconsistent cell seeding density. The typical coefficient of variation (CV) for the printing volume is 10%–20%.

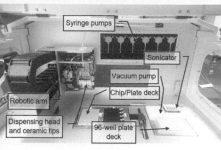

FIGURE 13.1 The picture of a microarray spotter (S+ MicroArrayer) and associated parts. The top image depicts the housing structure of the bioprinter, the main printing head, and all the large containers for fluids that clean the system and regulate the temperature. The bottom picture depicts the main printing area, including the printing and cleaning apparatuses and regions where samples are dispensed.

The printing head consists of solenoid valves and ceramic tips, which are connected to plastic tubes and syringe pumps (Figure 13.1). The solenoid valve is an electromechanical device consisting of a metal rod and solenoid coils. The metal rod moves up and down by a magnetic field generated by an electric current in solenoid coils, thereby regulating the opening and closing of fluid flow in the valve. Ceramic tips are corrosive resistant and do not react with biological samples that are loaded from a 96-well plate and dispensed on the microarray chip platforms. The volume dispensed during printing is governed by the duration of solenoid valves being open, the total pressure applied by syringe pumps through tubes filled with water, and the viscosity of printed samples. Thus, careful optimization of the three parameters is necessary for consistent cell printing in hydrogels. In general, longer open times, higher pressures, and hydrogels with lower viscosities allow for the printing of larger volumes. Typical dispensing volumes are as small as 30 nL when paired with smaller scale devices, such as the micropillar and microwell chips, and no more than 5 μL when paired with 384-pillar plates. The syringe pumps can aspirate biological samples of 250 μL–1 mL, depending on the size of the syringes. When large volumes are aspirated, multiple 96-well plates can be used to load the syringes. Since biological samples are pushed by syringe pumps through water, approximately 5–20 μL of air acts as a buffer to avoid direct contact of water to the biological samples, which prevents sample contamination or accidental dilution with water. During the printing of biological samples, the ceramic tips encounter multiple different samples directly from a 96-well plate (sample reservoir) and dispense the materials onto various chip surfaces, thus requiring rinsing and drying steps between each sample loading to avoid cross-contamination. In order to maintain stability of the materials and prevent evaporation, temperature controls are implemented as to prevent any premature gelation and denaturation of proteins, such as growth factors. Generally, this involves keeping the 96-well plate and the surface of microarray chips on which materials are printed at 4°C–10°C, and, if necessary, keep the printing head at 4°C–10°C as well. The temperature of chilling decks where microarray chips are loaded is determined depending on the relative humidity in the laboratory. Generally, 6°C–12°C is preferred, as this is slightly below the condensation temperature so that the surface of microarray chips is slightly moist to avoid sample evaporation after printing. The temperatures of the deck of the 96-well plate and the printing head are determined depending on the thermostability of samples and gelation temperature of thermosensitive hydrogels.

Prior to sample printing, all dispensing surfaces are checked under inspection cameras by measuring the alignment of surfaces on the chilling deck and the volume of samples dispensed. During sample printing, ceramic tips are rinsed, sonicated, and vacuumed to ensure no debris or excess material is dangling from the tips as to ensure the accuracy of the sample dispensing. Over 90% of the printing time is dedicated to cleaning the tips and the solenoid valves. In addition, pre-dispensation of the samples occurs as a way for the microarray spotter to adjust syringe pressure in the tubes and check clogging of the samples. Pre-dispensation can involve anywhere between five and one hundred droplets being dispensed before sample printing starts.

13.3.2 PLATFORMS AND SURFACE COATINGS FOR CREATING MINIATURIZED TISSUE REPLICAS

The microarray chip platforms used for cell printing include the micropillar chip, the microwell chip, the 384-pillar plate with a flat tip surface, and the 384-pillar plate with sidewalls and slits (384PillarPlate) (Figure 13.2). Each of these platforms is well suited for different applications and different sizes of cultures. Cellular microarrays have been developed on functionalized glass slides after appropriate surface chemistry to promote spot attachment. However, combinatorial media exposure and cell layering would be impossible on these platforms as the defined area for microarrays is limited and hydrogels are not paired easily with separate wells. Thus, functionalized glass slides are no longer used due to their user unfriendliness.

Micropillar and microwell chips are best suited for cultures that support only a few hundred cells. Owing to the size of the droplets and the surface area of the micropillars, cells will generally arrange themselves only into one or two layers on these pillars. The microwell chips can serve either as a recipient of growth media to nourish cells on the micropillar chips or as vessels for cells themselves. Cells deposited in the microwell chips are larger in droplet size; however, growth media can only go through the top of the well acting as potential barriers to delivering nutrients to the cells printed into the microwells.

The 384-pillar plate with a flat tip surface and the 384PillarPlate are two different platforms better suited for larger scale, multiplexed cultures than the micropillar and microwell chips, but are still well suited for microarray 3D bioprinting technology. These platforms are specifically designed to be paired with commercially available 384-well plates, which have been adopted as a standard platform for high-throughput compound screening in the pharmaceutical industry. The 384-pillar plate has been

FIGURE 13.2 Schematics and images of various microarray chip platforms used in microarray 3D bioprinting including the micropillar chip with encapsulated cells (top left), microwell chip with printed reagents (top center), images of both platforms next to a microscope glass slide for scale (top right), schematic of 384-pillar plate with flat tip surface and encapsulated cells and platform image (bottom left), and schematic of the 384PillarPlate and image of platform (bottom right).

used for 2D cell monolayer cultures for toxicology screens (Yu et al. 2018). The 384PillarPlate is a modified 384-pillar plate that has been manufactured by injection molding of polystyrene for creating tissue replicas and modeling diseases *in vitro*. The pillars have various shapes and structures with sidewalls and slits that have been tested for optimum layer-by-layer cell printing, formation of a gel for cell encapsulation, and analysis of 3D cells in different layers for high-content imaging (HCI) assays for miniaturized tissue regeneration (Yu et al. 2018). The unique sidewall and slit structure on the 384PillarPlate facilitate layered cell printing, ensure robust cell spot attachment for HCI and immunofluorescent assays, and minimize air bubble entrapment for rapidly creating miniature multicellular biological constructs. By sandwiching the 384PillarPlate containing tissue replicas with a 384-well plate containing biological samples such as growth media, enzymes, viruses, compounds, and reagents, various miniaturized biochemical- and cell-based assays can be performed on the chip platform.

The surface of all microarray chip platforms must be chemically modified to promote robust cell spot attachment. Polystyrene is ideal as its optical properties have little to no interference with fluorescent imaging. Amphiphilic functional polymers, such as poly(maleic anhydride-*alt*-1-octadecene) (PMA-OD) and poly(styrene-*co*-maleic anhydride) (PS-MA), are used to provide hydrophobic affinity to polystyrene on the chip surface and reactivity to hydrophilic polymers and hydrogels, such as poly-L-lysine (PLL), collagen, and Matrigel. For example, hydrophilic alginate and PuraMatrix spots with cells have been successfully attached on the hydrophobic microarray chip surface *via* PMA-OD and PLL chemistry (Joshi, Yu, et al. 2018; Roth et al. 2018; Figure 13.3). PMA-OD is strongly attached on the surface

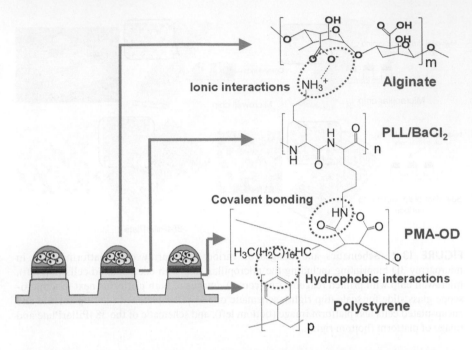

FIGURE 13.3 Surface chemistry of microarray chip platforms for cell encapsulation in alginate. Hydrophobic interactions between polystyrene chips and the octadecene end of PMA-OD allow for hydrogel attachment. The anhydride functional groups in PMA-OD bond to PLL that is co-printed with BaCl$_2$. This allows for attachment and polymerization of alginate to both anchor alginate and form a hydrogel.

of polystyrene through hydrophobic interactions. Maleic anhydride groups on PMA-OD can react with amine groups on PLL. PLL ionically binds to negatively charged alginate spots and creates robust spot attachment. For alginate, CaCl$_2$ or BaCl$_2$ is added to promote gelation. Plasma exposure under a vacuum can also work to negatively charge plastic and glass surfaces, and allow for hydrogel spot attachment {User's Manual for PDC-001-HP (115V) or PDC-002-HP (230V) High Power Expanded Plasma Cleaner [and Optional PDC-FMG (115V) or PDC-FMG-2 (230V) PlasmaFlo] 2015; Chen et al. 2013; Lee, Kwon et al. 2013}.

13.3.3 HYDROGELS FOR CELL ENCAPSULATION

Hydrogels are the most important parameter in microarray 3D bioprinting for characterizing 3D cell growth. As hydrogels form the main scaffold for these cells to grow in 3D, it is important to select a material that is mechanically stable and biocompatible to promote 3D cell proliferation. With hydrogels being rather viscous and the small volumes used in printing, it is important to optimize dispensing parameters, such as valve open time, nozzle diameter, and syringe pressure, to minimize errors associated with the droplet size of printed hydrogels. Common hydrogels that can be used for printing include salt-sensitive hydrogels (alginate, PuraMatrix)

(Joshi, Yu, et al. 2018; Roth et al. 2018), temperature-sensitive hydrogels (Matrigel, Geltrex, Collagen I, MaxGel) (Joshi, Yu, et al. 2018; Wong et al. 2007; Zarowna-Dabrowska et al. 2012), enzymatically catalyzed hydrogels (fibrinogen) (Zhang et al. 2017), and photopolymerized hydrogels [oxidized, methacrylated alginate (OMA), methacrylated collagen] (Klotz et al. 2016; Arya et al. 2016). These hydrogels can be supplemented with ECMs and growth factors to provide better cellular micro-environments. For example, alginate can be mixed with Matrigel and cells, and the mixture can be printed on the microarray chip platforms (Joshi, Yu, et al. 2018). The uses for various hydrogels are detailed in the applications section.

The parameters of hydrogel gelation include salt concentration, light intensity, temperature, and pH (Liang et al. 2011; Gaharwar et al. 2013; Xing et al. 2014; Andersen et al. 2014). The mechanical property of hydrogels can be changed by the concentration of hydrogels, which influences the viscosity of cell–hydrogel mixtures printed (Kang et al. 2017). Typical viscosity of 1% low-molecular-weight alginate with cells is below 100 centipoise (cP), which can be easily handled by the microarray spotter (Sun and Tan 2013). Both natural and synthetic hydrogels can be used for microarray 3D bioprinting (Roth et al. 2018; Joshi, Yu, et al. 2018). As discussed earlier, maintaining low temperature on both the dispensing head and the printed chip surface is critical for functional performance. Typical cell viability after printing is above 90%.

Alginate and OMA are polysaccharide-based hydrogels, which are structurally stable and biocompatible because they are not degraded by matrix metalloproteinases (MMPs) (Samorezov, Morlock, and Alsberg 2015; Pawar and Edgar 2012). These seaweed-based hydrogels initially showed their biocompatibility by encapsulating pancreatic islet cells for the study of diabetes (Rowley, Madlambayan, and Mooney 1999). Alginate is polymerized *via* ionic crosslinking between negatively charged carboxylic acid groups and divalent cations, such as $CaCl_2$ or $BaCl_2$, while OMA is functionalized alginate that is photopolymerized *via* covalent crosslinking of methacrylate groups using photoinitiators, such as Irgacure-2959 (Sigma-Aldrich) (Kang et al. 2017). Because alginate requires cations for polymerization, nontoxic salts such as $CaCl_2$ need to be printed on the microarray chip platforms prior to cell printing (Figure 13.3). Owing to the fact that near-UV light is required for photopolymerization, OMA does not require any pre-printing steps; however, UV-light exposure in the protocol generates free radicals and increases surface temperature, causing possible basal cytotoxicity that can be an issue if all parameters are not optimized for cell encapsulation (Xu et al. 2015).

In the case of Matrigel, Geltrex, and collagen, polymerization is initiated through elevated temperatures, typically at 37°C (Klouda and Mikos 2008; Van Vlierberghe, Dubruel, and Schacht 2011). This necessitates using a temperature-controlled dispensing head to prevent premature gelation within the solenoid valves and the tubes connected to the valves and the syringe pumps. These protein-based hydrogels generally contain active sites that promote the binding and adherence of various cell types (Jonker, Löwik, and Van Hest 2012; Censi et al. 2012). However, they are more susceptible to degradation by MMPs and generally take 20–30 min to gel as compared to OMA and alginate, which is resistant to MMPs and polymerizes in 1–4 min (Saldin et al. 2017; Skardal et al. 2012). Furthermore, because these hydrogels require the temperature to be 37°C for polymerization, the microarray chips with temperature-sensitive hydrogels must be kept in a humid chamber during the gelation period to prevent water evaporation. Another drawback to specifically using Matrigel and Geltrex is the undefined formulations associated with them. Matrigel and Geltrex consist of combinations of proteins that are proprietary to Engelbreth-Holm-Swarm mouse sarcoma cells. These protein formulations are not always consistent, which results in some variation in the behavior of cells in 3D culture.

13.3.4 COMPARISON TO OTHER MINIATURIZED 3D CELL CULTURE TECHNOLOGY

A wide range of cell types, including primary cells (e.g., cancer cells, stem cells, mature cells) and cell lines, can be mixed with hydrogels and printed on the microarray chip platforms for spheroid and organoid cultures. For successful cell printing, it is critical to maintain cells suspended in hydrogels for 5–10 min, which is the time necessary for dispensing. When the concentration of cells printed is too high or viscosity of hydrogels is too low, cells sometimes can block the tips and solenoid valves. Unlike other 3D cell culture platforms, including ULA well plates, Transwell inserts, hanging droplet plates, and microfluidic plates, which are relatively low throughput and require cumbersome steps for changing growth media, microarray chip platforms are amenable to high-throughput 3D tissue culture and HCI. This allows scientists and researchers to test multiple parameters simultaneously when optimizing cell culture conditions.

The key requirement associated with microarray 3D bioprinting involves the optimization procedures related to the cell printing, including the selection of hydrogels and the printing parameters. As discussed earlier, the printing volumes are very small, so that samples printed are susceptible to evaporation more than in regular cell culture or large-scale 3D printing conditions. In addition, depending on the cells used and the applications, the hydrogels used and the parameters for encapsulation (including hydrogel concentration, any reagents necessary for gelation, and time necessary for gelation) can vary significantly. Furthermore, surface chemistry is necessary to promote attachment between hydrogels and the chip surfaces that are suitable for subsequent imaging and analysis, as discussed in the next section. However, owing to the high-throughput nature of the microarray platforms, all of these parameters can be tested and optimized rather easily.

13.4 METHODS OF CELL STAINING, IMAGING, AND ANALYSIS

In general, imaging cells and spheroids in tissue constructs created by current 3D bioprinting poses significant challenges because the cells are not grown in a single focal plane and the size of the tissue is too large. The only viable options for imaging these cells involve cryosectioning and immunostaining, followed by confocal microscopy (Guillouzo and Guguen-Guillouzo 2008; Cui et al. 2012). In addition, quantitative analysis of cells printed in 3D is challenging due to high inconsistency and low reproducibility. Unlike relatively large 3D printed tissues, miniature tissue replicas on the microarray chip platforms allow the whole sample depth to fit within the focus depth of a normal objective (Sukumaran et al. 2009; Fernandes et al. 2010; Kwon et al. 2014;

Dolatshahi-Pirouz et al. 2014). Thus, the imaging area is clear and suitably thin, making imaging mini-tissue replicas easier than imaging large-scale 3D printed tissues.

Spheroids and tissue replicas cultured on the microarray chip platforms can be stained with various reagents for absorbance, fluorescence, and luminescence measurements. The 384-pillar plate with flat tips and the 384PillarPlate are compatible with conventional microtiter plate readers and automated fluorescence microscopes to rapidly acquire signals from cells. Unlike current 3D bioprinting approaches, microarray 3D bioprinting is ideal for HTS assays because of the small size of bioprinted tissues, which also decreases the amount of expensive human cells required and improves diffusion of nutrients and oxygen into the core of mini-tissue replicas. In this section, we explain the various methods used in imaging cells within bioprinted mini-tissues, including using fluorescent dyes, antibodies, and reporter genes. In addition, we will focus on the image analysis used to quantify cellular behavior at this small scale.

13.4.1 AUTOMATED FLUORESCENCE MICROSCOPE

Imaging of bioprinted mini-tissue replicas requires an automated fluorescence microscope that can adjust the positions around the pillar array given the large number of samples that need to be analyzed. In addition, because of the number of parameters that need to be measured, microscopes should be fitted with a variety of filters that can detect various fluorescent signals, specifically those corresponding with blue, green, orange, and red wavelengths. The stages used need to be able to fit both microscope slide-sized micropillar/microwell chips and 384-pillar plates. For example, the S+ scanner from Samsung Electro-Mechanics Co. (SEMCO) is equipped with a BrightLine® full-multiband filter set (DA/FI/TR/Cy5-A-000 from Semrock) for measuring blue, green, orange, and red fluorescent dyes simultaneously, and three single-band filter sets (DAPI-5060C-000, XF404, and TXRED-4040C-000) optimized for broad blue fluorophores, green fluorophores, and deep red fluorophores. This allows the user to obtain multi-color images from stained cells. To avoid excitation/emission spectrum overlapping among different fluorescent probes and to obtain clear separation of fluorescence signals, precautions should be taken when mixing multiple fluorescent dyes for cell staining. To obtain images of cell spots, cell colonies, individual cells, and small organelles, 4× and 20× objective lenses generally are used in combination with the full-multiband filter set. The time for image acquisition can be relatively short; the time to scan a single chip with 532 micropillars or 384 pillars is 10–30 min, depending on exposure time. While real-time image analysis is generally unnecessary on cells that are stained with fluorescent dyes or antibodies, cells with fluorescent biomarkers on 384-pillar plates do have the capability of being monitored *in situ* over an extended period.

13.4.2 HIGH-CONTENT IMAGING (HCI) ASSAYS WITH FLUORESCENT DYE STAINS

Stains can give insight into cues of what is occurring inside the cell. They involve the use of chemical dyes that bind to certain molecules and organelles within the cells which can be detected physically or chemically. Fluorescent stains can be used to label organelles or measure the activity of molecules within the cell *via* kinetic analysis (Joshi and Lee 2015). Common targets for fluorescent stains include the cell membrane, nucleus, mitochondria, endoplasmic reticulum (ER), and lysosomes (Wilson, Graham, and Ball 2014; Joshi and Lee 2015). A major advantage to specific organelle labeling is the ability to perform HCI assays on cells. While this is easier to do in 2D-cultured cells, HCI has been performed in 3D-cultured cells, giving scientists a versatile tool to measure multiple parameters (Joshi and Lee 2015).

Several established dyes used in HCI that are compatible with bioprinted mini-tissue replicas include acetoxymethyl calcein (calcein AM) and ethidium homodimer-1 for cell viability/cytotoxicity (Lee et al. 2008), Hoechst 33342 for changes in nuclear function, YO-PRO-1 and propidium iodide for apoptosis/necrosis (Kwon et al. 2014), tetramethyl rhodamine methyl ester (TMRM) for mitochondrial membrane potential (O'Brien et al. 2006), BODIPY 665 for oxidative stress (Tolosa, Gómez-Lechón, and Donato 2015), fluo-4 acetoxy methyl ester (Fluo-4 AM) for intracellular calcium ions (O'Brien et al. 2006), and monochlorobimane (mBCl)/ thiol green dye for glutathione (Joshi, Datar, et al. 2018; Lee 2016; Figure 13.4). In addition, clonogenic assays have been established to measure colony formation on the micropillar chip (Lee, Choi, Seo, Lee, Jeon, Ku, Kim, et al. 2014; Lee, Choi,

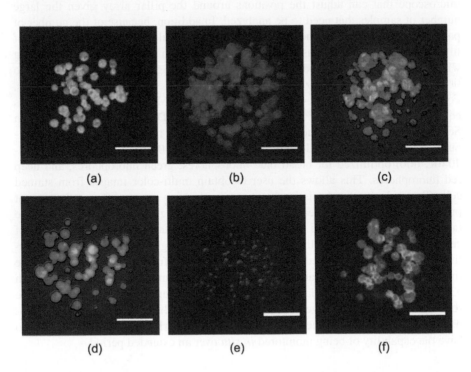

FIGURE 13.4 Representative images of 3D-cultured Hep3B human liver cells (60 nL) on the micropillar chip and stained with (a) calcein AM for cell viability, (b) TMRM for mitochondrial membrane potential, (c) Hoechst 33342 for nucleus morphology and cell count, (d) mBCl for glutathione level, (e) propidium iodide for cell viability and necrosis, and (f) Fluo-4 AM for intracellular calcium level. The scale bar is 200 μm.

Seo, Lee, Jeon, Ku, and Nam 2014). These assays will likely shed light on cellular processes that play pivotal roles in mechanistic toxicity.

13.4.3 IN-CELL IMMUNOFLUORESCENCE ASSAYS

While chemical dye stains can characterize specific organelles or quantify general behavior, they are generally toxic to cell health and are quite susceptible to photobleaching due to overexposure and membrane leakage over time. Applications in immunofluorescence allow for targeting specific protein markers in the cell. Because antibodies bind strongly to their target proteins, this method is widely used to detect the upregulation of disease targets. The disadvantage to using immunofluorescence is that antibodies are less likely to penetrate the cell membrane without appropriate permeabilization. Although antibodies bind strongly to their targeted antigen sites, there are occasionally issues with antibodies where binding is non-specific, leading to the detection of false positives. In addition, immunofluorescent stains

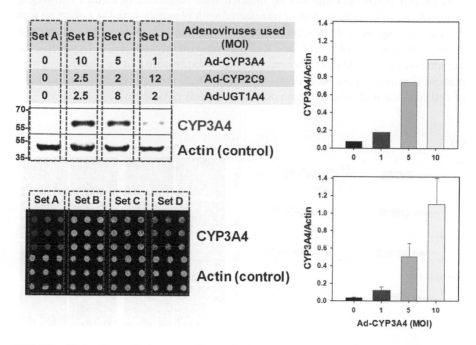

FIGURE 13.5 Controlled expression of drug-metabolizing enzyme CYP3A4 by co-transfecting different ratios of Ad-CYP3A4 (top left). Western blot analysis of THLE-2 cell monolayers co-expressing CYP3A4 (bottom left). In-cell immunofluorescence assay of THLE-2 cells co-expressing CYP3A4 on the micropillar chip. Three combinations of multiplicities of infection (MOIs) of the three recombinant adenoviruses at a total MOI of 15 (sets B, C, and D) were used to compare the co-expression levels of the three drug-metabolizing enzymes expressed in THLE-2 cells on the micropillar chip with those obtained from THLE-2 cell monolayers in six-well plates. The bar graphs represent the co-expression levels of CYP3A4 expressed in THLE-2 cell monolayers (top) and THLE-2 cells on the micropillar chip (bottom) (Kwon et al. 2014).

require aldehyde use during the fixation steps, which cause problems with toxicity. Immunofluorescence has been successfully demonstrated on the microarray chip platforms with cells overexpressing protein markers (Figure 13.5; Kwon et al. 2014; Fernandes et al. 2008). For example, the co-expression levels of three drug-metabolizing enzymes (CYP3A4, CYP2C9, and UGT1A4) expressed in THLE-2 cells on the micropillar chip were assessed using fluorescently labeled antibodies (Kwon et al. 2014; Gustafsson et al. 2014).

13.4.4 PROMOTER–REPORTER GENE ASSAYS

While HCI assays and in-cell immunofluorescence assays can provide useful information on mechanisms of toxicity and cell functions, they cannot be used for real-time cell imaging because of fluorophores leaching out, photobleaching over time, and cell membrane permeabilization. *In situ* real-time cell imaging on the microarray chip platforms can be performed by promoter–reporter gene assays to track cell behavior. Cells can be infected with recombinant viruses carrying a promoter–reporter gene system or transfected with plasmids with the system so that the cells can express fluorescently labeled proteins, such as blue, green, and red fluorescent proteins (BFP, GFP, and RFP), along with target proteins. Depending on the method of gene delivery chosen, the protein expression can be transient or permanent (Sekine, Takebe, and Taniguchi 2014). Thus, this approach allows tracking of cell function and differentiation over long periods of time.

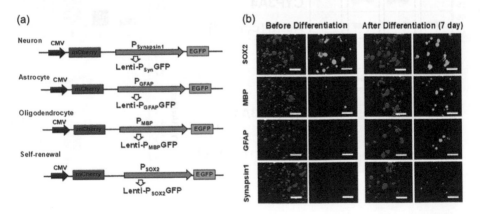

FIGURE 13.6 (a) Construction of lentiviral vectors carrying a promoter–reporter assay system for high-throughput assessment of ReNcell VM differentiation (Joshi, Yu, et al. 2018). (b) ReNcell VM cultured in 3D on the micropillar chip after infection with lentiviruses for self-renewal and differentiation. Lentivirus-infected ReNcell VM encapsulated in a mixture of 0.75% (w/v) alginate and 1 mg/mL Matrigel was printed on the micropillar chip. Differentiation was induced by removing the growth factors (EGF and bFGF) in the growth medium. Images were obtained before and 1 week after differentiation to determine the self-renewal and differentiation capability of ReNcell VM. Red-colored cells indicate the ReNcell VM infected with the lentiviruses, and the green-colored cells indicate ReNcell VM differentiation into respective lineages. The scale bar is 200 μm.

Recombinant lentiviruses carrying a promoter–reporter assay system have been constructed and infected into neural stem cells (NSCs) in bioprinted cellular microarrays to monitor cell differentiation over time on the microarray chip platforms (Joshi, Yu, et al. 2018; Figure 13.6). The CMV promoter-driven mCherry gene was used and constitutively expressed in ReNcell VM, and the enhanced green fluorescent protein (EGFP) gene was expressed under the control of cell-type-specific promoters. Four NSC-specific biomarkers, including sex-determining region Y box 2 (SOX2), synapsin 1, glial fibrillary acidic protein (GFAP), and myelin basic protein (MBP), were used to evaluate self-renewal, neuron differentiation, astrocyte differentiation, and oligodendrocyte differentiation, respectively.

13.4.5 IMAGE ACQUISITION AND PROCESSING

Given the large number of images generated from the microarray chips, it is essential to perform high-throughput image analysis. ImageJ and S+ Chip Analysis (in-house data analysis software) have been utilized to identify 3D-cultured cells in different hydrogel layers and investigate multiple parameters involved in cell functions and mechanisms of toxicity (Joshi, Yu, et al. 2018; Yu et al. 2018; Figure 13.7). This software automatically extracts fluorescent intensities from cells in each cell layer and then generates sigmoidal dose–response curves and calculates IC_{50} values for each test condition.

To quantify any assays conducted on the microarray chips, cell images obtained must be put through various analyses that will filter out noise and extract a signal that adequately represents any reactions occurring within the cells. First, fluorescently labeled cell images must be separated through different channels to extract out information on the colors of interest (Figure 13.7a). This is common for livedead assays performed with calcein AM and ethidium homodimer-1, or in situations

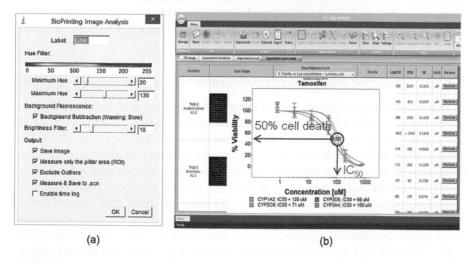

(a)   (b)

FIGURE 13.7 Image analysis software developed: (a) ImageJ plugin developed for data extraction from fluorescent images and (b) S+ Chip Analysis developed for plotting dose–response curves and calculating IC_{50} values of compounds.

where there are multiple stains. After splitting channels, background is subtracted from the image by removing low- and high-frequency noise. Following this, fluorescence is quantified relative to the size of the plate. ImageJ provides plenty of tools to analyze images, as does much software that is paired with fluorescent microscopes.

One of the major challenges to imaging in 3D is filtering out signals that are taken in various Z focus planes within a given tissue replica. This is especially true with fluorescently labeled cells, as expression of these markers can be so great as to either bleed into a different Z-position image or be eliminated during initial background subtraction, which necessitates a good choice of a gain value when obtaining images. Yu et al. have established the ability to acquire hepatic cell images from two different alginate layers containing Hoechst 33342 and TMRM-stained Hep3B cells in the microwell chip (Yu et al. 2017). For situations where fluorescence can bleed into different Z-positions, performing high-frequency background subtraction under the frequency domain of a fast Fourier transform (FFT) is generally enough (Driscoll and Danuser 2015). While this decreases the relative size examined for fluorescent cells in a given tissue replica, performing the analysis over the course of the whole mini-tissue will result in the normalization of the data, and the error will be corrected.

13.5 APPLICATIONS OF MINI-TISSUE REPLICAS ON MICROARRAY CHIP PLATFORMS

While robotic liquid dispensing technology has been implemented in wide use in terms of 2D cell culture, microarray 3D bioprinting technology is a recent advancement in high-throughput platforms. Within this technology, scientists can create miniaturized constructs of various human tissues, allowing for modeling of disease states associated with these tissues for predictive drug screening. In this section, we focus on several applications of spheroids and tissue replicas cultured on the microarray chips, with specific examples of modeling drug toxicity, personalized cancer screens, and generalized disease models.

13.5.1 Predictive Drug Toxicity Assessment

A major application of microarray 3D bioprinting is the use of tissue replicas for predictive screening of drug toxicity and efficacy (Lee et al. 2008; Kwon et al. 2014; Joshi, Datar, et al. 2018; Lee, Choi, Seo, Lee, Jeon, Ku, Kim, et al. 2014; Lee 2016). While HTS is widely used in assessing the potential for various compounds to effectively treat certain diseases, current HTS technology has mostly focused on using large-scale liquid dispensing technology to determine potential interactions between compounds and disease targets on 2D cell monolayers. As discussed earlier in this chapter, 2D cell models have distinct limitations to 3D cell models in terms of their genetic and phenotypic properties and lack important interactions with ECMs, which makes 3D cell models better suited for predicting toxicity and efficacy of drug candidates.

These deviations between 2D and 3D can manifest as changes in the experienced IC_{50} values associated with certain compounds (Joshi, Datar, et al. 2018), although mechanistic toxicology may be influenced by multiple parameters including gene expression levels (Lee, Choi, Seo, Lee, Jeon, Ku, Kim, et al. 2014). In addition, diffusion and transport of molecules including oxygen, nutrients, and compounds play an important role in 3D cell cultures (Haycock 2011; Justice, Badr, and Felder 2009). Owing to miniaturization of 3D cell cultures, spontaneous cell death in the core of bioprinted mini-tissue replicas (called "zonation") and diffusion limitation of compounds tested have not been observed (Fernandes et al. 2010; Joshi, Yu, et al. 2018). In addition, non-specific binding of compounds on the microarray chips is minimal because all the platforms are made of polystyrene (Sanyal 2014; Lee, Yi et al. 2013). Unlike 2D cell cultures where cell–ECM interactions are largely absent unless some ECM components are placed on the surface as to facilitate cells binding to the surface, cell–ECM interactions can be facilitated by supplementing ECM molecules directly in hydrogels (Ma et al. 2015). These unique features of 3D cells cultured on the microarray chips could potentially lead to more predictive outcomes of drug toxicity and efficacy.

13.5.2 PERSONALIZED CANCER SCREENING

While most HTS assays are performed on 2D cells overexpressing disease targets such as enzymes, receptors, and transporters, personalized anticancer drug screening can be performed on the microarray chip platforms with 3D-cultured cells from patients (Lee, Choi, Seo, Lee, Jeon, Ku, Kim, et al. 2014). The most important technical issue in predictive anticancer drug screening is to create tissue replicas that can mimic heterogeneous tumor tissues *in vivo*. Depending on the stage of tumor progression, underlying genetic mutations, and the microenvironment surrounding the cancerous tissue, the physiological conditions *in vivo* can be vastly different (Holle, Young, and Spatz 2016; Li et al. 2008). Recently, great progress has been made in creating tumor organoids derived from biopsy samples from patients (Nuciforo et al. 2018). Several groups recapitulated tumor tissues by mincing and trypsinizing biopsied tissues, generating primary cell aggregates and embedding them in Matrigel (Tsai et al. 2018). They successfully demonstrated that the outcomes from tumor organoid studies are more predictive to clinical data (Tsai et al. 2018). However, tumor organoid cultures are performed on relatively low-throughput platforms, such as petri dishes and six-well plates (Shroyer 2016). In addition, it is difficult to control the size of tumor organoids and diffusion of oxygen and nutrients in hydrogels due to variations in the locations of cells (Shroyer 2016; Skardal et al. 2015).

These issues can be addressed by using microarray 3D bioprinting technology, which has the potential to run biopsied samples against a suite of various treatment options. For example, cell aggregates in hydrogels can be printed rapidly and uniformly on the 384PillarPlate and cultured long term to create tumor organoids, which can be treated with anticancer drugs and stained with a suite of fluorescent dyes to identify mechanisms of drug action (Yu et al. 2018). The best working drugs and dosages identified from this assay could be potentially used for creating personalized drug regiments for cancer patients.

The major advantage microarray 3D bioprinting technology offers is the large suite of multiplexed assays that can be performed on patient samples. Scientists and clinicians can print mixtures of healthy and diseased cells in single hydrogel layer or print individual cell types in multiple hydrogel layers to better mimic cellular microenvironments. In addition, the size of tumor tissues can be confined to 1–5 μL total volume, which allows for better control of the diffusion of oxygen and nutrients. This in turn aids in experimental reproducibility and less variance in the data obtained from mini-tissue replicas.

13.5.3 TISSUE AND DISEASE MODELS

Recent advances in organoid cultures, including the brain, heart, lung, liver, intestine, and pancreas, have demonstrated great promise as human tissue replicas. They can mimic morphological features of human tissues and contain multiple cell types relevant to specific tissues *in vivo*, maintaining viability and function for several weeks (Picollet-D'hahan et al. 2017). While organoids represent a new direction in predictive *in vitro* efficacy and toxicity screening, there are several technical challenges to adopt organoids in disease modeling (Laurent et al. 2017). Current organoid cultures rely on the ability of pluripotent stem cells (PSCs) to self-organize into discrete tissue structures with a step-wise process that mimics normal organ development (McCauley and Wells 2017). This spontaneous differentiation of PSCs into multiple cell lineages, as well as clone-dependent differences, determines the structural complexity and cell-type diversity in organoids, leading to considerable organoid variations (Ranga, Gjorevski, and Lutolf 2014; Gjorevski, Ranga, and Lutolf 2014).

Since it takes so long for organoids to fully mature (typically 1–6 months), it is important to minimize batch-to-batch differences. These variations present obstacles for quantitative analysis and reproducibility of results. In addition, current methods of culturing and analyzing organoids are low throughput, which is a major obstacle to conduct rapid drug screening *in vitro*. As previously mentioned, cultures in petri dishes and bioreactors, as well as their structural and cellular characterization, including cryosectioning, immunostaining, and image collection, are labor-intensive and slow. In addition, the size of these organoids often exceeds the diffusion limit of oxygen and nutrients, leading to cell death in the organoid core despite its highly endothelialized structures. While several efforts have been made to generate uniform organoids under defined conditions, it has been difficult to establish standardized organoid culture and quality, such as lineage specificity, proportion of different cell types, and the relative size of organoids (Camp et al. 2017; Takai et al. 2016).

These problems represent a major gap in the development of high-throughput, predictive models for concordance between *in vitro* assays and *in vivo* responses and a significant opportunity for new technologies to fill this gap. Current 3D cell culture platforms are relatively low throughput, require cumbersome steps for changing growth media, and are not amenable to high-throughput, multicellular 3D tissue culture and high-content cell analysis. Although current 3D cell co-cultures partly mimic *in vivo* tissue structure, these approaches still lack the ability to provide "highly predictive" information on efficacy and toxicity *in vivo*, due to the lack of mimicking heterogeneous multicellular interactions and tissue structures, which are critical to model any disease state. We envision that engineering approaches such as robust cell printing on the microarray chip platforms with defined matrix and media components can play an important role in streamlining organoid culture and image analysis with minimal manual intervention for disease studies. For example, human liver organoids in biomimetic hydrogels can be precisely dispensed on the 384PillarPlate to replicate miniature liver tissues *in vitro* with intraluminal structures and mimic critical human hepatocyte functions, including metabolism, protein and bile acid production, and transport functions. In addition, miniaturized liver tissue constructs with key hepatic functions can be created on the 384PillarPlate by mimicking the microstructure of liver tissues *via* layered cell printing, which can be paired with 384-well plates containing additives such as growth factors or compounds to model inflammation reactions or other disease states in the liver. Ultimately, the goal of clinicians and researchers is to obtain a more complete profile of the disease by initially looking at healthy tissue models before accounting for disease state microenvironments. Mini-tissue replicas could be the best method for this purpose because of their high-throughput nature, which allows for accounting of any variation in disease states.

13.6 CONCLUSION

As scientists, doctors, and researchers seek to understand more about organ/tissue behavior, microarray 3D bioprinting brings a necessary tool to bridge that information gap. Bioprinted mini-tissue replicas allow for a combinatorial examination of tissues, specifically looking into multiple parameters that can affect cell proliferation, survival, and behavior. There are several steps that are needed to undertake microarray 3D bioprinting that require careful selection and optimization, including the choice of hydrogels and printing parameters. Once cell printing and maturation are completed, mini-tissue replicas can be exposed to test compounds and fluorescent imaging and analysis can quantify efficacy and toxicity of the compounds. This technology has a broad range of applications in tissue engineering, disease modeling, and drug discovery.

REFERENCES

Andersen, Therese, Jan Egil Melvik, Olav Gåserød, Eben Alsberg, and Bjørn E. Christensen. 2014. Ionically gelled alginate foams: Physical properties controlled by type, amount and source of gelling ions. *Carbohydrate Polymers* 99:249–56.

Arya, Anuradha D., Pavan M. Hallur, Abhijith G. Karkisaval, Aditi Gudipati, Satheesh Rajendiran, Vaibhav Dhavale, Balaji Ramachandran, Aravindakshan Jayaprakash, Namrata Gundiah, and Aditya Chaubey. 2016. Gelatin methacrylate hydrogels as biomimetic three-dimensional matrixes for modeling breast cancer invasion and chemoresponse in vitro. *ACS Applied Materials and Interfaces* 8 (34):22005–17.

Asai, Akihiro, Eitaro Aihara, Carey Watson, Reena Mourya, Tatsuki Mizuochi, Pranavkumar Shivakumar, Kieran Phelan, et al. 2017. Paracrine signals regulate human liver organoid maturation from IPSC. *Development* 144 (6):1056–64.

Bertassoni, Luiz E., Juliana C. Cardoso, Vijayan Manoharan, Ana L. Cristino, Nupura S. Bhise, Wesleyan A. Araujo, Pinar Zorlutuna, et al. 2014. Direct-write bioprinting of cell-laden methacrylated gelatin hydrogels. *Biofabrication* 6 (2):024105.

Bhatia, Sangeeta N., and Donald E. Ingber. 2014. Microfluidic organs-on-chips. *Nature Biotechnology* 32 (8):760–72.

Camp, J. Gray, Keisuke Sekine, Tobias Gerber, Henry Loeffler-Wirth, Hans Binder, Malgorzata Gac, Sabina Kanton, et al. 2017. Multilineage communication regulates human liver bud development from pluripotency. *Nature* 546:533–8.

Cavnar, Stephen P., Emma Salomonsson, Kathryn E. Luker, Gary D. Luker, and Shuichi Takayama. 2014. Transfer, imaging, and analysis plate for facile handling of 384 hanging drop 3D tissue spheroids. *Journal of Laboratory Automation* 19 (2):208–14.

Censi, Roberta, Piera Di Martino, Tina Vermonden, and Wim E. Hennink. 2012. Hydrogels for protein delivery in tissue engineering. *Journal of Controlled Release* 161 (2):680–92.

Chen, Yashao, Qiang Gao, Haiyan Wan, Jinhong Yi, Yanlin Wei, and Peng Liu. 2013. Surface modification and biocompatible improvement of polystyrene film by Ar, O_2 and Ar + O_2 plasma. *Applied Surface Science* 265:452–7.

Cui, Xiaofeng, Thomas Boland, Darryl D. D'Lima, and Martin K. Lotz. 2012. Thermal inkjet printing in tissue engineering and regenerative medicine. *Recent Patents on Drug Delivery & Formulation* 6 (2):1–13.

Dolatshahi-Pirouz, Alireza, Mehdi Nikkhah, Akhilesh K. Gaharwar, Basma Hashmi, Enrico Guermani, Hamed Aliabadi, Gulden Camci-Unal, et al. 2014. A combinatorial cell-laden gel microarray for inducing osteogenic differentiation of human mesenchymal stem cells. *Scientific Reports* 4:3896.

Driscoll, Meghan K., and Gaudenz Danuser. 2015. Quantifying modes of 3D cell migration. *Trends in Cell Biology* 25 (12):749–59.

Fernandes, Tiago G., Seok Joon Kwon, Shyam Sundhar Bale, Moo Yeal Lee, Maria Margarida Diogo, Douglas S. Clark, Joaquim M. S. Cabral, and Jonathan S. Dordick. 2010. Three-dimensional cell culture microarray for high-throughput studies of stem cell fate. *Biotechnology and Bioengineering* 106 (1):106–18.

Fernandes, Tiago G., Seok Joon Kwon, Moo Yeal Lee, Douglas S. Clark, Joaquim M. S. Cabral, and Jonathan S. Dordick. 2008. On-chip, cell-based microarray immunofluorescence assay for high-throughput analysis of target proteins. *Analytical Chemistry* 80 (17):6633–9.

Gaharwar, Akhilesh K., Christian Rivera, Chia Jung Wu, Burke K. Chan, and Gudrun Schmidt. 2013. Photocrosslinked nanocomposite hydrogels from PEG and silica nanospheres: Structural, mechanical and cell adhesion characteristics. *Materials Science and Engineering C* 33 (3):1800–7.

Gjorevski, Nikolche, Adrian Ranga, and Matthias P. Lutolf. 2014. Bioengineering approaches to guide stem cell-based organogenesis. *Development* 141 (9):1794–804.

Guillouzo, André, and Christiane Guguen-Guillouzo. 2008. Evolving concepts in liver tissue modeling and implications for in vitro toxicology. *Expert Opinion on Drug Metabolism & Toxicology* 4 (10):1279–94.

Gustafsson, Frida, Alison J. Foster, Sunil Sarda, Matthew H. Bridgland-Taylor, and J. Gerry Kenna. 2014. A correlation between the in vitro drug toxicity of drugs to cell lines that express human P450s and their propensity to cause liver injury in humans. *Toxicological Sciences* 137 (1):189–211.

Haycock, John. 2011. 3D cell culture: A review of current approaches and techniques. *Methods in Molecular Biology* 695:243–59.

van der Helm, Marinke W., Andries D. van der Meer, Jan C. T. Eijkel, Albert van den Berg, and Loes I. Segerink. 2016. Microfluidic organ-on-chip technology for blood-brain barrier research. *Tissue Barriers* 4 (1):e1142493.

Holle, Andrew W., Jennifer L. Young, and Joachim P. Spatz. 2016. In vitro cancer cell-ECM interactions inform in vivo cancer treatment. *Advanced Drug Delivery Reviews* 97:270–9.

Iwai, Soichi, Satoko Kishimoto, Yuto Amano, Akihiro Nishiguchi, Michiya Matsusaki, Akinori Takeshita, and Mitsuru Akashi. 2018. Three-dimensional cultured tissue constructs that imitate human living tissue organization for analysis of tumor cell invasion. *Journal of Biomedical Materials Research Part A* 12:1–9.

Jang, Mi, Pavel Neuzil, Thomas Volk, Andreas Manz, and Astrid Kleber. 2015. On-chip three-dimensional cell culture in phaseguides improves hepatocyte functions in vitro. *Biomicrofluidics* 9 (3):1–12.

Jo, Junghyun, Yixin Xiao, Alfred Xuyang Sun, Engin Cukuroglu, Hoang-Dai Tran, Jonathan Göke, Zi Ying Tan, et al. 2015. Midbrain-like organoids from human pluripotent stem cells contain functional dopaminergic and neuromelanin-producing neurons. *Cell Stem Cell* 19 (2):248–57.

Jonker, Anika M., Dennis W. P. M. Löwik, and Jan C. M. Van Hest. 2012. Peptide- and protein-based hydrogels. *Chemistry of Materials* 24 (5):759–73.

Joshi, Pranav, Akshata Datar, Kyeong-Nam Yu, Soo-Yeon Kang, and Moo-Yeal Lee. 2018. High-content imaging assays on a miniaturized 3D cell culture platform. *Toxicology in Vitro* 50:147–59.

Joshi, Pranav, and Moo-Yeal Lee. 2015. High content imaging (HCI) on miniaturized three-dimensional (3D) cell cultures. *Biosensors* 5 (4):768–90.

Joshi, Pranav, Kyeong Nam Yu, Soo Yeon Kang, Seok Joon Kwon, Paul S. Kwon, Jonathan S. Dordick, Chandrasekhar R. Kothapalli, and Moo Yeal Lee. 2018. 3D-cultured neural stem cell microarrays on a micropillar chip for high-throughput developmental neurotoxicology. *Experimental Cell Research* 370 (2):680–91.

Justice, Bradley A., Nadia A. Badr, and Robin A. Felder. 2009. 3D cell culture opens new dimensions in cell-based assays. *Drug Discovery Today* 14 (1):102–7.

Kang, Laura Hockaday, Patrick A. Armstrong, Lauren Julia Lee, Bin Duan, Kevin Heeyong Kang, and Jonathan Talbot Butcher. 2017. Optimizing photo-encapsulation viability of heart valve cell types in 3D printable composite hydrogels. *Annals of Biomedical Engineering* 45 (2):360–77.

Kermanizadeh, Ali, Mille Løhr, Martin Roursgaard, Simon Messner, Patrina Gunness, Jens M. Kelm, Peter Møller, Vicki Stone, and Steffen Loft. 2014. Hepatic toxicology following single and multiple exposure of engineered nanomaterials utilising a novel primary human 3D liver microtissue model. *Particle and Fibre Toxicology* 11 (56):1–15.

Klotz, Barbara J., Debby Gawlitta, Antoine J. W. B. Rosenberg, Jos Malda, and Ferry P. W. Melchels. 2016. Gelatin-methacryloyl hydrogels: Towards biofabrication-based tissue repair. *Trends in Biotechnology* 34 (5):394–407.

Klouda, Leda, and Antonios G. Mikos. 2008. Thermoresponsive hydrogels in biomedical applications. *European Journal of Pharmaceutics and Biopharmaceutics* 68 (1):34–45.

Kwon, Seok Joon, Dong Woo Lee, Dhiral A. Shah, Bosung Ku, Sang Youl Jeon, Kusum Solanki, Jessica D. Ryan, Douglas S. Clark, Jonathan S. Dordick, and Moo-Yeal Lee. 2014. High-throughput and combinatorial gene expression on a chip for metabolism-induced toxicology screening. *Nature Communications* 5:3739.

Laurent, Jérémie, Guillaume Blin, Francois Chatelain, Valérie Vanneaux, Alexandra Fuchs, Jérôme Larghero, and Manuel Théry. 2017. Convergence of microengineering and cellular self-organization towards functional tissue manufacturing. *Nature Biomedical Engineering* 1:939–56.

Lee, Dong Woo, Yeon Sook Choi, Yun Jee Seo, Moo Yeal Lee, Sang Youl Jeon, Bosung Ku, Sangjin Kim, Sang Hyun Yi, and Do Hyun Nam. 2014. High-throughput screening (HTS) of anticancer drug efficacy on a micropillar/microwell chip platform. *Analytical Chemistry* 86 (1):535–42.

Lee, Dong Woo, Yeon-Sook Choi, Yun Jee Seo, Moo-Yeal Lee, Sang Youl Jeon, Bosung
Ku, and Do-Hyun Nam. 2014. High-throughput, miniaturized clonogenic analysis of a
limiting dilution assay on a micropillar/microwell chip with brain tumor cells. *Small*
10 (24):5098–105.

Lee, Dong Woo, Sang Hyun Yi, Se Hoon Jeong, Bosung Ku, Jhingook Kim, and Moo Yeal
Lee. 2013. Plastic pillar inserts for three-dimensional (3D) cell cultures in 96-well
plates. *Sensors and Actuators, B: Chemical* 177 (February):78–85.

Lee, Jung Hwan, Jae Sung Kwon, Yong Hee Kim, Eun Ha Choi, Kwang Mahn Kim, and
Kyoung Nam Kim. 2013. The effects of enhancing the surface energy of a polystyrene
plate by air atmospheric pressure plasma jet on early attachment of fibroblast under
moving incubation. *Thin Solid Films* 547:99–105.

Lee, Moo-Yeal. 2016. Microarray Bioprinting Technology: Fundamentals and Practices.
Springer, New York.

Lee, Moo-Yeal, R. Anand Kumar, Sumitra M. Sukumaran, Michael G. Hogg, Douglas S.
Clark, and Jonathan S. Dordick. 2008. Three-dimensional cellular microarray for
high-throughput toxicology assays. *Proceedings of the National Academy of Sciences*
105 (January):59–63.

Li, Chun Li, Tao Tian, Ke Jun Nan, Na Zhao, Ya Huan Guo, Jie Cui, Jin Wang, and Wang
Gang Zhang. 2008. Survival advantages of multicellular spheroids vs. monolayers of
HepG2 cells in vitro. *Oncology Reports* 20 (6):1465–71.

Liang, Youyun, Jaehyun Jeong, Ross J. DeVolder, Chaenyung Cha, Fei Wang, Yen Wah Tong,
and Hyunjoon Kong. 2011. A cell-instructive hydrogel to regulate malignancy of 3D
tumor spheroids with matrix rigidity. *Biomaterials* 32 (35):9308–15.

Ma, Yufei, Yuan Ji, Guoyou Huang, Kai Ling, Xiaohui Zhang, and Feng Xu. 2015. Bioprinting
3D cell-laden hydrogel microarray for screening human periodontal ligament stem cell
response to extracellular matrix. *Biofabrication* 7 (4):044105.

McCauley, Heather A., and James M. Wells. 2017. Pluripotent stem cell-derived organoids: Using principles of developmental biology to grow human tissues in a dish.
Development 144 (6):958–62.

Nuciforo, Sandro, Isabel Fofana, Matthias S. Matter, Tanja Blumer, Diego Calabrese, Tujana
Boldanova, Salvatore Piscuoglio, et al. 2018. Organoid models of human liver cancers
derived from tumor needle biopsies. *Cell Reports* 24 (5):1363–76.

O'Brien, Peter James, William Irwin, Dolores Diaz, Elodie Howard Cofield, Cecile M.
Krejsa, Mark R. Slaughter, B. Gao, et al. 2006. High concordance of drug-induced
human hepatotoxicity with in vitro cytotoxicity measured in a novel cell-based model
using high content screening. *Archives of Toxicology* 80 (9):580–604.

Otsuka, Hidenori, Kohei Sasaki, Saya Okimura, Masako Nagamura, and Yuichi Nakasone.
2013. Micropatterned co-culture of hepatocyte spheroids layered on non-parenchymal
cells to understand heterotypic cellular interactions. *Science and Technology of
Advanced Materials* 14 (6):1–10.

Pawar, Siddhesh N., and Kevin J. Edgar. 2012. Alginate derivatization: A review of chemistry, properties and applications. *Biomaterials* 33 (11):3279–305.

Picollet-D'hahan, Nathalie, Monika E. Dolega, Delphine Freida, Donald K. Martin, and
Xavier Gidrol. 2017. Deciphering cell intrinsic properties: A key issue for robust organoid production. *Trends in Biotechnology* 35 (11):1035–48.

Proctor, William R., Alison J. Foster, Jennifer Vogt, Claire Summers, Brian Middleton,
Mark A. Pilling, Daniel Shienson, et al. 2017. Utility of spherical human liver microtissues for prediction of clinical drug-induced liver injury. *Archives of Toxicology*
91 (8):2849–63.

Ranga, Adrian, Nikolche Gjorevski, and Matthias P. Lutolf. 2014. Drug discovery through
stem cell-based organoid models. *Advanced Drug Delivery Reviews* 69–70:19–28.

Roth, Alexander D., Pratap Lama, Stephen Dunn, Stephen Hong, and Moo-Yeal Lee. 2018. Polymer coating on a micropillar chip for robust attachment of PuraMatrix peptide hydrogel for 3D hepatic cell culture. *Materials Science and Engineering C* 90:634–44.

Rowley, Jon A., Gerard Madlambayan, and David J. Mooney. 1999. Alginate hydrogels as synthetic extracellular matrix materials. *Biomaterials* 20 (1):45–53.

Saldin, Lindsey T., Madeline C. Cramer, Sachin S. Velankar, Lisa J. White, and Stephen F. Badylak. 2017. Extracellular matrix hydrogels from decellularized tissues: Structure and function. *Acta Biomaterialia* 49:1–15.

Samorezov, Julia E., Colin M. Morlock, and Eben Alsberg. 2015. Dual ionic and photo-crosslinked alginate hydrogels for micropatterned spatial control of material properties and cell behavior. *Bioconjugate Chemistry* 26 (7):1339–47.

Sanyal, Suparna. 2014. Culture and assay systems used for 3D cell culture. Corning.

Sekine, Keisuke, Takanori Takebe, and Hideki Taniguchi. 2014. Fluorescent labeling and visualization of human induced pluripotent stem cells with the use of transcription activator-like effector nucleases. *Transplantation Proceedings* 46 (4):1205–7.

Selby, Mike, Rene Delosh, Julie Laudeman, Chad Ogle, Russell Reinhart, Thomas Silvers, Scott Lawrence, et al. 2017. 3D models of the NCI60 cell lines for screening oncology compounds. *SLAS Discovery* 22 (5):473–83. doi:10.1177/2472555217697434.

Shroyer, Noah F. 2016. Tumor organoids fill the niche. *Cell Stem Cell* 18 (6):686–7.

Skardal, Aleksander, Mahesh Devarasetty, Christopher Rodman, Anthony Atala, and Shay Soker. 2015. Liver-tumor hybrid organoids for modeling tumor growth and drug response in vitro. *Annals of Biomedical Engineering* 43 (10):2361–73.

Skardal, Aleksander, Leona Smith, Shantaram Bharadwaj, Anthony Atala, Shay Soker, and Yuanyuan Zhang. 2012. Tissue specific synthetic ECM hydrogels for 3-D in vitro maintenance of hepatocyte function. *Biomaterials* 33 (18):4565–75.

Sukumaran, Sumitra M., Benjamin Potsaid, Moo-Yeal Lee, Douglas S. Clark, and Jonathan S. Dordick. 2009. Development of a fluorescence-based, ultra high-throughput screening platform for nanoliter-scale cytochrome P450 microarrays. *Journal of Biomolecular Screening* 14 (6):668–78.

Sun, Jinchen, and Huaping Tan. 2013. Alginate-based biomaterials for regenerative medicine applications. *Materials* 6 (4):1285–309.

Takai, Atsushi, Valerie Fako, Hien Dang, Marshonna Forgues, Zhipeng Yu, Anuradha Budhu, and Xin Wei Wang. 2016. Three-dimensional organotypic culture models of human hepatocellular carcinoma. *Scientific Reports* 6 (1):21174.

Tibbe, Martijn P., Anne M. Leferink, Albert van den Berg, Jan C. T. Eijkel, and Loes I. Segerink. 2018. Microfluidic gel patterning method by use of a temporary membrane for organ-on-chip applications. *Advanced Materials Technologies* 3 (3):1700200.

Tolosa, Laia, M. José Gómez-Lechón, and M. Teresa Donato. 2015. High-content screening technology for studying drug-induced hepatotoxicity in cell models. *Archives of Toxicology* 89 (7):1007–22.

Tsai, Susan, Laura McOlash, Katie Palen, Bryon Johnson, Christine Duris, Qiuhui Yang, Michael B. Dwinell, et al. 2018. Development of primary human pancreatic cancer organoids, matched stromal and immune cells and 3D tumor microenvironment models. *BMC Cancer* 18 (1):1–13.

User's Manual for PDC-001-HP (115V) or PDC-002-HP (230V) High Power Expanded Plasma Cleaner (and Optional PDC-FMG (115V) or PDC-FMG-2 (230V) PlasmaFlo). 2015. Ithaca, NY.

Van Vlierberghe, Sandra, Peter Dubruel, and Etienne Schacht. 2011. Biopolymer-based hydrogels as scaffolds for tissue engineering applications: A review. *Biomacromolecules* 12 (5):1387–408.

Wagner, Ilka, Eva-Maria Materne, Sven Brincker, Ute Süßbier, Caroline Frädrich, Mathias Busek, Frank Sonntag, et al. 2013. A dynamic multi-organ-chip for long-term cultivation and substance testing proven by 3D human liver and skin tissue co-culture. *Lab on a Chip* 13 (18):3538.

Wilson, Melinda S., James R. Graham, and Andrew J. Ball. 2014. Multiparametric high content analysis for assessment of neurotoxicity in differentiated neuronal cell lines and human embryonic stem cell-derived neurons. *Neurotoxicology* 42 (May):33–48.

Wong, Hoi Ling, Ming Xi Wang, Pik To Cheung, Kwok Ming Yao, and Barbara Pui Chan. 2007. A 3D collagen microsphere culture system for GDNF-secreting HEK293 cells with enhanced protein productivity. *Biomaterials* 28 (35):5369–80.

Xing, Qi, Keegan Yates, Caleb Vogt, Zichen Qian, Megan C. Frost, and Feng Zhao. 2014. Increasing mechanical strength of gelatin hydrogels by divalent metal ion removal. *Scientific Reports* 4:1–10.

Xu, Leyuan, Natasha Sheybani, W. Andrew Yeudall, and Hu Yang. 2015. The effect of photoinitiators on intracellular AKT signaling pathway in tissue engineering application. *Biomaterials Science* 3 (2):250–5.

Yu, Kyeong-Nam, S. Soo-Yeon Kang, Stephen Hong, and Moo-Yeal Lee. 2018. High-throughput metabolism-induced toxicity assays demonstrated on a 384-pillar plate. *Archives of Toxicology* 92 (8):2501–16.

Yu, Sean, Pranav Joshi, Yi Ju Park, Kyeong Nam Yu, and Moo Yeal Lee. 2017. Deconvolution of images from 3D printed cells in layers on a chip. *Biotechnology Progress* 34 (2):445–54.

Zarowna-Dabrowska, Alicja, Ekaterina O. McKenna, Maaike E. Schutte, Andrew Glidle, Li Chen, Carlos Cuestas-Ayllon, Damian Marshall, et al. 2012. Generation of primary hepatocyte microarrays by piezoelectric printing. *Colloids and Surfaces B: Biointerfaces* 89 (1):126–32.

Zhang, Kaile, Qiang Fu, James Yoo, Xiangxian Chen, Prafulla Chandra, Xiumei Mo, Lujie Song, Anthony Atala, and Weixin Zhao. 2017. 3D bioprinting of urethra with PCL/PLCL blend and dual autologous cells in fibrin hydrogel: An in vitro evaluation of biomimetic mechanical property and cell growth environment. *Acta Biomaterialia* 50:154–64.

Zhang, Lijie Grace, John P. Fisher, Kam W. Leong, Samuel C. Sklare, Theresa B. Phamduy, J. Lowry Curly, Yong Huang, and Douglas B. Chrisey. 2015. 3D Bioprinting and Nanotechnology in Tissue Engineering and Regenerative Medicine. Elsevier, Amsterdam.

14 Integration of the Immune System into Complex *In Vitro* Models for Preclinical Drug Development

Jason Ekert, Sunish Mohanan, Julianna Deakyne, Philippa Pribul Allen, Nikki Marshall, and Claire Jeong
GlaxoSmithKline

Spiro Getsios
Aspect Biosystems

CONTENTS

14.1 INTRODUCTION

Drug discovery has faced low success rates and remains a slow and expensive business. Lack of efficacy caused more than half of all drugs to fail in Phase II and Phase III clinical trials. Safety issues and limited therapeutic index lead to another third of drugs to fail. High clinical attrition rates continue to be high, and there is a pressing requirement for new cellular models that are more predictive, translate to the clinical setting, and better recapitulate *in vivo* biology and microenvironmental factors. An ideal *in vitro* system would contain all the pertinent cells of the model organ, the three-dimensional (3D) physiological or pathophysiological microenvironment, as well as the vascular perfusion or microfluidics to tie the whole system together. An important element often missing in these higher complexity systems is the immune component.

Immune cells are a vital component of every organ system and contribute to normal development, homeostasis, tissue repair, and immunity (Langhans 2018, Wynn, Chawla, and Pollard 2013). In addition, stromal immune cells play a role in disease development such as atherosclerosis, fibrosis, cancer, and neurodegeneration (Langhans 2018, Wynn, Chawla, and Pollard 2013). Immune cells can also modify and change the organ systems' response to drugs that will not be captured in model systems lacking immune components (Langhans 2018). Recent advances in cancer (oncology) treatments have led to the emergence of immunotherapies, which specifically capitalize on modulating the immune response to target cancer cells, again highlighting the importance in understanding and capturing the immune component in new complex models. This book chapter will detail the reasons behind the high attrition rates in pharmaceutical development, critical aspects that are required in developing complex *in vitro* models (CIVMs) for both efficacy and safety models. It will explore the incorporation of immune cells in detail for oncology and immune-inflammation diseases (i.e., inflammatory bowel disease, IBD). There is a section that will cover current *in vitro* safety models reviewing liver; gut; and absorption, distribution, metabolism, and excretion (ADME) models and addition of immune cells into these complex cellular models. The last section will examine immune system models and the future incorporation with disease or healthy models.

14.2 BACKGROUND IN CLINICAL SUCCESS IN DRUG DISCOVERY FOR ONCOLOGY AND INFLAMMATORY DISEASES

Current cellular models to identify a suitable candidate in early drug discovery use 2D and increasingly 3D cell culture but most often with immortalized cell lines that fail to generate cell–cell interactions and cell morphology in native microenvironments and lack an immune component. Animal models are used in supporting efficacy studies, pharmacokinetics (PK)/pharmacodynamics (PD), and toxicology studies after selecting the lead molecule in small and large molecule (antibody) development, but there are many issues including interspecies differences, including the immune system, that result in unexpected drug failure from no or increased side effects (secondary pharmacology) that couldn't be mimicked in the animal models that limit the therapeutic window for efficacy of the drug. The animal models also don't always fully mimic different patient populations for a particular disease indication to help stratify patients when entering later phase clinical trials.

There are several well-known examples of drug trials where successful animal models did not translate into clinical trial success such as TGN1412 (immunomodulatory humanized agonistic anti-CD28 monoclonal antibody), Hedgehog pathway antagonist IPI-926 (Saridegib), matrix metalloproteinases (MMPs) inhibitors, and many other therapeutic cancer vaccines. These failures were due to either lack of efficacy, safety issues, or unintended exaggerated pharmacology findings in patients (Mak, Evaniew, and Ghert 2014). Clinical experiences and failed *in vivo* models provide insights into translational weaknesses for both safety and efficacy (Mak, Evaniew, and Ghert 2014).

Immunocompetent animal models for preclinical development of immuno-oncology (I-O) drugs have either not been available or have been of limited effectiveness, as *in vivo* models do not closely resemble the immune system of cancer patients who enroll in immunotherapy trials. This is a major problem and requires the generation of I-O model systems that can be used to increase clinical relevance to perform preclinical drug screening, toxicity, and efficacy prediction. For agents targeting different immune effector cells, there has been an increase in the likelihood of cytokine-release syndrome (CRS) that has occurred with anti-CD20 (Coiffier et al. 2008, Goede et al. 2014) or anti-CD19 (Buie et al. 2015, Topp et al. 2015) antibodies and chimeric antigen receptor (CAR)-T therapy (Brudno and Kochenderfer 2016). While immune checkpoint inhibitors such as anti-PD-1 and anti-CTLA-4 can stimulate cellular immune effectors by blocking inhibitory signals, they can reduce tolerance and may lead to inflammation, tissue damage, and autoimmunity which have been reported as immune-related adverse events (IRAEs) in a percentage of patients (Michot et al. 2016). PD that is associated with the biochemical and physiological effect of an administered drug on a patient are still being developed for many I-O agents (Marshall and Djamgoz 2018).

For inflammatory diseases, recent drug failures in IBD, particularly anti-interleukin 17 (anti-IL17) in Crohn's disease (CD) and anti-IL13 in **Ulcerative colitis** (UC), together with the average performance of therapies that target anti-IL12p40, anti-IL23p19, and mucosal addressin cell adhesion molecule (MadCAM) highlight the central lack of understanding of the impact of pathogenic heterogeneity in the treatment of IBD. This is probably due to the etiology of IBD that is not fully understood but is likely influenced by numerous factors such as genetic predisposition, environmental factors, like the microbiota, and social behaviors, including smoking and diet. These factors are thought to mediate subsequent epigenetic and immunological changes that contribute to the heterogeneity in the pathogenesis leading to disease.

14.3 BETTER UNDERSTANDING THE IMMUNE COMPONENT COULD IMPROVE DRUG DEVELOPMENT SUCCESS

To improve success in drug development, there is a need to better mimic and understand the immune system in healthy and diseased settings along with having more predictive cellular models that can recapitulate this complex system. As the immune system is a key element of the disease process that is made up of the innate and adaptive immune system, it is important that immune components are included in models for drug development, and yet it is often missing or insufficiently incorporated at most stages of the drug development process.

There are immune aspects to all diseases but ironically the majority of human cell-based models lack an immune component. A large proportion of medicines (i.e., small molecules, biologics, cell-based/ CAR-T) target immune cells, but we have no predictive way of testing preclinically in human cells until first time in human (FTIH) and hence rely heavily on incongruous animal models. There has been a shotgun approach and consequently numerous failures for medicines that work for mice but not humans. Animal models have a role in immunology research, but it's important to note their limitations in various aspects of recapitulating human immunity. Some human diseases have no suitable animal model, and others are hindered by *in vivo* models that incompletely recapitulate key aspects of a human disease. Indeed, it's been constantly reported that preclinical studies have often been poorly predictive of response in humans (Ostrand-Rosenberg 2004, Pallardy and Hunig 2010, Linette et al. 2013). Murine and nonhuman primate (NHP) models are best for immunologic studies because of the availability of reagents and tools (Wagar, DiFazio, and Davis 2018). One way to complement or if needed to use instead of animal models and avoid trying to understand interspecies differences is to directly study human immune cells in more translationally relevant cellular models. Tissue models tend to focus on structural components (extracellular matrix (ECM), stromal cells, epithelial barrier, muscle, fat, vasculature–tumor cells) at the expense of resident or infiltrating immune cells that most likely drive disease progression.

14.3.1 SHIFTING FROM TRADITIONAL DRUG PROCESS TO INCORPORATE CIVMs WITH AN IMMUNE COMPONENT

A traditional hit identification and lead screening campaign is time consuming and costly as outlined in Figure 14.1. Before moving into a screening campaign, a target identification and validation approach is undertaken where different approaches can be taken to validate a target. These include perturbing the target function in healthy

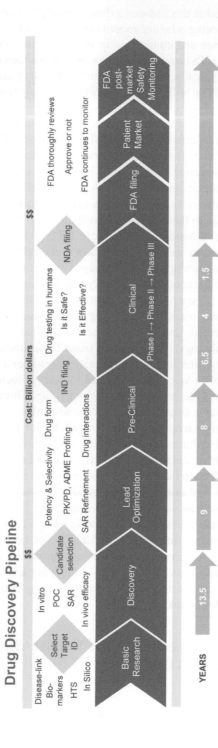

FIGURE 14.1 Drug discovery process depicted in a chevron diagram from basic research to post market monitoring (Hughes et al. 2011, Marx et al. 2016, Paul et al. 2010).

cells and looking to see whether this induces a disease state, or vice versa, working with a disease model and seeing whether modulating the target restores a healthy state. Once sufficient evidence and confidence in the target has been generated, a high-throughput screening (HTS) campaign occurs. It is an exhaustive approach as shown in Figure 14.1 where typically 1–2 million small compounds are screened to be a percentage of hits (1%–2%) that will be active in inhibiting or agonistic to the target. It comprises at the simple level an automated workstation that handles liquids, drugs, cells, and microtiter plates (384 or 1,536 wells) allowing multiple drugs and their efficacy to be tested on reporter cells to be assayed simultaneously. HTS in its current format has limited translatability from screened drugs to clinical drugs for patients and especially for immune-targeting compounds. These issues are due to simple cell culture techniques where nonrelevant overexpressing target cell lines are used that lack the capacity to mimic cell–cell and cell–matrix interactions in native tissue. The cell lines have been used due to their high reproducibility, ability to generate large quantities of cells, and ease of handling. To improve HTS, high-content screening assays have also been adopted to observe perturbation in the cells that are grown in a monolayer culture by using multiparametric analysis that consists of changes in morphology, cell features, and mitochondrial activity to better understand the mechanism. A model doesn't need to be a perfect model of the entire disease pattern which would be difficult, but it should mimic the vital component to study the biological and clinical question being addressed. Understanding the key attributes (gene expression, protein levels, and cell morphology) of a disease or tissue is paramount.

Safety and toxicity testing after identifying a series of leads or a lead compound typically involves a series of *in vivo* animal studies to assess mechanisms such as carcinogenicity, genetic toxicity, safety pharmacology, and immunotoxicology. There is, however, considerable *in vitro* and *in silico* work that occurs before and during *in vivo* testing. Cytotoxicity screening for cell health parameters can be implemented using 2D cell cultures and is suitable for high-throughput systems; however, to more faithfully recapitulate the complexity of cellular, or organ, systems, CIVMs are required to understand the mechanistic processes and more accurately predict toxic responses. *Ad hoc* investigations can also be undertaken to support toxicity findings in preclinical as well as clinical studies. In line with the overall drive in pharmaceutical development for *in vitro* and *in silico* alternatives to support the 3Rs (replacement, refinement, and reduction), CIVMs can be used in the safety space. In addition, the ability to carry out human *in vitro* modeling enables the generation of human safety data prior to clinical trials, which is particularly beneficial for immune-related targets for which preclinical safety testing in rodents and dogs will not generate species relevant toxicity data.

14.4   DESIGN CRITERIA FOR INCREASED TRANSLATIONAL RELEVANCE IN DRUG DISCOVERY

14.4.1   Cells

Immune cells can be obtained from several different sources that directly impact the human relevance of the cellular model and whether the model is an autologous (same tissue and blood) or allogenic (different donor source for tissue and blood) model. The most common source of immune cells that has been used extensively in drug development are immortalized cell lines such as the human Jurkat cell line for T cells, THP-1 cell line for macrophage, K562 cells that can be differentiated into dendritic cells (DCs), and NK-92 cell line for natural killer (NK) cells.

Most other experimental *in vitro* studies of primary immune cells have been restricted to those collected from peripheral blood, though discarded tissues and invasive sampling have expanded the options in studying human immune cells. Examining intact tissues can be an avenue for determining immune cell types that do not circulate within the blood system (e.g. tissue-resident memory T cells and macrophages, mesenchymal stromal cells, and lymphoid germinal center populations) and the study of immune cell infiltration in diseases with tissue- or organ-specific pathologies. Using human tissue will allow for the ability to capture and study the immune microenvironment. There are a number of obstacles to using primary immune cells that include limited numbers of starting cells, limited ability to expand immune cells, and expression of receptors that can be highly variable as differences in cell differentiation and activation are influenced by the local microenvironment.

In the future, generating immune cells on demand could occur from self-renewing human pluripotent stem cells (PSCs) and allow the generation of fully autologous cellular models. There are already established protocols for example for deriving macrophages, T cells (Vizcardo et al. 2018), B cells (Kawamura et al. 2017), and NK cells (Zeng et al. 2017) from PSCs, but there are still issues in developing the immune cells into fully functioning adult counterparts.

The cells used in recapitulating the tissue or disease should have a clear link to the disease. Immortalized cell lines have been the workhorse for cell culturing systems for drug development, but the relationship with the diseased cell type or tissue is questionable. The use of primary or patient-derived cells in recreating the CIVM model will raise concerns of donor variability or the ability to generate sufficient cell numbers, but there are numerous negative costs in using immortalized or cancerous cells, such as genetic abnormalities, cell proliferation rates, gene and protein levels, and drug response profiles. An alternative to primary or patient-derived cells is the use of induced pluripotent stem cell (iPSC)-derived cells with a disease-linked mutation or adult organoids that have been created from the diseased tissue or modified with CRISPR-Cas9 such as tumor organoids, intestinal IBD organoids, non-alcoholic steatohepatitis (NASH) liver organoids (Fujii, Clevers, and Sato 2019, Drost and Clevers 2017) or chronic obstructive pulmonary disorder (COPD) and cystic fibrosis in lung organoids (Sachs et al. 2019).

14.4.2 Tissue Architecture and Biological Readouts

When designing the CIVM model, there are number of choices in recapitulating the tissue architecture. There are a variety of options with increasing complexity from spheroid models (such as ultra-low attachment plates, hanging drop, matrix embedded, magnetic bioprinting, and microfluidic), microarrays like hydrogel micropads or inverted droplet arrays, organoids both healthy and diseased, 3D bioprinting technologies (Peng et al. 2017), microphysiological systems (MPSs) (Marx et al.

2016), and tissue slices that are probably the closest replica of a human organ physiologically.

In general, the design of CIVM will define the readout of a drug assay. Many 3D models rely on cell viability readouts that measure metabolic activity or cellular ATP as surrogate markers. Phenotyping using high-content imaging as a readout to deconvolute drug effects is an alternative approach where drug effects measuring apoptosis, cell death, cell health, and other phenotypic markers can be measured. Other endpoint analysis that are required when studying immune cells in the context of resident tissue cells includes cytokine analysis, flow cytometry or cytof where the number and type of cells can be determined.

14.5 EFFICACY MODELS

14.5.1 Solid Tumor Models

There are three compartments involved in the antitumor immune response for solid tumors that include the tumor microenvironment (TME) (tumor outer, inner, and core), blood vessel that is depicted in Figure 14.2, and lymph node/spleen. It is a

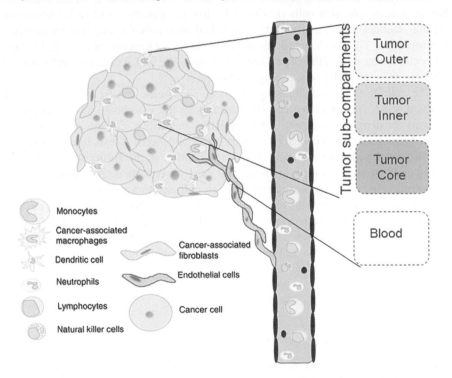

FIGURE 14.2 The solid TME. The illustration of the TME shows cancer cells along with cancer-associated fibroblasts in the tumor stroma, that attract monocytes through the leaky vascular network, that differentiate into cancer-associated macrophages and DCs. There are also other immune cell types including neutrophils, lymphoid cells, and NK cells. (Adapted from Nyga et al. 2016.)

challenge to build cellular models that incorporate all three compartments to perform assays in an HTS manner. A simplified model that recapitulates the TME would be most amenable for HTS while more complex models that include the vascular system and in the long term all three compartments.

The antitumor immune response can be dampened by the immunosuppressive microenvironment through several mechanisms that include ultimately either restricting immune cells entering the tumor, become inactivated within the tumor, or even promote further immunosuppression. Recapitulating the various mechanisms of immunosuppressive microenvironment is important when considering developing a translationally relevant tumor model (Cohen and Blasberg 2017).

14.5.1.1 Examples of Solid Tumor Models with Increasing Complexity and Human Relevance

14.5.1.1.1 *Spheroids*

Tumor spheroids can be formed by a considerable number of solid tumor cell lines using traditional methods such as ultra-low adherent plates, spinner flasks, magnetic bioprinting, and the hanging drop methods (Hoffmann et al. 2015, Tseng et al. 2015). The tumor spheroids recreate in an *ex vivo* setting either avascular tumor nodules or micrometastases. Tumor spheroids have been implemented in drug screening studies of tumor growth and proliferation (Madoux et al. 2017, Ekert et al. 2014, Hou et al. 2018). Several microfluidic techniques have also been developed to create tumor spheroids by either hydrodynamic trapping of cells in microwells or chamber structures (Liu et al. 2016, Baye, Galvin, and Shen 2017). More recently, tumor spheroid HTS assays have been created to assess CAR-T killing of tumor cells (Dillard et al. 2018, Bergeron and Gitschier 2018) or the effect of tumor fibroblasts on T cell infiltration (Koeck et al. 2017). The drawback with tumor spheroids so far is the use of cell lines that are not physiologically relevant and have low complexity as shown in Figure 14.3a. More recently, tumor spheroid studies have started to include patient samples in the drug screening phase, but this considerably limits the number of compounds that can be screened and there is an issue with donor-to-donor variability (Zhang et al. 2018).

14.5.1.1.2 *Organoids*

Tumor organoids have been derived from patient tumor samples from various tumor types that include lung, colorectal, ovarian, and pancreatic. Subsequently, similar culture protocols were developed for healthy and malignant tissue of the pancreas, stomach, prostate, liver, and lung (Dutta, Heo, and Clevers 2017). Tumor organoid cultures compared to tumor spheroids have increased complexity as shown in Figure 14.3b, more closely recapitulate morphological and genetic heterogeneous composition of the cancer cells in the original tumor. As the organoids can be genetically characterized and used for drug screening, it makes it possible to correlate the genetic background of a tumor with a drug response.

Tumor organoids for drug screening have been developed using an automated platform in a 384-well format for three-dimensional colon cancer organoid cultures derived from colon cancer patients. The study was able to show the possibility of using patient-derived tumor samples for high-throughput assays that was automated

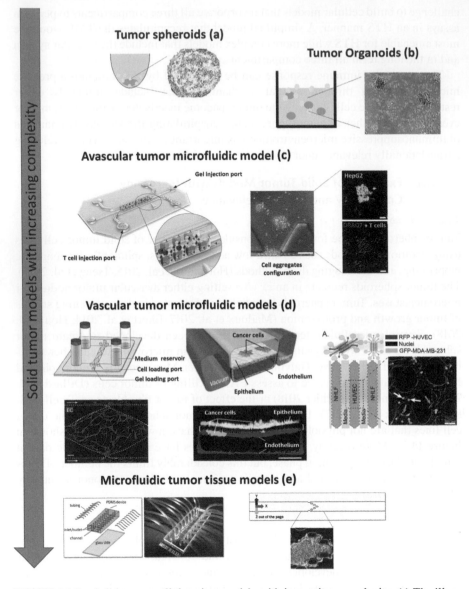

FIGURE 14.3 Solid tumor cellular vitro models with increasing complexity. (a) The illustration on the left shows tumor spheroids formed from tumor cells (green) or combined with fibroblasts that can be co-cultured with T cells (blue). The immunohistochemical image shows T cells (CD45+ cells in red) have infiltrated into the tumor (E-cadherin, brown) and fibroblast spheroid (Koeck et al. 2017). (b) Tumor organoids (green) can be co-cultured with autologous T cells (blue). The phase-contrast image shows tumor organoids in violet surrounded by T cells (unstained) (Dijkstra et al. 2018). (c) Schematics and fluorescent image of an avascular 3D tumor microfluidic model where HepG2 cells (green) are embedded in a collagen gel and T cells (blue) are in the flow compartment. (Scale bar: 100 μm [middle]; 50 μm [right]). (Pavesi et al. 2017).

(Continued)

by embedding single cells from colorectal tumor organoids in an ECM and being treated with compound and their integration as disease-specific models in drug discovery (Boehnke et al. 2016).

A similar assay but examining ovarian organoids was also successful in automating a microscopy-based assay to distinguish drug-induced cell death and cell proliferation inhibition. The compounds used in the organoid model were direct cytostatic or cytotoxic compounds. The study interestingly was able to reveal that drug effects on organoids correlated with patient-specific genomic alterations, while not present in 2D cell monolayers (Jabs et al. 2017). Both these studies further advanced tumor models that better recapitulate the TME and enhance predictivity and scalability of drug screening in the context of personalized oncology, but the models still lack an immune component.

Two recent studies have further advanced the *ex vivo* tumor organoid field to now start to understand and incorporate the immune component into organoid culture models. The first study was able to generate autologous tumor (organoids) reactive T cells from the same peripheral blood cells of a patient's organoids that were made from either colorectal cancer or non–small-cell lung cancer (NSCLC). The study showed the ability of the autologous T cells to efficiently kill matched tumor organoids (Dijkstra et al. 2018). The second study established a co-culture system for colorectal cancer and cytotoxic T cell (CTL) to model antigen-specific killing of tumor cells *ex vivo*. The T cells that were utilized in this study were $\alpha\beta$ T cells carrying a transgenic T-cell receptor (TCR) recognizing an HLA-A2-restricted Wilms tumor (WT)-1 derived peptide. Efficient killing of tumor organoids by the T cells was observed by co-localization of NucRed dead dye with T cells that expressed H2B-mNeonGreen and T cell infiltration into the epithelium of organoids could be detected. Enhanced tumor killing was observed with the addition of anti-PD1 and IFN-γ production in PD-L1 expressing IFN-γ stimulated organoids (Bar-Ephraim et al. 2018).

One recent study developed an alternative method where patient-derived organoids (PDOs) were cultured in an air–liquid interface (ALI) system. The PDOs that were propagated from diverse human tumors were able to maintain their complex histological TME architecture with tumor parenchyma, stroma plus functional

---

FIGURE 14.3 (CONTINUED) (d) Vascular tumor microfluidic models from left to right include a vascular micro-organ in the bottom left image (endothelial cells in red) (scale bar: 100 µm [bottom]. (Sobrino et al. 2016; Sontheimer-Phelps, Hassell, and Ingber 2019). In the middle is an illustration of a cross section of a two-channel MPS chip and a confocal image showing lung cancer cells (green) co-cultured with primary lung alveolar epithelial cells in the apical compartment and primary lung microvascular endothelial cells (red is anti-VE-cadherin) in the lower microchannel (scale bar: 200 µm) (Hassell et al. 2018). On the right is a diagram of an MPS three-gel channels' device. The fluorescent image displays the vascular network (red) with MDA-MB-231 tumor cells (green). Extravasated cells are indicated in orange, whereas non-extravasated cells are indicated in white (scale bar: 100 µm) (adapted from Song, Miermont, Lim, and Kamm, 2018). (e) Figures on the left show a microfluidic tumor tissue model showing compartment view and fully assembled view. Images on the right are an illustration of a channel and a 10×phase-contrast image of a sample trapped in the device. (Adapted from Holton et al. 2017.)

tumor-specific tumor-infiltrating lymphocytes (TILs). Importantly, the human PDOs were able produce a tumor cytotoxic response to immune checkpoint blockade molecules (i.e., anti-PD-1- and/or anti-PD-L1) (Neal et al. 2018). These methods are a step closer toward to rebuilding the TME *in vitro* but still lack other crucial immune cells that can cause immunosuppression such as Tregs, myeloid-derived suppressor cells (MDSCs), stromal and endothelial cells of the TME.

14.5.1.1.3   Microfluidic Model

Microfluidic models that recapitulate the TME can be separated into either avascular or vascularized tumor models as demonstrated in Figure 14.3c and d. In both formats of the microfluidic tumor models, the channels or microwells/chambers can be coated with various adhesion proteins and/or matrix materials to control cell adhesion or can be filled with ECM or scaffolds (decellularized tissue or several ECM proteins or biocompatible polymers) to promote either cell attachment, growth, migration, or morphogenesis in three dimensions. Tumor cells grown in the microfluidic models typically show increased resistance to chemotherapy. This was observed in a study using the Mimetas platform where the drug response of triple negative breast cancer cells was reduced in the microfluidic platform compared to 2D monolayer cultures. In the study by Pavesi et al. (2017), an avascular 3D microdevice was developed as seen in Figure 14.3d that contained human hepatocellular carcinoma cells and TCR-engineered T cells. In the avascular 3D microdevice antitumor efficacy of the TCR-engineered T cells was observed as the TCR–T cells showed increased ability to migrate and kill the tumor. Another study looking at immune checkpoint inhibitors was able to show PD-1 blockade using patient-derived organotypic tumor spheroids that retain autologous immune cells cultured in collagen hydrogels suspended in a 3D microfluidic device (Jenkins et al. 2018), while an alternative 3D microfluidic model was able to study the effects of monocytes on killing efficacy of Hepatitis B virus (HBV)-specific TCR–T cells that were flowed through a media channel demonstrating differences compared to a standard 2D assay (Lee et al. 2018).

To further enhance complexity, the addition of the vascular component as shown in Figure 14.3e can occur through lining the channels or chambers with endothelial cells, recreating a neovascular system or via 3D bioprinting (Kolesky et al. 2016, Sontheimer-Phelps, Hassell, and Ingber 2019). The tumor vasculature plays an important role in tumor biology. For cancer to grow and spread, it requires constant nutrient support supplied by a growing endothelial network that also modulates the TME. Traditional static transwell assays are inadequate in replicating tumor angiogenesis, metastatic cascade, and immune cell infiltration. Microfluidic systems have been fabricated to create microvasculature networks allowing for the complex interplay between tumor vasculature, cancer cells, and TME. The complex tumor platform could allow for (1) monitoring of immune cells and killing of tumor cells in high-resolution time-lapse imaging or (2) interrogating human immune-cancer cell interactions in ways that can't readily be achieved in animal models.

14.5.1.1.4   Ex Vivo Tissue

Ex vivo tumor tissue models maintain the vast cellular heterogeneity of the tumor stromal and immune component of the tissue when the tissue is extracted via tumor biopsy or resection. The *ex vivo* tumor tissue has the highest level of complexity compared to the other solid tumor models as outlined in Figure 14.3e. This allows the prospect to test and understand whether a certain patient population using an *ex vivo* tissue will respond to a compound or a cellular therapy such as a CAR-T or TCR therapy. It will allow to better predict what and how a future patient population will respond to cancer treatment. Maintaining the immune component is an important factor when evaluating immuno-oncology compounds so as to retain the autologous context of the tumor-infiltrating immune cells along with other key immune cells such as Tregs, DCs, and macrophage cells are important. The disadvantage of *ex vivo* models is the high cost and low-throughput nature of the assay that allows the testing of only few compounds and the high variability due to patient-to-patient tumor heterogeneity. The tissue slice, fragment, or biopsy can be kept in culture through either static (Majumder et al. 2015) or microfluidic devices (Moore et al. 2018, Holton et al. 2017) allowing the 3D architecture of the tumor and the TME plus the immune component to be retained (Moore et al. 2018, Holton et al. 2017).

14.5.2 INFLAMMATORY DISEASES WITH A FOCUS ON INFLAMMATORY BOWEL DISEASE (IBD)

The important roles of the gut include digestion, absorption of nutrients, and defense against invading pathogens. Disruption of the balance of these functions can have a significant impact on health and can lead to serious inflammatory or metabolic disorders. Maintenance of this homeostasis is regulated by the many specialized cells types present both making up the epithelial gut wall and ancillary cells such as cells of the immune system, which can be present in the proximate blood supply and infiltrating the gut wall. Sharkey et al. (2018) extensively review the details of many of the cell types of the gut and their importance for retaining the homeostasis and stimulating appropriate responses to commensal and invading microbes (Sharkey, Beck, and McKay 2018).

IBD is a chronic inflammatory disorder of the GI tract, the causes of which can be multifactorial consisting of environmental, genetic, microbial, and immunological elements. Immune homeostasis in the gut consists of a delicate balance of immune tolerance to self, food, and to the commensal organisms that inhabit the intestine along with the protective, inflammatory immune responses mounted against pathogenic organisms (Ahluwalia et al. 2018). The mucosal layer and intestinal epithelium serve as a physical barrier that separate the contents of the gut lumen and the mucosal immune system. When this barrier is compromised, the result is increased gut permeability causing a dysregulation of immune homeostasis and chronic inflammation. Innate cells including neutrophils, macrophages, DCs, and innate lymphoid cells are found to contribute to this ongoing inflammation. DCs, which help to link the innate and adaptive immune responses, are influenced by this inflammatory environment and can perpetuate activation and differentiation of T cells. IBD patients can also present with mucosal plasma B cell infiltrates with increases in immunoglobulins in the intestinal mucosa, in some cases, directed against commensal bacteria and food antigens (Macpherson et al. 1996, Hevia et al. 2014). T cells can home to and accumulate in the gut mucosa of IBD patients and different T cell subsets each playing a role in IBD pathogenesis. Genetic and epidemiological studies in human patients, as well as experimental studies in animal models, have identified numerous genetic and environmental risk factors for CD and UC but have not identified clear driver mutations that could lead to therapeutic opportunities.

The most widely used animal models for IBD are the dextran sodium sulfate (DSS) and hapten reagent 2,4,6-trinitrobenzene sulfonic acid (TNBS) rodent models, and these models display numerous disadvantages. The damage to the epithelial barrier through chemical stimulation and the subsequent immune reaction it causes does not take into account the dysregulation of the innate or adaptive immune system in IBD (Wirtz and Neurath 2007). To account for the complexity in IBD etiology, researchers have taken transcriptomic, proteomic, metabolomic, and metagenomic approaches in an attempt to understand global disease networks and to identify genes, proteins, and metabolites that may be involved in disease pathogenesis. Although these approaches have provided valuable insights, they have fallen short of identifying potential high-value therapeutic targets in IBD.

14.5.2.1 Examples of Models with Increasing Complexity and Human Relevance for IBD

Conventional intestinal models that have utilized the transwell system using Caco-2 cells or enterocyte cells cannot fully mimic the complex interactions with other cells, such as the immune system. These interactions may be of vital importance for the epithelial barrier function and don't fully recapitulate pathophysiological changes happening in an inflamed region of IBD. Transwell intestinal mucosa models that have utilized enterocyte cell lines (Caco-2, HT-29, or T84) and the addition of monocytes, DCs, and a pro-inflammatory signal (lipopolysaccharides (LPS) or IL-β) have resulted in an assay that can study the intestinal barrier in the state of inflammation and can differentiate therapeutic efficacy for different budesonide formulations (Leonard et al. 2012). Creating an intestinal-immune model that doesn't require fresh patient tissue and displays normal intestinal features would be a further advancement in human relevance for studying coordinated epithelial immune responses under normal or pathogenic conditions.

This has been achieved through the generation and culturing of human intestinal cells by a number of different protocols that include differentiation of iPSC to intestinal lineages (Munera and Wells 2017), developing so-called "organoids" or "mini-guts" (Nakamura 2019) or cloning as illustrated in Figure 14.4a. Enteroids and colonoids derived from LGR5$^+$ stem cells or LGR5$^+$-containing crypts from the small intestine and colon and iPSC-derived gut organoids, respectively, contain the four major human intestinal epithelial cell lineages with distinct roles in gut homeostasis and immune modulation (i.e., absorptive enterocytes, mucus-producing Goblet cells, hormone-producing enteroendocrine cells, and antimicrobial molecule-producing Paneth cells). However, the challenge with these gut organoid models is that they are inward facing where the epithelial layer in gut organoid models typically face the interior of the 3D structure, reducing the ability to perform experimental manipulation. There have been a few ways to overcome this technical challenge through static culture models (Blutt et al. 2017). A technique was developed to clone and propagate "ground state" human intestinal stem cells (ISCs) (Duleba et al. 2019,

Wang et al. 2015) on fibroblast feeder layers. The ISCs can be transferred to ALI cultures where the cells differentiate into intestine-like structures including all cell lineages such as enterocytes, goblet cells, enteroendocrine cells, and Paneth cells (Wang et al. 2015). Another method is where plating the gut organoids in a mono-layer in an ALI system generates open-faced organoids. A study utilized this method and went one step further incorporating an immune component where macrophages were seeded on the basolateral side of the transwell (Noel et al. 2017). Compared to enteroid alone cultures, the addition of macrophage cells leads to changes in enteroid morphology; enhanced maturation of the enteroid monolayers; and IL-8, IFN-γ, and IL-6 levels were substantially increased in the basolateral compartment (Noel et al. 2017). These methods allow the luminal surface of the epithelial layer to be available for experimentation from a monolayer of adult stem cells. These open-faced models provide an assembly for basolateral addition of applicable cell types, such as immune cells. This allows improved experimental evaluation of specific physical interactions, paracrine communication, and molecular mechanisms underlying host responses to apical stimuli to be performed in open-faced models. The open-faced models show high reproducibility and their suitability to study gut physiology.

To better recapitulate human complexity and more closely mimic the pathophysiology of IBD, a gut inflammation on a chip model has been developed that endeavored to recreate the hallmark signatures of IBD that include a leaky gut, gut microbiome imbalance, and a hyperactivated immunity (Shin and Kim 2018). It consisted of Caco-2 cells that were added on the luminal microchannel side and were treated with DSS to recapitulate a Sodium trimethylsilylpropanesulfonate (DDS)-induced inflammation model used in mice models. The intestinal cells were also under a mechanically dynamic microenvironment similar to human intestine that had been created in a previous gut-on-a-chip study. An *ex vivo* experimental environment was created to recapitulate an inflamed intestine whereby inflamed (DSS and LPS challenged) intestinal cells were challenged with *Escherichia coli* to mimic the microbiome on the lumen microchannel side and peripheral blood mononuclear cells (PBMCs) were added on the basolateral side. This induced increased production of pro-inflammatory cytokines into the basolateral compartment, the villi showed morphological damage and reduced villous height, reduced barrier integrity occurred, and there was recruitment and infiltration of PBMCs at the basolateral area of villi, all characteristics of an inflamed intestine (Shin and Kim 2018). An alternative high-throughput perfusable microfluidic platform that used Caco-2 cells demonstrated an intestinal tract epithelium with cellular polarization, tight junction formation, and expression of key intestinal receptors. The organ-on-a-chip model allows for 40 gut models to be grown in a tubular shape and allow accessibility from both the apical and basal sides (Trietsch et al. 2017). Further refinement of the organ-on-a-chip platform is required to understand whether it can mimic the pathophysiology of IBD using different stimuli like DSS and the addition of immune cells.

A more recent gut-on-a-chip study exchanged the immortalized Caco-2 cells with intestinal organoids by dissociating and culturing of the gut organoid cells on the permeable luminal membrane within the gut on a microfluidic device that incorporates flow and cyclic deformation. The gut organoid on a chip study also included in the parallel basolateral microchannel human intestinal endothelial cells

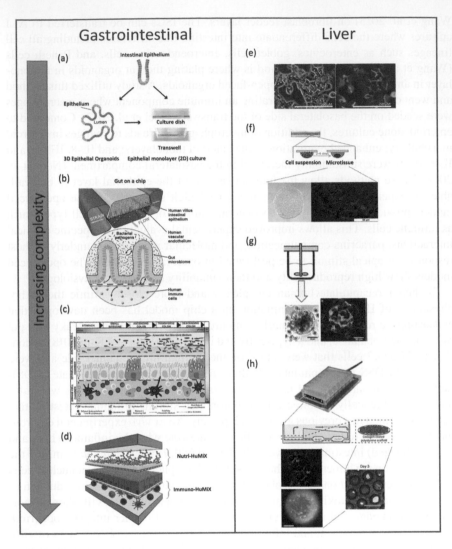

FIGURE 14.4 Gastrointestinal and liver cellular models with increasing complexity. Gastrointestinal models with increasing complexity are on the left. (a) Schematics of 3D epithelial organoids embedded in a matrix, organoids plated onto culture dish or transwells (adpated from Nakamura 2019), or (b) seeded in a gut-on-a-chip model. (Adapted from Bein et al. 2018.) (c) Diagram of an idealized *in vitro* gastrointestinal experimental model. (Adapted from Fritz et al. 2013.) (d) An illustration of a nutri-HuMiX and the immuno-HuMiX platforms. (Adapted from Giolla Eain et al. 2017.) Liver models with increasing complexity are on the right. (e) Polarized rat hepatocytes in collagen sandwich cultures with a phase contrast on the left and a fluorescent image of bile canalicular-like structures (green) shown on the right (Scale bar 200 μm), (f) hanging drop liver organoids (scale bar: 50 μm). (g) Stirrer bioreactor for 3D aggregates showing on the right hepatocyte spheroids in the bioreactor (scale bar = 50 μm) and fluorescent image of hepatocytes stained with albumin (red) and CK18 (green) (scale bar: 30 μm). (Adapted from Alepee et al. 2014.)

(Continued)

to further recapitulate the gut tissue architecture. This gut-on-a-chip model as shown in Figure 14.4b would allow for the introduction of immune cells on the basolateral side of the chip (Kasendra et al. 2018, Bein et al. 2018).

An alternative to integrating immune components into reconstituted intestinal tissues, gut organoids, or gut-on-a-chip models is to use intact intestinal tissue explants sourced from colonoscopy biopsies of healthy or IBD patients (Russo et al. 2016). These intestinal mucosal explants retain at least some resident immune cells that are distributed in an appropriate spatial manner for responding to exogenous stimuli or drug treatment, at least in a transient manner. Moreover, IBD patient biopsies with genetic and epigenetic marks of the disease exhibit elevated cytokine production and responsiveness such that artificial stimulation is no longer a prerequisite for modeling this inflammatory disease in a dish. IBD explants have been used in such a manner to test a variety of therapeutic approaches, including anti-TNF biologics that have some clinical efficacy in a subset of patients (Vadstrup et al. 2016). Thus, these *ex vivo* IBD tissue models may prove useful for patient stratification efforts to identify potential nonresponders. Adding a microfluidic element to these intestinal tissue explants increases their life span *ex vivo* and has recently been used to illustrate an important role of the enteric nervous system in modulating immune cell response to bacterial pathogens (Yissachar et al. 2017). While these *ex vivo* tissue models present pragmatic challenges for drug discovery efforts, including limited supply, donor heterogeneity, and inability to genetically manipulate the resident tissue cells, they remain an attractive preclinical tool that remains most proximal to the disease *in vivo*.

14.6 SAFETY MODELS

Over the past 60 years, the pharmaceutical R&D world has seen great scientific advancement in new technologies; however, the number of new drugs approved has halved every 9 years since 1950 (Scannell et al. 2012). Part of this discrepancy is due to the lack of predictive human responses in costly animal studies and the lack of complexity and complete human model systems for current *in vitro* assays (Scannell et al. 2012, Langhans 2018). Recently, in 2007, there was a call for change and an emphasis on developing more relevant *in vitro* human models in the NRC report, *Toxicity Testing in the 21st Century: A Vision and a Strategy* (Krewski et al. 2010). The immune system is a major component lacking in current CIVMs, and without recapitulating this system, it is nearly impossible to determine drug toleration, drug effectiveness, and also how disease state of the particular organ and genetic background of a patient will receive the drug. This section focuses on the importance of the immune component in the liver and gut organ system models for toxicology studies.

FIGURE 14.4 (CONTINUED) (h) A perfusion microfluidic device with hepatocytes embedded in a collagen-coated scaffold. The phase image in the bottom right is hepatocyte cells. The bottom left fluorescent images are an image of bile canalicular-like (bc in red) structures and the other image shows uptake of 5 (and 6)-carboxy-2′,7′-dichlorofluorescein diacetate (CDFDA) stain (green) into bile canaliculi. (Scale bar: 200 μm, [bottom left] and 30 μm [bottom right]). (Adapted from Ortega-Prieto et al. 2018, Rowe et al. 2018.)

14.6.1 LIVER

Drug-induced liver injury (DILI) is difficult to predict and treat because of all the different factors and pathways involved that can lead to hepatotoxicity. The liver itself is very complex and, in addition to hepatocytes (80%), includes various non-parenchymal cells such as endothelial, stellate and Kupffer cells (Alepee et al. 2014). Immune cells and their responses within the liver play important roles in disease pathogenesis, resistance, and therapy effectiveness (Lau and Thomson 2003). For example, during drug metabolism, when the liver metabolizes a drug to form certain toxic metabolites or protein–drug adducts, this can create a neoantigen signal and elicit an immune response (Funk and Roth 2017). However, without a functioning immune component in a model, this reaction would never be captured and investigated. Figure 14.4e–h shows the various complex culture systems that have already been utilized in an attempt to more accurately recapitulate the liver organ system including sandwich cultures, solid scaffolds, perfused bioreactor systems, multicellular cultures, and microfluidic/chip-based systems (Alepee et al. 2014, Ortega-Prieto et al. 2018, Rowe et al. 2018). However, to lower attrition rates, it has become more imperative to develop a model that encompasses not only the pathways of metabolic enzymes and transporters but also includes the interplay of the immune system in order to increase predictivity and success in the safety space (Funk and Roth 2017, Parker and Picut 2005).

The innate immune response has liver tissue-specific components involved in the production of acute-phase proteins, nonspecific phagocytosis and cell killing as well as disposal of molecules after the whole response (Parker and Picut 2005). Hepatic sinusoids are involved in the nonspecific phagocytosis response and include five cell types: endothelial cells, Kupffer cells (macrophages), pit cells (liver-specific NK cells), fat-storing cells, and DCs (Parker and Picut 2005). The Kupffer cells are the liver's resident macrophages and contribute to normal liver physiology by participating in acute and chronic responses of the liver to toxic compounds. Pit cells represent the liver-specific more mature NK cells and morphologically resemble large granular lymphocytes. These cells play a role in defending the liver against invading tumor cells. Pit cells have critical cell–cell interactions with both endothelial and Kupffer cells and depend on Kupffer cells to proliferate and also kill tumor cells.

The liver also functions in the adaptive or specific immune response. The liver deletes activated T cells after an immune response, builds tolerance to antigens, aids in proliferation of T lymphocytes, and disposes waste molecules accumulated from an immune response. DCs are antigen-presenting cells that can migrate from the liver to other organs (e.g., regional lymphoid organs) to aid in the immune response (Lau and Thomson 2003). Increasing evidence indicates that the microenvironment where the DCs mature or activate greatly influences their phenotype, function, and effects on other immune modulators (e.g., T cells) (Lau and Thomson 2003). Along with this idea, different disease states such as hepatitis B, C, and hepatocellular carcinoma will have different profiles of immune cells. For example, there are larger populations of dysfunctional DCs in the liver in these pathogenic states, and these differences alone could affect how drug therapies are received or tolerated by the patient. Additionally, hepatic T cells express different percentages of surface markers compared to T cells in the blood, lymph nodes, or spleen (Lau and Thomson 2003). Therefore, it is critical to add these types of immune cells in current 3D models because it would allow for the exploration of the inflammatory response and also the cross talk between epithelial cells and immune cells during infection, drug treatments, and different immune disease states pertinent to that organ.

Recent experiments have looked into co-culturing hepatocytes with the Kupffer cells or the liver's resident macrophages at different ratios in micropatterned plates to see if this control led to benefits of hepatocyte function. The Kupffer cells naturally establish cell-to-cell contacts with the hepatocytes and work to phagocytose foreign entities as well as clean up the different stages of cells (e.g., apoptotic cells, old cells) (Zinchenko et al. 2006). In addition, the Kupffer cells can secrete cytokines that can control hepatocyte function.

14.6.2 GUT

In terms of pharmaceutical development, the gut is important for absorption of drugs and has metabolic functions. Gut-related toxicity not only impacts on health of the individual but can also influence drug PK/PD, downstream availability, and efficacy. The immune system is involved in many inflammatory conditions of the gut and can be an important component of safety studies both through monitoring detrimental activation of the immune system or if drug treatment is negatively impacting immune system and downregulating healthy responses. Similar to gaps in current liver-immune CIVMs, resident immune cells are critical to the homeostasis and disease state of the gut.

Safety applications of a gut *in vitro* model can range from recapitulating a well-characterized toxicity on a specific cell type, to investigating the mechanisms of pathology in an *in vivo* like system, or ranking drug candidates on their effect on barrier integrity. Relevant constituents could consist of relevant cell types such as stem cells and proliferating cells, also a functioning crypt to villus axis of differentiation, spatial organization and architecture, and tight junctions. Of particular interest for a model of gut, in the safety space and beyond, is the ability to work with systems promoting cellular polarization and an air–liquid interface. This enables modeling of drug treatments from both oral and systemic administration through exposure from the apical or basolateral side, respectively, as well as from an immune component point of view being able to introduce immune cells from the blood (basolateral) side.

Several gut CIVMs illustrated in Figure 14.4a–d have been developed including cell culture insert, gastric/intestinal organoids, and perfusion models such as organs-on-a-chip (Giolla Eain et al. 2017, Ramadan et al. 2013, Fritz et al. 2013, Short, Costacurta, and Williams 2017, Kasendra et al. 2018). Successful implementation of an *in vitro* gut organoid model of the intestine for safety assessment was reported by Wagoner et al. (2014) and reviewed in Morgan et al. (2018) whereby a cross-species organoid investigation was conducted to assess species intestine sensitivity to bromodomain (BRD) inhibitors (Wagoner et al. 2014, Morgan et al. 2018). 3D organoids of intestinal stem cell tissues were used to determine the relative sensitivity of rat, dog, and human intestinal cells to a panel of BRD4 inhibitors. Clinical translatability was demonstrated, and the results enabled progression of a candidate.

Additionally, a microfluidic model has been used for screening chemotherapy drugs (Pocock et al. 2017), the system enabled increased mechanistic analysis of drug permeability, which is a challenge for drug development in this area. Mechanistic analysis of toxicities was made possible by the use of *in vitro* colon models which examined anticancer drug combinations through assessment of diarrheal profiles (Moisan et al. 2018). Using the cellular model, it was possible to delineate effects on chloride channel activity and barrier integrity.

Ewart et al. (2018) agree in their review that MPSs and technology have the chance to enhance safety assessment and advise that due to compatibility issues between different cell types within multiorgan devices/models using induced PSCs from single donors could be a practical solution particularly when comprising parts of the immune system (Ewart et al. 2018).

Many *in vitro* GI models are reported to display innate immune-like responses such as the release of cytokines and chemokines, which *in vivo* would signal to immune cells to mount an appropriate response. The capacity for this response by the epithelial cells provides evidence for the potential for immune cell interaction and communication upon introduction of immune cells; however, this does not constitute a true immune component. Immune cells, such as macrophages, of which there is a large population in the gut and T cells (also known as intraepithelial lymphocytes in the mucosal gut) are key immune cells considered relevant, and PBMCs containing a mixed population of cells are readily compatible with fluidic CIVMs. Smith et al. (2011) characterized resident gut mucosal macrophages and found differences in expression of proteins in comparison to blood monocytes such as no CD14, a co-receptor for LPS, and reduced IgA and IgG receptors, indicating the importance of working with appropriate immune cells (Smith et al. 2011).

The gut-on-a-chip model illustrated in Figure 14.4b and discussed in the Section 14.5.2.1 shows further increased complexity (Bein et al. 2018, Kasendra et al. 2018), and the addition of commensal and pathogenic microbes and immune cells has been achieved in this model. A highly differentiated intestinal mucosal model including epithelial and endothelial cells was developed by Salerno-Goncalves et al, (2011), exhibiting many features of *in vivo* tissues (Salerno-Goncalves, Fasano, and Sztein 2011). Lymphocytes were introduced and persisted for in excess of 7 days colocalized with endothelial cells and fibroblasts in the ECM. Functional studies were carried out to assess the response to treatment with bacterial pathogens, and both the morphological and cytokine responses were representative of an *in vivo* response. Duell et al. (2011) summarize other epithelial immune cell co-culture systems which may be applicable to studying infectious diseases and emphasize that cross talk between epithelial cells and macrophages, in particular, is important for driving the responses of both cell types (Duell et al. 2011). An *in vitro* transwell co-culture of intestinal epithelial cells and macrophages using Caco-2 cells (seeded on apical side of membrane) and differentiated THP-1 cells (seeded into the basolateral media) was chosen to maximize both applicability to a human exposure scenario and accessibility through use of cell lines (Kampfer et al. 2017). Homeostasis was determined through stable transepithelial electrical resistance (TEER) measurements, lactate dehydrogenase (LDH) release, and lack of epithelial cell or macrophage activation, demonstrating feasibility in mimicking the intestine in a healthy state;

furthermore, this system could be stimulated into a controlled inflammatory state (Kampfer et al. 2017). There is much evidence *in vitro* of variations in responses depending on the presence of immune cells, such as the ability for greater discrimination between pathogenic and nonpathogenic bacteria (Parlesak et al. 2004, Haller et al. 2000).

Learnings from the wider use of immune competent models for investigations into microbial responses can be applied to the safety space and drug treatment testing to better reproduce gut toxicological responses.

14.6.3 Absorption, Distribution, Metabolism, and Excretion (ADME)

Allied with toxicity testing to evaluate drug impact is the study of how the drug is distributed in the body, termed PK. PK is evaluated through four mechanisms: the extent of ADME. Many *in vitro* models are available to identify ADME properties, systemic toxicity, and hazard identification for target organs. A report coming from an European Partnership for Alternative Approaches to Animal Testing (EPAA) workshop recommended that mechanism-based toxicity assays should be further developed and validated, and that known toxic and adverse effects should be defined for the kidney, heart, lung, CNS, immune system, adrenal and thyroid glands (Schroeder et al. 2011). Collation of this information on known substances may help develop quantitative structure–activity relationship (QSAR) models and new assays. Hepatocyte cell culture is an important *in vitro* system applicable to ADME testing, with the most valuable having metabolically active cells, as metabolites are important factors to take into account with ADME (Ukairo et al. 2013).

Generally, more CIVMs would be more effective at recapitulating *in vivo* conditions that support the ADME assessment mechanisms and allow human prediction of PK. Specifically, addition of the gut microbiota would be beneficial as the gut bacteria are involved in many aspects of ADME properties that can affect drug disposition as well as be involved to some degree in metabolizing the drug. Biological factors such as an immune reaction could affect a precise PK prediction; therefore, Vellonen et al. (2014) suggest that the benefits and limitations of the model in use must be understood.

Kimura et al. (2018) and Abaci and Shuler (2015) review the role for organ-on-a-chip technology for predicting PK through their increased cellular activity and physiological functions and where linked multiple organ-on-a-chip systems that create a body-on-a-chip may more accurately observe the linked ADME PK processes.

14.7 FUTURE DIRECTION OF *IN VITRO* IMMUNE SYSTEM ON A CHIP MODELS

The demand for *in vitro* systems to model human immune system has increased tremendously recently with advancements in immuno-oncology and complex biologicals. There has also been progress in developing artificial bone marrow and lymph node models for drug development and mechanistic study applications. de la Puente et al. (2015) created a 3D-engineered bone marrow derived from bone marrow cells from the multiple myeloma patient samples. More complex efforts have been undertaken in the lymphoid tissue modeling space and creation of artificial lymph nodes (aLNs). Efforts to develop a human aLN has been reported by several labs (Tomei et al. 2009, Kobayashi, Kato, and Watanabe 2011, Sardi, Lubitz, and Giese 2016, Purwada et al. 2019), but none of these studies have been able to replicate a complete LN *in vitro*. These models if established successfully can be used to assess immunomodulation, immunogenicity and immunotoxicity of biopharmaceuticals. In order to develop aLN models, it is critical to understand the complex lymph node architecture and cell types. Lymph node anatomy consists of cortex, paracortex, and medulla. Functional architecture of the LNs is comprised of a stationary antigen-presenting DC population mixed with highly dynamic lymphocytes. These cells are embedded within a stromal environment composed of non-hematopoietic cells such as stromal cells and endothelial cells which are critical in regulating structural organization and trafficking of immune cells and cytokines. Lymphocytes enter LN from the blood circulation through specialized endothelial cell junctions and interact with APCs. Once activated, T lymphocytes can undergo clonal proliferation. They can regulate B cell-rich areas of the lymph node and facilitate an effective B-lymphocyte response.

There is an opportunity for future drug development efforts to combine the lymph node with a disease or safety model to create a multi-organ-on-a-chip platform. Single organ-on-chip models that replicate a single organ do not allow the study of a drug's systemic effect or replicate the movement of immune cells. There has been the demonstration of multi-organ-on-a-chip systems such as a liver-gut, liver-skin, pancreatic-liver, liver-kidney, or heart-lung on a chip that has demonstrated cross talk between the two or more organs, but none have incorporated an immune tissue into their models (Prantil-Baun et al. 2018). Existing microfluidic models of tumor immunity, immune inflammation, or safety models have focused primarily on events in the tumor, inflamed or designated tissue, or the nearby vasculature. A recent paper demonstrated the feasibility of creating a two-organ microfluidic tumor-lymph node system by using murine tissue. Interestingly, the co-cultured lymph node-tumor murine slices showed signs of being more immunosuppressed than those co-cultured with healthy tissue by reduced IFN-γ secretion, suggesting that the chip may possibly display features of tumor-immune interactions. Developing a two-compartment human system where there is continuously recirculating flow to transport secreted signals and immune cells between the immune compartment and the healthy or diseased tissue would start to recapitulate *in vivo* responses and allow the study of communication between the organs, mechanistic study of immune physiology to drugs, permit experiments to be undertaken in cancer metastasis, education of immune cells in the lymph tissue, and immune cell homing to the tumor.

14.8   CONCLUSION

This chapter was intended to capture the complexity and challenges of the current preclinical drug discovery process and its shortcomings in improving clinical attrition due to safety or efficacy concerns. We outlined the different requirements that include picking the most relevant immune cells and tissue cells with a tissue architecture that recapitulates the tissue when designing complex cellular models that has an immune component. We went into great details into explaining the various cellular models for Oncology and Immuno-Inflammation with increasing physiological relevance and how an immune component could be incorporated into these cellular models. The majority of these models have been characterized but require extensive qualification to better understand the limitations and the right context of use. We turned our attention in the middle section of the chapter to safety models and focused on the need to incorporate the immune cells into complex liver, gut, and ADME models. The addition of immune cells into complex safety cellular models could help to capture events that are idiosyncratic events. Finally, increasing the adoption of CIVMs with an integrated immune component in drug development either as a single CIVM or moving to a two-tissue CIVM with an immune tissue should ultimately help to decrease both the high cost and the high attrition rate associated with drug development especially since the immune system is a major contributor to drug failure in clinical trials due to safety and toxicity concerns.

REFERENCES

Abaci, H. E., and M. L. Shuler. 2015. Human-on-a-chip design strategies and principles for physiologically based pharmacokinetics/pharmacodynamics modeling. *Integr Biol (Camb)* 7 (4):383–91.

Ahluwalia, B., L. Moraes, M. K. Magnusson, and L. Ohman. 2018. Immunopathogenesis of inflammatory bowel disease and mechanisms of biological therapies. *Scand J Gastroenterol* 53 (4):379–89.

Alepee, N., A. Bahinski, M. Daneshian, B. De Wever, E. Fritsche, A. Goldberg, J. Hansmann, T. Hartung, J. Haycock, H. Hogberg, L. Hoelting, J. M. Kelm, S. Kadereit, E. McVey, R. Landsiedel, M. Leist, M. Lubberstedt, F. Noor, C. Pellevoisin, D. Petersohn, U. Pfannenbecker, K. Reisinger, T. Ramirez, B. Rothen-Rutishauser, M. Schafer-Korting, K. Zeilinger, and M. G. Zurich. 2014. State-of-the-art of 3D cultures (organs-on-a-chip) in safety testing and pathophysiology. *ALTEX* 31 (4):441–77.

Bar-Ephraim, Y. E, K. Kretzschmar, P. Asra, E. de Jongh, K. E Boonekamp, J. Drost, J. van Gorp, A. Pronk, N. Smakman, I. J. Gan, Z. Sebestyen, J. H. Kuball, R. G. J. Vries, and H. Clevers. 2018. Modelling cancer immunomodulation using epithelial organoid cultures. *bioRxiv*:377655. doi:10.1101/377655.

Baye, J., C. Galvin, and A. Q. Shen. 2017. Microfluidic device flow field characterization around tumor spheroids with tunable necrosis produced in an optimized off-chip process. *Biomed Microdevices* 19 (3):59.

Bein, A., W. Shin, S. Jalili-Firoozinezhad, M. H. Park, A. Sontheimer-Phelps, A. Tovaglieri, A. Chalkiadaki, H. J. Kim, and D. E. Ingber. 2018. Microfluidic organ-on-a-chip models of human intestine. *Cell Mol Gastroenterol Hepatol* 5 (4):659–68.

Bergeron, A. B., and H. J. Gitschier. 2018. CAR-T cell screening in tumor spheroids using corning spheroid microplates and the KILR cytotoxicity assay. Corning. www.corning.com/media/worldwide/cls/documents/applications/CLS-AN-447%20DL.pdf.

Blutt, S. E., J. R. Broughman, W. Zou, X. L. Zeng, U. C. Karandikar, J. In, N. C. Zachos, O. Kovbasnjuk, M. Donowitz, and M. K. Estes. 2017. Gastrointestinal microphysiological systems. *Exp Biol Med (Maywood)* 242 (16):1633–42.

Boehnke, K., P. W. Iversen, D. Schumacher, M. J. Lallena, R. Haro, J. Amat, J. Haybaeck, S. Liebs, M. Lange, R. Schafer, C. R. Regenbrecht, C. Reinhard, and J. A. Velasco. 2016. Assay establishment and validation of a high-throughput screening platform for three-dimensional patient-derived colon cancer organoid cultures. *J Biomol Screen* 21 (9):931–41.

Brudno, J. N., and J. N. Kochenderfer. 2016. Toxicities of chimeric antigen receptor T cells: Recognition and management. *Blood* 127 (26):3321–30.

Buie, L. W., J. J. Pecoraro, T. Z. Horvat, and R. J. Daley. 2015. Blinatumomab: A first-in-class bispecific T-cell engager for precursor B-cell acute lymphoblastic leukemia. *Ann Pharmacother* 49 (9):1057–67.

Cohen, I. J., and R. Blasberg. 2017. Impact of the tumor microenvironment on tumor-infiltrating lymphocytes: Focus on breast cancer. *Breast Cancer (Auckl)* 11:1178223417731565.

Coiffier, B., S. Lepretre, L. M. Pedersen, O. Gadeberg, H. Fredriksen, M. H. van Oers, J. Wooldridge, J. Kloczko, J. Holowiecki, A. Hellmann, J. Walewski, M. Flensburg, J. Petersen, and T. Robak. 2008. Safety and efficacy of ofatumumab, a fully human monoclonal anti-CD20 antibody, in patients with relapsed or refractory B-cell chronic lymphocytic leukemia: A phase 1–2 study. *Blood* 111 (3):1094–100.

de la Puente, P., B. Muz, R. C. Gilson, F. Azab, M. Luderer, J. King, S. Achilefu, R. Vij, and A. K. Azab. 2015. 3D tissue-engineered bone marrow as a novel model to study pathophysiology and drug resistance in multiple myeloma. *Biomaterials* 73:70–84.

Dijkstra, K. K., C. M. Cattaneo, F. Weeber, M. Chalabi, J. van de Haar, L. F. Fanchi, M. Slagter, D. L. van der Velden, S. Kaing, S. Kelderman, N. van Rooij, M. E. van Leerdam, A. Depla, E. F. Smit, K. J. Hartemink, R. de Groot, M. C. Wolkers, N. Sachs, P. Snaebjornsson, K. Monkhorst, J. Haanen, H. Clevers, T. N. Schumacher, and E. E. Voest. 2018. Generation of tumor-reactive T cells by co-culture of peripheral blood lymphocytes and tumor organoids. *Cell* 174 (6):1586–98 e12.

Dillard, P., H. Koksal, E. M. Inderberg, and S. Walchli. 2018. A spheroid killing assay by CAR T cells. *J Vis Exp* (142). doi:10.3791/58785.

Drost, J., and H. Clevers. 2017. Translational applications of adult stem cell-derived organoids. *Development* 144 (6):968–75.

Duell, B. L., A. W. Cripps, M. A. Schembri, and G. C. Ulett. 2011. Epithelial cell coculture models for studying infectious diseases: Benefits and limitations. *J Biomed Biotechnol* 2011:852419.

Duleba, M., Y. Qi, R. Mahalingam, A. A. Liew, R. Neupane, K. Flynn, F. Rinaldi, M. Vincent, C. P. Crum, K. Y. Ho, J. K. Hou, J. S. Hyams, F. A. Sylvester, F. McKeon, and W. Xian. 2019. An efficient method for cloning gastrointestinal stem cells from patients via endoscopic biopsies. *Gastroenterology* 156 (1):20–3.

Dutta, D., I. Heo, and H. Clevers. 2017. Disease modeling in stem cell-derived 3D organoid systems. *Trends Mol Med* 23 (5):393–410. doi:10.1016/j.molmed.2017.02.007.

Ekert, J. E., K. Johnson, B. Strake, J. Pardinas, S. Jarantow, R. Perkinson, and D. C. Colter. 2014. Three-dimensional lung tumor microenvironment modulates therapeutic compound responsiveness in vitro--implication for drug development. *PLoS One* 9 (3):e92248.

Ewart, L., E. M. Dehne, K. Fabre, S. Gibbs, J. Hickman, E. Hornberg, M. Ingelman-Sundberg, K. J. Jang, D. R. Jones, V. M. Lauschke, U. Marx, J. T. Mettetal, A. Pointon, D. Williams, W. H. Zimmermann, and P. Newham. 2018. Application of microphysiological systems to enhance safety assessment in drug discovery. *Annu Rev Pharmacol Toxicol* 58:65–82.

Fritz, J. V., M. S. Desai, P. Shah, J. G. Schneider, and P. Wilmes. 2013. From meta-omics to causality: Experimental models for human microbiome research. *Microbiome* 1 (1):14.

Fujii, M., H. Clevers, and T. Sato. 2019. Modeling human digestive diseases with CRISPR-Cas9-modified organoids. *Gastroenterology* 156 (3):562–76.

Funk, C., and A. Roth. 2017. Current limitations and future opportunities for prediction of DILI from in vitro. *Arch Toxicol* 91 (1):131–42. doi:10.1007/s00204-016-1874-9.

Giolla Eain, M. M., J. Joanna Baginska, K. Greenhalgh, J. V. Fritz, F. Zenhausern, and P. Paul Wilmes. 2017. Engineering solutions for representative models of the gastrointestinal human-microbe interface. *Engineering* 3 (1):60–5.

Goede, V., K. Fischer, R. Busch, A. Engelke, B. Eichhorst, C. M. Wendtner, T. Chagorova, J. de la Serna, M. S. Dilhuydy, T. Illmer, S. Opat, C. J. Owen, O. Samoylova, K. A. Kreuzer, S. Stilgenbauer, H. Dohner, A. W. Langerak, M. Ritgen, M. Kneba, E. Asikanius, K. Humphrey, M. Wenger, and M. Hallek. 2014. Obinutuzumab plus chlorambucil in patients with CLL and coexisting conditions. *N Engl J Med* 370 (12):1101–10.

Haller, D., C. Bode, W. P. Hammes, A. M. Pfeifer, E. J. Schiffrin, and S. Blum. 2000. Non-pathogenic bacteria elicit a differential cytokine response by intestinal epithelial cell/leucocyte co-cultures. *Gut* 47 (1):79–87.

Hassell, B. A., G. Goyal, E. Lee, A. Sontheimer-Phelps, O. Levy, C. S. Chen, and D. E. Ingber. 2018. "Human organ chip models recapitulate orthotopic lung cancer growth, therapeutic responses, and tumor dormancy in vitro." Cell Rep 23 (12):3698.

Hevia, A., P. Lopez, A. Suarez, C. Jacquot, M. C. Urdaci, A. Margolles, and B. Sanchez. 2014. Association of levels of antibodies from patients with inflammatory bowel disease with extracellular proteins of food and probiotic bacteria. *Biomed Res Int* 2014:351204.

Hoffmann, O. I., C. Ilmberger, S. Magosch, M. Joka, K. W. Jauch, and B. Mayer. 2015. Impact of the spheroid model complexity on drug response. *J Biotechnol* 205:14–23.

Holton, A. B., F. L. Sinatra, J. Kreahling, A. J. Conway, D. A. Landis, and S. Altiok. 2017. Microfluidic biopsy trapping device for the real-time monitoring of tumor microenvironment. *PLoS One* 12 (1):e0169797.

Hou, S., H. Tiriac, B. P. Sridharan, L. Scampavia, F. Madoux, J. Seldin, G. R. Souza, D. Watson, D. Tuveson, and T. P. Spicer. 2018. Advanced development of primary pancreatic organoid tumor models for high-throughput phenotypic drug screening. *SLAS Discov* 23 (6):574–84.

Hughes, J. P., S. Rees, S. B. Kalindjian, and K. L. Philpott. 2011. Principles of early drug discovery. *Br J Pharmacol* 162 (6):1239–49.

Jabs, J., F. M. Zickgraf, J. Park, S. Wagner, X. Jiang, K. Jechow, K. Kleinheinz, U. H. Toprak, M. A. Schneider, M. Meister, S. Spaich, M. Sutterlin, M. Schlesner, A. Trumpp, M. Sprick, R. Eils, and C. Conrad. 2017. Screening drug effects in patient-derived cancer cells links organoid responses to genome alterations. *Mol Syst Biol* 13 (11):955.

Jenkins, R. W., A. R. Aref, P. H. Lizotte, E. Ivanova, S. Stinson, C. W. Zhou, M. Bowden, J. Deng, H. Liu, D. Miao, M. X. He, W. Walker, G. Zhang, T. Tian, C. Cheng, Z. Wei, S. Palakurthi, M. Bittinger, H. Vitzthum, J. W. Kim, A. Merlino, M. Quinn, C. Venkataramani, J. A. Kaplan, A. Portell, P. C. Gokhale, B. Phillips, A. Smart, A. Rotem, R. E. Jones, L. Keogh, M. Anguiano, L. Stapleton, Z. Jia, M. Barzily-Rokni, I. Canadas, T. C. Thai, M. R. Hammond, R. Vlahos, E. S. Wang, H. Zhang, S. Li, G. J. Hanna, W. Huang, M. P. Hoang, A. Piris, J. P. Eliane, A. O. Stemmer-Rachamimov, L. Cameron, M. J. Su, P. Shah, B. Izar, M. Thakuria, N. R. LeBoeuf, G. Rabinowits, V. Gunda, S. Parangi, J. M. Cleary, B. C. Miller, S. Kitajima, R. Thummalapalli, B. Miao, T. U. Barbie, V. Sivathanu, J. Wong, W. G. Richards, R. Bueno, C. H. Yoon, J. Miret, M. Herlyn, L. A. Garraway, E. M. Van Allen, G. J. Freeman, P. T. Kirschmeier, J. H. Lorch, P. A. Ott, F. S. Hodi, K. T. Flaherty, R. D. Kamm, G. M. Boland, K. K. Wong, D. Dornan, C. P. Paweletz, and D. A. Barbie. 2018. Ex vivo profiling of PD-1 blockade using organotypic tumor spheroids. *Cancer Discov* 8 (2):196–215.

Kampfer, A. A. M., P. Urban, S. Gioria, N. Kanase, V. Stone, and A. Kinsner-Ovaskainen. 2017. Development of an in vitro co-culture model to mimic the human intestine in healthy and diseased state. *Toxicol In Vitro* 45 (Pt 1):31–43.

Kasendra, M., A. Tovaglieri, A. Sontheimer-Phelps, S. Jalili-Firoozinezhad, A. Bein, A. Chalkiadaki, W. Scholl, C. Zhang, H. Rickner, C. A. Richmond, H. Li, D. T. Breault, and D. E. Ingber. 2018. Development of a primary human Small Intestine-on-a-Chip using biopsy-derived organoids. *Sci Rep* 8 (1):2871.

Kawamura, F., M. Inaki, A. Katafuchi, Y. Abe, N. Tsuyama, Y. Kurosu, A. Yanagi, M. Higuchi, S. Muto, T. Yamaura, H. Suzuki, H. Noji, S. Suzuki, M. A. Yoshida, M. Sasatani, K. Kamiya, M. Onodera, and A. Sakai. 2017. Establishment of induced pluripotent stem cells from normal B cells and inducing AID expression in their differentiation into hematopoietic progenitor cells. *Sci Rep* 7 (1):1659.

Kimura, H., Y. Sakai, and T. Fujii. 2018. Organ/body-on-a-chip based on microfluidic technology for drug discovery. *Drug Metab Pharmacokinet* 33 (1):43–8.

Kobayashi, Y., K. Kato, and T. Watanabe. 2011. Synthesis of functional artificial lymphoid tissues. *Discov Med* 12 (65):351–62.

Koeck, S., J. Kern, M. Zwierzina, G. Gamerith, E. Lorenz, S. Sopper, H. Zwierzina, and A. Amann. 2017. The influence of stromal cells and tumor-microenvironment-derived cytokines and chemokines on CD3(+)CD8(+) tumor infiltrating lymphocyte subpopulations. *Oncoimmunology* 6 (6):e1323617.

Kolesky, D. B., K. A. Homan, M. A. Skylar-Scott, and J. A. Lewis. 2016. Three-dimensional bioprinting of thick vascularized tissues. *Proc Natl Acad Sci U S A* 113 (12):3179–84.

Krewski, D., D. Acosta, Jr., M. Andersen, H. Anderson, J. C. Bailar, 3rd, K. Boekelheide, R. Brent, G. Charnley, V. G. Cheung, S. Green, Jr., K. T. Kelsey, N. I. Kerkvliet, A. A. Li, L. McCray, O. Meyer, R. D. Patterson, W. Pennie, R. A. Scala, G. M. Solomon, M. Stephens, J. Yager, and L. Zeise. 2010. Toxicity testing in the 21st century: A vision and a strategy. *J Toxicol Environ Health B Crit Rev* 13 (2–4):51–138.

Langhans, S. A. 2018. Three-dimensional in vitro cell culture models in drug discovery and drug repositioning. *Front Pharmacol* 9:6. doi:10.3389/fphar.2018.00006.

Lau, A. H., and A. W. Thomson. 2003. Dendritic cells and immune regulation in the liver. *Gut* 52 (2):307–14.

Lee, S. W. L., G. Adriani, E. Ceccarello, A. Pavesi, A. T. Tan, A. Bertoletti, R. D. Kamm, and S. C. Wong. 2018. Characterizing the role of monocytes in T cell cancer immunotherapy using a 3D microfluidic model. *Front Immunol* 9:416.

Leonard, F., H. Ali, E. M. Collnot, B. J. Crielaard, T. Lammers, G. Storm, and C. M. Lehr. 2012. Screening of budesonide nanoformulations for treatment of inflammatory bowel disease in an inflamed 3D cell-culture model. *ALTEX* 29 (3):275–85.

Linette, G. P., E. A. Stadtmauer, M. V. Maus, A. P. Rapoport, B. L. Levine, L. Emery, L. Litzky, A. Bagg, B. M. Carreno, P. J. Cimino, G. K. Binder-Scholl, D. P. Smethurst, A. B. Gerry, N. J. Pumphrey, A. D. Bennett, J. E. Brewer, J. Dukes, J. Harper, H. K. Tayton-Martin, B. K. Jakobsen, N. J. Hassan, M. Kalos, and C. H. June. 2013. Cardiovascular toxicity and titin cross-reactivity of affinity-enhanced T cells in myeloma and melanoma. *Blood* 122 (6):863–71.

Liu, W., C. Tian, M. Yan, L. Zhao, C. Ma, T. Li, J. Xu, and J. Wang. 2016. Heterotypic 3D tumor culture in a reusable platform using pneumatic microfluidics. *Lab Chip* 16 (21):4106–20.

Macpherson, A., U. Y. Khoo, I. Forgacs, J. Philpott-Howard, and I. Bjarnason. 1996. Mucosal antibodies in inflammatory bowel disease are directed against intestinal bacteria. *Gut* 38 (3):365–75.

Madoux, F., A. Tanner, M. Vessels, L. Willetts, S. Hou, L. Scampavia, and T. P. Spicer. 2017. A 1536-well 3D viability assay to assess the cytotoxic effect of drugs on spheroids. *SLAS Discov* 22 (5):516–24.

Majumder, B., U. Baraneedharan, S. Thiyagarajan, P. Radhakrishnan, H. Narasimhan, M. Dhandapani, N. Brijwani, D. D. Pinto, A. Prasath, B. U. Shanthappa, A. Thayakumar, R. Surendran, G. K. Babu, A. M. Shenoy, M. A. Kuriakose, P. Bergthold, P. Horowitz, M. Loda, R. Beroukhim, S. Agarwal, S. Sengupta, M. Sundaram, and P. K. Majumder. 2015. Predicting clinical response to anticancer drugs using an ex vivo platform that captures tumour heterogeneity. *Nat Commun* 6:6169.

Mak, I. W., N. Evaniew, and M. Ghert. 2014. Lost in translation: Animal models and clinical trials in cancer treatment. *Am J Transl Res* 6 (2):114–8.

Marshall, H. T., and M. B. A. Djamgoz. 2018. Immuno-oncology: Emerging targets and combination therapies. *Front Oncol* 8:315.

Marx, U., T. B. Andersson, A. Bahinski, M. Beilmann, S. Beken, F. R. Cassee, M. Cirit, M. Daneshian, S. Fitzpatrick, O. Frey, C. Gaertner, C. Giese, L. Griffith, T. Hartung, M. B. Heringa, J. Hoeng, W. H. de Jong, H. Kojima, J. Kuehnl, M. Leist, A. Luch, I. Maschmeyer, D. Sakharov, A. J. Sips, T. Steger-Hartmann, D. A. Tagle, A. Tonevitsky, T. Tralau, S. Tsyb, A. van de Stolpe, R. Vandebriel, P. Vulto, J. Wang, J. Wiest, M. Rodenburg, and A. Roth. 2016. Biology-inspired microphysiological system approaches to solve the prediction dilemma of substance testing. *ALTEX* 33 (3):272–321.

Michot, J. M., C. Bigenwald, S. Champiat, M. Collins, F. Carbonnel, S. Postel-Vinay, A. Berdelou, A. Varga, R. Bahleda, A. Hollebecque, C. Massard, A. Fuerea, V. Ribrag, A. Gazzah, J. P. Armand, N. Amellal, E. Angevin, N. Noel, C. Boutros, C. Mateus, C. Robert, J. C. Soria, A. Marabelle, and O. Lambotte. 2016. Immune-related adverse events with immune checkpoint blockade: A comprehensive review. *Eur J Cancer* 54:139–48.

Moisan, A., F. Michielin, W. Jacob, S. Kronenberg, S. Wilson, B. Avignon, R. Gerard, F. Benmansour, C. McIntyre, G. Meneses-Lorente, M. Hasmann, A. Schneeweiss, M. Weisser, and C. Adessi. 2018. Mechanistic investigations of diarrhea toxicity induced by anti-HER2/3 combination therapy. *Mol Cancer Ther* 17 (7):1464–74.

Moore, N., D. Doty, M. Zielstorff, I. Kariv, L. Y. Moy, A. Gimbel, J. R. Chevillet, N. Lowry, J. Santos, V. Mott, L. Kratchman, T. Lau, G. Addona, H. Chen, and J. T. Borenstein. 2018. A multiplexed microfluidic system for evaluation of dynamics of immune-tumor interactions. *Lab Chip* 18 (13):1844–58.

Morgan, P., D. G. Brown, S. Lennard, M. J. Anderton, J. C. Barrett, U. Eriksson, M. Fidock, B. Hamren, A. Johnson, R. E. March, J. Matcham, J. Mettetal, D. J. Nicholls, S. Platz, S. Rees, M. A. Snowden, and M. N. Pangalos. 2018. Impact of a five-dimensional framework on R&D productivity at AstraZeneca. *Nat Rev Drug Discov* 17 (3):167–81.

Munera, J. O., and J. M. Wells. 2017. Generation of gastrointestinal organoids from human pluripotent stem cells. *Methods Mol Biol* 1597:167–77.

Nakamura, T. 2019. Recent progress in organoid culture to model intestinal epithelial barrier functions. *Int Immunol* 31 (1):13–21.

Neal, J. T., X. Li, J. Zhu, V. Giangarra, C. L. Grzeskowiak, J. Ju, I. H. Liu, S. H. Chiou, A. A. Salahudeen, A. R. Smith, B. C. Deutsch, L. Liao, A. J. Zemek, F. Zhao, K. Karlsson, L. M. Schultz, T. J. Metzner, L. D. Nadauld, Y. Y. Tseng, S. Alkhairy, C. Oh, P. Keskula, D. Mendoza-Villanueva, F. M. De La Vega, P. L. Kunz, J. C. Liao, J. T. Leppert, J. B. Sunwoo, C. Sabatti, J. S. Boehm, W. C. Hahn, G. X. Y. Zheng, M. M. Davis, and C. J. Kuo. 2018. Organoid modeling of the tumor immune microenvironment. *Cell* 175 (7):1972–88 e16.

Noel, G., N. W. Baetz, J. F. Staab, M. Donowitz, O. Kovbasnjuk, M. F. Pasetti, and N. C. Zachos. 2017. A primary human macrophage-enteroid co-culture model to investigate mucosal gut physiology and host-pathogen interactions. *Sci Rep* 7:45270.

Nyga, A., J. Neves, K. Stamati, M. Loizidou, M. Emberton, and U. Cheema. 2016. "The next level of 3D tumour models: immunocompetence." *Drug Discov Today* 21 (9):1421–28.

Ortega-Prieto, A. M., J. K. Skelton, S. N. Wai, E. Large, M. Lussignol, G. Vizcay-Barrena, D. Hughes, R. A. Fleck, M. Thursz, M. T. Catanese, and M. Dorner. 2018. 3D microfluidic liver cultures as a physiological preclinical tool for hepatitis B virus infection. *Nat Commun* 9 (1):682.

Ostrand-Rosenberg, S. 2004. Animal models of tumor immunity, immunotherapy and cancer vaccines. *Curr Opin Immunol* 16 (2):143–50. doi:10.1016/j.coi.2004.01.003.

Pallardy, M., and T. Hunig. 2010. Primate testing of TGN1412: Right target, wrong cell. *Br J Pharmacol* 161 (3):509–11.

Parker, G. A., and C. A. Picut. 2005. Liver immunobiology. *Toxicol Pathol* 33 (1):52–62.

Parlesak, A., D. Haller, S. Brinz, A. Baeuerlein, and C. Bode. 2004. Modulation of cytokine release by differentiated CACO-2 cells in a compartmentalized coculture model with mononuclear leucocytes and nonpathogenic bacteria. *Scand J Immunol* 60 (5):477–85.

Paul, S. M., D. S. Mytelka, C. T. Dunwiddie, C. C. Persinger, B. H. Munos, S. R. Lindborg, and A. L. Schacht. 2010. How to improve R&D productivity: The pharmaceutical industry's grand challenge. *Nat Rev Drug Discov* 9 (3):203–14.

Pavesi, A., A. T. Tan, S. Koh, A. Chia, M. Colombo, E. Antonecchia, C. Miccolis, E. Ceccarello, G. Adriani, M. T. Raimondi, R. D. Kamm, and A. Bertoletti. 2017. A 3D microfluidic model for preclinical evaluation of TCR-engineered T cells against solid tumors. *JCI Insight* 2 (12). doi:10.1172/jci.insight.89762.

Peng, W., P. Datta, B. Ayan, V. Ozbolat, D. Sosnoski, and I. T. Ozbolat. 2017. 3D bioprinting for drug discovery and development in pharmaceutics. *Acta Biomater* 57:26–46.

Pocock, Kyall, Ludivine Delon, Vaskor Bala, Shasha Rao, Craig Priest, Clive Prestidge, and Benjamin Thierry. 2017. Intestine-on-a-chip microfluidic model for efficient in vitro screening of oral chemotherapeutic uptake. *ACS Biomater Sci Eng* 3 (6):951–9.

Prantil-Baun, R., R. Novak, D. Das, M. R. Somayaji, A. Przekwas, and D. E. Ingber. 2018. Physiologically based pharmacokinetic and pharmacodynamic analysis enabled by microfluidically linked organs-on-chips. *Annu Rev Pharmacol Toxicol* 58:37–64.

Purwada, A., S. B. Shah, W. Beguelin, A. August, A. M. Melnick, and A. Singh. 2019. Ex vivo synthetic immune tissues with T cell signals for differentiating antigen-specific, high affinity germinal center B cells. *Biomaterials* 198:27–36.

Ramadan, Q., H. Jafarpoorchekab, C. Huang, P. Silacci, S. Carrara, G. Koklu, J. Ghaye, J. Ramsden, C. Ruffert, G. Vergeres, and M. A. Gijs. 2013. NutriChip: Nutrition analysis meets microfluidics. *Lab Chip* 13 (2):196–203.

Rowe, C., M. Shaeri, E. Large, T. Cornforth, A. Robinson, T. Kostrzewski, R. Sison-Young, C. Goldring, K. Park, and D. Hughes. 2018. Perfused human hepatocyte microtissues identify reactive metabolite-forming and mitochondria-perturbing hepatotoxins. *Toxicol In Vitro* 46:29–38.

Russo, I., P. Zeppa, P. Iovino, C. Del Giorno, F. Zingone, C. Bucci, A. Puzziello, and C. Ciacci. 2016. The culture of gut explants: A model to study the mucosal response. *J Immunol Methods* 438:1–10.

Sachs, N., A. Papaspyropoulos, D. D. Zomer-van Ommen, I. Heo, L. Bottinger, D. Klay, F. Weeber, G. Huelsz-Prince, N. Iakobachvili, G. D. Amatngalim, J. de Ligt, A. van Hoeck, N. Proost, M. C. Viveen, A. Lyubimova, L. Teeven, S. Derakhshan, J. Korving, H. Begthel, J. F. Dekkers, K. Kumawat, E. Ramos, M. F. van Oosterhout, G. J. Offerhaus, D. J. Wiener, E. P. Olimpio, K. K. Dijkstra, E. F. Smit, M. van der Linden, S. Jaksani, M. van de Ven, J. Jonkers, A. C. Rios, E. E. Voest, C. H. van Moorsel, C. K. van der Ent, E. Cuppen, A. van Oudenaarden, F. E. Coenjaerts, L. Meyaard, L. J. Bont, P. J. Peters, S. J. Tans, J. S. van Zon, S. F. Boj, R. G. Vries, J. M. Beekman, and H. Clevers. 2019. Long-term expanding human airway organoids for disease modeling. *EMBO J* 38 (4). doi:10.15252/embj.2018100300.

Salerno-Goncalves, R., A. Fasano, and M. B. Sztein. 2011. Engineering of a multicellular organotypic model of the human intestinal mucosa. *Gastroenterology* 141 (2):e18–20.

Sardi, M., A. Lubitz, and C. Giese. 2016. Modeling human immunity in vitro: Improving artificial lymph node physiology by stromal cells. *Appl In Vitro Toxicol* 2 (3):143–50.

Scannell, J. W., A. Blanckley, H. Boldon, and B. Warrington. 2012. Diagnosing the decline in pharmaceutical R&D efficiency. *Nat Rev Drug Discov* 11 (3):191–200.

Schroeder, K., K. D. Bremm, N. Alepee, J. G. Bessems, B. Blaauboer, S. N. Boehn, C. Burek, S. Coecke, L. Gombau, N. J. Hewitt, J. Heylings, J. Huwyler, M. Jaeger, M. Jagelavicius, N. Jarrett, H. Ketelslegers, I. Kocina, J. Koester, J. Kreysa, R. Note, A. Poth, M. Radtke, V. Rogiers, J. Scheel, T. Schulz, H. Steinkellner, M. Toeroek, M. Whelan, P. Winkler, and W. Diembeck. 2011. Report from the EPAA workshop: In vitro ADME in safety testing used by EPAA industry sectors. *Toxicol In Vitro* 25 (3):589–604.

Sharkey, K. A., P. L. Beck, and D. M. McKay. 2018. Neuroimmunophysiology of the gut: Advances and emerging concepts focusing on the epithelium. *Nat Rev Gastroenterol Hepatol* 15 (12):765–84.

Shin, W., and H. J. Kim. 2018. Intestinal barrier dysfunction orchestrates the onset of inflammatory host-microbiome cross-talk in a human gut inflammation-on-a-chip. *Proc Natl Acad Sci U S A* 115 (45):E10539–47.

Short, S. P., P. W. Costacurta, and C. S. Williams. 2017. Using 3D organoid cultures to model intestinal physiology and colorectal cancer. *Curr Colorectal Cancer Rep* 13 (3):183–91.

Smith, P. D., L. E. Smythies, R. Shen, T. Greenwell-Wild, M. Gliozzi, and S. M. Wahl. 2011. Intestinal macrophages and response to microbial encroachment. *Mucosal Immunol* 4 (1):31–42.

Sobrino, A., D. T. Phan, R. Datta, X. Wang, S. J. Hachey, M. Romero-Lopez, E. Gratton, A. P. Lee, S. C. George, and C. C. Hughes. 2016. "3D microtumors in vitro supported by perfused vascular networks." *Sci Rep* 6:31589.

Song, J., A. Miermont, C. T. Lim, and R. D. Kamm. 2018. "A 3D microvascular network model to study the impact of hypoxia on the extravasation potential of breast cell lines." *Sci Rep* 8 (1):17949.

Sontheimer-Phelps, A., B. A. Hassell, and D. E. Ingber. 2019. Modelling cancer in microfluidic human organs-on-chips. *Nat Rev Cancer* 19 (2):65–81.

Tomei, A. A., S. Siegert, M. R. Britschgi, S. A. Luther, and M. A. Swartz. 2009. Fluid flow regulates stromal cell organization and CCL21 expression in a tissue-engineered lymph node microenvironment. *J Immunol* 183 (7):4273–83.

Topp, M. S., N. Gokbuget, A. S. Stein, G. Zugmaier, S. O'Brien, R. C. Bargou, H. Dombret, A. K. Fielding, L. Heffner, R. A. Larson, S. Neumann, R. Foa, M. Litzow, J. M. Ribera, A. Rambaldi, G. Schiller, M. Bruggemann, H. A. Horst, C. Holland, C. Jia, T. Maniar, B. Huber, D. Nagorsen, S. J. Forman, and H. M. Kantarjian. 2015. Safety and activity of blinatumomab for adult patients with relapsed or refractory B-precursor acute lymphoblastic leukaemia: A multicentre, single-arm, phase 2 study. *Lancet Oncol* 16 (1):57–66.

Trietsch, S. J., E. Naumovska, D. Kurek, M. C. Setyawati, M. K. Vormann, K. J. Wilschut, H. L. Lanz, A. Nicolas, C. P. Ng, J. Joore, S. Kustermann, A. Roth, T. Hankemeier, A. Moisan, and P. Vulto. 2017. Membrane-free culture and real-time barrier integrity assessment of perfused intestinal epithelium tubes. *Nat Commun* 8 (1):262.

Tseng, H., J. A. Gage, T. Shen, W. L. Haisler, S. K. Neeley, S. Shiao, J. Chen, P. K. Desai, A. Liao, C. Hebel, R. M. Raphael, J. L. Becker, and G. R. Souza. 2015. A spheroid toxicity assay using magnetic 3D bioprinting and real-time mobile device-based imaging. *Sci Rep* 5:13987.

Ukairo, O., C. Kanchagar, A. Moore, J. Shi, J. Gaffney, S. Aoyama, K. Rose, S. Krzyzewski, J. McGeehan, M. E. Andersen, S. R. Khetani, and E. L. Lecluyse. 2013. Long-term stability of primary rat hepatocytes in micropatterned cocultures. *J Biochem Mol Toxicol* 27 (3):204–12.

Vadstrup, K., E. D. Galsgaard, J. Gerwien, M. K. Vester-Andersen, J. S. Pedersen, J. Rasmussen, S. Neermark, M. Kiszka-Kanowitz, T. Jensen, and F. Bendtsen. 2016. Validation and optimization of an ex vivo assay of intestinal mucosal biopsies in Crohn's disease: Reflects inflammation and drug effects. *PLoS One* 11 (5):e0155335.

Vellonen, K. S., M. Malinen, E. Mannermaa, A. Subrizi, E. Toropainen, Y. R. Lou, H. Kidron, M. Yliperttula, and A. Urtti. 2014. A critical assessment of in vitro tissue models for ADME and drug delivery. *J Control Release* 190:94–114.

Vizcardo, R., N. D. Klemen, S. M. R. Islam, D. Gurusamy, N. Tamaoki, D. Yamada, H. Koseki, B. L. Kidder, Z. Yu, L. Jia, A. N. Henning, M. L. Good, M. Bosch-Marce, T. Maeda, C. Liu, Z. Abdullaev, S. Pack, D. C. Palmer, D. F. Stroncek, F. Ito, F. A. Flomerfelt, M. J. Kruhlak, and N. P. Restifo. 2018. Generation of tumor antigen-specific iPSC-derived thymic emigrants using a 3D thymic culture system. *Cell Rep* 22 (12):3175–90.

Wagar, L. E., R. M. DiFazio, and M. M. Davis. 2018. Advanced model systems and tools for basic and translational human immunology. *Genome Med* 10 (1):73.

Wagoner, M., J. Kelsal, M. Hattersley, K. Hickling, J. Pederson, J. Harris, N. Keirstead, D. Heathcote, and P. Newham. 2014. Bromodomain and extraterminal (BET) domain inhibitors induce a loss of intestinal stem cells and villous atrophy. *Toxicol Lett* 229:S75–6.

Wang, X., Y. Yamamoto, L. H. Wilson, T. Zhang, B. E. Howitt, M. A. Farrow, F. Kern, G. Ning, Y. Hong, C. C. Khor, B. Chevalier, D. Bertrand, L. Wu, N. Nagarajan, F. A. Sylvester, J. S. Hyams, T. Devers, R. Bronson, D. B. Lacy, K. Y. Ho, C. P. Crum, F. McKeon, and W. Xian. 2015. Cloning and variation of ground state intestinal stem cells. *Nature* 522 (7555):173–8.

Wirtz, S., and M. F. Neurath. 2007. Mouse models of inflammatory bowel disease. *Adv Drug Deliv Rev* 59 (11):1073–83.

Wynn, T. A., A. Chawla, and J. W. Pollard. 2013. Macrophage biology in development, homeostasis and disease. *Nature* 496 (7446):445–55.

Yissachar, N., Y. Zhou, L. Ung, N. Y. Lai, J. F. Mohan, A. Ehrlicher, D. A. Weitz, D. L. Kasper, I. M. Chiu, D. Mathis, and C. Benoist. 2017. An intestinal organ culture system uncovers a role for the nervous system in microbe-immune crosstalk. *Cell* 168 (6):1135–48 e12.

Zeng, J., S. Y. Tang, L. L. Toh, and S. Wang. 2017. Generation of off-the-shelf natural killer cells from peripheral blood cell-derived induced pluripotent stem cells. *Stem Cell Reports* 9 (6):1796–812.

Zhang, Z., H. Wang, Q. Ding, Y. Xing, Z. Xu, C. Lu, D. Luo, L. Xu, W. Xia, C. Zhou, and M. Shi. 2018. Establishment of patient-derived tumor spheroids for non-small cell lung cancer. *PLoS One* 13 (3):e0194016.

Zinchenko, Y. S., L. W. Schrum, M. Clemens, and R. N. Coger. 2006. Hepatocyte and kupffer cells co-cultured on micropatterned surfaces to optimize hepatocyte function. *Tissue Eng* 12 (4):751–61.

Index

A

Printed and bound by CPI Group (UK) Ltd, Croydon, CR0 4YY

17/10/2024

01775660-0012